D0842923

PLANT BIOTECHNOLOGY AND GENETICS

Published by John Wiley & Sons, Inc., Hoboken, New Jersey
Published simultaneously in Canada

For general information on our other products and services or for technical support, please contact our Customer Care Department within the United States at (800) 762-2974, outside the United States at (317) 572-3993 or fax (317) 572-4002.

Wiley also publishes its books in variety of electronic formats. Some content that appears in print may not be available in electronic format. For more information about Wiley products, visit our web site at www.wiley.com.

Library of Congress Cataloging-in-Publication Data:

Stewart, C. Neal.
Plant biotechnology and genetics: principles, techniques and applications/
C. Neal Stewart, Jr.
p. cm.
Includes index.
ISBN 978-0-470-04381-3 (cloth/cd)
1. Plant biotechnology. 2. Plant genetics. 3. Transgenic plants. I. Title.
TP248.27.P55S74 2008
660.6′5—dc22

2008002719

Printed in the United States of America

10 9 8 7 6 5 4 3 2 1

This book is dedicated to both the pioneers and the students of plant biotechnology

CONTENTS

Preface xvii

Foreword to Plant Biotechnology and Genetics xix

Contributors xxiii

1. Plant Agriculture: The Impact of Biotechnology 1
Graham Brookes

1.0 Chapter Summary and Objectives 1
1.0.1 Summary 1
1.0.2 Discussion Questions 1
1.1 Introduction 1
1.2 Biotechnology Crops Plantings 2
1.3 Why Farmers Use Biotech Crops 4
1.3.1 Herbicide-Tolerant Crops 7
1.3.2 Insect-Resistant Crops 7
1.3.3 Conclusion 8
1.4 How the Adoption of Plant Biotechnology Has Impacted the Environment 8
1.4.1 Environmental Impacts from Changes in Insecticide and Herbicide Use 9
1.4.2 Impact on Greenhouse Gas (GHG) Emissions 12
1.5 Conclusions 14
References 19

2. Mendelian Genetics and Plant Reproduction 21
Matthew D. Halfhill and Suzanne I. Warwick

2.0 Chapter Summary and Objectives 21
2.0.1 Summary 21
2.0.2 Discussion Questions 21
2.1 Genetics Overview 22
2.2 Mendelian Genetics 25
2.2.1 Law of Segregation 28
2.2.2 Law of Independent Assortment 28
2.3 Mitosis and Meiosis 30
2.3.1 Mitosis 31
2.3.2 Meiosis 32
2.3.3 Recombination 32
2.3.4 Cytogenetic Analysis 33

2.4 Plant Reproductive Biology 34
2.4.1 History of Research 34
2.4.2 Mating Systems 35
2.4.2.1 Sexual Reproduction 35
2.4.2.2 Asexual Reproduction 38
2.4.2.3 Mating Systems Summary 38
2.4.3 Hybridization and Polyploidy 39
2.5 Conclusion 41
References 45

3. Plant Breeding **47**
Nicholas A. Tinker

3.0 Chapter Summary and Objectives 47
3.0.1 Summary 47
3.0.2 Discussion Questions 47
3.1 Introduction 48
3.2 Central Concepts in Plant Breeding 49
3.2.1 Simple versus Complex Inheritance 49
3.2.2 Phenotype versus Genotype 51
3.2.3 Mating Systems, Varieties, Landraces, and Pure Lines 52
3.2.4 Other Topics in Population and Quantitative Genetics 55
3.2.5 The Value of a Plant Variety Depends on Many Traits 56
3.2.6 Varieties Must Be Adapted to Environments 56
3.2.7 Plant Breeding Is a Numbers Game 57
3.2.8 Plant Breeding Is an Iterative and Collaborative Process 57
3.2.9 Diversity, Adaptation, and Ideotypes 58
3.2.10 Other Considerations 61
3.3 Objectives for Plant Breeding 62
3.4 Methods of Plant Breeding 63
3.4.1 Methods of Hybridization 63
3.4.1.1 Self-Pollinated Species 64
3.4.1.2 Outcrossing Species 69
3.4.1.3 Synthetic Varieties 72
3.4.1.4 Hybrid Varieties 72
3.4.2 Clonally Propagated Species 74
3.5 Breeding Enhancements 74
3.5.1 Doubled Haploidy 74
3.5.2 Marker-Assisted Selection 75
3.5.3 Mutation Breeding 76
3.5.4 Apomixis 77
3.6 Conclusions 77
References 82

4. Plant Development and Physiology **83**
Glenda E. Gillaspy

4.0 Chapter Summary and Objectives 83
4.0.1 Summary 83
4.0.2 Discussion Questions 83
4.1 Plant Anatomy and Morphology 84

4.2 Embryogenesis and Seed Germination 85
4.2.1 Gametogenesis 85
4.2.2 Fertilization 88
4.2.3 Fruit Development 88
4.2.4 Embryogenesis 89
4.2.5 Seed Germination 91
4.2.6 Photomorphogenesis 91
4.3 Meristems 92
4.3.1 Shoot Apical Meristem 92
4.3.2 Root Apical Meristem and Root Development 94
4.4 Leaf Development 96
4.4.1 Leaf Structure 96
4.4.2 Leaf Development Patterns 97
4.5 Flower Development 98
4.5.1 Floral Evocation 98
4.5.2 Floral Organ Identity and the ABC Model 99
4.6 Hormone Physiology and Signal Transduction 101
4.6.1 Seven Plant Hormones and Their Actions 101
4.6.2 Plant Hormone Signal Transduction 103
4.6.2.1 Auxin and GA Signaling 104
4.6.2.2 Cytokinin and Ethylene Signaling 105
4.6.2.3 Brassinosteroid Signal Transduction 105
4.7 Conclusions 106
References 110

5. Tissue Culture: The Manipulation of Plant Development **113**
Vinitha Cardoza

5.0 Chapter Summary and Objectives 113
5.0.1 Summary 113
5.0.2 Discussion Questions 113
5.1 Introduction 113
5.2 History 114
5.3 Media and Culture Conditions 115
5.3.1 Basal Media 115
5.3.2 Growth Regulators 116
5.4 Sterile Technique 118
5.4.1 Clean Equipment 118
5.4.2 Surface Sterilization of Explants 118
5.5 Culture Conditions and Vessels 119
5.6 Culture Types and Their Uses 120
5.6.1 Callus Culture 120
5.6.1.1 Somaclonal Variation 122
5.6.2 Cell Suspension Culture 122
5.6.2.1 Production of Secondary Metabolites and Recombinant Proteins Using Cell Culture 122
5.6.3 Anther/Microspore Culture 123
5.6.4 Protoplast Culture 123
5.6.4.1 Somatic Hybridization 124
5.6.5 Embryo Culture 124
5.6.6 Meristem Culture 124

5.7 Regeneration Methods of Plants in Culture 125
5.7.1 Organogenesis 125
5.7.1.1 Indirect Organogenesis 125
5.7.1.2 Direct Organogenesis 125
5.7.2 Somatic Embryogenesis 126
5.7.2.1 Synthetic Seeds 127
5.8 Rooting of Shoots 127
5.9 Acclimation 128
5.10 Conclusions 128
Acknowledgments 128
References 132

6. Molecular Genetics of Gene Expression **135**
Maria Gallo and Alison K. Flynn

6.0 Chapter Summary and Objectives 135
6.0.1 Summary 135
6.0.2 Discussion Questions 135
6.1 The gene 135
6.1.1 DNA Coding for a Protein via the Gene 135
6.1.2 DNA as a Polynucleotide 136
6.2 DNA Packaging into Eukaryotic Chromosomes 136
6.3 Transcription 140
6.3.1 Transcription of DNA to Produce Messenger RNA (mRNA) 140
6.3.2 Transcription Factors 143
6.3.3 Coordinated Regulation of Gene Expression 144
6.3.4 Chromatin as an Important Regulator of Transcription 144
6.3.5 Regulation of Gene Expression by DNA Methylation 146
6.3.6 Processing to Produce Mature mRNA 146
6.4 Translation 148
6.4.1 Initiation of Translation 150
6.4.2 Translation Elongation 152
6.4.3 Translation Termination 152
6.5 Protein Postranslational Modification 152
References 156

7. Recombinant DNA, Vector Design, and Construction **159**
Mark D. Curtis

7.0 Chapter Summary and Objectives 159
7.0.1 Summary 159
7.0.2 Discussion Questions 159
7.1 DNA Modification 160
7.2 DNA Vectors 163
7.2.1 DNA Vectors for Plant Transformation 166
7.2.2 Components for Efficient Gene Expression in Plants 167
7.3 Greater Demands Lead to Innovation 170
7.3.1 Site-Specific DNA Recombination 171

7.3.1.1 Gateway Cloning 172
7.3.1.2 Creator™ Cloning 175
7.3.1.3 Univector (Echo™) Cloning 175
7.4 Vector Design 177
7.4.1 Vectors for High-Throughput Functional Analysis 177
7.4.2 Vectors for RNA Interference (RNAi) 179
7.4.3 Expression Vectors 179
7.4.4 Vectors for Promoter Analysis 180
7.4.5 Vectors Derived from Plant Sequences 181
7.4.6 Vectors for Multigenic Traits 183
7.5 Targeted Transgene Insertions 184
7.6 Safety Features in Vector Design 186
7.7 Prospects 188
References 190

8. Genes and Traits of Interest for Transgenic Plants 193
Kenneth L. Korth

8.0 Chapter Summary and Objectives 193
8.0.1 Summary 193
8.0.2 Discussion Questions 193
8.1 Introduction 194
8.2 Identifying Genes of Interest via Genomic Studies 194
8.3 Traits for Improved Crop Production 197
8.3.1 Herbicide Resistance 197
8.3.2 Insect Resistance 200
8.3.3 Pathogen Resistance 202
8.4 Traits for Improved Products and Food Quality 205
8.4.1 Nutritional Improvements 205
8.4.2 Modified Plant Oils 207
8.4.3 Pharmaceutical Products 208
8.4.4 Biofuels 209
8.5 Conclusions 210
References 216

9. Marker Genes and Promoters 217
Brian Miki

9.0 Chapter Summary and Objectives 217
9.0.1 Summary 217
9.0.2 Discussion Questions 217
9.1 Introduction 218
9.2 Definition of Marker Genes 218
9.2.1 Selectable Marker Genes: An Introduction 218
9.2.2 Reporter Genes: An Introduction 222
9.3 Promoters 224
9.4 Selectable Marker Genes 227
9.4.1 Conditional Positive Selectable Marker Gene Systems 227

9.4.1.1 Selection on Antibiotics 228
9.4.1.2 Selection on Herbicides 229
9.4.1.3 Selection Using Nontoxic Metabolic Substrates 229
9.4.2 Nonconditional Positive Selection Systems 230
9.4.3 Conditional Negative Selection Systems 230
9.4.4 Nonconditional Negative Selection Systems 230
9.5 Nonselectable Marker Genes or Reporter Genes 231
9.5.1 β-Glucuronidase 231
9.5.2 Luciferase 232
9.5.3 Green Fluorescent Protein 232
9.6 Marker-Free Strategies 233
9.7 Conclusions 237
References 242

10. Transgenic Plant Production **245**
John Finer and Taniya Dhillon

10.0 Chapter Summary and Objectives 245
10.0.1 Summary 245
10.0.2 Discussion Questions 245
10.1 Overview 246
10.2 Basic Components for Successful Gene Transfer to Plant Cells 246
10.2.1 Visualizing the General Transformation Process 246
10.2.2 DNA Delivery 247
10.2.3 Target Tissue Status 248
10.2.4 Selection and Regeneration 248
10.3 Agrobacterium 249
10.3.1 History of Our Knowledge of *Agrobacterium* 249
10.3.2 Use of the T-DNA Transfer Process for Transformation 251
10.3.3 Optimizing Delivery and Broadening the Range of Targets 253
10.3.4 Agroinfiltration 254
10.3.5 Arabidopsis Floral Dip 254
10.4 Particle Bombardment 255
10.4.1 History of Particle Bombardment 255
10.4.2 The Fate of Introduced DNA 257
10.4.3 The Power and Problems of Direct DNA Introduction 259
10.4.4 Improvements in Transgene Expression 259
10.5 Other Methods 260
10.5.1 The Need for Additional Technologies 260
10.5.2 Protoplasts 261
10.5.3 Whole-Tissue Electroporation 262
10.5.4 Silicon Carbide Whiskers 262
10.5.5 Viral Vectors 263
10.5.6 Laser Micropuncture 263
10.5.7 Nanofiber Arrays 263
10.6 The Rush to Publish 265
10.6.1 Controversial Reports of Plant Transformation 265
10.6.1.1 DNA Uptake in Pollen 265
10.6.1.2 *Agrobacterium*-Mediated Transformation of Maize Seedlings 265

10.6.1.3 Pollen Tube Pathway 266
10.6.1.4 Rye Floral Tiller Injection 266
10.6.1.5 Electrotransformation of Germinating Pollen Grain 267
10.6.1.6 *Medicago* Transformation via Seedling Infiltration 267
10.6.2 Criteria to Consider: Whether My Plant Is Transgenic 268
10.6.2.1 Resistance Genes 268
10.6.2.2 Marker Genes 268
10.6.2.3 Transgene DNA 269
10.7 A Look to the Future 269
References 272

11. Transgenic Plant Analysis **275**
Janice Zale

11.0 Chapter Summary and Objectives 275
11.0.1 Summary 275
11.0.2 Discussion Questions 275
11.1 Introduction 276
11.2 Directionally Named Analyses: As the Compass Turns 276
11.3 Initial Screens: Putative Transgenic Plants 277
11.3.1 Screens on Selection Media 277
11.3.2 Polymerase Chain Reaction 278
11.3.3 Enzyme-Linked Immunosorbent Assays (ELISAs) 279
11.4 Definitive Molecular Characterization 280
11.4.1 Intact Transgene Integration 280
11.4.2 Determining the Presence of Intact Transgenes or Constructs 284
11.4.3 Transgene Expression: Transcription 284
11.4.3.1 Northern Blot Analysis 284
11.4.3.2 Quantitative Real-Time Reverse Transcriptase (RT)-PCR 286
11.4.4 Transgene Expression: Translation: Western Blot Analyses 286
11.5 Digital Imaging 287
11.6 Phenotypic Analysis 287
11.7 Conclusions 288
References 288

12. Regulations and Biosafety **291**
Alan McHughen

12.0 Chapter Summary and Objectives 291
12.0.1 Summary 291
12.0.2 Discussion Questions 291
12.1 Introduction 291
12.2 History of Genetic Engineering and its Regulation 293
12.3 Regulation of GE 296
12.3.1 United States 296
12.3.1.1 USDA 297
12.3.1.2 FDA 297
12.3.1.3 EPA 298
12.3.2 EU 299
12.3.3 Canada 300

12.3.4 International Perspectives 301
12.4 Conclusions 302
References 309

13. Field Testing of Transgenic Plants **311**
Detlef Bartsch, Achim Gathmann, Christiane Saeglitz, and Arti Sinha

13.0 Chapter Summary and Objectives 311
13.0.1 Summary 311
13.0.2 Discussion Questions 311
13.1 Introduction 312
13.2 Environmental Risk Assessment (Era) Process 312
13.2.1 Initial Evaluation (ERA Step 1) 313
13.2.2 Problem Formulation (ERA Step 2) 313
13.2.3 Controlled Experiments and Gathering of Information (ERA Step 3) 313
13.2.4 Risk Evaluation (ERA Step 4) 313
13.2.5 Progression through a Tiered Risk Assessment 313
13.3 An Example Risk Assessment: The Case of Bt Maize 314
13.3.1 Effect of Bt Maize Pollen on Nontarget Caterpillars 315
13.3.2 Statistical Analysis and Relevance for Predicting Potential Adverse Effects on Butterflies 317
13.4 Proof of Safety versus Proof of Hazard 319
13.5 Proof of Benefits: Agronomic Performance 319
13.6 Conclusions 320
References 323

14. Intellectual Property in Agricultural Biotechnology: Strategies for Open Access **325**
Alan B. Bennett, Cecilia Chi-Ham, Gregory Graff, and Sara Boettiger

14.0 Chapter Summary and Objectives 325
14.0.1 Summary 325
14.0.2 Discussion Questions 325
14.1 Introduction 326
14.2 Intellectual Property Defined 327
14.3 Intellectual Property in Relation to Agricultural Research 329
14.4 Development of an "Anticommons" in Agricultural Biotechnology 331
14.4.1 Transformation Methods 331
14.4.2 Selectable Markers 332
14.4.3 Constitutive Promoters 332
14.4.4 Tissue- or Development-Specific Promoters 332
14.4.5 Subcellular Localization 333
14.5 Freedom to Operate (FTO) 333
14.6 Strategies for Open Access 336
14.7 Conclusions 338
References 339

15. Why *Transgenic* Plants Are So Controversial **343**
Douglas Powell

15.0 Chapter Summary and Objectives 343
15.0.1 Summary 343
15.0.2 Discussion Questions 343
15.1 Introduction 343
15.1.1 The Frankenstein Backdrop 344
15.1.2 Agricultural Innovations and Questions 344
15.2 Perceptions of Risk 345
15.3 Responses to Fear 347
15.4 Feeding Fear: Case Studies 348
15.4.1 Pusztai's Potatoes 349
15.4.2 Monarch Butterfly Flap 349
15.5 How Many Benefits are Enough 350
15.6 Continuing Debates 351
15.6.1 Process versus Product 351
15.6.2 Health Concerns 352
15.6.3 Environmental Concerns 353
15.6.4 Consumer Choice 353
15.7 Business and Control 353
15.8 Conclusions 354
References 354

16. The Future of Plant Biotechnology **357**
C. Neal Stewart, Jr. and David W. Ow

16.0 Chapter Summary and Objectives 357
16.0.1 Summary 357
16.0.2 Discussion Questions 357
16.1 Introduction 357
16.2 Site-Specific Recombination Systems to Provide Increased Precision 359
16.2.1 Removal of DNA from Transgenic Plants or Plant Parts 361
16.2.2 More Precise Integration of DNA 362
16.3 Zinc-Finger Nucleases 363
16.4 The Future of Food (and Fuel and Pharmaceuticals) 364
16.5 Conclusions 365
References 368

Index **371**

PREFACE

Serendipity.

One thing led to another.

My department at the University of Tennessee decided to offer a plant biotechnology concentration to the Plant Science undergraduate major. I thought that was a really good idea. But we were missing a key course—a capstone course to integrate plant biotechnology genetics and breeding. I think that plant biotechnology only makes sense in the backdrop of genetics and breeding. So I volunteered to teach such a course. I soon found out that not only were we missing the course, but also the world was missing a textbook to support such a course. Plenty of good textbooks on plant biotechnology are available, but the levels or contents were not what I envisioned for the course we needed. Some were too advanced, and others were too applied with not enough of the basic science. At around the same time, Wiley must have seen the same thing because they asked me to edit a plant biotechnology textbook.

As you have gathered by now, this is that book. It takes the student on a tour from basic plant biology and genetics to breeding and principles and applications of plant biotechnology. Toward the end of the book, we diverge from science to perceptions and patents, which are arguably as important as the science in delivering agricultural biotechnology products to people who need them. I asked some of the leading scientists in the field, many of whom teach in biotechnology, to write chapters of this textbook. I think that seeing several points of view is more interesting to the reader than if I'd written the entire text myself (at least it is more interesting to me).

One of my favorite aspects of this book, and one that makes it fairly unique, is the autobiographical segments that accompany chapters. I asked many of the fathers and mothers of plant biotechnology to author these things I call "life boxes"; to tell their stories and give advice about science and life. In addition to my "elders," I also asked several scientists in the prime of their careers to share their stories. As I expected, their stories have a different feel to them because they lack deep retrospection, but they look more toward the future—futures they hope to contribute and live out. The one person who was too ill at the time to make a contribution was Norman Borlaug. As he is the most famous plant breeder of all time (and I think who will ever live), I could not foresee this book without his life box, and so I asked his biographer to boil his own book down to just a few paragraphs. Finally, on the other end of the spectrum, the book ends with life boxes from two graduate students. They have lived but a very short time in this exciting field, but their stories tell of dreams and future contributions that could change the face of agriculture and science.

I look back to when I was in college—in the early 1980s when Mary-Dell Chilton and colleagues were transforming the first plant. I was hardly a serious student. I was more interested in rebuilding engines than building transgenic plants. I did not set out to be a plant biotechnologist. Likewise, although I did not set out to teach a plant biotechnology course a few years ago nor did I seek to construct a textbook on the subject, it was serendipitous that it all came together in this product. The rest of the story is up to the biotechnology student and researcher to make new things happen that will cause revisions to this text to be necessary. I am counting on that.

This science called "plant biotechnology" is far from static, and that is what makes it exciting. I hope the reader catches a glimpse of the excitement from each of the chapter authors and decides to change the world into a better place for us all. I am counting on that too.

En route to teaching my plant biotechnology course for the first time my colleagues and I have made a set of lecture Power Point sites that are freely accessible for any student or instructor at http://plantsciences.utk.edu/pbg

NEAL STEWART

Knoxville, Tennessee
December 11, 2007

FOREWORD TO PLANT BIOTECHNOLOGY AND GENETICS

An international (but widely unnoticed) race took place in the mid-1970s to understand how *Agrobacterium tumefaciens* caused plant cells to grow rapidly into a gall that produced its favorite substrates—called opines. Belgian, German, Australian, French, and U.S. groups were at the forefront of different aspects of the puzzle. By 1977, it was clear that gene transfer from the bacterium to its plant host was the secret, and that the genes from the bacterium were functioning to alter characteristics of the plant cells. Participants in the race as well as observers began to speculate that we might exploit the capability of this cunning bacterium in order to get plants to produce our favorite substrates. Small startup companies and multinational corporations took notice and began to work with *Agrobacterium* and other means of gene transfer to plants. One by one the problems were dealt with, and each step in the use of *Agrobacterium* for the genetic engineering of a tobacco plant was demonstrated.

As I look back to those early experiments, I see that we have come a long way since the birth of plant biotechnology, which most of us who served as midwives would date from the Miami Winter Symposium of January 1983. The infant technology was weak and wobbly, but its viability and vitality were already clear. Its growth and development were foreseeable although not predictable in detail. I thought that the difficult part was behind us, and now (as I used to predict at the end of my lectures) the main challenge would be thinking of what genes we might use to bring about desired changes in crop plants. Unseen at that early date were the interesting problems, some technical and some of other kinds, to be encountered and overcome.

To my surprise, one of the biggest challenges turned out to be tobacco, which worked so well that it made us cocky. Tobacco was the guinea pig of the plant kingdom in 1983. This plant has an uncanny ability to reproduce a new plant from (almost) any of its cells. We practiced our gene-transfer experiments on tobacco cells with impunity, and we could coax transgenic plants to develop from almost any cell into which *Agrobacterium* had transferred our experimental gene. This ease of regeneration of tobacco did not prepare us for the real world, whose principal food crops (unlike tobacco) were monocots—corn, wheat, rice, sorghum, and millet—to which the technology would ultimately need to be applied. Regeneration of these monocot plants from certain rare cells would be needed, and gene transfer to those very cells must be achieved. This process took years of research, and solutions were unique for each plant. In addition, much of the work was performed in small or large biotech companies, which sought to block competitors by applying for patent protection on methods they developed. Thus, still other methods had to be developed if licensing was not an option.

Another challenge we faced was bringing about expression of the "transgenes" we introduced into the plant cell. We optimistically supposed that any transgene, if given a plant gene promoter, would function in plants. After all, in 1983 the first gene everyone tried, the one coding for neomycin phosphotransferase II, had worked beautifully! The gene encoding a *Bacillus thuringiensis* insecticidal protein (nicknamed Bt, among other things, in the lab) was to teach us humility. Considerable ingenuity was needed to figure out why the Bt gene refused to express properly in the plant, and what to do about it. In the end, we learned to avoid many problems by using an artificial copy of this Bt gene constructed from plant-preferred codons. Although we thought of the genetic code as universal, as a practical matter, correct and fluent gene translation turned out to require, where a choice of codons was provided, that we use the plant's favorites.

An entirely new problem was how to determine product safety. Once the transgenic plant was performing properly, how should it be tested for any unforeseen properties that might conceivably make it harmful, toxic, allergenic, weedy (i.e., a pest in subsequent crops grown in the field), or disagreeable in any other way one could imagine? Ultimately, as they gained experience with these new products, regulatory agencies developed protocols for testing transgenic plants. The transgene must be stable, the plant must produce no new material that looks like an allergen, and the plant must have (at least) the original nutritional value expected of that food. In essence, it must be the same familiar plant you start with except for the (predicted) new trait encoded by the transgene. And of course the protein encoded by the transgene must be safe—for consumption by humans or animals if it is food or feed, and by non-target organisms in the environment likely to encounter it. Plants made by traditional plant breeding using "wide crossing" to bring in a desired gene from a distant (weedy or progenitor) relative are more likely to have unexpected properties than are transgenic plants. That is because unwanted and unknown genes will always be linked to the desirable trait sought in the wide cross.

The final problem—one still unsolved in many parts of the world—is that the transgenic plant, once certified safe and functional, must be accepted by consumers. Here I speak as an aging but fond midwife looking at this adolescent technology that I helped to birth. I find that we are now facing a new kind of challenge, one on which all of the science discussed here seems to have surprisingly little impact.

Many consumers oppose transgenic plants as something either dangerous or unethical, possibly both. These opponents are not likely to inform themselves about plant biotechnology by reading materials such as you will find assembled between the covers of this book. But many are at least curious about this unknown thing that they oppose. I hope that many of you who read this book will become informed advocates of plant biotechnology. Talk to the curious. Replace suspicion, where you can, with information. Replace doubt with evidence. I do not think, however, that in order to spread trust, it is necessary to teach everyone about this technology. People are busy. They will not expend the time and energy to inform themselves in depth. I think that you only need to convince people that *you* have studied this subject in detail, that you have read this book, that you harbor no bias, and that you think that it is safe and natural, as I believe you will.

I have invested most of my career in developing and exploiting the technology for putting new genes into plants. My greatest hope is to see wide—at least wider—acceptance of transgenic plants by consumers during my lifetime. Transgene integration by plants is a natural phenomenon, so much so that we are still trying to figure out exactly how Mother Nature does it. *Agrobacterium* was a microbial genetic engineer long before I began studying DNA. Plant biotechnology has already made significant and positive environmental contributions, as you will discover in the very first chapter of this book. It has the potential to be a powerful new tool for plant breeders, one that they will surely need in facing the challenges of rapid climate change, flood and drought, global warming, as well as the new pests and diseases that these changes may bring. The years ahead promise to be very challenging and interesting. I think that this book will serve you readers well as you prepare for your various roles in meeting those challenges. Enjoy your travels through these chapters and beyond, and I sincerely hope that your journey may turn out to be as interesting and rewarding as mine has been.

MARY-DELL CHILTON

Syngenta Biotechnology
Research Triangle Park, North Carolina

CONTRIBUTORS

Detlef Bartsch, BVL, Bundesamt für Verbraucherschhutz und Lebensmittelsicherheit (Federal Office of Consumer Protection and Food Safety), Mauerstrasse 39-42, D-10117 Berlin, Germany (Detlef.Bartsch@bvl.bund.de)

Alan B. Bennett, Public Intellectual Property Resource for Agriculture, Department of Plant Sciences, Plant Reproductive Biology Building, Extension Center Circle, One Shields Avenue, University of California, Davis, CA 95616 (abbennett@ucdavis.edu)

Sara Boettiger, Public Intellectual Property Resource for Agriculture, Department of Plant Sciences, Plant Reproductive Biology Building, Extension Center Circle, One Shields Avenue, University of California, Davis, CA, 95616-8631 (sara.hearn@ucop.edu)

Graham Brookes, PG Economics Ltd, Wessex Barn, Dorchester Road, Frampton, Dorchester, Dorset DT2 9NB, United Kingdom (graham.brookes@btinternet.com)

Vinitha Cardoza, BASF Plant Science LLC, 26 Davis Drive, Durham, NC 27709 (vinitha.cardoza@BASF.com)

Cecilia Chi-Ham, Public Intellectual Property Resource for Agriculture, Department of Plant Sciences, Plant Reproductive Biology Building, Extension Center Circle, One Shields Avenue, University of California, Davis, CA 95616 (clchiham@ucdavis.edu)

Mark D. Curtis, Institute of Plant Biology, University of Zurich, Zollikerstrasse 107, 8008 Zurich, Switzerland (mcurtis@botinst.unizh.ch)

Taniya Dhillon, Department of Horticulture and Crop Science, OARDC/Ohio State University, 1680 Madison Avenue, Wooster, OH 44691 (dhillon.18@osu.edu)

John Finer, Department of Horticulture and Crop Science, OARDC/Ohio State University, 1680 Madison Avenue, Wooster, OH 44691 (Finer.1@osu.edu)

Alison K. Flynn, Veterinary Medical Center, University of Florida, Gainesville, FL (Flynn-LurieA@vetmed.ufl.edu)

Maria Gallo, Agronomy Department, Cancer/Genetics Research Complex, Room 303, 1376 Mowry Road, PO Box 103610, University of Florida, Gainesville, FL 32610-3610 (mgm@ufl.edu)

Achim Gathmann, BVL, Bundesamt für Verbraucherschhutz und Lebensmittelsicherheit (Federal Office of Consumer Protection and Food Safety), Mauerstrasse 39-42, D-10117 Berlin, Germany (achim.gathmann@bvl.bund.de)

Glenda E. Gillaspy, Department of Biochemistry, 542 Latham Hall, Virginia Tech, Blacksburg, VA 24061 (gillaspy@vt.edu)

Gregory Graff, Public Intellectual Property Resource for Agriculture, Department of Plant Sciences, Plant Reproductive Biology Building, Extension Center Circle, One Shields Avenue, University of California, Davis, CA, 95616-8631 (ggraff@are.berkeley.edu)

Matthew D. Halfhill, Department of Biology, Saint Ambrose University, Davenport, IA 52803 (HalfhillMatthewD@sau.edu)

Kenneth L. Korth, Department of Plant Pathology, 217 Plant Science Building, University of Arkansas, Fayetteville, AR 72701 (kkorth@uark.edu)

Alan McHughen, Batchelor Hall 3110, University of California, Riverside, CA 92521-0124 (alanmc@ucr.edu)

Brian Miki, Agriculture and Agri-Food Canada, 960 Carling Avenue, Ottawa, Ontario, Canada K1A 0C6 (mikib@agr.gc.ca)

David W. Ow, Plant Gene Expression Center, USDA-ARS/UC Berkeley, 800 Buchanan Street, Albany, CA 94710 (david_ow@berkeley.edu)

Douglas Powell, International Food Safety Network, Department of Diagnostic Medicine/Pathobiology, Kansas State University, Manhattan, KS 66506 (dpowell@ksu.edu)

Christiane Saeglitz, Institute of Environmental Research (Biology V), Aachen University (RWTH), Worringerweg 1, D-52056 Aachen, Germany (saeglitz@rwth-aachen.de)

Arti Sinha, Department of Biology, Carleton University, 1233 Colonel By Drive, Ottawa, Ontario K1S5B7, Canada (biology@carleton.ca)

C. Neal Stewart, Jr., Department of Plant Sciences, 2431 Joe Johnson Drive, Room 252 Ellington Plant Sciences, University of Tennessee, Knoxville, TN 37996-4561 (nealstewart@utk.edu)

Nicholas (Nick) Tinker, Agriculture and Agri-Food Canada, 960 Carling Avenue, Ottawa, Ontario, Canada K1A 0C6 (tinkerna@agr.gc.ca)

Suzanne I. Warwick, Agriculture and Agri-Food Canada, Eastern Cereal and Oilseeds Research Centre, K.W. Neatby Bldg., C.E.F., Ottawa, Ontario, Canada K1A 0C6 (warwicks@agr.gc.ca)

Janice Zale, Department of Plant Sciences, University of Tennessee, Knoxville, TN 37996 (jzale@utk.edu)

CHAPTER 1

Plant Agriculture: The Impact of Biotechnology

GRAHAM BROOKES

PG Economics Ltd, Frampton, Dorchester, United Kingdom

1.0. CHAPTER SUMMARY AND OBJECTIVES

1.0.1. Summary

Since the first stably transgenic plant produced in the early 1980s and the first commercialized transgenic plant in 1995, biotechnology has revolutionized plant agriculture. More than a billion acres of transgenic cropland has been planted worldwide, with over 50 trillion transgenic plants grown in the United States alone. In the United States, over half of the corn and cotton and three-quarters of soybean produced are transgenic for insect resistance, herbicide resistance, or both. Biotechnology has been the most rapidly adopted technology in the history of agriculture and continues to expand in much of the developed and developing world.

1.0.2. Discussion Questions

1. What biotechnology crops are grown and where?
2. Why do farmers use biotech crops?
3. How has the adoption of plant biotechnology impacted on the environment?

1.1. INTRODUCTION

The year 2005 saw the tenth commercial planting season of genetically modified (GM) crops, which were first widely grown in 1996. In 2006, the billionth acre of GM crops was planted somewhere on Earth. These milestones provide an opportunity to critically assess the impact of this technology on global agriculture. This chapter therefore examines

Plant Biotechnology and Genetics: Principles, Techniques, and Applications, Edited by C. Neal Stewart, Jr.

specific global socioeconomic impacts on farm income and environmental impacts with respect to pesticide usage and greenhouse gas (GHG) emissions of the technology.[1]

1.2. BIOTECHNOLOGY CROPS PLANTINGS

Although the first commercial GM crops were planted in 1994 (tomato), 1996 was the first year in which a significant area [1.66 million hectares (ha)] of crops were planted containing GM traits. Since then there has been a dramatic increase in plantings, and by 2005/06, the global planted area reached approximately 87.2 million ha.

Almost all of the global GM crop area derives from soybean, maize (corn), cotton, and canola (Fig. 1.1).[2] In 2005, GM soybean accounted for the largest share (62%) of total GM crop cultivation, followed by maize (22%), cotton (11%), and canola (5%). In terms of the share of total global plantings to these four crops accounted for by GM crops, GM traits accounted for a majority of soybean grown (59%) in 2005 (i.e., non-GM soybean accounted for 41% of global soybean acreage in 2005). For the other three main crops, the GM shares in 2005 of total crop production were 13% for maize, 27% for cotton, and 18% for canola (i.e., the majority of global plantings of these three crops continued to be non-GM in 2005). The trend in plantings of GM crops (by crop) from 1996 to 2005 is shown in Figure 1.2. In terms of the type of biotechnology trait planted, Figure 1.3 shows that GM

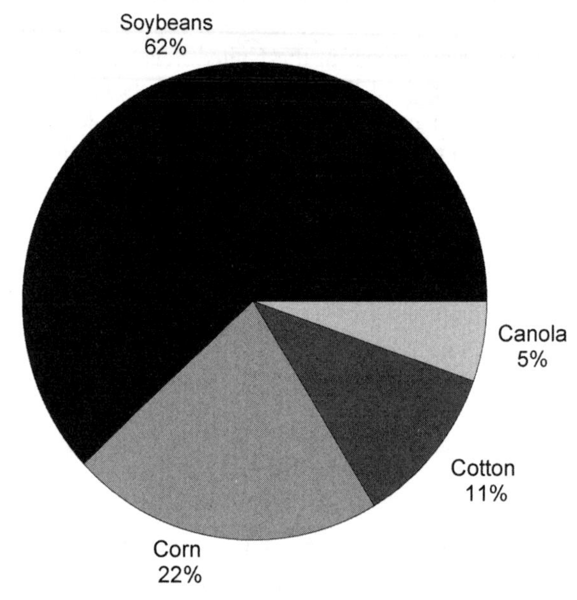

Figure 1.1. Global GM crop plantings in 2005 by crop (base area: 87.2 million ha). (*Sources*: ISAAA, Canola Council of Canada, CropLife Canada, USDA, CSIRO, ArgenBio.)

[1]Brookes G, Barfoot P (2007): Gm crops: The first ten years—global socio-economic and environmental impacts. *AgbioForum* **9**:1–13.

[2]In 2005 there were also additional GM crop plantings of papaya (530 ha) and squash (2400 hectares) in the United States.

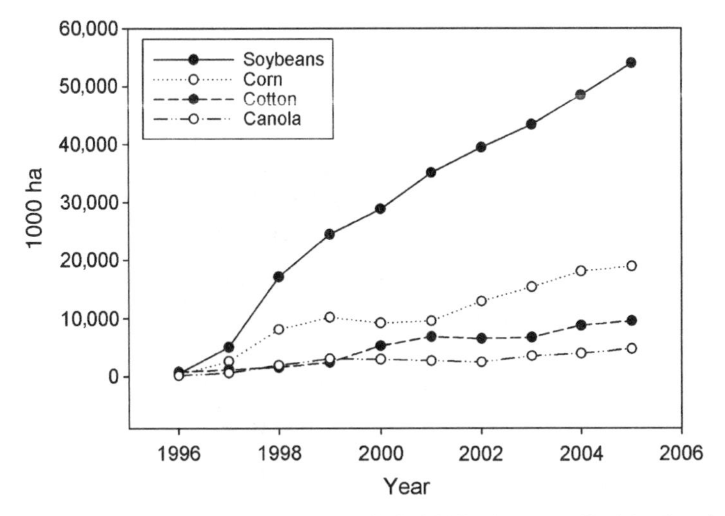

Figure 1.2. Global GM crop plantings by crop 1996–2005. (*Sources*: ISAAA, Canola Council of Canada, CropLife Canada, USDA, CSIRO, ArgenBio.)

herbicide-tolerant soybean dominate, accounting for 58% of the total, followed by insect-resistant (largely Bt) maize and cotton with respective shares of 16% and 8%.[3] In total, herbicide tolerant crops (GM HT) account for 76%, and insect resistant crops (GM IR) account for 24% of global plantings. Finally, looking at where biotech crops have been grown, the United States had the largest share of global GM crop plantings in 2005

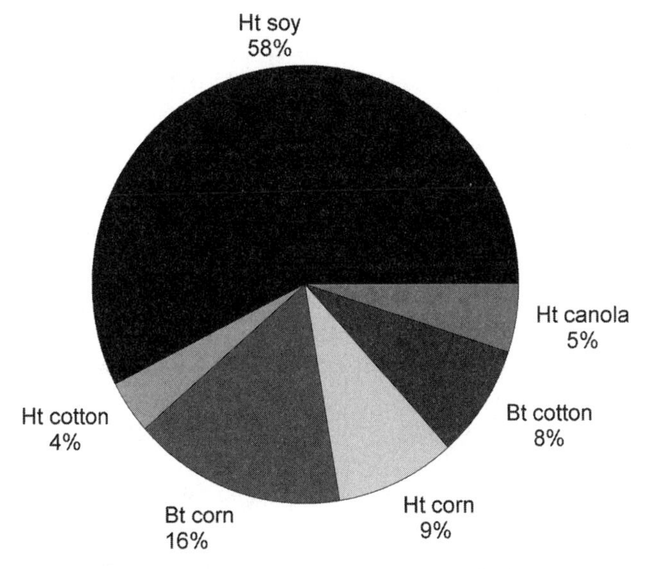

Figure 1.3. Global GM crop plantings by main trait and crop: 2005. (*Sources*: Various, including ISAAA, Canola Council of Canada, CropLife Canada, USDA, CSIRO, ArgenBio.)

[3]The reader should note that the total number of plantings by trait produces a higher global planted area (93.9 million ha) than the global area by crop (87.2 million ha) because of the planting of some crops containing the stacked traits of herbicide tolerance and insect resistance (e.g., a single plant with two biotech traits).

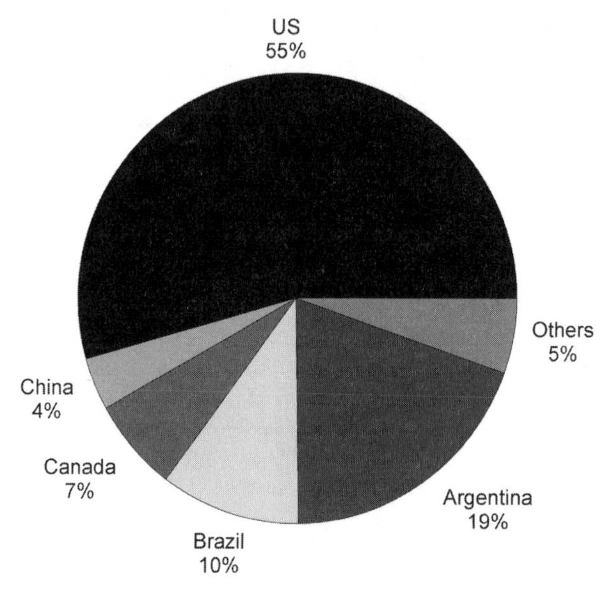

Figure 1.4. Global GM crop plantings 2005 by country. (*Sources*: ISAAA, Canola Council of Canada, CropLife Canada, USDA, CSIRO, ArgenBio.)

(55%: 47.4 million ha), followed by Argentina (16.93 million ha: 19% of the global total). The other main countries planting GM crops in 2005 were Canada, Brazil, and China (Fig. 1.4).

1.3. WHY FARMERS USE BIOTECH CROPS

The primary driver of adoption among farmers (both large commercial and small-scale subsistence) has been the positive impact on farm income. The adoption of biotechnology has had a very positive impact on farm income derived mainly from a combination of enhanced productivity and efficiency gains (Table 1.1). In 2005, the direct global farm income benefit from GM crops was $5 billion. If the additional income stemming from second crop soybeans in Argentina is considered,[4] this income gain rises to $5.6 billion. This is equivalent to having added between 3.6% and 4.0% to the value of global production of the four main crops of soybean, maize, canola, and cotton, a substantial impact. Since 1996, worldwide farm incomes have increased by $24.2 billion or $27 billion inclusive of second-crop soybean gains in Argentina directly because of the adoption of GM crop technology.

The largest gains in farm income have arisen in the soybean sector, largely from cost savings, where the $2.84 billion additional income generated by GM HT soybean in 2005 has been equivalent to adding 7.1% to the value of the crop in the GM-growing countries, or adding the equivalent of 6.05% to the $47 billion value of the global soybean crop in 2005. These economic benefits should, however, be placed within the context of a significant increase in the level of soybean production in the main

[4]The adoption of herbicide-tolerant soybean has facilitated the adoption of no and reduced tillage production practices, which effectively shorten the production season from planting to harvest. As a result, it has enabled many farmers in Argentina to plant a crop of soybean immediately after a wheat crop in the same season (hence the term *second-crop soybean*). In 2005, about 15% of the total soybean crop in Argentina was second-crop.

TABLE 1.1. Global Farm Income Benefits from Growing GM Crops 1996–2005 (million US $)

Trait and Crop	Increase in Farm Income, 2005	Increase in Farm Income, 1996–2005	Farm Income Benefit in 2005 as % of Total Value of Production of These Crops in GM-Adopting Countries	Farm Income Benefit in 2005 as % of Total Value of Global Production of These Crops
GM herbicide-tolerant soybean	2281 (2842)	11,686 (14,417)	5.72 (7.1)	4.86 (6.05)
GM herbicide-tolerant maize	212	795	0.82	0.39
GM herbicide-tolerant cotton	166	927	1.16	0.64
GM herbicide-tolerant canola	195	893	9.45	1.86
GM insect-resistant maize	416	2,367	1.57	0.77
GM insect-resistant cotton	1,732	7,510	12.1	6.68
Others	25	66	N/A	N/A
Totals	5027 (5588)	24,244 (26,975)	6.0 (6.7)	3.6 (4.0)

Notes: Others = virus-resistant papaya and squash, rootworm resistant maize. Figures in parentheses include second-crop benefits in Argentina. Totals for the value shares exclude "other crops" (i.e., relate to the four main crops of soybeans, maize, canola and cotton). Farm income calculations are net farm income changes after inclusion of impacts on costs of production (e.g., payment of seed premia, impact on crop protection expenditure). (N/A = not applicable.)

GM-adopting countries. Since 1996, the soybean area and production in the leading soybean producing countries of the United States, Brazil, and Argentina increased by 58% and 65%, respectively.

Substantial gains have also arisen in the cotton sector through a combination of higher yields and lower costs. In 2005, cotton farm income levels in the GM-adopting countries increased by $1.9 billion and since 1996, the sector has benefited from an additional $8.44 billion. The 2005 income gains are equivalent to adding 13.3% to the value of the cotton crop in these countries, or 7.3% to the $26 billion value of total global cotton production. This is a substantial increase in value-added terms for two new cotton seed technologies.

Significant increases to farm incomes have also resulted in the maize and canola sectors. The combination of GM IR and GM HT technology in maize has boosted farm incomes by over $3.1 billion since 1996. An additional $893 million has been generated in the North American canola sector.

Overall, the economic gains derived from planting GM crops have been of two main types: (1) increased yields (associated mostly with GM insect-resistant technology) and (2) reduced costs of production derived from less expenditure on crop protection (insecticides and herbicides) products and fuel.

Table 1.2 summarizes farm income impacts in key GM-adopting countries highlighting the important direct farm income benefit arising from growing GM HT soybeans in

TABLE 1.2. GM Crop Farm Income Benefits during 1996–2005 in Selected Countries (million US $)

Country	GM HT Soybean	GM HT Maize	GM HT Cotton	GM HT Canola	GM IR Maize	GM IR Cotton	Total
USA	7570	771	919	101	1957	1627	12,945
Argentina	5197	0.2	4.0	N/A	159	29	5389.2
Brazil	1367	N/A	N/A	N/A	N/A	N/A	1367
Paraguay	132	N/A	N/A	N/A	N/A	N/A	132
Canada	69	24	N/A	792	145	N/A	1031
South Africa	2.2	0.3	0.2	N/A	59	14	75.7
China	N/A	N/A	N/A	N/A	N/A	5168	5168
India	N/A	N/A	N/A	N/A	N/A	463	463
Australia	N/A	N/A	4.1	N/A	N/A	150	154.1
Mexico	N/A	N/A	N/A	N/A	N/A	55	55
Philippines	N/A	N/A	N/A	N/A	8	N/A	8
Spain	N/A	N/A	N/A	N/A	28	N/A	28

Note: Argentine GM HT soybeans includes second crop soybeans benefits.

Argentina, GM IR cotton in China, and a range of GM cultivars in the United States. It also illustrates the growing level of farm income benefits obtained in developing countries such as South Africa, Paraguay, India, the Philippines, and Mexico from planting GM crops.

In terms of the division of the economic benefits, it is interesting to note that farmers in developing countries derived the majority of the farm income benefits in 2005 (55%) relative to farmers in developed countries (Table 1.3). The vast majority of these income gains for developing country farmers have been from GM IR cotton and GM HT soybean.[5]

Examination of the cost farmers pay for accessing GM technology relative to the total gains derived shows (Table 1.4) that across the four main GM crops, the total cost was equal to about 26% of the total farm income gains. For farmers in developing countries the total cost is equal to about 13% of total farm income gains, while for farmers in

TABLE 1.3. GM Crop Farm Income Benefits, 2005: Developing Versus Developed Countries (million US $)

Crop	Developed	Developing[a]	% Developed	% Developing
GM HT soybean	1183	1658	41.6	58.4
GM IR maize	364	53	86.5	13.5
GM HT maize	212	0.3	99.9	0.1
GM IR cotton	354	1378	20.4	79.6
GM HT cotton	163	3	98.4	1.6
GM HT canola	195	0	100	0
GM VR papaya and squash	25	0	100	0
Totals	2496	3092	45	55

[a]Developing countries include all countries in South America.

[5]The author acknowledges that the classification of different countries into "developing" or "developed country" status affects the distribution of benefits between these two categories of country. The definition used here is consistent with the definition used by others, including the International Service for the Acquisition of Agri-Biotech Applications (ISAAA) [see the review by James C (2006) Global Status of GM Crops 2006 ISAAA Brief No 35.].

TABLE 1.4. Cost of Accessing GM Technology[a] (in % Terms) Relative to Total Farm Income Benefits, 2005

Crop	All Farmers	Developed Countries	Developing Countries
GM HT soybean	21	32	10
GM IR maize	44	43	48
GM HT maize	38	38	81
GM IR cotton	21	41	13
GM HT cotton	44	43	65
GM HT canola	47	47	N/A
Totals	26	38	13

[a]Cost of accessing the technology is based on the seed premia paid by farmers for using GM technology relative to its conventional equivalent.

developed countries the cost is about 38% of the total farm income gain. Although circumstances vary among countries, the higher share of total gains derived by farmers in developing countries relative to farmers in developed countries reflects factors such as weaker provision and enforcement of intellectual property rights.

In addition to the tangible and quantifiable impacts on farm profitability presented above, there are other important, more intangible (difficult to quantify) impacts of an economic nature. Many studies on the impact of GM crops have identified the factors listed below as being important influences for adoption of the technology.

1.3.1. Herbicide-Tolerant Crops

- This method provides increased management flexibility due to a combination of the ease of use associated with broad-spectrum, postemergent herbicides like glyphosate (often referred to by its more commonly known brand name of Roundup) and the increased/longer time window for spraying.
- In a conventional crop, postemergent weed control relies on herbicide applications before the weeds and crop are well established. As a result, the crop may suffer "knockback" to its growth from the effects of the herbicide. In the GM HT crop, this problem is avoided because the crop is tolerant to the herbicide and spraying can occur at a later stage when the crop is better able to withstand any possible knockback effects.
- This method facilitates the adoption of conservation or no-tillage systems. This provides for additional cost savings such as reduced labor and fuel costs associated with plowing.
- Improved weed control has contributed to reduced harvesting costs—cleaner crops have resulted in reduced times for harvesting. It has also improved harvest quality and led to higher levels of quality price bonuses in some regions (e.g., Romania).
- Potential damage caused by soil-incorporated residual herbicides in follow-on crops has been eliminated.

1.3.2. Insect-Resistant Crops

- For production risk management/insurance purposes, this method eliminates the risk of significant pest damage.
- A "convenience" benefit is derived because less time is spent walking through the crop fields to survey insects and insect damage and/or apply insecticides.

- Savings in energy use are realized—associated mainly with less frequent aerial spraying.
- There are savings in machinery use (for spraying and possibly reduced harvesting times).
- The quality of Bt maize is perceived as superior to that of non-Bt maize because the level of fungal (*Fusarium*) damage, which leads to mycotoxin presence in plant tissues, is lower with Bt maize. As such, there is an increasing body of evidence that *Fusarium* infection levels and mycotoxin levels in GM insect resistant maize are significantly (5–10-fold) lower than those found in conventional (nonbiotech) crops. This lower mycotoxin contamination in turn leads to a safer food or feed product for consumption.
- There Health and safety for farmers and farmworkers is improved (handling and use of pesticides is reduced).
- The growing season is shorter (e.g., for some cotton growers in India), which allows some farmers to plant a second crop in the same season (notably maize in India). Also some Indian cotton growers have reported commensurate benefits for beekeepers as fewer bees are now lost to insecticide spraying.

1.3.3. Conclusion

It is important to recognize that these largely intangible benefits are considered by many farmers as the primary reasons for adoption of GM technology, and in some cases farmers have been willing to adopt for these reasons alone, even when the measurable impacts on yield and direct costs of production suggest marginal or no direct economic gain. As such, the estimates of the farm level benefits presented above probably understate the real value of the technology to farmers. For example, the easier and more convenient weed control methods and facilitation of no/reduced tillage practices were cited as the most important reason for using GM herbicide-tolerant soybean by US farmers when surveyed by the American Soybean Association in 2001.

With respect to the nature and size of GM technology adopters, there is clear evidence that farm size has not been a factor affecting use of the technology. Both large and small farmers have adopted GM crops. Size of operation has not been a barrier to adoption. In 2005, 8.5 million farmers, more than 90% of whom were resource-poor farmers in developing countries, were using the technology globally. This is logical. The benefit is in the seed, which must be planted by both small and large farmers.

The significant productivity and farm income gains identified above have, in some countries (notably Argentina), also made important contributions to income and employment generation in the wider economy. For example, in Argentina, the economic gains resulting from the 140% increase in the soybean area since 1995 are estimated to have contributed to the creation of 200,000 additional agriculture-related jobs (Trigo et al. 2002) and to export-led economic growth.

1.4. HOW THE ADOPTION OF PLANT BIOTECHNOLOGY HAS IMPACTED THE ENVIRONMENT

The two key aspects of environmental impact of biotech crops examined below are decreased insecticide and herbicide use, and the impact on carbon emissions and soil conservation.

1.4.1. Environmental Impacts from Changes in Insecticide and Herbicide Use

Usually, changes in pesticide use with GM crops have traditionally been presented in terms of the volume (quantity) of pesticide applied. While comparisons of total pesticide volume used in GM and non-GM crop production systems can be a useful indicator of environmental impacts, it is an imperfect measure because it does not account for differences in the specific pest control programs used in GM and non-GM cropping systems. For example, different specific chemical products used in GM versus conventional crop systems, differences in the rate of pesticides used for efficacy, and differences in the environmental characteristics (mobility, persistence, etc.) are masked in general comparisons of total pesticide volumes used.

To provide a more robust measurement of the environmental impact of GM crops, the analysis presented below includes an assessment of both pesticide active-ingredient use and the specific pesticides used via an indicator known as the *environmental impact quotient* (EIQ). This universal indicator, developed by Kovach et al. 1992 and updated annually, effectively integrates the various environmental impacts of individual pesticides into a single *field value per hectare*. This index provides a more balanced assessment of the impact of GM crops on the environment as it draws on all of the key toxicity and environmental exposure data related to individual products, as applicable to impacts on farmworkers, consumers, and ecology, and provides a consistent and comprehensive measure of environmental impact. Readers should, however, note that the EIQ is an indicator only and therefore does not account for all environmental issues and impacts.

The EIQ value is multiplied by the amount of pesticide active ingredient (AI) used per hectare to produce a field EIQ value. For example, the EIQ rating for glyphosate is 15.3. By using this rating multiplied by the amount of glyphosate used per hectare (e.g., a hypothetical example of 1.1 kg applied per hectare), the field EIQ value for glyphosate would be equivalent to 16.83/ha. In comparison, the field EIQ/ha value for a commonly used herbicide on corn crops (atrazine) is 22.9/ha.

The EIQ indicator is therefore used for comparison of the field EIQ/ha values for conventional versus GM crop production systems, with the total environmental impact or load of each system, a direct function of respective field EIQ/ha values, and the area planted to each type of production (GM vs. non-GM).

The EIQ methodology is used below to calculate and compare typical EIQ values for conventional and GM crops and then aggregate these values to a national level. The level of pesticide use in the respective areas planted for conventional and GM crops in each year was compared with the level of pesticide use that probably would otherwise have occurred if the whole crop, in each year, had been produced using conventional technology (based on the knowledge of crop advisers). This approach addresses gaps in the availability of herbicide or insecticide usage data in most countries and differentiates between GM and conventional crops. Additionally, it allows for comparisons between GM and non-GM cropping systems when GM accounts for a large proportion of the total crop planted area. For example, in the case of soybean in several countries, GM represents over 60% of the total soybean crop planted area. It is not reasonable to compare the production practices of these two groups as the remaining non-GM adopters might be farmers in a region characterized by below-average weed or pest pressures or with a tradition of less intensive production systems, and hence, below-average pesticide use.

TABLE 1.5. Impact of Changes in Use of Herbicides and Insecticides from Growing GM Crops Globally, 1996–2005

Trait	Change in Volume of Active Ingredient Used (million kg)	Change in Field EIQ Impact[a]	% Change in AI use in GM-Growing Countries	% Change in Environmental Impact in GM-Growing Countries
GM herbicide-tolerant soybean	−51.4	−4,865	−4.1	−20.0
GM herbicide-tolerant maize	−36.5	−845	−3.4	−4.0
GM herbicide-tolerant cotton	−28.6	−1,166	−15.1	−22.7
GM herbicide-tolerant canola	−6.3	−310	−11.1	−22.6
GM insect-resistant maize	−7.0	−403	−4.1	−4.6
GM insect-resistant cotton	−94.5	−4,670	−19.4	−24.3
Totals	−224.3	−12,259	−6.9	−15.3

[a]In terms of million field EIQ/ha units.

GM crops have contributed to a significant reduction in the global environmental impact of production agriculture (Table 1.5). Since 1996, the use of pesticides was reduced by 224 million kg of active ingredient, constituting a 6.9% reduction, and the overall environmental impact associated with pesticide use on these crops was reduced by 15.3%. In absolute terms, the largest environmental gain has been associated with the adoption of GM HT soybean and reflects the large share of global soybean plantings accounted for by GM soybean. The volume of herbicide use in GM soybean decreased by 51 million kg since 1996, a 4.1% reduction, and the overall environmental impact decreased by 20%. It should be noted that in some countries, such as in Argentina and Brazil, the adoption of GM HT soybean has coincided with increases in the volume of herbicides used relative to historic levels. This net increase largely reflects the facilitating role of the GM HT technology in accelerating and maintaining the switch away from conventional tillage to no/low-tillage production systems, along with their inherent environmental benefits (discussed below). This net increase in the volume of herbicides used should, therefore, be placed in the context of the reduced GHG emissions arising from this production system change (see discussion below) and the general dynamics of agricultural production system changes.

Major environmental gains have also been derived from the adoption of GM insect-resistant (IR) cotton. These gains were the largest of any crop on a per hectare basis. Since 1996, farmers have used 95.5 million kg less insecticide in GM IR cotton crops (a 19.4% reduction), and reduced the environmental impact by 24.3%. Important environmental gains have also arisen in the maize and canola sectors. In the maize sector, pesticide

TABLE 1.6. Reduction in "Environmental Impact" from Changes in Pesticide Use Associated with GM Crop Adoption by Country, 1996–2005, Selected Countries (% Reduction in Field EIQ Values)

Country	GM HT Soybean	GM HT Maize	GM HT Cotton	GM HT Canola	GM IR Maize	GM IR Cotton
USA	29	4	24	38	5	23
Argentina	21	NDA	NDA	N/A	0	4
Brazil	6	N/A	N/A	N/A	N/A	N/A
Paraguay	13	N/A	N/A	N/A	N/A	N/A
Canada	9	5	N/A	22	NDA	N/A
South Africa	7	0.44	6	N/A	2	NDA
China	N/A	N/A	N/A	N/A	N/A	28
India	N/A	N/A	N/A	N/A	N/A	3
Australia	N/A	N/A	4	N/A	N/A	22
Mexico	N/A	N/A	N/A	N/A	N/A	NDA
Spain	N/A	N/A	N/A	N/A	30	N/A

Note: Zero impact for GM IR maize in Argentina is due to the negligible (historic) use of insecticides on the Argentine maize crop. (NDA = no data available.)

use decreased by 43 million kg and the environmental impact decreased because of reduced insecticide use (4.6%) and a switch to more environmentally benign herbicides (4%). In the canola sector, farmers reduced herbicide use by 6.3 million kg (an 11% reduction) and the environmental impact has fallen by 23% because of a switch to more environmentally benign herbicides.

The impact of changes in insecticide and herbicide use at the country level (for the main GM-adopting countries) is summarized in Table 1.6.

In terms of the division of the environmental benefits associated with less insecticide and herbicide use for farmers in developing countries relative to farmers in developed countries, Table 1.7 shows that in 2005, the majority of the environmental benefits associated with lower insecticide and herbicide use have been for developing-country farmers. The vast majority of these environmental gains have been from the use of GM IR cotton and GM HT soybeans.

TABLE 1.7. GM Crop Environmental Benefits from Lower Insecticide and Herbicide Use in 2005: Developing versus Developed Countries

	Percent of Total Reduction in EI[a]	
Crop	Developed Countries	Developing Countries[b]
GM HT soybean	53	47
GM IR maize	92	8
GM HT maize	99	1
GM IR cotton	15	85
GM HT cotton	99	1
GM HT canola	100	0
Totals	46	54

[a]Environmental impact.
[b]"Developing countries", include all countries in South America.

1.4.2. Impact on Greenhouse Gas (GHG) Emissions

Reductions in the level of GHG emissions from GM crops are from two principal sources:

1. GM crops contribute to a reduction in fuel use from less frequent herbicide or insecticide applications and a reduction in the energy use in soil cultivation. For example, Lazarus and Selley (2005) estimated that one pesticide spray application uses 1.045 liters (L) of fuel, which is equivalent to 2.87 kg/ha of carbon dioxide emissions. In this analysis we used the conservative assumption that only GM IR crops reduced spray applications and ultimately GHG emissions. In addition to the reduction in the number of herbicide applications there has been a shift from conventional tillage to no/reduced tillage. This has had a marked effect on tractor fuel consumption because energy-intensive cultivation methods have been replaced with no/reduced tillage and herbicide-based weed control systems. The GM HT crop where this is most evident is GM HT soybean. Here, adoption of the technology has made an important contribution to facilitating the adoption of reduced/no-tillage (NT) farming (CTIC 2002). Before the introduction of GM HT soybean cultivars, NT systems were practiced by some farmers using a number of herbicides and with varying degrees of success. The opportunity for growers to control weeds with a nonresidual foliar herbicide as a "burndown" preseeding treatment, followed by a postemergent treatment when the soybean crop became established, has made the NT system more reliable, technically viable, and commercially attractive. These technical advantages, combined with the cost advantages, have contributed to the rapid adoption of GM HT cultivars and the near-doubling of the NT soybean area in the United States (and also a $\geq$5-fold increase in Argentina). In both countries, GM HT soybean crops are estimated to account for 95% of the NT soybean crop area. Substantial growth in NT production systems has also occurred in Canada, where the NT canola area increased from 0.8 to 2.6 million ha (equal to about half of the total canola area) between 1996 and 2005 (95% of the NT canola area is planted with GM HT cultivars). Similarly, the area planted to NT in the US cotton crop increased from 0.2 to 1 million ha over the same period (86% of which is planted to GM HT cultivars). The increase in the NT cotton area has been substantial from a base of 200,000 ha to over 1.0 million ha between 1996 and 2005. The fuel savings resulting from changes in tillage systems are drawn from estimates from studies by Jasa (2002) and CTIC (2002). The adoption of NT farming systems is estimated to reduce cultivation fuel usage by 32.52 L/ha compared with traditional conventional tillage and 14.7 L/ha compared with (the average of) reduced tillage cultivation. In turn, this results in reductions in CO_2 emissions of 89.44 and 40.43 kg/ha, respectively.
2. The use of reduced/no-tillage[6] farming systems that utilize less plowing increase the amount of organic carbon in the form of crop residue that is stored or sequestered in the soil. This carbon sequestration reduces carbon dioxide emissions to the environment. Rates of carbon sequestration have been calculated for cropping systems using

[6]*No-tillage farming* means that the ground is not plowed at all, while *reduced tillage* means that the ground is disturbed less than it would be with traditional tillage systems. For example, under a no-tillage farming system, soybean seeds are planted through the organic material that is left over from a previous crop such as corn, cotton, or wheat. No-tillage systems also significantly reduce soil erosion and hence deliver both additional economic benefits to farmers, enabling them to cultivate land that might otherwise be of limited value and environmental benefits from the avoidance of loss of flora, fauna, and landscape features.

TABLE 1.8. Impact of GM Crops on Carbon Sequestration Impact in 2005; Car Equivalents

Crop/Trait/ Country	Permanent CO_2 Savings from Reduced Fuel Use (million kg CO_2)	Average Family Car Equivalents Removed from Road per Year from Permanent Fuel Savings	Potential Additional Soil Carbon Sequestration Savings (million kg CO_2)	Average Family Car Equivalents Removed from Road per Year from Potential Additional Soil Carbon Sequestration
US: GM HT soybean	176	78,222	2,195	975,556
Argentina: GM HT soybean	546	242,667	4,340	1,928,889
Other countries: GM HT soybeans	55	24,444	435	193,333
Canada: GM HT canola	117	52,000	1,083	481,520
Global GM IR cotton	68	30,222	0	0
Totals	962	427,556	8,053	3,579,298

Note: It is assumed that an average family car produces 150 g CO_2/km. A car does an average of 15,000 km/year and therefore produces 2250 kg of CO_2 per year.

normal tillage and reduced tillage, and these were incorporated in our analysis on how GM crop adoption has significantly facilitated the increase in carbon sequestration, ultimately reducing the release of CO_2 into the atmosphere. Of course, the amount of carbon sequestered varies by soil type, cropping system, and ecoregion. In North America, the International Panel on Climate Change estimates that the conversion from conventional tillage to no-tillage systems stores between 50 and 1300 kg C/ha annually (average 300 kg C/ha per year). In the analysis presented below, a conservative savings of 300 kg C/ha per annum was applied to all no-tillage agriculture and 100 kg C/ha^{-1} year^{-1} was applied to reduced-tillage agriculture. Where some countries aggregate their no/reduced-tillage data, the reduced-tillage saving value of 100 kg C/ha^{-1} year^{-1} was used. One kilogram of carbon sequestered is equivalent to 3.67 kg of carbon dioxide. These assumptions were applied to the reduced pesticide spray applications data on GM IR crops, derived from the farm income literature review, and the GM HT crop areas using no/reduced tillage (limited to the GM HT soybean crops in North and South America and GM HT canola crop in Canada[7]).

[7]Because of the likely small-scale impact and/or lack of tillage-specific data relating to GM HT maize and cotton crops (and the US GM HT canola crop), analysis of possible GHG emission reductions in these crops have not been included in the analysis. The no/reduced-tillage areas to which these soil carbon reductions were applied were limited to the increase in the area planted to no/reduced tillage in each country since GM HT technology has been commercially available. In this way the authors have tried to avoid attributing no/reduced-tillage soil carbon sequestration gains to GM HT technology on cropping areas that were using no/reduced-tillage cultivation techniques before GM HT technology became available.

Table 1.8 summarizes the impact on GHG emissions associated with the planting of GM crops between 1996 and 2005. In 2005, the permanent CO_2 savings from reduced fuel use associated with GM crops was 0.962 billion kg. This is equivalent to removing 430,000 cars from the road for a year.

The additional soil carbon sequestration gains resulting from reduced tillage with GM crops accounted for a reduction in 8.05 billion kg of CO_2 emissions in 2005. This is equivalent to removing nearly 3.6 million cars from the roads per year. In total, the carbon savings from reduced fuel use and soil carbon sequestration in 2005 were equal to removing 4 million cars from the road (equal to 17% of all registered cars in the UK).

1.5. CONCLUSIONS

GM technology has to date delivered several specific agronomic traits that have overcome a number of production constraints for many farmers. This has resulted in improved productivity and profitability for the 8.5 million GM-adopting farmers who have applied the technology to over 87 million ha in 2005.

Since the mid-1990s, this technology has made important positive socioeconomic and environmental contributions. These have arisen despite the limited range of GM agronomic traits commercialized thus far, in a small range of crops.

GM technology has delivered economic and environmental gains through a combination of their inherent technical advances and the role of technology in the facilitation and evolution of more cost-effective and environmentally friendly farming practices. More specifically:

- The gains from the GM IR traits have mostly been delivered directly from the technology (through yield improvements, reduced production risk, and decreased insecticide use). Thus, farmers (mostly in developing countries) have been able to improve their productivity and economic returns while also practicing more environmentally friendly farming methods.
- The gains from GM HT traits have come from a combination of direct benefits (mostly cost reductions to the farmer) and the facilitation of changes in farming systems. Thus, GM HT technology (especially in soybean) has played an important role in enabling farmers to capitalize on the availability of a low-cost, broad-spectrum herbicide (glyphosate) and in turn, facilitated the move away from conventional to low/no-tillage production systems in both North and South America. This change in production system has made additional positive economic contributions to farmers (and the wider economy) and delivered important environmental benefits, notably reduced levels of GHG emissions (from reduced tractor fuel use and additional soil carbon sequestration).

The impact of GM HT traits has, however, contributed to increased reliance on a limited range of herbicides, and this raises questions about the possible future increased development of weed resistance to these herbicides. For example, some degree of reduced effectiveness of glyphosate (and glufosinate) against certain weeds has already occurred. To the extent to which this may occur in the future, there will be an increased need to include low-dose applications of other herbicides in weed control programs (commonly used in conventional production systems), which may, in turn, marginally reduce the level of net environmental and economic gains derived from the current use of GM technology.

LIFE BOX 1.1. NORMAN E. BORLAUG

Norman E. Borlaug, Retired, President of the Sasakawa Africa Association and Distinguished Professor of Agriculture at Texas A&M Univeristy; Laureate, Winner, Nobel Peace Prize, 1970; Recipient, Congressional Gold Medal 2007

Norman Borlaug

The following text is excerpted from the book by biographer Leon Hesser, *The Man Who Fed the World: Nobel Peace Prize Laureate Norman Borlaug and His Battle to End World Hunger*, Durban House Dallas, Texas (2006):

> From the day he was born in 1914, Norman Borlaug has been an enigma. How could a child of the Iowa prairie, who attended a one-teacher, one-room school; who flunked the university entrance exam; and whose highest ambition was to be a high school science teacher and athletic coach, ultimately achieve the distinction as one of the hundred most influential persons of the twentieth century? And receive the Nobel Peace Prize for averting hunger and famine? And eventually be hailed as the man who saved hundreds of millions of lives from starvation—more than any other person in history?
>
> Borlaug, ultimately admitted to the University of Minnesota, met Margaret Gibson, his wife to be, and earned B.S., M.S., and Ph.D. degrees. The latter two degrees were in plant pathology and genetics under Professor E. C. Stakman, who did pioneering research on the plant disease rust, a parasitic fungus that feeds on phytonutrients in wheat, oats, and barley. Following three years with DuPont, Borlaug went to Mexico in 1944 as a member of a Rockefeller Foundation team to help increase food production in that hungry nation where rust diseases had taken their toll on wheat yields.
>
> Dr. Borlaug initiated three innovations that greatly increased Mexico's wheat yields. First, he and his Mexican technicians crossed thousands of varieties to find a select few that were resistant to rust disease. Next, he carried out a "shuttle breeding" program to cut in half the time it took to do the breeding work. He harvested seed from a summer crop that was grown in the high altitudes near Mexico City, flew to Obregon to plant the seed for a winter crop at sea level. Seed from that crop was flown back to near Mexico City and planted for a summer crop. Shuttle breeding not only worked, against the advice of fellow scientists, but serendipitously the varieties were widely adapted globally because it had been grown at different altitudes and latitudes and during different day lengths.
>
> But, there was a problem. With high levels of fertilizer in an attempt to increase yields, the plants grew tall and lodged. For his third innovation, then, Borlaug crossed his rust-resistant varieties with a short-strawed, heavy tillering Japanese variety. Serendipity squared. The resulting seeds were responsive to heavy applications of fertilizer without lodging. Yields were six to eight times higher than for traditional varieties in Mexico. It was these varieties, introduced in India and Pakistan

in the mid-1960s, which stimulated the Green Revolution that took those countries from near-starvation to self-sufficiency. For this remarkable achievement, Dr. Borlaug was awarded the Nobel Peace Prize in 1970.

In 1986, Borlaug established the World Food Prize, which provides $250,000 each year to recognize individuals in the world who are deemed to have done the most to increase the quantity or quality of food for poorer people. A decade later, the World Food Prize Foundation added a Youth Institute as a means to get young people interested in the world food problem. High school students are invited to submit essays on the world food situation. Authors of the 75 best papers are invited to read them at the World Food Prize Symposium in Des Moines in mid-October each year. From among these, a dozen are sent for eight weeks to intern at agricultural research stations in foreign countries. By the summer of 2007, approximately 100 Youth Institute interns had returned enthusiastically from those experiences and all are on track to become productively involved. This is an answer to Norman Borlaug's dream.

Borlaug has continually advocated increasing crop yields as a means to curb deforestation. In addition to his being recognized as having saved millions of people from starvation, it could be said that he has saved more habitat than any other person.

When Borlaug was born in 1914, the world's population was 1.6 billion. During his lifetime, population has increased four times, to 6.5 billion. Borlaug is often asked, "How many more people can the Earth feed?" His usual response: "I think the Earth can feed 10 billion people, IF, and this is a big IF, we can continue to use chemical fertilizer and there is public support for the relatively new genetic engineering research in addition to conventional research."

To those who advocate only organic fertilizer, he says, "For God's sake, let's use all the organic materials we can muster, but don't tell the world that we can produce enough food for 6.5 billion people with organic fertilizer alone. I figure we could produce enough food for only 4 billion with organics alone."

One of Borlaug's dreams, through genetic engineering, is to transfer the rice plant's resistance to rust diseases to wheat, barley, and oats. He is deeply concerned about a recent outbreak of rust disease in sub-Saharan Africa which, if it gets loose, can devastate wheat yields in much of the world.

Since 1984, Borlaug has served each fall semester at Texas A&M University as distinguished professor of international agriculture. In 1999, the university's Center for Southern Crop Improvement was named in his honor.

As President of the Sasakawa Africa Association (SAA) since 1986, Borlaug has demonstrated how to increase yields of wheat, rice, and corn in sub-Saharan Africa. To focus on food, population and agricultural policy, Jimmy Carter initiated Sasakawa-Global 2000, a joint venture between the SAA and the Carter Center's Global 2000 program.

Norman Borlaug has been awarded more than fifty honorary doctorates from institutions in eighteen countries. Among his numerous other awards are the U.S. Presidential Medal of Freedom (1977); the Rotary International Award (2002); the National Medal of Science (2004); the Charles A. Black Award for contributions to public policy and the public understanding of science (2005); the Congressional Gold Medal (2006); and the Padma Vibhushan, the Government of India's second highest civilian award (2006).

The Borlaug family includes son William, daughter Jeanie, five grandchildren and four great grandchildren. Margaret Gibson Borlaug, who had been blind in recent years, died on March 8, 2007 at age 95.

LIFE BOX 1.2. MARY-DELL CHILTON

Mary-Dell Chilton, Scientific and Technical Principal Fellow, Syngenta Biotechnology, Inc.; Winner of the Rank Prize for Nutrition (1987), and the Benjamin Franklin Medal in Life Sciences (2001); Member, National Academy of Sciences

Mary-Dell Chilton in the Washington University (St. Louis) Greenhouse 1982 with tobacco, the white rat of the plank kingdom.

I entered the University of Illinois in the fall of 1956, the autumn that Sputnik flew over. My major was called the "Chemistry Curriculum," and was heavy on science and light on liberal arts. When I entered graduate school in 1960 as an organic chemistry major, still at the University of Illinois, I took a minor in microbiology (we were required to minor in something. . .). To my astonishment I found a new love: in a course called "The Chemical Basis of Biological Specificity" I learned about the DNA double helix, the genetic code, bacterial genetics, mutations and bacterial transformation. I was hooked! I found that I could stay in the Chemistry Department (where I had passed prelims, a grueling oral exam) and work on DNA under guidance of a new thesis advisor, Ben Hall, a professor in physical chemistry. When Hall took a new position in the Department of Genetics at the University of Washington, I followed him. This led to a new and fascinating dimension to my education. My thesis was on transformation of *Bacillus subtilis* by single-stranded DNA,

As a postdoctoral fellow with Dr. Brian McCarthy in the Microbiology Department at the University of Washington, I did further work on DNA of bacteria, mouse, and finally maize. I became proficient in all of the then-current DNA technology. During this time I married natural products chemist Prof. Scott Chilton and we had two sons to whom I was devoted. But that was not enough. It was time to start my career!

Two professors (Gene Nester in microbiology and Milt Gordon in biochemistry) and I (initially as an hourly employee) launched a collaborative project on *Agrobacterium tumefaciens* and how it causes the plant cancer "crown gall." In hindsight it was no accident that we three represented at least three formal disciplines (maybe four or five, if you count my checkered career). Crown gall biology would involve us in plants, microbes, biochemistry, genetics, protein chemistry, natural products chemistry (in collaboration with Scott) and plant tissue culture. The multifaceted nature of the problem bound us together.

My first task was to write a research grant application to raise funds for my own salary. My DNA hybridization proposal was funded. Grant money flowed in the wake of Sputnik. Our primary objective was to determine whether DNA transfer from the bacterium to the plant cancer cells was indeed the basis of the disease, as some believed and others disputed. We disputed this continually amongst ourselves, often switching sides! This was the start of a study that has extended over my entire career. While we hunted for bacterial DNA, competitors in Belgium discovered that virulent strains of *Agrobacterium* contained enormous plasmids (circular DNA molecules) which we now know as Ti (tumor-inducing) plasmids. Redirecting our analysis, we found that gall cells contained not the whole Ti plasmid but a sector of it large enough to encompass 10–20 genes.

Further studies in several laboratories world-wide showed that this transferred DNA, T-DNA, turned out to be in the nuclei of the plant cells, attached to the plant's own chromosomal DNA. It was behaving as if it were plant genes, encoding messenger RNA and proteins in the plant. Some proteins brought about the synthesis of plant growth hormones that made the plant gall grow. Others caused the plant to synthesize, from simple amino acids and sugars or keto acids, derivatives called opines, some of which acted as bacterial hormones, inducing conjugation of the plasmid from one *Agrobacterium* to another. The bacteria could live on these opines, too, a feat not shared by most other bacteria. Thus, a wonderfully satisfying biological picture emerged. We could envision *Agrobacterium* as a microscopic genetic engineer, cultivating plant cells for their own benefit.

At that time only a dreamer could imagine the possibility of exploiting *Agrobacterium* to put genes of our choice into plant cells for crop improvement. There were many obstacles to overcome. We had to learn how to manipulate genes on the Ti plasmid, how to remove the bad ones that caused the plant cells to be tumorous and how to introduce new genes. We had to learn what defined T-DNA on the plasmid. It turned out that *Agrobacterium* determined what part of the plasmid to transfer by recognizing a 25 basepair repeated sequence on each end. One by one, as a result of research by several groups around the world, the problems were solved. The Miami Winter Symposium in January 1983 marked the beginning of an era. Presentations by Belgian, German and two U.S. groups, including mine at Washington University in St. Louis, showed that each of the steps in genetic engineering was in place, at least for (dicotyledonous) tobacco and petunia plants. Solutions were primitive by today's standards, but in principle it was clear that genetic engineering was feasible; *Agrobacterium* could be used to transform a number of dicots.

I saw that industry would be a better setting than my university lab for the next step: harnessing the Ti plasmid for crop improvement. When a Swiss multinational company, CIBA–Geigy, offered me the task of developing from scratch an agricultural biotechnology lab to be located in North Carolina where I had grown up, it seemed tailor made for me. I joined this company in 1983. CIBA–Geigy and I soon found that we had an important incompatibility: while I was good at engineering genes into (dicotyledonous) tobacco plants, the company's main seed business was (monocotyledonous) hybrid corn seed. Nobody knew whether *Agrobacterium* could transfer T-DNA. This problem was solved and maize is now transformable by either *Agrobacterium* or the "gene gun" technique. Our company was first to the market with Bt maize.

The company underwent mergers and spinoffs, arriving at the new name of

Syngenta a few years ago. My role also evolved. After 10 years of administration, I was allowed to leave my desk and go back to the bench. I began working on "gene targeting," which means finding a way to get T-DNA inserts to go where we want them in the plant chromosomal DNA, rather than random positions it goes of its own accord.

Transgenic crops now cover a significant fraction of the acreage of soybeans and corn. In addition, transgenic plants serve as a research tool in plant biology. *Agrobacterium* has already served us well, both in agriculture and in basic science. New developments in DNA sequencing and genomics will surely lead to further exploitation of transgenic technology for the foreseeable future.

REFERENCES

American Soybean Association Conservation Tillage Study (2001) (http://www.soygrowers.com/ctstudy/ctstudy_files/frame.htm).

Brookes G, Barfoot P (2007): GM crops: The first ten years—global socio-economic and environmental impacts. *AgbioForum* **9**:1–13.

Conservation Tillage and Plant Biotechnology (CTIC) (2002): *How New Technologies Can Improve the Environment by Reducing the Need to Plough* (http://www.ctic.purdue.edu/CTIC/Biotech.html).

James C (2006): *Global Status of Transgenic Crops, Various Global Review Briefs from 1996 to 2006*. International Service for the Acquisition of Agri-Biotech Applications (ISAAA).

Jasa P (2002): *Conservation Tillage Systems*, Extension Engineer, Univ Nebraska.

Kovach J, Petzoldt C, Degni J, Tette J (1992): *A Method to Measure the Environmental Impact of Pesticides*. New York's Food and Life Sciences Bulletin, NYS Agricultural Experiment Station, Cornell Univ, Geneva, NY, p 139, 8 pp annually updated (http://www.nysipm.cornell.edu/publications/EIQ.html).

Lazarus W, Selley R (2005): *Farm Machinery Economic Cost Estimates for 2005*, Univ Minnesota Extension Service.

Trigo et al. (2002): *Genetically Modified Crops in Argentina Agriculture: An Opened Story*. Libros del Zorzal Buenos Aires, Argentina.

CHAPTER 2

Mendelian Genetics and Plant Reproduction

MATTHEW D. HALFHILL

Saint Ambrose University, Department of Biology, Davenport, Iowa

SUZANNE I. WARWICK

Agriculture and Agri-Food Canada, Eastern Cereal and Oilseeds Research Centre, Ottawa, Ontario, Canada

2.0. CHAPTER SUMMARY AND OBJECTIVES

2.0.1. Summary

Flowering plants (angiosperms) and conifers (gymnosperms) are diverse organisms that have conquered the terrestrial world and made the planet green (Fig. 2.1). Angiosperms are the most important crop and horticultural plants, while gymnosperms are important in forestry. These plants have sundry methods of reproduction ranging from vegetative propagation to sex by cross-fertilization, which sets them apart from the relatively mundane world of animal reproduction. With the incredible diversity of reproduction methods, plants maintain genetic variation in various ways. Gregor Mendel, the nineteenth-century monk, was the first person to demonstrate the inheritance of genes (even though he did not know what genes were in the molecular sense) using the garden pea plant. His research is the basis of inheritance theory and practice.

2.0.2. Discussion Questions

1. What is a gene, and why are there multiple viable definitions?
2. How does the discrete nature of chromosomes impact sexual reproduction in plants?
3. What would be the consequence of sexual reproduction if mitosis were the only form of cell division?
4. How do the reproductive features of plants regulate the degree of inbreeding?

Plant Biotechnology and Genetics: Principles, Techniques, and Applications, Edited by C. Neal Stewart, Jr.

Figure 2.1. In many ecosystems on Earth, plants change the color of the land to shades of green.

2.1. GENETICS OVERVIEW

The field of genetics impacts all aspects of the science of biology, but individual disciplines within biology utilize different types of genetic information. In order to discuss plant reproduction specifically, several universal genetic definitions must be introduced. In its simplest definition, the field of *genetics* is the study of genes. *DNA* (deoxyribonucleic acid) is the genetic material in organisms that stores all the information that encodes for life. The sequence of *nucleotides* (DNA building blocks: A, C, G, T) stores the instructions to produce proteins and information that allows for the regulation of the genetic material. The *DNA sequence* serves as a type of software or programming language that the cell uses to produce and regulate all the necessary products for life. DNA exists as a double helix, and each nucleotide is paired with its complementary base making a *base pair* (*a*denine with *t*hymine, *c*ytosine with *g*uanine). For this chapter, a *gene* is a contiguous sequence of DNA that contains regulatory regions and a sequence that encodes for a protein. Many sequences in the genome of an organism are outside this definition of the gene, and in fact, much of a plant's DNA would not be considered as part of a gene. At the next level of genetic organization are the *chromosomes*, which are discrete molecules of DNA and associated proteins that reside within the nucleus. The chromosome-associated proteins help package and condense DNA for packing into the nucleus of a cell. The *genome* of an organism is the entire sequence of DNA inclusive of all the chromosomes. DNA is also present within certain cellular organelles: the mitochondria and chloroplasts. Plants therefore contain three distinct genomes (the nuclear, mitochondrial, and chloroplast genomes), and this chapter focuses specifically on the DNA contained within the nucleus. If we draw an analogy comparing genetics to the structure of this book, nucleotides are similar to letters that form three-letter words. Genes are similar to sentences, and chromosomes are similar to chapters. The genome is similar to the complete book, and a library would be a collection of different species (see Chapter 6 for detailed explanation on molecular genetics).

Molecular, cellular, organismal, population, and evolutionary studies all have genetic components, and build on traditional knowledge about genes. For molecular research, the DNA sequence of a gene and its presence and role within the genome are critically important. The sequence itself determines how a gene functions and impacts on the final characteristics of the organism. In larger-scale research such as population and evolutionary studies, both the transcribed DNA within a gene and the DNA that falls outside genes (spacer regions) may be used to describe population structure. Often in a comparative study, the sequences within the genes are highly conserved, that is, too similar in makeup, and are therefore noninformative with respect to deciphering genetic relatedness. In this respect, variable genetic information outside the genes is often more useful for large-scale population studies. These DNA sequences are often used in various types of DNA fingerprinting procedures to elucidate differences between populations. It should be noted that there are differences of opinion on basic definitions of critical terms such as "gene." Unlike our definition, some scientists/researchers refer to the gene as simply the coding region (without the DNA responsible for regulating gene expression). Others have a broader view of the gene to encompass nearly any stretch of DNA. Genetics is a dynamic field whose terminology can be confusing—almost like a rapidly evolving language.

For plant reproduction, the most important genetic level is the chromosome, since chromosomes are the largest units of DNA passed from parents to offspring (progeny). In other words, this chapter is the story of chromosomes. In plants as in all eukaryotes (organisms with a nucleus), chromosomes are linear pieces of DNA that have a single centromere and two arms (Fig. 2.2a). The *centromere* is the constricted region of the chromosome and serves as a connection between the chromosome arms. Centromeres also play an important role in cell division, which is discussed later in the chapter. The genes exist mainly on the chromosome arms. Different plant species vary widely in

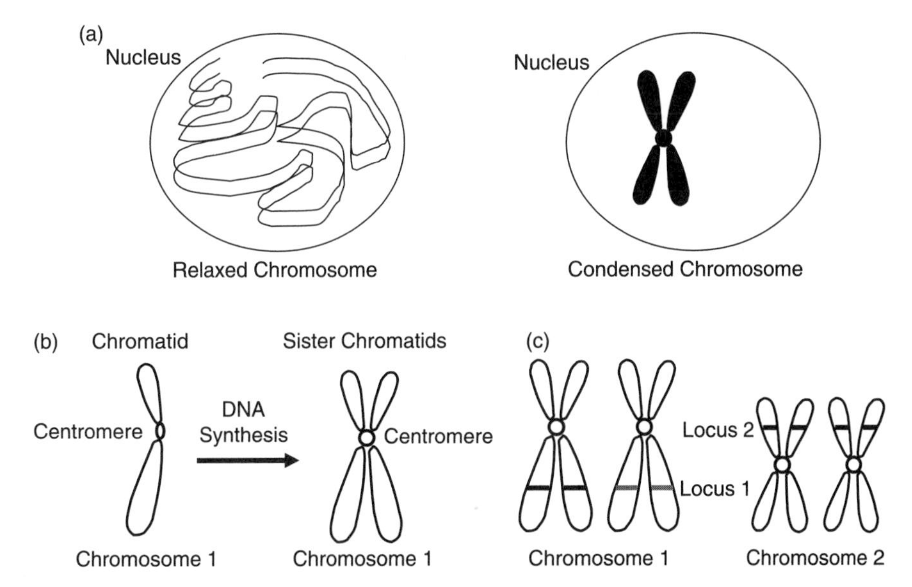

Figure 2.2. Chromosomes have several physical states during the life of a cell: (a) chromosome physical states; (b) chromosome conformations; (c) homologous chromosomes.

chromosome number, and this number often defines a species as being different from another. The number of chromosomes within a nucleus is defined as the *ploidy* of the cell. For example, the model plant *Arabidopsis thaliana* has a total of 10 chromosomes (5 pairs), while the crop plant soybean (*Glycine max*) has 40 chromosomes (20 pairs). Some plants have tremendously large genomes. For example, some lilies have hundreds of chromosomes. Chromosomes vary in length (i.e., in the number of nucleotides that make up the DNA molecule) and therefore vary in size when visualized under the microscope. Each chromosome has hundreds to thousands of genes contained within the sequence of DNA, along with sequences between the genes. This connecting DNA has been historically called "junk DNA," but current research is discovering that intergenic DNA sequences may play several critical roles such as regulating how genes and chromosomes interact at higher levels.

To understand biotechnology and genetics, it is essential to define and understand how chromosomes exist within the nucleus. Chromosomes are organized in two different basic physical structures during the life of the cell. During most of the cells' adult life, the chromosome exists in a relaxed state, where the DNA is loosely wrapped around chromosomal proteins (Fig. 2.2a). This physical state allows the DNA to be read (transcribed and translated) so that the appropriate proteins are produced. As the chromosomes prepare for cell division, they become tightly wound around chromosomal proteins and are described as being in the condensed state (Fig. 2.2a). Chromosomes can be visualized under the light microscope only when they are condensed. During different points in the cell cycle, chromosomes may be in different conformations. Initially after cell division, a chromosome exists as a single molecule of double-stranded DNA with a single centromere, called a *chromatid* (Fig. 2.2b). After the DNA synthesis phase of the cell cycle, the chromosome exists as two molecules of identical double-stranded DNA connected at the single centromere. The two DNA molecules within a chromosome are called *sister chromatids*, and they stay connected until they are separated by one of the types of cell division. DNA synthesis does not represent a change in the total chromosome number, as chromosome numbers remain the same during the lifetime of the plant. A single chromosome then may exist in either a prereplicated (one chromatid) or replicated state (two sister chromatids). The different states of chromosomal arrangements within the life of a cell will be important as we describe cell division and sexual reproduction.

Most cells in a plant have two copies of each chromosome, which are called *homologous chromosomes* or a *chromosome pair* (Fig. 2.2c). Generally speaking, one of the individual chromosomes in a pair is derived from the maternal parent and one from the paternal parent. Gender identity and parenting is sometimes confusing to think about in plants that have the ability to *self-fertilize* (when the same plant's pollen fertilizes the ovum), but one of the homologous chromosomes comes from the pollen and one from the ovum even if all the chromosomes come from the same plant. Hermaphrodites (organisms with both male and female organs) and selfing are considered to be anomalies in the animal kingdom but are frequent among plants. As we will discuss later in this chapter, plants have a wide array of reproductive strategies to achieve the pairing of the chromosomes.

Most adult plant cells have two copies of all chromosomes, and the ploidy level is defined as the *diploid* state ($2N$). In order to sexually reproduce, the total chromosome number is divided in half, and this reduced chromosome number in the sexual gametes is defined as the *haploid* state (N). During most of an angiosperm plant's life, the diploid *sporophyte* stage dominates and produces diploid cells during cell division.

In the small reproductive structures (pollen grains and ovaries), the haploid *gametophyte* stage is present and gives rise to haploid sex cells. Even with the diversity of chromosome numbers observed among plant species, eukaryotic chromosomes function under the same rules during cell division. During normal cell division (*mitosis*) in the sporophyte, the chromosome number is maintained in the diploid state. During gametophyte production (*meiosis*), the two copies of each chromosome separate from one another and produce cells with half the normal number of chromosomes. All the variations of reproductive mode are simply complexities of how the two homologous chromosomes come together during the process of reproduction.

2.2. MENDELIAN GENETICS

Gregor Mendel, a member of the Augustinian monastery in what is now the current Czech Republic, was the first person to describe how chromosomes are transmitted between generations (Fig. 2.3). Mendel combined what are now considered typical plant breeding procedures, such as keeping accurate records of the characteristics that appeared in the offspring of selected parents and the control of pollination of the experimental plants, with statistics to describe how traits behave over generations. The molecular basis of genetics was not understood in the 1800s, but Mendel observed and recorded the phenotypic traits within the plants that he grew on the monastery grounds. The *phenotype* is the physical appearance of an organism, and the *genotype* is the underlying genetic makeup of an organism. Using pea plants (*Pisum sativum*), Mendel was able to track the segregation of traits over generations, and thus indirectly described the laws of how chromosomes act within cells. He accurately described the cellular process of chromosomal segregation

Figure 2.3. Gregor Mendel was the father of genetics.

without the benefit of knowing what was occurring within the nucleus or that chromosomes existed. Gregor Mendel's work in genetics was relatively obscure in his own day but was "rediscovered" in the twentieth century (see Bateson 1909, Sutton 1903).

Mendel's choice of working with peas was a good one, since the pea plants he used differed from one another in several relatively simple phenotypic traits. Seed shape and color, pod shape and color, plant height, and flower position were the traits that he traced over generations of sexual reproduction (Mendel 1866). The pea plants had different variants for a given trait (Fig. 2.4). For example, some of the pea plants had yellow seeds, while others had green seeds. Each trait that Mendel followed was controlled by a single gene, and the traits themselves were often discrete. That is, seeds could be scored as either yellow or green, and not a mixed or splotched variant that was in between the original parents. *Mendelian traits* are controlled by a single gene, and therefore the protein product from a single gene directly leads to the characteristic phenotype. This is one of the most important concepts in plant biotechnology since all transgenic plants produced to date have traits controlled by single transgenes. Mendelian traits may have multiple different versions that make different proteins with varying characteristics, but the gene that controls the trait is at a single location within a chromosome in the genome called a *locus* (Fig. 2.2c). The different versions of each gene are called *alleles*, and differ from one another in the sequence of DNA at that chromosomal locus. Mendelian traits are also characterized by *discrete variation*, where the different phenotypes of the trait can be broken into obvious categories. In the example of pea plant height, tall versus short plant type is determined by the genotype at a single genetic locus that controls height.

As you will see throughout this book, most traits are more complex than Mendelian traits because they are controlled by the gene products of many genes, and hence are called *polygenic traits*. Polygenic traits exhibit *continuous variation*, where the trait can show a wide range of phenotypes. *Multifactorial traits* are controlled by multiple genes and the environment in which the plant is grown. Multifactorial traits also exhibit continuous variation, and will vary with the environmental conditions. Polygenic and multifactorial traits will be discussed specifically in Chapter 3 of this book. The traits that Mendel followed had two specific characteristics; they had discrete variation and were controlled by the action of a single gene.

Mendel was very observant, and was a good botanist. His choice of peas was fortuitous in that peas normally self-fertilize, which made all of his interpretations of transmission

Trait	Dominant	Recessive
Seed color	Yellow—*Y*	Green—*y*
Seed form	Round—*R*	Wrinkled—*r*
Seed coat color	Grey—*A*	White—*a*
Pod form	Inflated—*V*	Restricted—*v*
Pod color	Green—*G*	Yellow—*g*
Flower position	Axial along stem—*F*	Terminal on top—*f*
Stem length	Tall—*L*	Short—*l*

Figure 2.4. Traits of the pea plant used by Mendel to discover the genetic laws of segregation and independent assortment. Each trait had two phenotypes: one controlled by a dominant allele and one by a recessive allele.

genetics much simpler than would be the case if he'd picked plants that were normally (or even partially) *outcrossers*. He used plant lines that would only generate plants of a single type when the plants were allowed to self-fertilize. These plants were *homozygous* for that trait, which meant that the two homologous chromosomes had the same allele. When homozygous plants are selfed, the resulting progeny are always homozygous. Mendel's method of tracking segregation was based on crossing plants that were homozygous and differed for the phenotypic trait of interest. For example, he would cross (instead of selfing) plants that were homozygous yellow and homozygous green for seed color, and then record the phenotypic ratio in progeny of each subsequent generation.

By crossing different homozygotes, Mendel generated plants whose two homologous chromosomes each had a different allele of the gene (Fig. 2.5a). The condition of having two different alleles in a single gene is called *heterozygous*. All the plants generated from the initial cross (*F_1 hybrids* or F_1 generation) would have the same genotype, but could have either one of two different parental phenotypes. In the heterozygous plants, Mendel discovered that certain variants of a trait appeared to mask or cover the expression of other variants. A variant that would cover the other type was termed *dominant*, while the phenotype that would disappear was called *recessive*. When we write allele names, we often use uppercase letters for dominant alleles and lowercase letters for recessive alleles. Today we understand that dominant alleles have a sequence of DNA that encodes for a functional protein, while many recessive alleles have changes in the DNA sequence, called *mutations*, which render the encoded protein nonfunctional. Therefore, in a heterozygous plant, functional and nonfunctional proteins are produced, and the plant has the phenotype of the dominant allele from the functional protein. In Mendel's experiments, he would see that the dominant trait would mask the expression of the recessive trait.

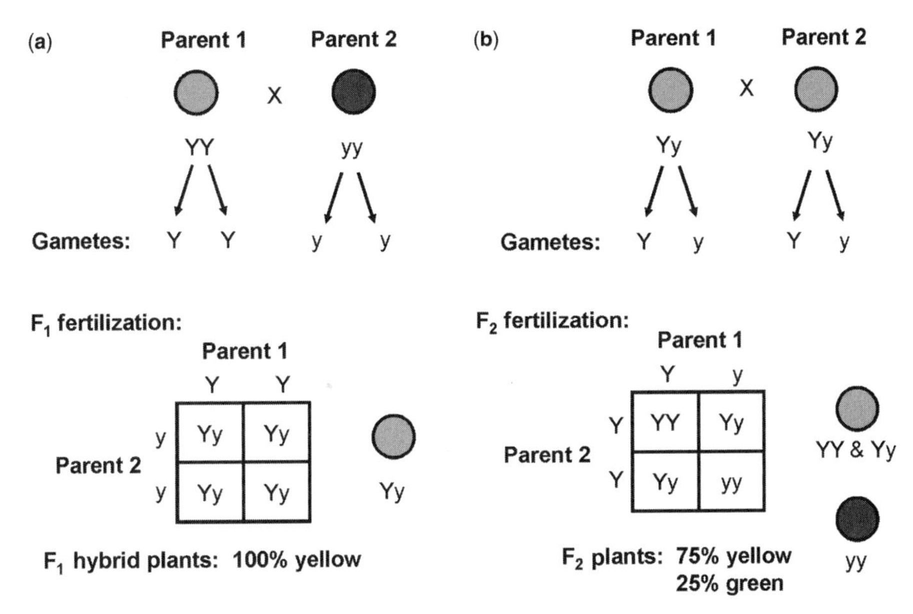

Figure 2.5. A monohybrid crossing system involving a single-gene model where the two alleles segregate from one another in the production of gametes: (a) monohybrid cross; (b) F1 self-fertilization.

After crossing the homozygous parents and generating a heterozygous hybrid plant (F_1), Mendel would allow the hybrid plant to self-fertilize. In the subsequent F_2 plants or *F_2 generation*, plants with the recessive trait would reappear (Fig. 2.5b). Mendel realized that the recessive allele was not replaced or destroyed by the dominant allele, but its phenotype was just masked in the heterozygous individuals. With his intricate recordkeeping of counting the plants with different phenotypes, Mendel observed that the dominant plants occurred in 75% of individual F_2 plants, while recessive plants occurred at a frequency of 25%. Mendel's crosses may be visualized in a graphical table called a *Punnett square* that depicts the number and variety of genetic combinations in a genetic cross. The latter was named after Reginald Punnett, who worked with William Bateson to confirm experimentally the findings of Gregor Mendel. Their investigations of the exceptions to Mendel's rules led to the discovery of genetic linkage in the pea, discussed later in this chapter. Using a Punnett square, the possible genotypes of the gametes from each parent are placed on adjacent axes, and the matrix within the Punnett square represents all possible outcomes from sexual reproduction.

Using his crossing data, Mendel realized that plants contained two copies of genetic material. Although he did not know that each plant had two different sequences of DNA on the two homologous chromosomes, he could predict the expected segregation frequencies over all the traits that he tracked over multiple generations. The fundamental process that Mendel discovered was that plants contained two versions of every gene, and that those genes were discrete particles that could separate from one another over the generations.

2.2.1. Law of Segregation

In his crosses using single traits, or *monohybrid crosses*, Mendel described the first of his genetic laws explaining how traits are passed between generations. He didn't know that DNA was controlling the traits he was observing, but we will state his law on the basis of current knowledge that DNA is genetic material and is stored in chromosomes. Because dominant and recessive alleles segregate from one another in progeny derived from heterozygous plants, he described the *law of segregation*, which states that two homologous chromosomes separate from one another during the production of sex cells. In practical terms, this means that half of the sex cells will be produced with one allele and half with the other allele in a heterozygous plant.

2.2.2. Law of Independent Assortment

Mendel also crossed plants that differed at multiple traits at the same time. When plants that differed at two traits were crossed, or were *dihybrid crosses*, Mendel determined that the traits segregated independently from one another (Fig. 2.6). This phenomenon was described in the *law of independent assortment*, where chromosomes from different homologous chromosome pairs separate independently from one another during the production of sex cells. Chromosomes are independent molecules of DNA, and only homologous chromosomes pair with one another during gamete production. Therefore, nonhomologous chromosomes will divide completely randomly into the daughter cells.

It is an interesting historical fact that the traits that Mendel studied were controlled by genes on different chromosomes. This is often deemphasized when discussing Mendel's work and it should not be, because if the genes had been on the same chromosome, his

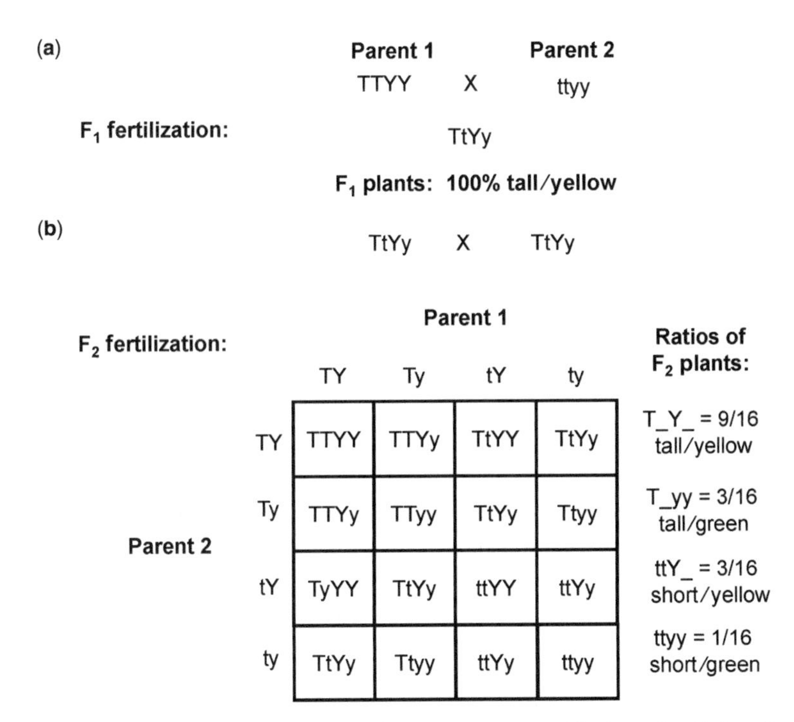

Figure 2.6. A dihybrid crossing system involving a two-gene model where the alleles of two genes independently assort from one another in the production of gametes: (a) dihybrid cross; (b) F_1 self-fertilization.

results would have been different. *Genetic linkage*, or the fact that genes on the same chromosome tend to be inherited together, would have caused linked alleles and corresponding traits to remain together rather than segregate independently. He did not understand it at the time, but Mendel's traits were each controlled by a single gene on a completely different chromosome, which allowed them to segregate in the patterns he observed.

There were numerous experiments on the crossing of different species or varieties of plants during the eighteenth and nineteenth centuries; the primary intention was to obtain new and improved varieties of fruits and vegetables. Knight (1799) and Goss (1824) in the United Kingdom both worked on edible pea (*Pisum sativum*)—in fact, made the same crosses as Mendel—and each observed the same general segregation patterns, but did not record the numbers as did Mendel. Knight chose pea, because of its short generation time, the numerous varieties available, and the self-fertilizing habit, which obviated the need to protect flowers from insects carrying pollen. Presumably, Mendel had the same goals and rationale.

Mendel's laws (Bateson 1909) have served as the basis for all fields of genetics. Of course, once DNA structure was described by Watson and Crick in 1953, the age of modern genetics began (Watson and Crick 1953a, 1953b, Watson 1968). Even though the mechanisms as to how DNA could store genetic information was not known, Mendel's principles still correctly described how genes were transferred between generations. Mendel's important work illustrates that comprehensive knowledge on a subject is not needed to make an important contribution in science. To continue our discussion of plant reproduction, we must describe the two types of cell division that separate chromosomes from one another during the life of the cell.

2.3. MITOSIS AND MEIOSIS

Mendel's observations and subsequent research prompted cell biologists to study the movement of chromosomes during the process of cell division. Plant growth and sex cell production are the result of two different types of cell division: chromosome copying (mitosis) and chromosome reducing (meiosis). Most cells in a plant and any other complex organism undergo an exact copying process in which the original chromosome number remains the same. This process that allows simple plant growth is called *mitosis*, in which a cell divides into two exact copies of the original (Fig. 2.7). In mitosis, the chromosome number is maintained in each daughter cell as a result of the division of sister chromatids at the centromere. In order to proceed through sexual reproduction, cells must undergo the process of *meiosis*, a form of cell division where the resultant cells have half (haploid) the total number of chromosomes (Fig. 2.8). If the chromosome number were not reduced in sex cells (*gametes*), the number of chromosomes would double after each generation of sexual reproduction. This, of course, is not the case, as each plant species generally retains its chromosome number over generations. Meiosis allows for two haploid cells to join during fertilization to reconstitute the two copies of each chromosome in the progeny. Mitosis and meiosis are the two processes by which a cell may divide, and each process has a different goal according to the total number of chromosomes required in the daughter cells.

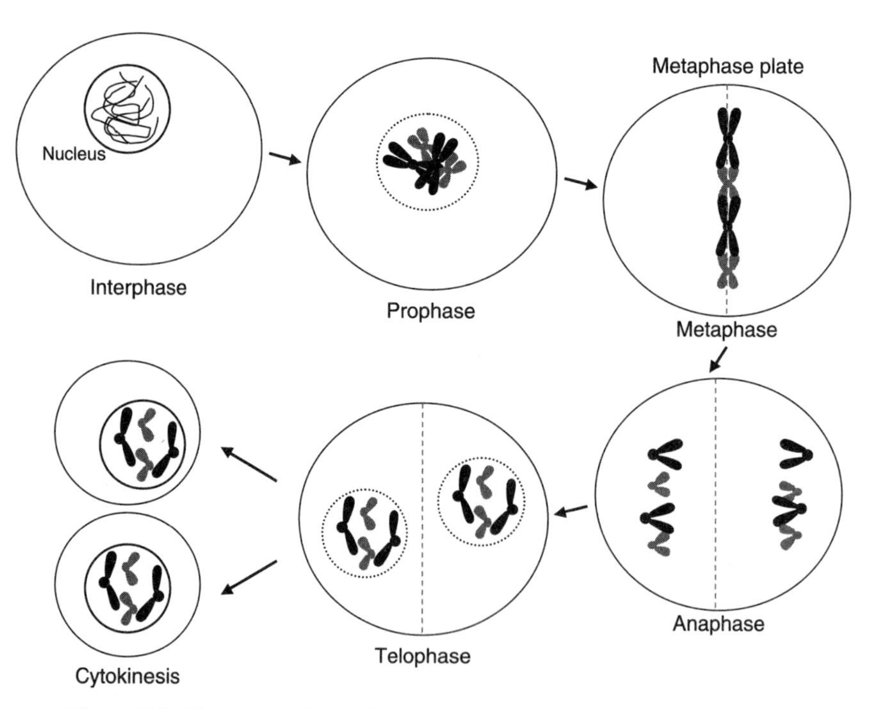

Figure 2.7. The stages of mitosis based on arrangement of the chromosomes.

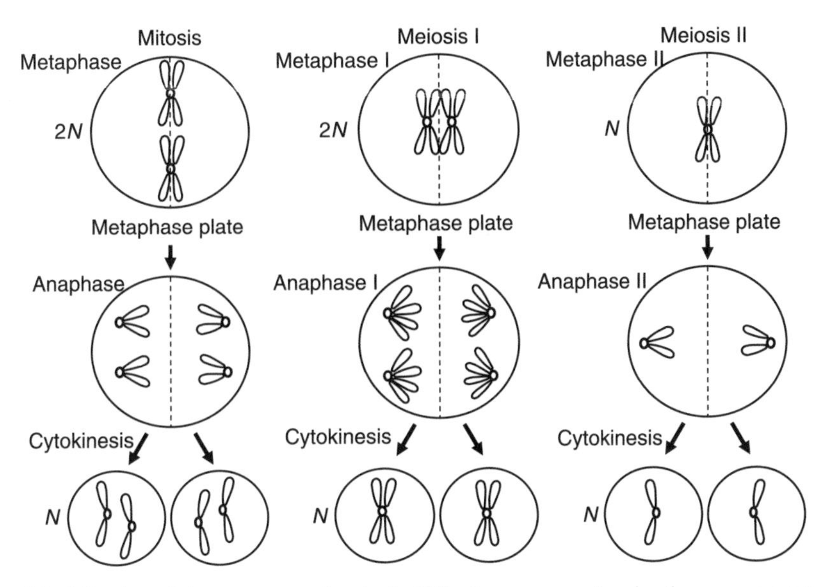

Figure 2.8. Mitosis and the two steps of meiosis differ from one another by the arrangements of the homologous chromosomes prior to cell division.

2.3.1. Mitosis

The goal of mitosis is to maintain the complete number of chromosomes during cell division. Mitosis is a highly ordered process, because chromosome loss during cell division would be detrimental to the adult plant. Mitosis can be broken into five basic steps, each defined by the organizational state of the chromosomes (Fig. 2.7).

The chromosomes are in the relaxed state throughout most of the life of the cell, called *interphase*, which is the period of cellular life when the cell grows and prepares its chromosomes for cell division. During the synthesis phase (*S phase*) of interphase, the chromosomes replicate their DNA and form the sister chromatids. As the cell enters mitosis, the chromosomes condense into the tightly wound state and the nucleus breaks down, which are characteristics of *prophase*. The chromosomes appear in a disorganized mass that can be seen under the light microscope. The cellular machinery that performs the actual work of cell division involves a group of proteins called the *mitotic spindle apparatus*, but we will focus on the state of the chromosomes during mitosis in this chapter. As the chromosomes become organized along the middle of the cell, they enter *metaphase*. During metaphase, the chromosomes line up at the center of the cell with each of the sister chromatids on opposite side of the *metaphase plate*. The centromere sits directly on the middle line, and is broken in half and pulled to the opposite ends of the cell during *anaphase*. The chromosomes appear as small V shapes, with the centromere pulled to the opposite poles with chromosome arms lagging behind. During this phase, the cell transiently has a 4*N* chromosomal number, because the centromeres between the sister chromatids are broken, producing two chromosomes. When the chromosomes reach the opposite ends of the cell, the nuclear membranes re-form, which characterizes *telophase*. At this point, the two sister chromatids from all the chromosomes have been separated from one another, and the cell can divide by a process called *cytokinesis* into two daughter cells that have the exact same DNA. During mitosis, the chromosomes are broken at the centromere and the two daughter cells each acquire a complete copy of the cell's genome.

2.3.2. Meiosis

Meiosis is the type of cell division used to make sex cells or gametes. The goal of meiosis is to generate haploid cells, which have half the number of chromosomes as the original cell. Meiosis is a two-step process, where the original cell undergoes two divisions in order to make haploid cells. In the first division (I), homologous chromosomes line up together and separate from one another to generate haploid cells. In the second meiotic division (II), sister chromatids of each chromosome divide in a process identical to mitosis. It can be said that meiosis simply adds a reductive division to separate the homologous chromosomes, and then goes through a mitotic division of remaining chromosomes.

The two meiotic divisions proceed in stepwise fashion similar to that described above for mitosis, with the condensation of the chromosomes, alignment in the center of the cell, pulled to opposite poles, followed by cell division. The differences occur in how the homologous chromosomes interact with one another (Fig. 2.8). In the first meiotic division, the homologous chromosomes find one another and form a structure called the *tetrad*. During prophase I, the homologous chromosomes interact with one another, which allows for the transfer of genetic material between the homologous chromosomes in a process known as *crossing over* (or *crossover*) or *recombination*. Recombination in this fashion generates diversity when the homologous chromosomes swap DNA. Metaphase I is also different in meiosis, as the homologous chromosomes in the tetrad straddle the metaphase plate, with each chromosome on one side. During anaphase I, complete homologous chromosomes, each with their two sister chromatids, are pulled to the opposite poles of the cell. The centromere remains completely intact as each separate homologous chromosome is pulled to the opposite end of the cell. After cell division, each daughter cell has only one of each homologous chromosome and therefore only half of the genetic material. The first meiotic division results in a reduction of genetic material by half.

The second meiotic division is exactly like mitosis, but with half the genetic material per cell, with the chromosomes lining up at the metaphase plate with the sister chromatids on each side of the cell. The centromeres are then broken, and the sister chromatids are pulled to opposite ends of the cell. This division results in two cells with identical genetic material, which is exactly the same process as mitosis, except with a haploid number of chromosomes. Meiosis and mitosis are similar processes but differ in how the chromosomes are pulled apart. In mitosis, the complete genome is retained in the daughter cells, while meiosis reduces the genome size in half by separating the homologous chromosomes. Therefore, growth is achieved by mitosis as numerous exact copies of the diploid cells are made, allowing for each cell to function in the adult plant. Meiosis prepares for sexual reproduction by generating haploid cells, which will be combined by the process of fertilization with other haploid cells to reconstitute the normal number of two homologous chromosomes.

2.3.3. Recombination

Recombination, or the crossing over of DNA between chromosomes during meiosis, a process first documented in Drosophila (Bridges 1916), is a critically important process that generates genetic diversity in plant species. If recombination did not occur, each chromosome would be essentially static and "immortal," with the same alleles always linked together on the same piece of DNA. The only changes that could occur in the DNA sequence would be caused by mutation, and each mutation would stay on the same piece of DNA forever. If this were the case, then plant improvement via breeding would be impossible. In both nature and agriculture the "goal" is to combine advantageous alleles together within the same breeding line to improve a plant for natural or agricultural settings. Without recombination, the target of selection would be the chromosome with the allele of interest, and there would be a limited number of chromosome combinations from which to make selections. Luckily for crop breeders, mutation is not the only process that generates genetic diversity.

Recombination allows for alleles to be shuffled during every meiotic division (Fig. 2.9). It has been estimated that crossing over occurs during every meiotic division for each chromosome, and therefore the lifespan of any chromosomal sequence is actually only one generation. This allows for different alleles at different chromosomal loci to reshuffle and land on the same chromosome. Crop breeders rely on this process because they attempt to select for recombination events that liberate the specific allele from a genetic background to improve the crop line without having to select for chromosomes. Often, crop plants have been highly selected to obtain a group of alleles that help the crop perform well under specific agricultural conditions. A single new allele may improve the crop, but the breeder needs to retain all the original genes of that crop. The process of recombination allows the breeder to try to find specific recombination events where the one allele has crossed over to join all the other original crop-selected alleles (see the next chapter for an in-depth description of plant breeding).

2.3.4. Cytogenetic Analysis

Scientific methods to observe chromosomes have improved greatly since Mendel outlined the laws that describe chromosome movement across generations. The easiest way to observe chromosomes is via chromosome staining during mitosis. Many readers can remember back to their high school biology classes where they observed stained onion (*Allium cepa*) root tips with the microscope. In these lab exercises, condensed chromosomes were stained with a DNA-specific dye (a fuchsin-based DNA-specific stain developed by Feulgen in 1914), and the different stages of cellular mitotic division determined by observing the patterns of the chromosomes in each cell. Chromosome viewing by simple light microscopy is, however, limited to those plant species with large chromosomes in which single layers of actively dividing cells can easily be attained. These conditions are not common to most tissue types in adult plants.

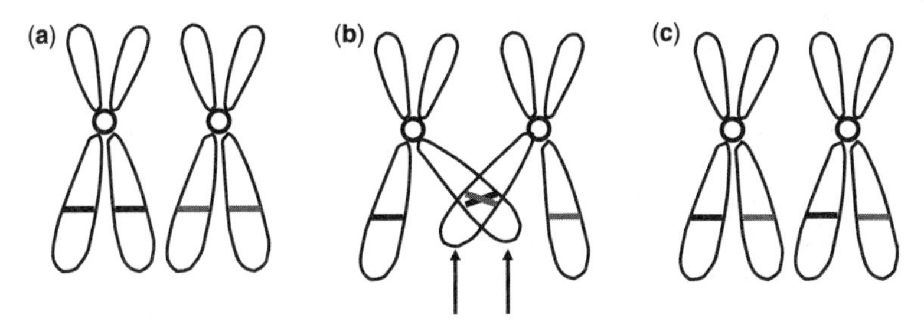

Figure 2.9. Recombination occurs when homologous chromosomes trade DNA sequences, thus generating genetic diversity.

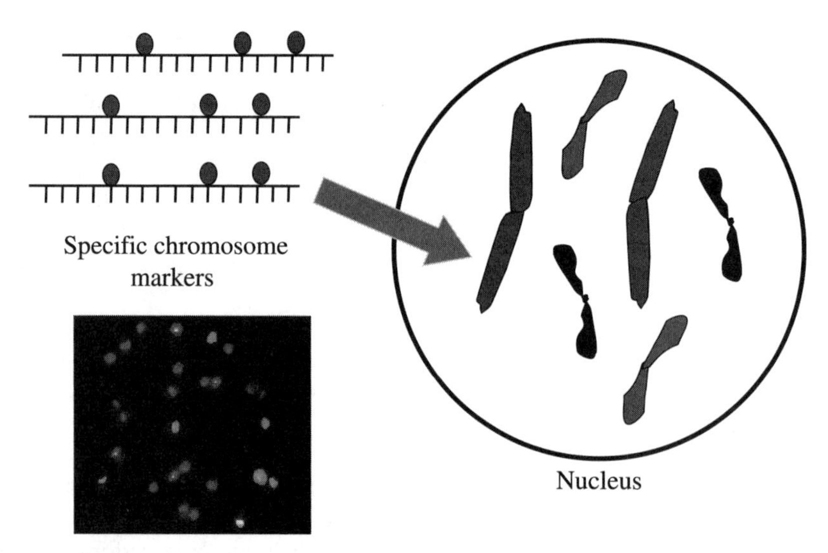

Figure 2.10. Fluorescent in situ hybridization (FISH) shows the physical location of a specific transgene or DNA. The inset (bottom left; courtesy of Chris Pires) shows *Brassica napus* mitotic metaphase chromosomes stained blue with two different centromere probes (red and green). See color insert.

More advanced cytogenetic techniques to observe chromosomes have been developed since the mid-1950s, and are now being combined with molecular tools in the field of plant genomics research. *Fluorescence* in situ *hybridization* (FISH) is a method that utilizes small fluorescently labeled DNA fragments to paint different chromosomes (Fig. 2.10). In this technique, nuclear DNA is fixed to the surface of a slide preparation and the labeled DNA fragments bind to chromosomes with homologous complementary sequences. Since the chromosomes are still in the nucleus, it is said to be in situ, or in the original location. *Flow cytometry* is a technique to determine the total amount of DNA within a cell. Although this is not a direct way to visualize chromosomes, it allows researchers to determine (along with chromosome number) *genome size*, that is, how much genetic material is present in a cell, which has implications during hybridization between species.

2.4. PLANT REPRODUCTIVE BIOLOGY

2.4.1. History of Research

When it comes to sex, angiosperms have evolved many ways of doing it and indeed of doing without it. Sexuality in plants (reviewed in Stuessy 1989) was first demonstrated experimentally over 300 years ago by a German botanist and physician, Rudolph Jakob Camerarius. In his 1694 book *Epistolae de Sexu Plantarum* (*Letter on the Sexuality of Plants*), he identified the stamen and pistil as the male and female organs, and the pollen as the fertilizing agent (Camerarius 1694). By the mid-1700s the role of insects in pollination was well accepted, and in 1793 another German, Sprengel, provided elaborate details on the floral adaptations of 500 or more species to insect pollinators (Sprengel 1793). Charles Darwin was also interested in pollination and plant mating systems from an evolutionary perspective, and one of his books outlining *The Effects of Cross and Self Fertilisation in the Vegetable Kingdom* in 1876, introduced the idea of self-incompatibility systems in plants (Darwin 1876). Plant mating systems have continued to fascinate botanists and geneticists since that time. Plant reproduction is clearly important to biotechnological improvements to agriculture, as it directly or indirectly affects the quality and quantity of all crop products.

2.4.2. Mating Systems

2.4.2.1. Sexual Reproduction. Traditional sexual reproduction is the best place to begin the discussion of plant mating systems. Seed production by sexual reproduction involves the transfer of pollen from an anther to the stigma of the pistil, followed by germination and growth of the pollen tube. The movement of nuclei in the pollen tube through the style to the embryo sac and the union of functional male and female gametes complete sexual reproduction in plants. Pollination vectors, such as insects or wind, are responsible for the transfer of pollen, but mating systems determine whether the pollen grain can germinate on a receptive stigma and penetrate the style. Mating systems are classified according to the source of pollen that is responsible for fertilization. Self-fertilization or *selfing* (also known as *autogamy*) occurs when the pollen that effects fertilization is produced on the same plant as the female gamete with which it unites. Cross-pollination or *outcrossing* (*xenogamy*) occurs when the pollen of one plant is responsible for fertilization of the female gamete of another plant.

The mating system of a plant species is also classified according to the relative frequency of self- versus cross-pollination in their seed production. There is a continuum of variation among species, ranging from complete selfing to obligate outcrossers, with those species demonstrating both characteristics often referred to as having a *mixed mating system.* Most crops have been bred and selected for selfing, but can also be outcrossed. This situation enables "true" seed to be produced by selfing in which the progeny are genetically very similar to the parent. "Homozygosity begets homozygosity." This situation also allows plant breeders to "shuffle" genomes from outcrossing when needed. The predominant mechanism of pollination for a species is an important factor in determining the breeding method used to develop the cultivar (see Chapter 3). For example, hybrid seed production is more readily accomplished in an outcrossing species than in a selfing species. The formation of homozygous lines occurs naturally in a self-pollinating species, but artificial self/sib-pollination must be practiced in outcrossing species to obtain homozygous genotypes. Both flower morphology and development, as discussed in more detail below, can influence rates of self- and cross-pollination.

2.4.2.1.1. Selfing (Autogamy) versus Outcrossing (Xenogamy). Some plants have natural mechanisms that encourage self-pollination. One such mechanism, in which pollination takes place while the flower is still closed, is known as *cleistogamy*, and is a process that can occur even in self-incompatible species (Fig. 2.11). *Homogamy*, the synchronous maturation of stamens and stigma, also facilitates self-pollination.

The effects of repeated self-fertilization, first documented in maize at the turn of the (nineteenth–twentieth) century, has been confirmed for many crop species. Repeated self-fertilization will yield complete homozygosity in a few generations unless the heterozygous state is favored by selection. In an heterozygous diploid, the dominant allele can shelter recessive alleles that would be deleterious in the homozygous state. Self-fertilization quickly results in the segregation of lethal or sublethal types as homozygous recessives are produced. Further selfings rapidly separate the material into uniform lines,

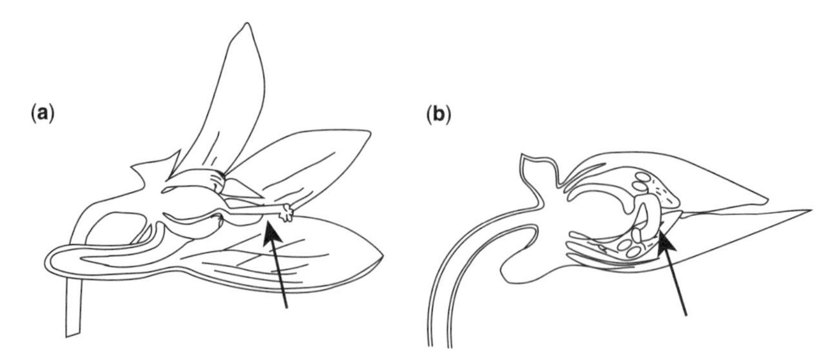

Figure 2.11. Cleistogamous flowers (b) are fertilized prior to the opening of sepals and petals, which ensures that the plant is self-pollinated. A noncleistogamous flower is shown in (a). (Adapted from Briggs and Walters 1997).

often called *pure lines*. Some of the surviving lines may be characterized by reduced vigour and fertility, a condition known as *inbreeding depression*. If pure lines originating from different parental stocks are crossed together, *hybrid vigor* (i.e., *heterosis*) may be demonstrated. Outcrossing thus avoids the deleterious effects of inbreeding depression, and promotes heterozygosity, genetic variability, and genetic exchange. Plants species have therefore evolved a wide variety of natural mechanisms that favor cross-pollination; and scientists have needed to invent an alphabet soup to describe the myriad of mating syndromes observed (see Richards 1986, Stuessy 1989). Several of these, including protandry; protogyny; chasmogamy; heterostyly; imperfect flowers on monoecious, dioecious, or polygamous plants; and incompatibility, are discussed in somewhat greater detail below.

2.4.2.1.2. Sex Distribution within a Flower and within a Plant. Plants are the ultimate hermaphrodites—most are bisexual with male and female organs together in one flower (also referred to as a "perfect flower"), but there are many ways in which sex organs are distributed within a flower, within a plant, and within a plant population. Some plants have separate male (*staminate*) flowers and female (*pistillate*) flowers on a single plant and are termed *monoecious* (e.g., maize) In other species the male and female flowers occur on separate plants (known as *dioecy*), or can have a mixture of male, female, and perfect flowers on the same plants (termed *mixed polygamous*). Sex determination in such plants is under genetic control, with monoecy in maize, for example, under the control of a set of genes known as the *tasselseed* loci. A number of different mechanisms have been identified that establish the sexuality of dioecious plants, including the presence of heteromorphic *sex chromosomes* with males having XY and females XX chromosomes, or varying X : autosome ratios similar to that found in *Drosophila* (Bridges 1925). Even when both male and female organs occur in the same flower, the timing of sexual expression can vary. Sometimes pollen is shed before the stigma is receptive in a process known as *protandry*, or a stigma can mature and cease to be receptive before pollen is shed (*protogyny*).

2.4.2.1.3. Self-Incompatability Genetic Systems. Many plant species have a genetic self-incompatibility (SI) mechanism that promotes outcrossing and is defined as "the inability of a fertile hermaphrodite seed plant to produce zygotes after self-pollination." SI mechanisms are estimated to occur in more than half of all angiosperm species. The effectiveness of SI in promoting outbreeding is believed to be one of the most important factors that ensured the evolutionary success of flowering plants, an idea first promoted by Darwin. It is a genetically controlled phenomenon, and in many cases the control is by a single locus known as the *S locus*. This locus often has up to several hundred alleles in some species. The SI mechanism promotes outcrossing by arresting "self" pollen tubes as determined by the genotype at the S locus (Fig. 2.12). SI is based on the

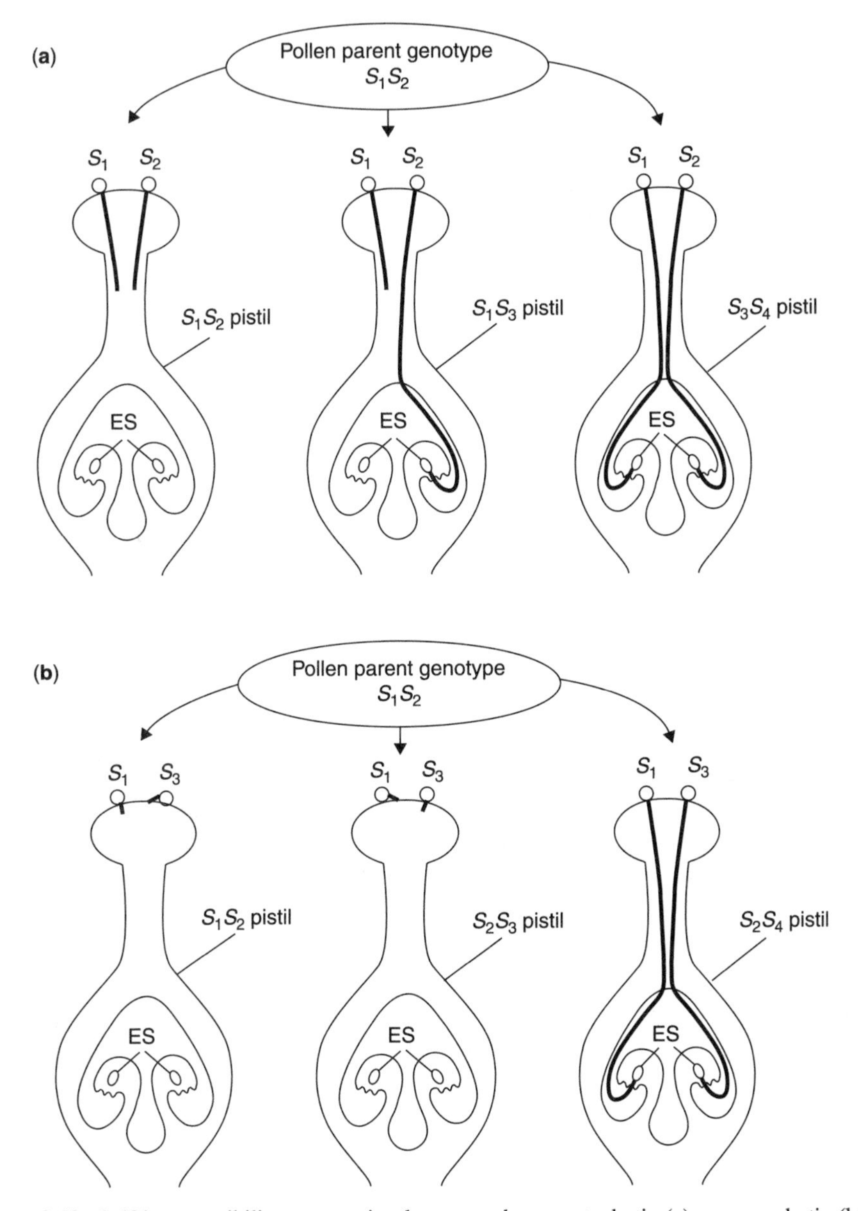

Figure 2.12. Self-incompatibility systems in plants may be gametophytic (a) or sporophytic (b). In gametophytic self-incompatibility, the pollen grain will not grow and fertilize ovules if the female plant has the same self-incompatibility (S) alleles. In sporophytic self-incompatibility, the diploid parent prevents germination of pollen grains that share an allele with the parent. (Adapted from Briggs and Walters 1997).

ability of the pistil to discern the presence of self-pollen and to inhibit the germination or subsequent development of self-related, but not genetically unrelated, pollen. There are two types of SI mechanisms, *gametophytic* and *sporophytic* (Fig. 2.12); these differ in whether the haploid pollen genotype or the diploid pollen parent genotype, respectively, determines the success of pollination. These are important traits for controlling pollinations and are much sought after in breeding programs.

2.4.2.1.4. Male Sterility. The ability to produce hybrid seed has been of fundamental importance to modern agricultural practice. "Hybrid vigor" has increased the yield in maize since the mid-1960s. The genetic approach to the production of F_1 hybrid seed was made possible by the exploitation of various male sterility mechanisms. *Male sterility* refers to the failure of a plant to produce functional pollen by either genetic or cytoplasmic mechanisms. *Cytoplasmic male sterility* (CMS) is a maternally inherited trait that suppresses the production of viable pollen grains. It is a common trait reported in hundreds of species of higher plants. The CMS phenotype (female parent) is used commercially in the production of F_1 hybrid seed by preventing self-fertilization of the seed parent, in such crops as maize, sorghum, rice, sugarbeet, and sunflower. The use of CMS lines as female parents also requires the introduction of nuclear fertility restorer genes from the pollen parent, so that male fertile F_1 hybrids can be produced. Novel sources of CMS and fertility restorer genes are very important to plant breeders and the traits can be introduced via biotechnological means.

2.4.2.2. Asexual Reproduction. Plants can also reproduce by asexual means, resulting in the multiplication of genetically identical individuals. An individual reproducing asexually is referred to as a *clone* and the process as *cloning*. Potatoes and cranberries are two agricultural plants that are propagated by asexual reproduction. Asexual reproduction in seed plants can be divided into two main classes; *vegetative propagation*, which can occur through plant parts other than seed (bulbs, corms, rhizomes, stolon, tubers, etc.), and *apomixis*, which can be defined as the production of fertile seeds in the absence of sexual fusion of gametes or "seeds without sex." Sexual fusion presupposes a reductional meiosis if the ploidy level is to remain stable. During apomixes, the embryo may develop from either an *N* (haploid) egg cell or from a 2*N* (diploid) egg cell. In the latter type, known as *agamospermy*, a full reductional meiosis is usually absent and chromosomes do not segregate. Another rarer form of apomixis is that in which the embryo plant arises from tissue surrounding the embryo sac. These "adventitious" embryos occur, for example, in citrus crops.

2.4.2.3. Mating Systems Summary. Having discussed the three main modes of reproduction—selfing, outcrossing, and apomixis—we may now examine the advantages and disadvantages of different mating systems (reviewed in Briggs and Walters 1997). One possible advantage of repeated self-fertilization is that well-adapted genotypes can be replicated with little change. A further advantage, especially in extreme or marginal habitats, where relying on crossing between plants might be hazardous or even result in total failure, is that self-fertilization is an assured method of producing progeny. Outcrossing, on the other hand, avoids the deleterious effects of inbreeding depression, the main disadvantage of repeated selfing, and promotes heterozygosity, genetic variability, and genetic exchange. There are, however, costs to the plant, compared with selfers, as more biomass has to be employed in producing flowers, nectar, and so on. Other disadvantages to an obligate outcrosser are that if only one genotype is present in an area, the plant may not be able to reproduce sexually, or reproduction may be rendered uncertain or unlikely by environmental factors. With outcrosssing, each generation produces new variability, and although most progeny may be fit and well adapted, some progeny may be less fit and constitute "genetic load" to the population. The third method of reproduction—apomixis—factilitates the production of a large number of well-adapted plants of the maternal genotype with little or no genetic load. Apomixis offers the possibility of reproduction by seed in plants with "odd" or unbalanced chromosome numbers, as such plants are unable to produce viable gametes at meiosis and are likely to be totally or partially seed-sterile. Seed apomixis, for example, provides all the advantages of the seed habit (dispersal of propagules and a potential means of survival through unfavorable seasons). Apomicts are often of polyploid and hybrid origin, and therefore this reproductive mode can potentially serve as a means of preserving high heterozygosity. Apomixis, like selfing, would also appear to be important at the edge of the range of a species allowing populations to persist in areas in which various factors may limit or exclude the possibility of sexual reproduction. Given that all three reproductive modes have advantages and disadvantages depending on environmental circumstances, it is not too surprising to learn that plants often have highly flexibile mating systems, reproducing by several means, rather than relying on only a single reproductive mode.

The mating system of a plant species will influence the way in which the genetic diversity present in the species is distributed within and among its populations—specifically, its *population genetic structure*. In outcrossing species, higher levels of genetic diversity are found within than among populations. The opposite is true for predominantly selfing species where greater among-population (i.e., interpopulation) differentiation is expected. Knowledge of a plant's mating system is important in conservation of its genetic diversity in a seed genebank, or for efficient screening of populations of wild species as source of traits for crop improvement in plant breeding programs. More populations of a selfing species would be needed in order to capture the true diversity of a species.

2.4.3. Hybridization and Polyploidy

Although we think of species as discrete and static breeding entities, examples can be found throughout the angiosperms where different species have the capacity to cross with another. Plants are champions at interspecific hybridization. *Hybridization*, or the process of sexual reproduction between members of different species or biotypes within a species, produces plants that have genetic material from both parents. In most cases, the initial hybridization event results in hybrid plants that are haploid for each genome or in other words, have a single homologous chromosome from each parental chromosome set (Fig. 2.13). As homologous chromosomes are normally paired during metaphase I, the presence of only one of each homologous chromosome pair, can disrupt normal meiotic function. In fact, most of the gametes produced in hybrids are abnormal, leading to sterility to reduced viability of pollen or eggs in the hybrid plant. Although hybrids can be made from the crossing of many different species, hybridization of normal haploid gametes rarely generates plants that are fully fertile.

In some cases, sex cells are produced that have more than just one of each homologous chromosome. *Nondisjunction*, when homologous chromosomes fail to separate during meiosis, sometimes generates gametes that have complete sets of chromosomes from the

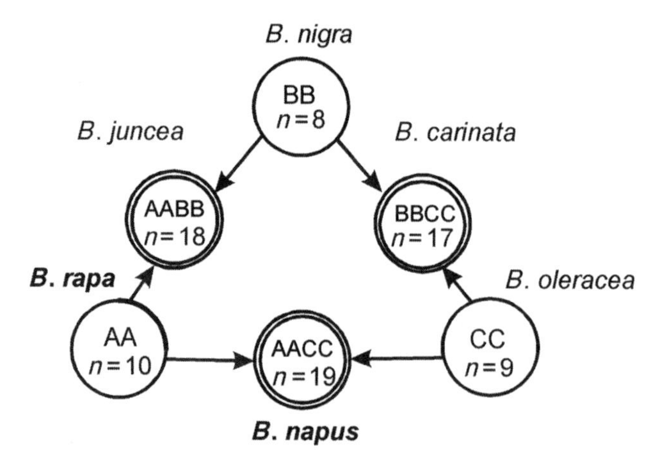

Figure 2.13. Triangle of U (1935) shows the relationships between several diploid and polyploid crop species within the *Brassica* genus.

parent species, called *unreduced gametes*. If two unreduced gametes fertilize one another, the resultant hybrid would have the complete genome of each parental species. In this case, meiosis can function normally, and the hybrid plant may represent a new species with a unique chromosome number. Species that contain multiple genomes or multiple sets of chromosomes beyond the diploid level are called *polyploids*. Again, among the myriad of organismal types, plants are champions at polyploid production—and indeed many plant species are polyploids.

Polyploidy may arise in two ways: by the doubling of a homologous set of chromosomes (autopolyploidy) or by combining two complete sets of chromosomes from genetically different parent plants (allopolyploidy). An autotetraploid contains four sets of homologous chromosomes, and pairing between the four homologous chromosomes is often irregular, with chromatids showing random segregation during gamete formation. In an *allotetraploid*, on the other hand, the parental chromosomes in each of the two sets of homologous chromosomes tend to pair with each other as they would in the parental plants, thus contributing to the stability and fertility of such plants. Several natural allopolyploids are known, and several have been created in the plant breeding field.

Hybridization is an important process that has occurred in the development of many of our agricultural crops. Many polyploid crop plants have been produced by either the combination of unreduced gametes or the doubling of the chromosomes after hybridization of haploid gametes. Canola (*Brassica napus*), which is used for vegetable cooking oil, is composed of the complete genomes of two different species (*B. rapa*, genome AA and *B. oleracea*, genome CC); similar polyploid origins have been confirmed for two other *Brassica* crops *B. juncea* and *B. carinata* (Fig. 2.13). Bread wheat, *Triticum aestivum*, was produced from the hybridization between three different species. In this case, each progenitor species donated their complete diploid genome (AA, BB, DD genomes, respectively) to making a species with three complete sets of chromosomes and a very large "new" wheat genome (AABBDD).

Polyploidization is undoubtedly a frequent mode of diversification and speciation in plants. More recent studies indicate that most plants have undergone one or more episodes of polyploidization (i.e., increase in the whole DNA complement beyond the diploid level)

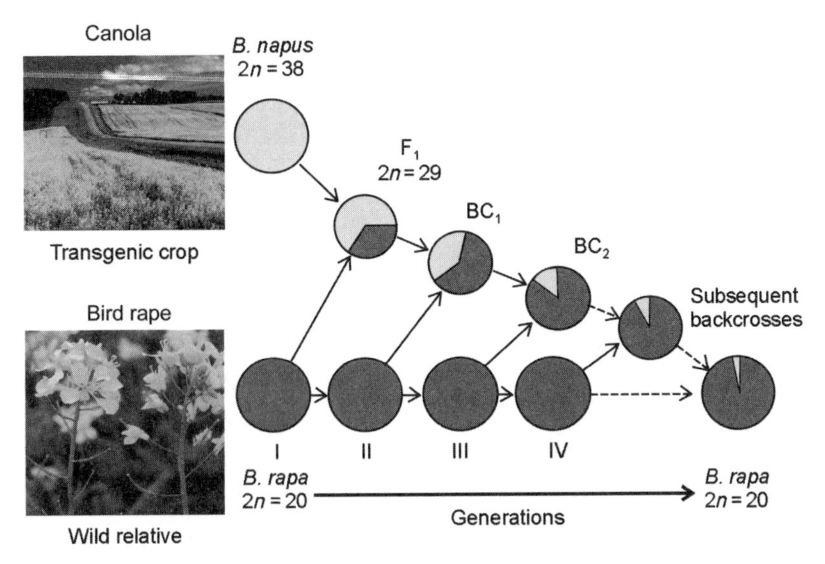

Figure 2.14. Hybridization and genetic integration between closely related species allows for the incorporation of genetic material from one species to another.

during their evolution (Soltis et al. 2004). Hybrid speciation is another important phenomenon. Interspecific hybridization and subsequent introgression of the portion of the genome of one species into that of another (Fig. 2.14) have often been recognized as a source of genetic variation and genetic novelties, and in some cases successful hybridization events have promoted rapid speciation radiation. The complexities of plant genetics can be traced to reproductive biology and mating systems in plants, which is an area of research that is very active and dynamic.

2.5. CONCLUSION

After biotechnologists introduce or manipulate genes in plants, if all goes as expected, the new genes should be part of the genomic fabric and behave like normal plant genes. Therefore, they should follow the laws of Mendelian genetics and be passed on to future generations like other genes of the particular species. Therefore, it is important for the plant biotechnologist to know botany and basic genetics. Transgenes also become part of breeding programs, which is why understanding the fate of transgenes in new plant cultivars is important—the subject of the next chapter.

LIFE BOX 2.1. RICHARD A. DIXON

Richard A. Dixon, Professor and Director, Plant Biology Division, Samuel Roberts Noble Foundation; Member, National Academy of Sciences

Rick Dixon relaxing at a faculty retreat, Quartz Mountain, Oklahoma (May 2007).

I first became interested in plant natural products as an undergraduate at Oxford. I was reading Biochemistry, and the course was quite heavily weighted towards physical biochemistry, an area I found hard because of my lack of mathematical prowess. Faced with the choice of either whole animal physiology or plant biochemistry as an elective, I jumped at the latter, a decision that determined the future course of my career. I had been excited by organic chemistry at an early age, and was fascinated to learn how plants "do" organic chemistry during the synthesis of natural products and lignin. This was before the era of molecular biology, and our understanding depended mostly on the results of in vivo labeling studies coupled with in vitro enzymology. I always remember my first lecture from Vernon Butt, in which he outlined current views on how the monolignol units of lignin are formed. It all seemed so beautiful and logical, although my group and others were later to show that it is actually more complex than envisaged at the time. This new understanding had to wait until we had the necessary genetic and genomic tools.

I decided to stay on in the Botany School at Oxford to work on my D.Phil. with Keith Fuller. Keith had suggested a project on galactomannan mobilization in alfalfa, but, when I returned from the summer vacation to start this project, we discovered that four papers, reporting essentially everything we were planning to do, had just appeared in the literature. Keith suggested I might instead look at how plants make bioactives in cell culture. I was disappointed at being "scooped" on my planned project (although better early than later!), and did not realize at the time that agreeing to the back-up plan was the defining moment in my career. Using the isoflavonoid phytoalexin phaseollin from bean as a model, I established conditions for turning on isoflavonoid metabolism in cell cultures. When Chris Lamb joined the lab as a postdoc we set up a collaboration that lasted nearly 20 years, in which we used the phytoalexin induction system as a model for studying microbially-induced gene expression in plants using the new tools of molecular genetics.

After two years of postdoctoral work in Cambridge and nine years of teaching and research at the University of London, I moved to become director of the newly formed Plant Biology Division at the Noble Foundation in Ardmore, Oklahoma, in 1988. During the first eight years of my tenure at Noble, I continued to work primarily on plant–microbe interactions. The Noble Foundation's major mission is to assist farmers and ranchers reach their production goals through basic and applied science and demonstration, and, during the previous years, I had hired a number of excellent principal investigators in the plant–microbe interaction field. I therefore decided to move away from the plant–microbe focus, and to concentrate my research on those natural product pathways that impacted forage quality, the health of ruminant animals, and human health. This was another decision, dictated by circumstances, that has paid dividends. The work I initiated on the biosynthesis and metabolic engineering of lignin and proanthocyanidins has been rewarding as basic science, has moved towards commercialization through a long-term research collaboration with Forage Genetics International, and has had important implications for plant metabolic engineering in relation to lignocellulosic bioenergy crops (lignin) and human health (proanthocyanidins). This is certainly more than I envisaged when I first decided that the plant–microbe field was too crowded and that quieter pastures might profitably be grazed!

Based on my personal experiences, my advice to young scientists would be to always stick with what you are passionate about, always try to work with people who are smarter than you are, and never turn down opportunities to adapt your program to emerging applications. It is also critical to get away from the lab and clean out your brain (regularly!). I have had a passion, since the age of 10, for studying, collecting and cultivating cacti. I also love hiking, particularly in mountains. The photograph shows me indulging both of these passions in the Quartz Mountains of Southwestern Oklahoma (although I have to admit that this was during a short break at a faculty retreat!).

LIFE BOX 2.2. MICHAEL L. ARNOLD

Michael L. Arnold, Professor of Genetics, University of Georgia

Mike Arnold with *Iris nelsonii*; Vermilion Parish, Louisiana.

From Whence I Come

In regard to my career as an evolutionary biologist, I start the clock with the Fall [1975] semester of my freshman year at Texas Tech University. During this time period, I fell in love with research science—sometimes to the detriment of my participation in classes! My initial plan was to work with a parasitologist who specialized in organisms dug from the rotting remains of farm animals. However, this professor stood me up for several scheduled meetings and so I turned instead to a plant evolutionary biologist, Professor Raymond Jackson, and an animal evolutionary biologist, Professor Robert Baker, as my first two mentors. Their patience and encouragement helped me to not only finish the lab work for several research projects, but to see the research published in scholarly journals as well. This taught me the love of discovery and creation—discovery of facts about the natural, evolving world and creation of word pictures in order to explain what had been discovered. Their careful tutelage gave me the understanding of how to pursue research projects. Because my earliest training was in both botany and zoology, it has been natural for me to emphasize tests for common evolutionary patterns between plants and animals that may reveal common underlying processes. This emphasis is reflected both by the breadth of organisms on which my students, post-doctoral associates and I have worked (everything from fruit flies to fungi and fruit bats to Louisiana Irises) and the synthetic treatments we have produced—e.g., the two books *Natural Hybridization and Evolution*, 1997 and *Evolution Through Genetic Exchange*, 2006.

(Re)Turning to Plants

Though, as indicated above, my colleagues and I have examined many types of organisms, 20 years ago I did make a decision to focus most of my research efforts on plant taxa. Several factors led to this decision, two of which related to my earliest training in evolutionary botany and zoology. I had learned quickly, that testing many of the hypotheses in which I was interested—especially those associated with the processes of genetic exchange, speciation and adaptation—required taxa that would allow a dual approach of experimental manipulations and surveys of natural populations. Most plant and animal groups (and for that matter, many bacterial and viral assemblages) provide opportunities to examine naturally occurring populations for the purpose of estimating evolutionary processes such as genetic exchange via introgressive hybridization and/or lateral transfer. However, few animal clades allow the type of direct assessments possible in studies of plant species (e.g. through reciprocal transplantations into both experimental and natural environments). In addition, my interest in testing the descriptiveness of the web-of-life metaphor (i.e., that emphasizes the importance of genetic exchange in the evolution of organisms) led me to choose plants over animals. Thus, evolutionary biologists consider plants to be paradigms of such processes as introgressive hybridization, hybrid speciation and adaptive trait transfers.

Has Our Work Affected Plant Biotechnology?

I believe that the work carried out by my colleagues and myself has impacted the field of plant biotechnology in several ways. However, all of the effects from this work can likely be traced back to our emphasis on studies of population level phenomena. In the early 1990s, when we began our research into reticulate evolution, plant evolutionary biology was characterized by systematic treatments (i.e., studies that defined the relationships of species). Indeed, many decades had passed since the appearance of the wealth of publications by such workers as Edgar Anderson and Ledyard Stebbins on the population-level phenomena associated with genetic exchange between plant lineages. With few exceptions—e.g., see many publications of Verne Grant and Don Levin—the study of plant evolution had emphasized pattern over process. In contrast, our work was designed to emphasize process over pattern. For example, we have asked how the processes of introgressive hybridization, hybrid speciation, lateral exchange, and adaptive trait transfer have affected the evolutionary patterns reflected in present-day bio diversity. This process-over-pattern focus has led to the application of our findings by plant biotechnologists, particularly when they are considering the effect that gene exchange might have on development and control of bio-engineered products. One example of this can be seen in the interest that we have generated by highlighting the observation common to the vast majority of hybridizing plant and animal taxa (as well as for those organisms exchanging genes via viral recombination and lateral exchange), that hybrid genotypes demonstrate a range of fitness estimates that are often affected by the environment. This key observation leads to an array of expectations concerning the challenges faced in forming hybrid lineages—both under [i.e., under both] natural and experimental conditions. Furthermore, the observation of a wide range of hybrid fitness should also lead to caution during the generation of predictions concerning the effects on natural ecosystems from the introduction of bio-engineered plant lineages.

To Where Are We Going?

I am reminded of the Old Testament mandate that states that prophets, once proven inaccurate, were to be stoned. In that context, I offer the following suggestion concerning one direction I believe studies of genetic exchange (of which I do consider myself a student) and plant biotechnology (of which I do not) should be progressing. The analyses of genetic exchange, across all taxonomic categories, are entering an exciting phase. The definition of the genomic architecture of related organisms allows the dissection of the causal factors that affect the transfer of specific loci. Given such information, it is possible to state with some certainty, which loci are prevented and which loci are facilitated in their transfer between organisms belonging to divergent evolutionary lineages. However, a more difficult, and much more significant, inference is needed. Specifically, it is necessary to define the "why" behind a transfer (or lack of transfer). In other words, what is the specific effect on the organism that causes either an increase in the fitness of hybrid genotypes (leading to genetic transfer) or a decrease in the fitness of hybrid genotypes (resulting in no transfer) when certain combinations of loci are present? The degree to which we are able to address and answer this question, will be the degree to which we are able to test hypotheses concerning such fundamentally important processes as (i) the effect of genetic exchange on hybrid lineage formation and the transfer of adaptations and (ii) the impact of genetic exchange between bio-engineered plants and wild relatives on both crop production and natural ecosystems.

REFERENCES

Anderson E (1953): Introgressive hybridization. *Biological Reviews* **28**:280–307.

Bateson W (1909): *Mendel's Principles of Heredity.* Cambridge Univ. Press, Cambridge, UK.

Bridges CB (1916): Nondisjunction as a proof of the chromosome theory of heredity. *Genetics* **1**: 1–52, 107–163.

Bridges CB (1925): Sex in relation to chromosomes and genes. *Am Nat* **59**:127–137.

Briggs D, Walters SM (1997): *Plant Variation and Evolution*, 3rd ed. Cambridge Univ. Press, Cambridge, UK.

Camerarius RJ (1694): Epistola ad M.B. Valentini de sexu plantarum. In Ostwald's Klassiker der exakten Naturwissenschaften, No. 105, 1899. Verlag von Wilhelm Engelmann, Leipzig.

Darwin CR (1876): *The Effects of Cross and Self Fertilisation in the Vegetable Kingdom.* John Murray, London.

Goss J (1824): On the variation in the colour of peas, occasioned by cross-impregnation. *Trans Hort Soc Lond* **5**:234.

Grant V (1981): *Plant Speciation* 2nd ed. Columbia University Press, New York.

Knight TA (1799): An account of some experiments on the fecundation of vegetables. *Phil Trans Roy Soc Lond 195–204.*

Levin DA (1993): Local speciation in plants: the rule not the exception. *Syst. Botany* **18**:197–208.

Mendel G (1866): Experiments in plant hybridization (translation). In *Classical Papers in Genetics*, Peters JA, (ed), 1959. Prentice-Hall, Englewood Cliffs, NJ.

Richards AJ (1986): *Plant Breeding Systems.* George Allen & Unwin, London.

Soltis DE, Soltis PS, Pires JC, Kovarik A, Tate J, Madrodiev E (2004): Recent and recurrent polyploidy in Tragopogon (Asteraceae): Cytogenetic, genomic, and genetic comparisons. *Biol J Linn Soc* **82**:485–501.

Sprengel CK (1793): Das entdeckte Geheimniss der Natur im Bau und in der Befruchtung der Blumen, [*The Secret of Nature in the Form and Fertilization of Flowers Discovered*]. Berlin.

Stebbins GL (1959): The role of hybridization in evolution. *Proc Amer Phil Soc* **103**:231–251.

Stuessy TF (1989): *Plant Taxonomy.* Columbia Univ. Press, New York, Oxford.

Sutton WS (1903): The chromosomes in heredity. *Biol Bull* **4**:231–251.

UN (1935): Genome analysis in *Brassica* with special reference to the experimental formation of *B. napus* and peculiar mode of fertilization. *Jpn J Bot* **7**:389–452.

Watson JD (1968): *The Double Helix.* Atheneum, New York.

Watson JD, Crick FHC (1953a): Genetical implications of the structure of deoxyribonucleic acid. *Nature* **171**:964–969.

Watson JD, Crick FHC (1953b): Molecular structure of nucleic acids. A structure for deoxyribose nucleic acid. *Nature* **171**:737–738.

CHAPTER 3

Plant Breeding

NICHOLAS A. TINKER

Agriculture and Agri-Food Canada, Ottawa, Ontario, Canada

3.0. CHAPTER SUMMARY AND OBJECTIVES

3.0.1. Summary

Breeding modifies plant genetics in thousands of genes at a time by hybridizing plant types then selecting on traits (and genes) of interest. An iterative process, the plant breeder makes crosses to accumulate genes and desirable traits into genetic backgrounds of choice. Breeding relies on the principles of Mendelian genetics and the use of statistical methods. In practice, biotechnology is nearly always combined with plant breeding for crop improvement.

3.0.2. Discussion Questions

1. Describe how plant breeding is both an art and a science.
2. Is seed color a qualitative or quantitative trait?
3. List six factors that can affect the distribution of quantitative trait phenotypes that will appear in a given population.
4. What proportion of plants in an F_6 generation are heterozygous at a given locus?
5. What is the probability that five segregating loci will all be homozygous in the F_6 generation?
6. What is the difference between a landrace and a pure-line plant variety?
7. The pedigree and the SSD methods are two strategies for developing pure-line varieties. List some factors that might influence your choice of one *vs.* the other.
8. Using the terms *homozygous*, *heterozygous*, *homogeneous*, and *heterogeneous*, describe each of the following: (a) a modern maize hybrid, (b) a synthetic alfalfa variety, (c) a mass-selected population of maize, (d) a landrace of wheat, and (e) a modern variety of wheat.

*This chapter is the work of the Department of Agriculture and Agri-Food (Canada) (AAAF).

Plant Biotechnology and Genetics: Principles, Techniques, and Applications, Edited by C. Neal Stewart, Jr.

3.1. INTRODUCTION

Plant breeders enjoy quoting a famous US president, who wrote that "The greatest service which can be rendered any country is, to add an useful plant to its culture."[1] Whether or not you agree, it must be acknowledged that the creation of new and better plant varieties is among the most useful and visible outcomes of biotechnology. Whether it is a noble service might depend on whether you do it for fun, for profit, or for the good of humanity, but most breeders will confess to all three motives. Plant breeding has been credited with helping to triple the productivity of modern agriculture, and it has been a fundamental part of international humanitarian achievements (Hoisington et al. 1999). But you do not have to "think big" to be excited about plant breeding. Admire the colors on your next plate of food, taste the subtle flavors in your next bite of fruit, feel the strength and softness of your cotton shirt, or smell your favorite rose—these characteristics are all derived from unique characters of different plant varieties. You spend less of your income to eat a much better variety of foods than your ancestors did, largely because of plant breeding. In future you might live longer or healthier because of the varieties of plants used to make your breakfast cereal. What fun it would be to create those varieties, or even just to understand how they are created!

Plant breeding is a skill that requires advanced learning and practical experience. Many universities and corporations worry that trained plant breeders are becoming scarce in relation to ongoing demands. Most of the modern concepts in biotechnology that are introduced in this book can be viewed either as enhancements to plant breeding, or as innovations that can be rendered useful only through plant breeding. The most groundbreaking achievements in biotechnology still need to be packaged in plants that are productive, disease-free, tasty, and nutritious. These qualities depend on the coordinated expression and complex interactions of thousands of plant genes and gene products. We have learned many things about many genes, but we may never know enough to fine-tune all of the genes required to make a plant variety that is adapted and competitive under modern agricultural production practices. Thus, we continue to depend on plant breeding as the cornerstone of commercialization and technology transfer.

Plant breeding was described by Nikolai Vavilov,[2] the famous Russian scientist, as "evolution directed by man." Thus, the job of a plant breeder is to replace natural selection with artificial selection, such that combinations of traits can be assembled into plant varieties that would not otherwise be found in nature. While correct, this definition hides many of the dimensions in which a breeder must work to produce successful plant varieties. There are two primary interventions that a breeder makes: the deliberate hybridization of

[1]From *Memoir, Correspondence, and Miscellanies, from the papers of Thomas Jefferson*, edited by Thomas Jefferson Randolf, 1829, p. 144.

[2]Vavilov and Lysenko: In the plant sciences, Russia has produced one of the most influential plant scientists as well as one of the most notorious. Nikolai Ivanovich Vavilov (1887–1943) is credited with several important discoveries in genetics, including the demonstration that the center of diversity of a plant species is an indication of its center of origin. He also assembled one of the largest and most diverse collections of plant germplasm in the world, now housed at the Vavilov Institute of Plant Industry in St. Petersburg. Trofim Denisovich Lysenko (1898–1976) was an experimentalist who claimed the discovery of many agricultural methods that now seem absurd. To his credit, he also studied some phenomena such as vernalization that we now recognize as important physiological mechanisms. Unfortunately, his claims of rapid and phenomenal success were popularized to the extent that he garnered great political influence in Stalin's Soviet Union. When he was put in charge of the Academy of Agricultural Sciences of the Soviet Union, he was able to silence or imprison his critics, including Vavilov, who died in prison in 1943.

specific parents, and the selection (or elimination) of progeny. This seemingly simple iterative process is elaborated by many factors: knowledge of what traits are important, knowledge of genetic control, knowledge of how environment affects traits, and knowledge of strategies to reduce the sheer numbers of progeny that must be examined. On top of this, a breeder must be a communicator, a team builder, an extension worker, an expert in commercialization, and a specialist in legal, ethical, and social issues. Plant breeding is often described as being an art as well as a science. While there are deterministic principles to discover and apply, there is often more than one acceptable result, and more than one way to achieve the same result. Plant breeders sometimes claim to recognize another breeder's "handiwork" by the way a variety looks in the field, and often, they find that the most efficient use of time and resources is to walk through a field and identify plants that "just look right."

While studying the topic of plant breeding, you might think of numerous analogies that help you conceptualize the process. For example, you might draw a parallel between a good plant variety and a favorite song; they are both dependent on many subtle characteristics, and although their quality may be widely acknowledged, appreciation of this quality is varied. In some ways, breeding is also similar to the iterative trial-and-error process that investors use to build strong and diverse investment portfolios, and principles of genetic selection have even been applied with great success to areas such as this. But the processes of genetic recombination and gene expression are unique to DNA-based organisms, and no analogy can completely replace the concepts that must be learned to become a successful plant breeder.

This chapter introduces some fundamental concepts of plant breeding, and describes some generic strategies that are typically used to breed plant species that have a variety of mating systems. In the "real world," every plant species presents unique challenges and opportunities, and it is beyond the scope of this chapter to discuss strategies used for specific crops. The emphasis in this chapter has been placed on describing the underlying concepts of plant breeding, which will help you understand and appreciate literature that is more detailed or specific. You are encouraged to look at some of the references listed at the end of this chapter to see how breeding is typically applied in plant species that interest you.

3.2. CENTRAL CONCEPTS IN PLANT BREEDING

Prior to reading this chapter, you should have studied the previous chapter, and you should have a good working knowledge of plant genetics and reproduction. The concepts introduced in this section will build on that knowledge. The following paragraphs introduce key concepts that collectively determine most of the decisions and strategies of a breeding program.

3.2.1. Simple versus Complex Inheritance

The previous chapter introduced Mendelian genetics—undeniably, the most important concept that a breeder must understand. The discovery of Mendelian principles was made in a plant species (pea) using traits that might be important in a pea breeding program (color, height, and starch content). These traits are considered *qualitative* (having discrete values such as green or yellow, tall or short) and *monogenic* (controlled by single genes). Such traits are also described as showing *simple inheritance*. However, many other traits that a plant breeder works with—such as fruit weight, maturity date, and grain yield—are *quantitative* (measured on a continuous scale) and *polygenic* (controlled by many genes). Such traits are also described as showing *complex inheritance*.

Figure 3.1 illustrates how a quantitative, polygenic trait can still have underlying Mendelian inheritance. In this illustration, the size of a melon fruit is determined by the type of alleles that are present at two different genetic loci. This type of mathematical simplification is commonly used to develop or test models that can help explain the numbers of genes and the types of gene action that are involved in quantitative trait expression. Although these are mathematical assumptions, models such as this can often approximate underlying biological phenomena. For example, the "capital" alleles in Figure 3.1 could represent gene promoters that trigger higher expression of fruit development factors, and the "small" alleles are less effective versions of these gene promoters.

The distinction between simple and complex inheritance is a common source of confusion. We say that green versus yellow is a simple monogenic trait, because it is often determined by one of two alternate alleles at a single genetic locus. But there are probably numerous other genes that might influence the intensity of the green or yellow color, and there are probably other gene loci that could mutate to block the production of the green

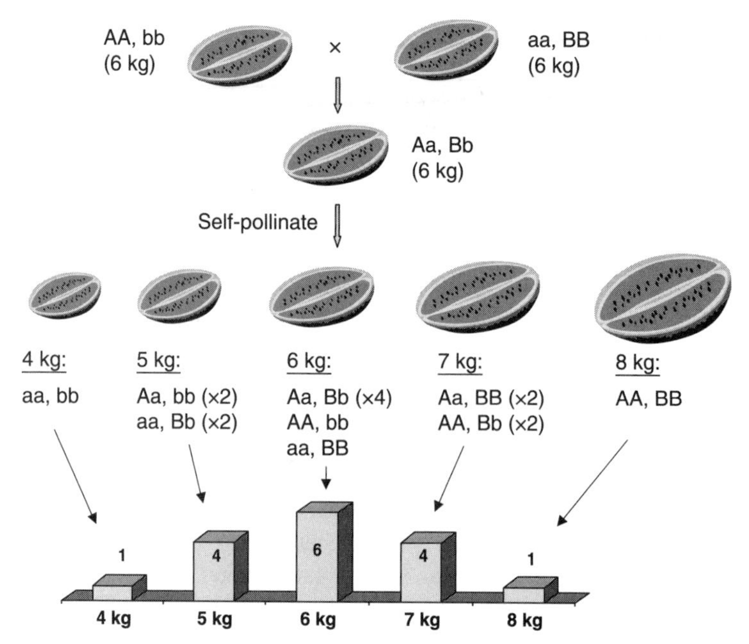

Figure 3.1. A hypothetical melon-breeding scenario that illustrates quantitative inheritance. Alleles at two loci (A and B) are represented by lower- versus uppercase letters. Assume that the allele represented by the capital letter increases melon weight relative to the small allele, such that the average weight of melons (in kg) produced by a variety is determined by 2 times the total number of capital alleles plus the number of small alleles. Nine different genotypes will be present in the F_2 generation. If the two loci are not linked on the same chromosome, their expected segregation ratios will be as depicted in Figure 2.6. However, because some different genotypes result in the same melon size, only five sizes of melons will be produced, as shown. The expected proportions of each melon size in a random F_2 population is depicted by the histogram at the bottom of the figure.

color. These other differences may or may not be noticeable, and they may or may not be present within a specific set of germplasm. Thus, in some populations, seed color could be a polygenic trait. Indeed, a pea breeder might make a cross between a yellow-seeded variety and a green-seeded variety, then try to select progeny that had even greener seeds than those of the original green parent. Why? Because the yellow-seeded variety might contain alleles at loci other than the primary seed color locus that are capable of enhancing the green color in the presence of the green alleles at the primary locus. Thus, seed color can be either qualitative and monogenic (simple inheritance), or quantitative and polygenic (complex inheritance) or both, depending on circumstance and on the germplasm being investigated.

3.2.2. Phenotype versus Genotype

An important term used throughout this book is *phenotype*, which simply means "what something looks like." We often speak about the phenotype of a specific trait, in which case it takes on units of measurement. For example, the phenotype of a quantitative trait such as seed weight in wheat might range between 30 and 80 mg. The term *phenotype* is also used to distinguish what a plant looks like from its *genotype* (what genes are present) or *genotypic value* (what we would expect the phenotype to be if we could predict it exactly from the genotype). A fundamental concept in plant breeding is that genotypic value is something that we try to measure and predict. If we could identify or control all the unpredictable effects of error and environment, then the phenotype of a plant (P) would be equal to its genotype (G) plus the effects caused by error and environment (E). Virtually all of the fancy equations that you will see in plant breeding books are derivations of this basic formula

$$P = G + E$$

or more precisely

$$P = \Sigma\, G + E$$

where Σ indicates that genetic effects may be summed over multiple genes, as they are in Figure 3.1.

The equations above refer to the genotypic or phenotypic values of a single plant or observation. However, breeders work with populations of many plants, and they often summarize a set of observations by calculating the *variance*, which is simply a mathematical formalization of *variability*, and genetic variability is the key to creating varieties through artificial selection. The basic breeding equation can also be written to describe a population of plants in terms of phenotypic variance (V_{P}), genetic variance (V_{G}) and environmental variance (V_{E}), such that

$$V_{\mathrm{P}} = V_{\mathrm{G}} + V_{\mathrm{E}}$$

It is imperative for any breeder to understand the relative proportion of genetic variance that contributes to phenotypic variance for a given trait. This concept is formalized using the term *heritability* (H), which, in its simplest form, is measured as

$$H = \frac{V_{\mathrm{G}}}{V_{\mathrm{P}}}$$

Since V_P is always greater than or equal to V_G, the heritability of a trait can range from 0 to 1. If H is equal to one, then all variance is caused by genetic effects, and the breeder will be very successful at selecting better plants. Such is the case for the imaginary melon trait illustrated in Figure 3.1. However, if H is zero, then V_G must also be zero, and there is no possibility of selecting plants that are genetically superior because all variation is environmental. Most traits that breeders work with show intermediate levels of heritability, between zero and one.

3.2.3. Mating Systems, Varieties, Landraces, and Pure Lines

The fundamental output of plant breeding is the *plant variety*, which is sometimes referred to as a *cultivar* (i.e., a cultivated variety). However, the genetic makeup of a variety, and the way in which it is produced, maintained, and released depends critically on the type of *mating system* found in the species to which the variety belongs. Many plants can tolerate *self-pollination* (or self-fertilization), and some of the most important crop species (including most grain and oilseed crops) are naturally self-pollinated. An important exception is maize (corn), which can tolerate self-pollination, but is normally *cross-pollinated* (or cross-fertilized). Other plants cannot tolerate self-pollination, and have specific genetic mechanisms to prevent this (see Chapter 2). Plants that normally cross-pollinate are subject to continual recombination and selection after varieties are released, and thus strategies for breeding and variety release can be quite different from those used in self-pollinating species.

For plant species that normally cross-pollinate, we often assume that mating occurs at random. In reality, this is seldom the case because plants that are near to each other are more likely to pollinate each other. Nevertheless, the assumption of random mating allows the development of theories that often give good approximations of reality. The most important theory regarding random mating is the *Hardy–Weinberg law*, which predicts the frequency of genotypes that will occur according to the frequency of alleles. Assume that there are two alleles, "A" and "a," at a given locus, and that the alleles are at frequencies p and q, where p must equal $(1 - q)$. The law states that the frequencies of genotypes, as represented below, can be predicted as

$$AA : Aa : aa = p^2 : 2\,pq : q^2$$

An important property of this law is that these frequencies are achieved after just one generation of random mating (the proof of this theory is shown in many textbooks). An important application of this theory is to identify whether random mating is occurring, or if other factors such as selection or mixing of populations (immigration/emigration) are occurring.

Plant species that are highly self-pollinated usually exist in a *homozygous* state (i.e., alleles exist in identical pairs at most loci). To understand why, consider what happens when a *hybrid* is formed, through either a chance pollination or a deliberate hybridization by a breeder. Figure 3.2 shows a cross between two homozygous genotypes. The product of this mating (a hybrid) will be *heterozygous* at any locus that differs between the parents, and all progeny will be identical. However, a mixture of genotypes will exist in the F_2 generation and beyond. Each generation of selfing reduces the level of heterozygosity by 50%, such that the proportion of homozygotes (P_{homo}) at a particular locus in generation F_X can be predicted as

$$P_{\text{homo}} = 1 - \tfrac{1}{2}(X - 1).$$

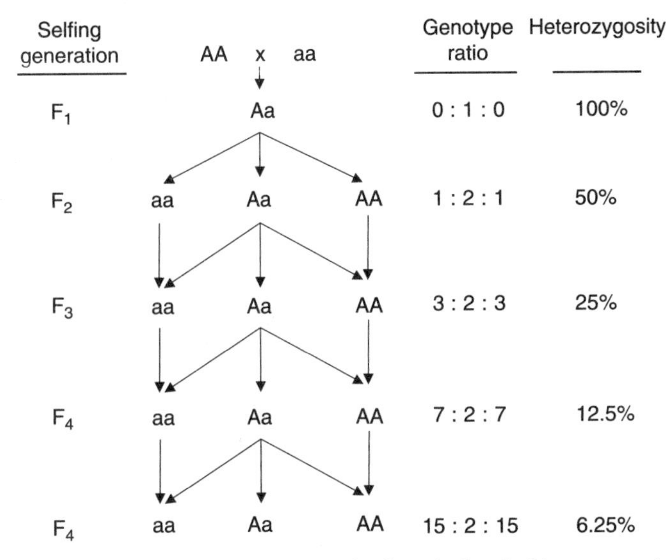

Figure 3.2. In repeated self-pollination with no selection, the level of heterozygosity is reduced by one-half with each selfing generation. This is because only half of the progeny from the heterozygous genotypes will still be heterozygous, while all the progeny from homozygous genotypes will be homozygous. Thus, the population gradually approaches complete homozygosity, but with a mixture of homozygous genotypes.

This prediction applies to a single locus, and to the average level of homozygosity after X generations of selfing. Thus, after just four generations of selfing, the average level of homozygosity and the probability that a given locus is homozygous are 94%. However, when N loci are considered simultaneously, and if all loci assort independently, then the probability that all N loci will be homozygous is equal to $(P_{\mathrm{homo}})^N$. Thus, if a large number of loci are segregating, then, even after many generations of selfing, there remains a high probability that at least one of those loci remains heterozygous. This prediction has important consequences in the breeding and release of plant varieties that are regarded as homozygous—it means that there are always a few loci that may segregate within the variety. Such phenomena sometimes turn up when a variety is grown in a novel environment where no one has ever tested it before.

Prior to the development of modern breeding methods (and even afterward, for various reasons), plant varieties in self-pollinating species were propagated as mixtures of homozygous lines called *landraces*. Each landrace typically arose from generations of bulked selections from a farmer's field. Gradually, through selection of desirable types (e.g., large ears of corn) or elimination of undesirable types (e.g., those with seed that fell off during harvest: shattering), a farmer might develop a particularly useful landrace and share it with friends. Landraces often took on the name of a farmer, a region of origin, a defining characteristic, or a combination thereof (e.g., "Swedish giant"). It was probably rare for a landrace to originate from a single plant, because there was no knowledge that this might be beneficial, and because this would have required careful multiplication of seed in isolation from other crops before there was adequate seed to plant a crop for harvest. In a landrace that had been grown for many generations, most plants would be homozygous, but the landrace would remain as a *heterogeneous* mixture of different genotypes.

In 1903 a Danish biologist, Wilhelm Johannsen, reported an important finding that has provided the foundation for modern breeding methods. He showed that progeny grown from a single plant selected from a mixture of inbred lines would produce progeny that were consistently different from those of another plant from the same mixture. Thus, he could create a large-seeded variety and a small-seeded variety through single-plant selections from the same mixture. Importantly, he also observed that further selections within progeny that were derived from the same single plant were not effective. This is because each selection represented a pure homozygous line, and all subsequent variation observed within a selected line was due to differences in environment, and not to genetic differences. A variety selected and multiplied from a single homozygous plant is known as a "pure line," and the alleles or traits possessed by this line are said to be "fixed," meaning that further selection is neither necessary nor possible. These observations, as illustrated in Figure 3.3, are known as *Johannsen's pure-line theory*. It is also noteworthy that these observations were probably the first time that a clear distinction was made between genotype and phenotype—an important step beyond Mendel's laws.

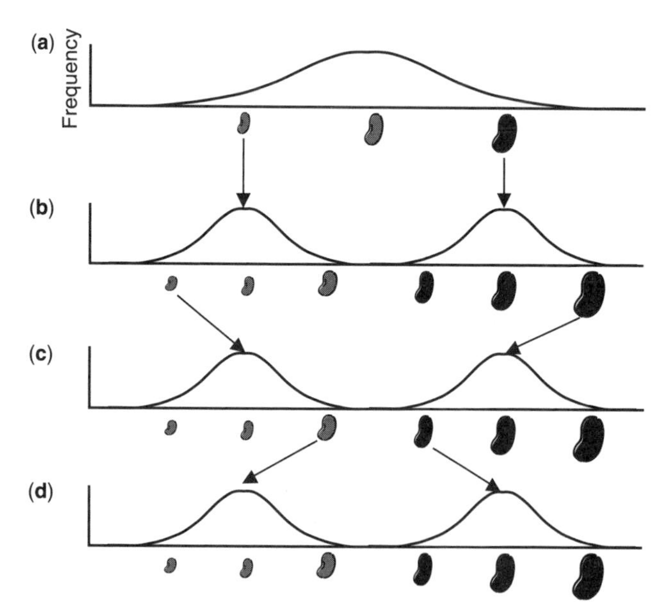

Figure 3.3. Development of pure lines from a mixture of homozygous, heterogeneous beans. Panels (a) through (d) show histograms representing the frequency of different bean sizes in various populations. Panel (a) could represent a landrace or a population derived by repeated selfing of progeny from a hybridized plant. Selection of single beans from the original population results in new populations (b) that have different average bean sizes. Further selection within these populations is not effective (c, d). In this illustration bean color is a qualitative trait that shows no environmental variation, whereas seed size is a quantitative trait that shows environmental variation. The dark-colored beans on the right represent a pure line, and all phenotypic variation for seed size within that line is environmental. A Danish biologist, Wilhelm Johannsen, conducted a similar series of experiments in 1903, and developed what we now call the *pure-line theory*.

3.2.4. Other Topics in Population and Quantitative Genetics

Refer back to Figure 3.1, and try to imagine factors that might complicate the situation that is illustrated. Rather than two genes affecting this trait, there could be dozens of genes. Different genes could have effects of different magnitudes, and there could also be an environmental influence. Rather than being *unlinked* (assorting independently), some of the loci might be *linked* together on the same chromosomes. This would result in some combinations of parental alleles being more frequent than others. The effects of individual alleles may not be completely *additive* as they are in this figure (i.e., the genotypic value is found by adding up the individual genotypic effects at each locus). If the genotypic value of the heterozygote is not equal to the average of the homozygotes, we refer to this as *dominance* (meaning that one allele has a dominant effect over another). If alleles at different loci interact (i.e., if the total genotypic value is different from the sum of the genotype values of the individual loci), we call this *epistasis*. Many of these factors will tend to produce a histogram of phenotypes that is more continuous (smoother) than the distribution shown in Figure 3.1, but they can also cause the shape of the distribution to deviate from the normal (bell-shaped) distribution that results when all genes have uniform, additive effects. All of these concepts are simplest to study in a diploid plant. However, many crop plants such as potato and strawberry are not diploids, but rather *polyploids*. Inheritance in polyploids is considerably more complex than in diploids because there can be more than two different alleles at a given locus, and they can interact in many different ways.

There is an entire field called *quantitative genetics* that is dedicated to the study and prediction of genetic effects that underlie quantitative traits, and any serious study of plant breeding must be accompanied by further study of quantitative genetics [e.g., see the text by Wrike and Weber (1986)]. An excellent introduction to many modern concepts in quantitative genetics is provided by Barton and Keightley (2002). Quantitative genetics builds on the topic of *population genetics* (the study of gene flow in populations), and many *curricula* separate these topics into different courses of study.

The study of quantitative genetics has been given a significant boost since the mid-1980s by the discovery of molecular markers, and the ability to produce high-density molecular maps of where these markers and genes lie within plant chromosomes. When mapped molecular markers are segregating in the same population as a quantitative trait, it is often possible to find discrete relationships between map locations and individual genes that control the quantitative trait. This procedure, known as *quantitative trait locus* (QTL) *analysis*, is the key to understanding the genetic control of many complex traits. It is also the concept that lies at the heart of marker-assisted breeding (the use of molecular markers to assist in the selection of linked traits). A detailed discussion of QTL analysis is provided by Paterson (1998), but an Internet search of "QTL + your favorite plant species" may direct you to primary literature regarding the discovery of QTL in your species of choice.

3.2.5. The Value of a Plant Variety Depends on Many Traits

If a melon breeder had nothing to worry about besides fruit size, then melon breeders might have finished their jobs long ago, and/or melons might now be approaching the size of small cars. However, plant varieties are often bought and sold in an open market, and the value of a plant variety is subject to complex and changing industry and consumer preferences. Some of these preferences are mentioned in Section 3.3. There can be no perfect melon variety, but a given market might be driven by the need for melons that are large (but not too large), oblong, sweet, and seedless, and that grow on compact plants that are resistant to insects and disease. These and other characteristics are controlled by many different genes. Thus, the perfect plant variety is a distant and moving target, determined by thousands of genes, dozens of which may be segregating in a given population.

3.2.6. Varieties Must Be Adapted to Environments

Why does a plant variety selected in the tropics not perform well in temperate climates? Many environmental factors influence how a given variety will perform, and *genotype–environment* interaction ($G \times E$) is an essential concept for breeders to understand. Some environmental factors that interact with genotype include soil type, soil fertility, amount of rainfall, temperature, length of growing season, production methods, and daylength. Some factors such as daylength are predictable, and much is known about how plants respond to daylength. Plants such as soybean require short days to initiate flowering, and there are specific genes that determine when a plant will flower at a given latitude. Other plants, such as oat, require long days to flower, and may flower only in high latitudes during the summer unless specific alleles of a "daylength sensing" gene are absent. However, many other factors that affect $G \times E$ are not so well understood. Moreover, many environmental factors such as rainfall are unpredictable, so it is important to select varieties that perform well in a range of environments. This is why plant varieties are tested in numerous locations over a period of at least 2 years before they are sold commercially. If a variety performs well in one year at one location, it may perform poorly the next year or at a different location (e.g., Figure 3.4). Only through multiyear, multilocation testing can we predict how a variety responds to different environments, and whether it will deliver as promised. Related to this is the concept of "stability." If a variety performs consistently in many different environments, we say that it is stable. But a stable variety may not be the top-performing variety in any given environment. Whether to release unstable varieties that perform extremely well in a few environments or stable varieties

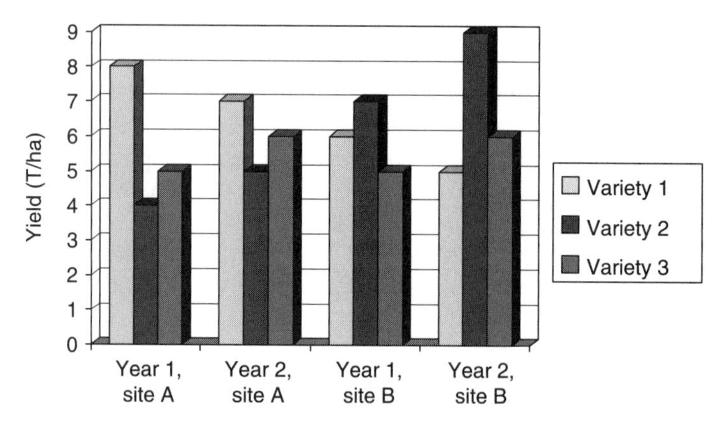

Figure 3.4. An illustration of $G \times E$ interaction. Plant variety 1 performs best at site A, whereas variety 2 performs best at site B. Both varieties show variation in performance over years. The fact that the ranking of varieties changes from site to site means that this is a *crossover interaction.* Variety 3 shows performance that is more consistent across environments, so it is described as being more stable than the other two varieties.

with average performance in many environments is an ongoing debate among plant breeders.

3.2.7. Plant Breeding Is a Numbers Game

Before discussing breeding strategies in more detail, it is important to put in context the scale on which a plant breeder must work. As illustrated above, a breeder may be working with multiple objectives that require the selection of many genes. The breeder will seldom know exactly how many important genes are segregating in a population, but there may be information about some of the genes. An example might be a population in which the breeder knows that there are a few specific genes affecting disease resistance, height, and flowering time. Suppose that there are only two genes affecting each of these traits, for a total of six genes, all segregating in a population derived from a biparental cross as shown in Figure 3.1. In the F_2 generation, the probability of a specific homozygote at each locus is $\frac{1}{4}$. If all six genes assort independently, then the probability of a specific genotype that is homozygous at all six loci is one in 4^6, or 1 in 4096. A breeder who wishes to be reasonably certain of recovering this genotype in the F_2 generation would need to grow many thousands of progeny. Given the fact that many other unknown (or unpredicted) genes will segregate, and that the true genotype is often obscured by the environment, it is not unusual for a breeder to evaluate hundreds or thousands of progeny from a given cross, and to work with many crosses simultaneously. Breeders remark that finding the perfect variety is like winning the lottery. The fact that they often "win something" is a result of "buying many tickets," but the elusive jackpot may never be won.

3.2.8. Plant Breeding Is an Iterative and Collaborative Process

A common depiction of plant breeding is that it is an ongoing process of gradual improvement, often represented by a gradually upward-sloping graph of yield versus time (e.g., see Fig. 3.5).[3] The sloping line represents the average of many plant varieties released in a given year. The measured performance may be historical, in which case it will reflect changing cultural practice and fluctuation due to "good or bad years," or it may be based on a modern experiment in which the performance of older "retired" varieties are tested together with new varieties in the same environment. While the typical graph represents yield, many other objectives are selected simultaneously (Section 3.2.5). Therefore, the one-dimensional progress shown by the graph in Figure 3.5 does not accurately represent what has been achieved, nor does it account for the fact that objectives and cultural practices change over the years, such that perfection is a moving target.

One might ask "Why not make the perfect cross and select the perfect pure line and be done with it?" The first answer is that the perfect cross cannot possibly contain all the best alleles. Disease resistance may come from one parental source, high protein from another, stem strength from another, and so on. In fact, the perfect parents have probably not been

[3]An interesting thing about this study, reported by Duvick and Cassman (1999), was that corn varieties showed no improvement if they were grown using the same cultural practices (wide rows) that were used in 1930 to accommodate the driving of horses between rows. Yet, old varieties performed poorly under the modern practice of narrow rows. Thus, the genetic gain that was achieved was accompanied by a trend toward narrow rows—a good example of genotype–environment interaction.

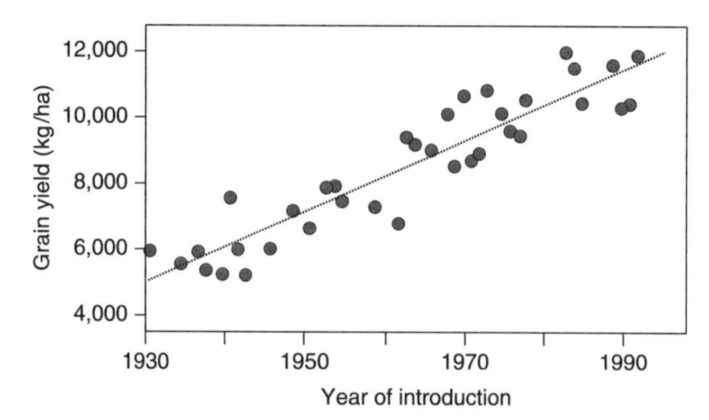

Figure 3.5. Yield of hybrid corn varieties versus year of release. Data were obtained from Duvick and Cassman (1999), based on field experiments conducted at a plant density of 79,000 plants per hectare at three locations in central Iowa in 1994.

identified for all traits of interest. The perfect variety might require recombining alleles from hundreds of different germplasm sources. The second answer is that the number of progeny required to generate a segregating population that contained the perfect combination of alleles would be prohibitive. So the breeder who set out to make a perfect variety would be busy for many decades, while his/her colleagues were busy releasing very good varieties.

Whether working in a cross-pollinating species or a self-pollinating species, the breeder needs to alternate between crossing and selection. Selection is done between crossing generations in order to increase the probability of success, and to release interim varieties. Crossing is done following selection, either to introduce new material, or to recombine existing material. The pedigree of most modern varieties shows a history of crosses that have been made (e.g., see Fig. 3.6). What may not be obvious in Figure 3.6 is that each cross is followed by selection, such that the final outcome is not a random result of the crosses that have been made. Whether it is done intentionally in a systematic process, or ad hoc in an ongoing breeding program, this iterative process of crossing and selection is called *recurrent selection*. Importantly, plant breeders use material from other breeding programs in their crosses. Legal and ethical principles allow most released plant varieties to be used for crossing purposes in any breeding program. Furthermore, many breeders actively exchange unreleased germplasm with each other, knowing that reciprocal exchange of germplasm has the net result of increasing the scale of their own program. Therefore, it is very rare to find a pedigree such as that shown in Figure 3.6 that does not contain material from many different breeding programs, and often from different countries.

3.2.9. Diversity, Adaptation, and Ideotypes

Why does natural or artificial selection not favor a single genotype? Where does genetic variation come from, and why does genetic diversity remain in the presence of intense natural or artificial selection? It is quite clear that genetic diversity originates through mutations in DNA sequence, but when, and how often, do these mutations occur? Why has it been possible, for example, to continually select for higher oil in a population of oats without ever exhausting the genetic diversity (see Fig. 3.7)? Is it because of new

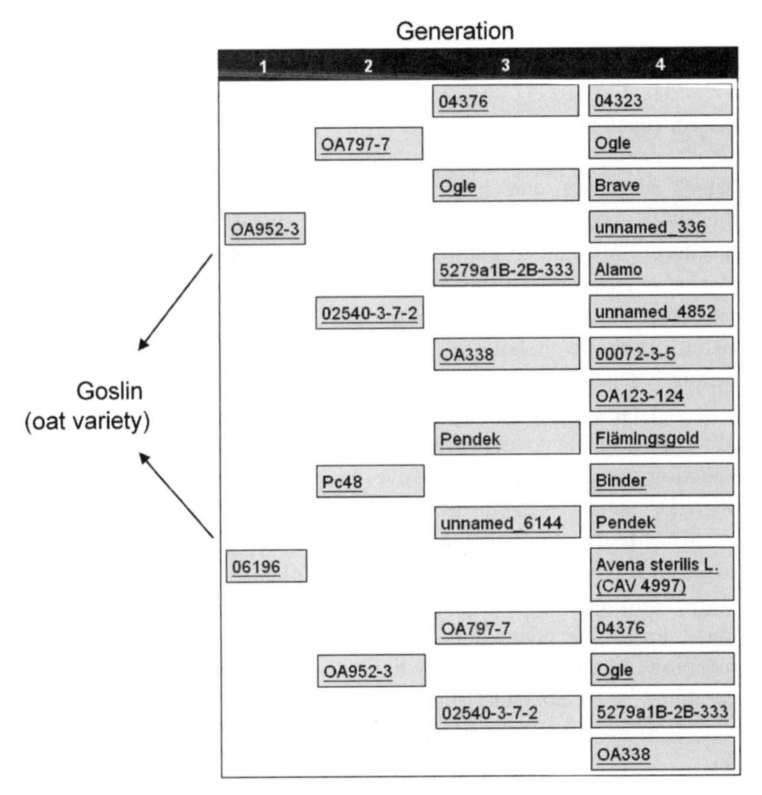

Figure 3.6. The pedigree of an oat variety named "Goslin." The parents of the cross from which Goslin was selected are shown on the left in column 1, grandparents and great-grandparents are shown in columns 2 and 3, and so on. Lines identified by numbers (e.g., OA952-3) were probably elite breeding lines that did not become varieties. This pedigree tree was drawn using an online database (http://avena.agr.gc.ca) that records pedigrees of historical oat varieties for many generations. See color insert.

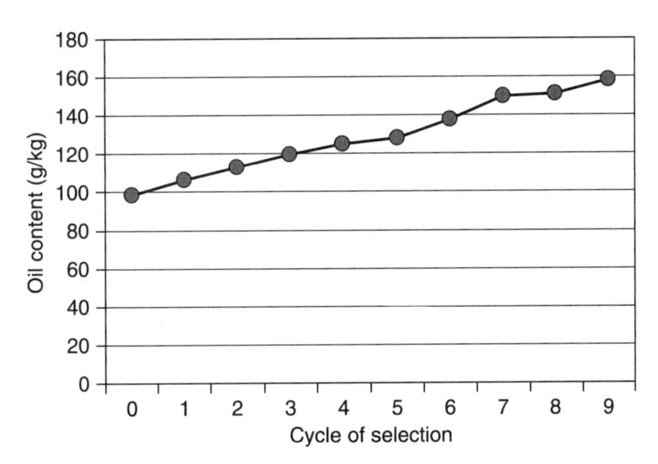

Figure 3.7. Mean oil content for oat lines representing nine cycles of recurrent selection evaluated in three Iowa locations in 1992 [from Frey and Holland (1999)].

mutations, the unmasking of suppressed genes, the gradual uncovering of rare epistatic gene interactions, or all of these? There have been many interesting debates about these questions, as reviewed by Orr (2005), and literature on this topic makes a fascinating and important side topic for plant breeders.

The famous geneticist Sewall Wright (1889–1988) introduced the *shifting balance theory*, in which adaptation and diversity are dependent partially on random population drift (Wright 1982). Although this specific theory is still debated, it is based on two concepts that are highly relevant to plant breeding: fitness surfaces and adaptive peaks (see Fig. 3.8). A *fitness surface* is a theoretical representation of genotypic value, given an underlying genotype. An *adaptive peak* represents the genetic coordinates on that surface that produce an optimum phenotype. Adaptive peaks may be local, or global. Part of Wright's shifting balance theory related to how selection was capable of moving a population from one adaptive peak to another, given that selection favors "going uphill." But for a breeder, it is possible to deliberately "go downhill" if it is apparent that this will move a population toward a higher adaptive peak. An example might be the deliberate selection of larger seeds, smaller pods, and a reduced number of pods per plant. Individually, these traits would result in poorer yield or adaptation, and the breeder might spend years producing seemingly worthless plants. But once all three traits have been recombined into just the right genotype, the breeder may release a plant variety that achieves a quantum-leap in adaptation. This concept was first formalized under the name "ideotype breeding" by Donald (1968), where an ideotype was defined as a plant with a particular combination of characteristics that have not yet been observed together, but that are predicted to be genetically achievable and are theorized to provide superior yield or adaptation.

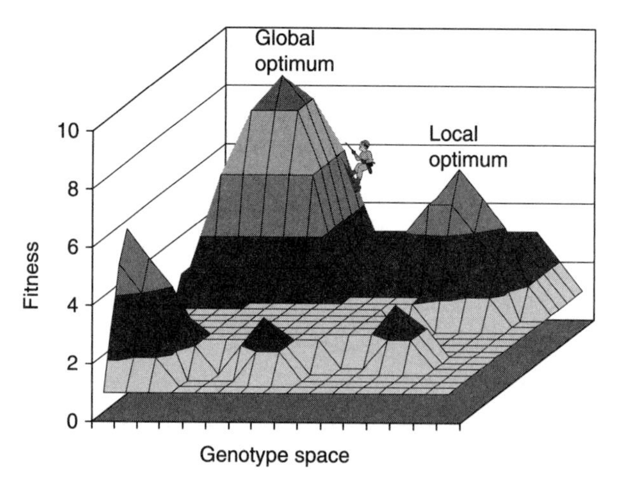

Figure 3.8. In Sewall Wright's shifting balance theory, a genotype or population is defined by coordinates in N-dimensional space, and a fitness value forms a surface in the $(N + 1)$th-dimension. Here, genotype coordinates are defined in two dimensions on the ground beneath a mountainous fitness surface (the third dimension). The coordinates of a given population can be changed by selection, but only in small increments. Direct selection tends to move a population toward coordinates where fitness is highest, but that may be only a local peak. Applying this concept to plant breeding, we see that to find genotypes beneath a global peak, we need to create and explore a large genotype space (i.e., genetic diversity) or to know exactly where we are going (an ideotype).

The important thing to remember about ideotypes, or fitness surfaces, is that both are dependent on a specific environment. If the environment changes (e.g., if a stress is added or removed), then the value of an ideotype or the shape of a fitness surface could change dramatically. A genotype that was near a global peak on the previous surface may now be in a valley. Given that different agricultural production techniques are really just different environments, it is not difficult to realize why varieties that were adapted to one production practice may not be adapted to another. A variety that is short may produce high yields only when fertilizer is applied, or a variety that does not produce branches may produce high yields only when planted in high-density stands.

By visualizing plant breeding as a mountaineering expedition on a complex and constantly changing fitness surface, you can easily see why different breeders can have vastly different approaches and philosophies. Whether a breeder believes that there is a higher mountain, or not, can determine long-term *vs.* short-term success.

3.2.10. Other Considerations

Lack of space in this chapter prevents detailed discussion of many additional topics that make plant breeding challenging and interesting. What follows are some general statements about additional factors that the interested reader may wish to pursue.

- Some crops exhibit *polyploidy* (doubling, tripling, or quadrupling of basic chromosome number). This can lead to nondiploid chromosome pairing, which complicates normal diploid inheritance. Polyploidy can also be induced artificially in order to create "artificial species" that may be more vigorous, or to combine characters from two different species.
- Variations in chromosome number (*aneuploidy*) or chromosome structure (*translocations, inversions, duplications*, and *deletions*) can affect genetic inheritance. These phenomena may also be induced artificially for genetic studies.
- *Xenia*, the expression of pollen genes in the tetraploid endosperm or embryo of a seed, can complicate the selection of seed traits. Seeds normally show the phenotype of the parent plant, but carry the genotype of a genetically different offspring.
- Breeders are important stakeholders in efforts to maintain biodiversity through *in situ* and *ex situ* collections of germplasm.
- Interspecific hybridization has been used to transfer alleles controlling traits such as disease resistance, which may not be present in the normal germplasm of a species.
- Male sterility, and various methods of pollen control, may be useful in the production of hybrid varieties, but have also been used in normal recurrent crossing programs.
- Plant breeders must manage large amounts of data, and they may need to share data with other researchers. Electronic data management systems are becoming increasingly important in plant breeding.
- Plant breeders must also be statisticians. There is a large body of literature concerning the optimization of field plot techniques, and the statistical analysis of test results.
- Resistance to disease is an ongoing battle between plant breeders and the organisms that cause disease (a variant of the "evolutionary arms race"). Many pest organisms mutate very quickly, and mutations that overcome new types of resistance are selected quickly in crop monocultures of a single plant variety. One strategy to overcome this

is the development of *multiline varieties*, which contain a mixture of resistance types. Another method is to "pyramid" or "stack" multiple sources of resistance into a single variety.

3.3. OBJECTIVES FOR PLANT BREEDING

The overall value of a plant variety is determined by many small and subtle characteristics that are quantitative and polygenic in addition to a few major characteristics that may be qualitative and monogenic. It is not difficult to draw a parallel in human traits. Think of someone whom you admire; you might like this person because you have a fondness for a certain eye color, hair color, height, or other characteristic, but most likely it would be because of a combination of many subtle traits (involving both appearance and personality) that are controlled by numerous genes but also influenced by environment. Beauty is in the eye of the beholder, or in this case, in the eye of the plant breeder.

Many breeding objectives fall into two general categories of traits. We often categorize certain traits as *agronomic traits* or *input traits*, because they relate to production practices and to the amount of raw material that can be harvested. Such traits include crop yield, pest resistance, height, flowering time, susceptibility to lodging (falling down), seed vigor, and seed dormancy. Crop yield includes many component traits, such as seed size, seeds per pod, pods per branch, and branches per plant. Some breeders prefer to select according to component traits rather than on final yield, but the value of a plant variety is almost always judged for its potential to produce high yields per unit area. The second general category is described as *output traits*, which include anything related to the composition or quality of what gets harvested. Examples are the composition and content of protein or oil; the relative proportions of oil, starch, and protein; and the composition of secondary compounds that may have value relative to human health and industrial use. Many output traits are extremely complex. These include traits related to the use of a plant product in processing.

For every cultivated plant species, there are different breeding objectives. Often there are several different sets of objectives, sometimes conflicting, for the same species. An example of this is a barley, which may be used for animal feed or malt production. A major objective for animal feed is high protein content, whereas malt production requires low protein content. The breeding objectives for crops such as malting barley or bread wheat are dependent on markets that have evolved very specific industrial processes that require dependable and uniform grain. Thus, a variety that is merely high-yielding and pest-resistant will not suffice. Other crops have highly diverse objectives that may be driven by many different markets. Soybean, for example, is used in the manufacture of many different food products, and each product has very specific requirements such as taste, texture, chemical composition, seed size, and seed coat color.

Horticultural crops can have many interesting and diverse objectives. The market for gardening varieties is driven by diversity because gardeners like to try new things. In contrast, the market for commercial varieties of horticultural crops is driven by the need for conformity and uniformity, and for improvement of specific high-value characteristics. For example, processing tomatoes need to deliver maximum amounts of soluble solids, but there would be little tolerance for a variety that had a different taste or color. The market for fresh shipping tomatoes has requirements for produce that is both durable in transit and attractive in appearance (and, some would argue, tasteless).

Some breeding objectives can produce interesting challenges. Consider the objective of reduced seed content in grapes or watermelons. Where does the seed come from to grow the next crop? In these cases, breeders can "trick" the plant by producing hybrid varieties that have an unbalanced number of chromosomes such that they cannot undergo proper meiosis to produce viable seed. Another interesting challenge is the incorporation of traits that we cannot measure directly. For example, it is desirable to incorporate multiple sources of disease resistance in a single plant variety so that it is more difficult for a pathogen to mutate and overcome the resistance. But if there are two genes at different loci that both confer resistance, how do you know that they are both there? In this case, one solution is to identify genetic markers that are linked to each resistance gene, so that the markers can be selected instead of the resistance.

Breeding objectives can change suddenly and unpredictably. An example was the sudden appearance in the 1990s of devastating levels of a fungal pathogen called *Fusarium* in wheat. Now every major wheat breeding program is attempting to introduce new sources of resistance to *Fusarium* disease. Another example is the development of a USDA health claim in the late 1990s around β-glucan in oat. While breeders were aware of this factor, and had initiated selections for higher β-glucan, it was not until the development of this health claim that a particular value was placed on varieties that had elevated levels of this trait. These points illustrate that breeders must be constantly aware of changing market forces and agricultural conditions. Indeed, it is not unusual for breeders to help drive some market forces through their knowledge of traits that are available, and through awareness of economic factors affecting industry and producers.

3.4. METHODS OF PLANT BREEDING

It has been claimed that breeding is a continuous cycle of recurrent mating and selection. There is rarely a startpoint or an endpoint in a breeding program; rather, it is a continuous pipeline that must be kept filled for continual delivery of new and better plant varieties. Breeders try to release improved varieties every year, but today's varieties may be the result of planning and crossing that began a decade ago. Add to this the fact that breeders mix and match various breeding methods, depending on objectives, and that they constantly modify and update their strategies, and you can understand why it is difficult to write down a simple "recipe" for successful plant breeding. Nevertheless, several core strategies have been developed, and most breeders adopt and adapt one or more of these strategies depending on plant characteristics, breeding objectives, resources, and personal preference. While the breeding systems that are described below appear to have a beginning and an end, you must remember that many cycles at different stages will be running simultaneously in a given breeding program, and that material from one system can become starting material for another system.

3.4.1. Methods of Hybridization

Most breeding methods incorporate sexual hybridization as a method of generating new genetic variability. Hybridization may occur naturally, as in the case of out-crossing species, or it may require tedious manipulation of flowers in the case of a self-pollinating species. In special cases, sexual hybridization has been used as a method to combine traits from species that are rarely cross-fertile. The methods for hybridizing most self-pollinating species involve emasculation (removal of stamens) and the introduction of pollen from another plant. The timing of these steps, and the methods by which they are best done, are critical. Outcrossing species may also require controlled hybridization in specific breeding methods, particularly if hybrid varieties are developed (see Section 3.4.3.4). Even varieties that are developed through random mating require special considerations. For example, alfalfa is poorly pollinated by honeybees, which do not trigger a special floral mechanism that transfers pollen to the bee, but they are efficiently pollinated by certain wild bees, which may be artificially reared near plots that are used for breeding or seed production. Fehr and Hadley (1980) have compiled a comprehensive reference source on methods of hybridization in crop plants that discusses technical details as well as many related issues such as environmental factors that affect the timing of flowering and fertilization.

3.4.2. Self-Pollinated Species

Most self-pollinated species are grown as varieties derived from a pure line (see Section 3.2.3). Therefore, the overall objective of the following strategies is to recombine as many desirable genes as possible into a single homozygous genotype. All of the following strategies involve one or more hybridizations followed by generations of selfing and selection. The key differences among these strategies are whether crossing is repeated, when selections are made, and how many selfed progeny are made from each selected plant. All systems generally culminate in the same final steps for variety testing and release.

3.4.2.1. Pedigree Breeding. The pedigree breeding method (Fig. 3.9) requires detailed record keeping. Selections are made in every generation except for the F_1, because it is assumed that all F_1 plants from a cross are genetically identical. A breeder would choose this method primarily because it allows elimination of poor lines at an early stage in the breeding program, thus leaving more room to increase the number of lines that can be tested from promising families. An additional benefit is that, by recording information about the performance of lines as well as their parents and families, the breeder ensures that selections can incorporate all three types of information. For example, a breeder may notice that one family has susceptibility to disease, while another family from the same cross appears to be completely resistant. This might lead to speculation that the first family was derived from a parent that was segregating for disease resistance, while the second family was derived from a parent where the resistance was fixed. This is useful information, since individual lines sometimes escape disease infection even if they do not carry genetic resistance. This information might allow the breeder to favor selections within the resistant family.

3.4.2.2. Single-Seed Descent. A primary criticism of the pedigree method is that it requires a lot of time and resources to keep records about material that will simply be discarded. Another criticism is that the performance of progeny in early generations may be enhanced by the effects of dominance, which is lost in later generations, and also that favorable gene interactions (epitasis) may not be evident until later generations. In other words, a good line in an early generation may give poor progeny in late generations, or a poor line in an early generation may give good progeny in late generations. The single-seed descent (SSD) method (Fig. 3.10) addresses all the concerns mentioned above. Rather than select lines and families in early generations, a large F_2 population is created, and one random line is developed from each F_2. Thus, the pedigree of each F_2 line is represented by

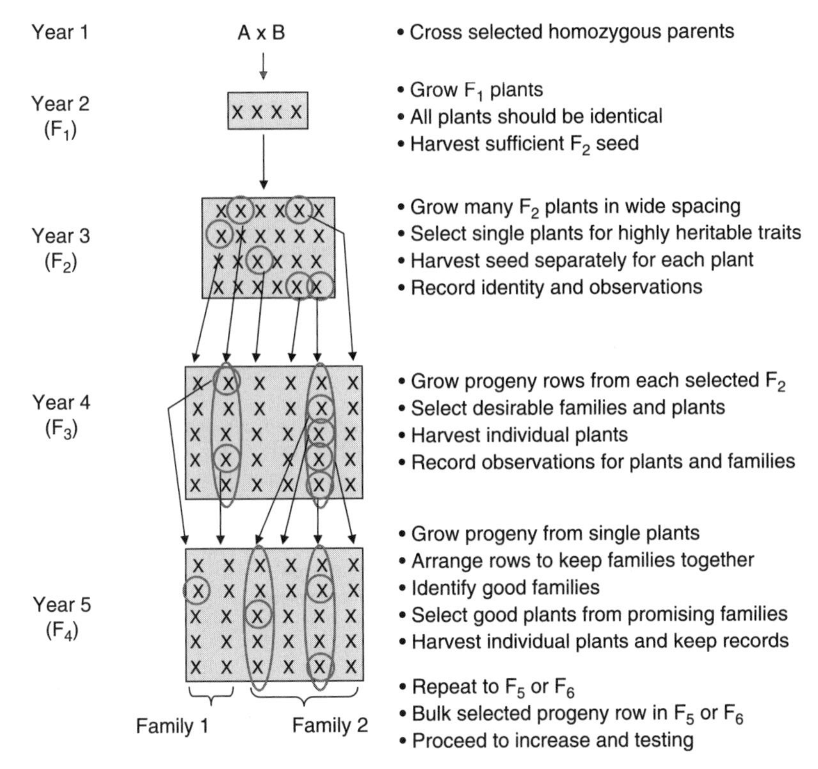

Figure 3.9. The pedigree breeding method is used in self-pollinated species to derive pure-line varieties when it is desirable to practice selection in early generations.

exactly one random line in each following generation by taking a single seed from each F_2 family in every segregating generation. The result is that maximum genetic diversity is preserved until late generations when selection will commence, and no recordkeeping is required. The SSD method can also be used to derive populations of nearly inbred lines (NILs). These populations are useful in genetic experiments because segregation can be considered random such that "good" lines can be contrasted with "bad" lines to identify genetic determinants. However, it is this same feature that leads to the primary criticism of SSD as a breeding method: poor material that might easily be removed in early generations continues to occupy space and resources in the breeding program.

3.4.2.3. Bulk Breeding Methods. A *bulk breeding method* is any method whereby generations are advanced by bulking and planting seed from the previous generation. However, if all seed from a given generation is harvested, then there will likely be too much seed to plant in the following generation, so some seed must be discarded or held in reserve. The SSD method is actually a special type of bulk breeding whereby each generation is advanced by saving only one seed per plant from the previous generation. In other methods of bulk breeding, the reduction of seed is achieved randomly, or through a selection process that is applied uniformly but indiscriminately (e.g., harvesting the earliest-ripening plants, or sieving to keep the largest seed). Some breeders favor bulk breeding as a method of generation advance because it is extremely simple. However, several issues must be considered. If the intention is to preserve a random nonselected population,

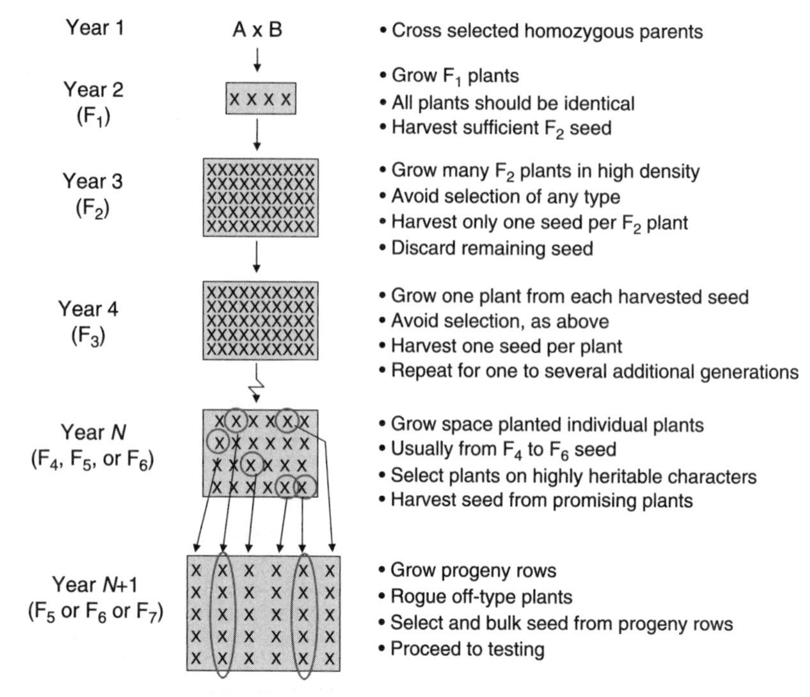

Figure 3.10. The single-seed descent (SSD) breeding method is used in self-pollinated species to derive pure-line varieties when it is desirable to select from random homozygous lines in an advanced generation.

then some lines will be represented by greater numbers of progeny than other lines simply by random chance; thus, the total genetic variance that is preserved is less than that of a true SSD population. Lines that are preserved are unlikely to be random. Plants that have more seed will be disproportionately represented, and plants that compete poorly may be lost completely from the lineage. This may sound like a desirable way to favor lines that are more adapted. In fact, some breeders deliberately practice bulk selection in the presence of some artificial selection. Examples include favoring tall plants by mechanically harvesting only the tops of the tallest plants, or conversely, penalizing tall plants by applying herbicide using a rope-wick prior to harvest. Many creative methods have been developed to apply selection during bulk generation advance. However, it must be remembered that plants that produce more seed during generation advance might actually produce less seed in a competitive community of identical genotypes, or they may simply produce seed that is smaller and less desirable.

3.4.2.4. Backcross Breeding. As illustrated in Figure 3.11, the backcross breeding method is quite different from the methods discussed so far. It involves much smaller populations and greater numbers of hybridizations. The objective of a backcrossing program is to preserve as many genes as possible in an inbred parent that has proven adaptation to a given environment, while introducing new alleles at just one or a few loci from an unadapted parent. The former is called the *recurrent parent*, and the latter is called the *donor parent*. Often, a backcross strategy is used when an unadapted genotype is found to have

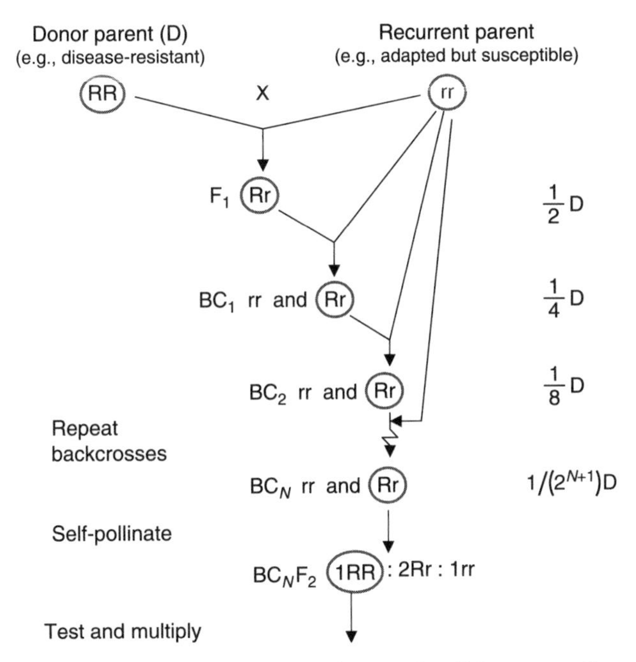

Figure 3.11. The backcross breeding method is used to transfer alleles at a small number of loci from a donor parent into the genetic background of a reciprocal parent. Each generation of backcrossing reduces the proportion of alleles from the donor (D) parent by half $(\frac{1}{2})$, as shown on the right.

disease resistance that is not present in adapted varieties, but it may be used to introduce any simply inherited trait from a parent that is otherwise undesirable.

After an initial cross between the donor and recurrent parent, repeated backcrosses are made to the recurrent parent. Each time a backcross is made, the progeny receive half their alleles from each parent; thus, the proportion of alleles that remain from the donor parent are reduced by 50% each generation, and after N backcrosses, the proportion of alleles remaining from the donor is formulated as $1/(2^{N+1})$.[4]

After the first hybridization, every backcross will produce a mixture of genotypes at the locus where we wish to introduce alleles from the donor. Thus, each generation must produce an adequate number of progeny such that it is possible to identify the heterozygous genotype carrying the donor allele. While Figure 3.11 shows an example of backcrossing alleles at a single locus, it is also possible to backcross alleles at two or more loci, but larger numbers of progeny must be made to identify backcross parents that contain donor alleles at all loci.

3.4.2.5. Increasing Seed, Testing, and Releasing of Pure-Line Varieties.

At the end of these breeding procedures (typically at the F_5, F_6, or BC_5 stage), most breeders will consider that individual plants are adequately homozygous to form a pure line that

[4]This is true only for loci that are not linked to the locus under selection. Donor parent alleles that are closely linked to the locus under selection will be highly favored. This phenomenon is called "linkage drag."

could become a plant variety, and the final stages of variety development will begin. Several additional things must now be accomplished:

1. Additional selection among inbred lines must be done. This selection will be among a decreasing number of lines that are increasingly elite. This selection may result in the complete elimination of all lines from a given cross—perhaps in favor of keeping additional lines from a cross that turned out more favorably.
2. Varietal purity must be monitored and maintained.
3. An inbred line may need to meet industry established standards for registration as an "official" plant variety. These standards include descriptions, uniformity, uniqueness, and sometimes merit in relation to competing varieties. Increasingly, plant variety protection, which restricts unauthorized use of the variety, is sought in addition (or as an alternative) to variety registration.
4. Seed must be increased to a level where there is enough to sell or to distribute to a seed producer.

These requirements may vary among different countries, states, or provinces, and they are often unique among different species. However, the general principles are the same, and so is that fact that they are often addressed simultaneously. Table 3.1 lists the typical final steps in the birth of a plant variety, and how each of the abovementioned requirements might be met.

TABLE 3.1. Steps[a] Involved in Final Stages of Variety Development in a Self-Pollinated Species

Step	Description	Activities
Progeny rows	Single row, nonreplicated	Select promising rows, rogue off-types, bulk seed from characteristic plants
Home tests	One site, two replicates, 1 year: material from home program + standard check variety	Detailed selection, may include characters such as quality that require increased amounts of seed; rogue off-types and bulk remaining seed
Preliminary tests	Three sites, three replicates, 1 year: material from home program + standard check variety + other current varieties and possibly lines from collaborating breeders	Detailed and final selection of several lines to enter into variety registration trials; identify defining characters, rogue off-types, bulk seed
Breeders' seed rows	Nonreplicated plots grown in isolation to increase seed for supply of potential variety	Identify defining characters, rogue off-types, bulk seed
Registration tests	Six sites, four replicates, 2 years: cooperative tests that include several best lines from each breeder + standard check varieties	Selection of lines to support for registration (often competitive among breeders); apply for plant variety protection on probable winners (year 2)

[a]These steps must address several requirements simultaneously: testing and selection, varietal purity, registration and/or protection, and seed increase and distribution.

Although not all breeders would consider this a separate stage in a breeding program, the steps listed in Table 3.1 technically begin after the last generation in which a family can be traced to the progeny from a single plant (the *founding generation*). If that were the F_5 generation, then we would consider a resulting variety to be F_5-derived. Since homozygosity continues to increase with each generation of selfing, breeders often use the notation $F_{X:Y}$, where X denotes the founding generation and Y designates generations of selfing and bulking that follow the founding generation. Varieties developed from early founding generations (i.e., a relatively heterozygous founding plant) can show a considerable amount of heterogeneity, especially in molecular traits that do not undergo selection. *Rogueing* (the culling of undesirable plants or off-types) is done in generations following the founding generation, and this reduces the amount of heterogeneity. However, it is common for phenotypic variation to show up within a variety, sometimes by surprise, when a new environment is encountered. For this reason, varieties may need to be described in terms of their range in characteristics, and descriptions based on molecular traits are increasingly favored.

Although they are not technically part of a breeding program, many jurisdictions have testing and recommendation procedures that enable comparison of plant varieties from different breeding programs with one another under different conditions (years, locations, and management practices) culminating in reports and recommendations that allow agricultural producers to select varieties that are most adapted to their conditions. The agricultural producer has a job that is parallel to that of the plant breeder: making the final selection among numerous plant varieties that are available for production on the farm.

3.4.3. Outcrossing Species

Since pure lines cannot easily be maintained in a naturally outcrossing species, the development and release of varieties in an outcrossing species is quite different from that in a self-pollinated species. Rather than identifying the "perfect genotype," the objective is to identify the perfect set of genes that work happily together in a random mating population. Some outcrossing plant species, such as rye, can tolerate a high degree of inbreeding, and can be effectively bred and grown as if they were self-pollinated species. The primary difference is the increased need for isolation (to prevent uncontrolled outcrossing) during variety development and seed production. However, other species do not tolerate inbreeding, or they have specific mechanisms to prevent it. Matings between different plants often produce offspring that are more fit than the parents, a concept called *hybrid vigor* or *heterosis*. Thus, maintaining a heterogeneous population in a random mating state is beneficial. However, there must still be some opportunity to select the breeding population such that it produces relatively uniform progeny that have desired and predictable characteristics.

3.4.3.1. Mass Selection. Historically, varieties of outcrossing species were improved in the same way that landraces of self-pollinated species were improved: by saving seed in bulk, or by saving seed from selected plants. Either way, selection would have taken place. For example, we know that selection for seeds that do not fall off the plant before manual harvest (called "shattering") was one of the earliest traits selected in the process of crop domestication. This most likely happened simply because genotypes with seeds that shattered were rapidly eliminated as soon as early agronomists started planting crops intentionally from seed that was harvested for food. Other traits, such as lack of seed dormancy, seed size, early flowering, and height, were probably selected in similar ways, whereas traits such as fruit flavor may have required a more deliberate effort to propagate favored genotypes. All traditional selection such as this is termed *mass selection*

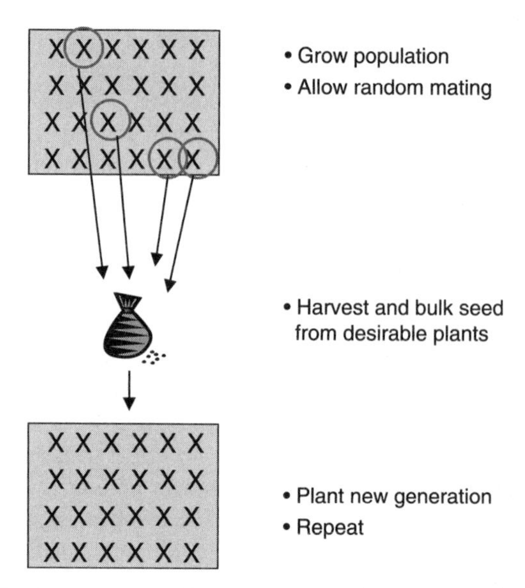

Figure 3.12. Mass selection, as practiced in an outcrossing species, is a traditional method of breeding that is still used to improve base populations from which parents may be chosen for other breeding methods.

(Fig. 3.12). The primary difference between a traditional open-pollinated variety and a traditional landrace is that the former undergoes continual random mating. This difference is the reason why mass selection has fallen from favor in self-pollinated crops, where continual recombination does not take place. But mass selection is still a viable method to improve a cross-pollinated species. With modern knowledge about genetic diversity, a mass selection strategy will now try to reconcile the intensity of selection with the need to maintain diversity. Mass selection is often used as a strategy for continual population improvement in crops such as maize, although it is now more likely to be used to improve a base population that will serve as a source of germplasm for other breeding strategies. Mass selection can also allow introduction of new or exotic germplasm that will recombine with an elite population.

3.4.3.2. Recurrent Selection. The term *recurrent selection* has been used earlier in this chapter to refer to any strategy where selection is alternated with recombination (Section 3.2.8). In fact, mass selection is technically a type of recurrent selection, because recombination occurs with every generation. However, special recurrent selection strategies have been devised for cross-pollinated species whereby selection and intermating are more discrete, and controlled.

An important modification over mass selection has been the development of methods whereby plant selection is based on the performance of their progeny. This is highly relevant if one wishes to favor genes that increase the fitness of the population. Mass selection merely saves plants that have a desirable phenotype, but there is no guarantee that the alleles controlling this phenotype will be expressed in the same way when they are mixed with other alleles in the following generation. Figure 3.13 shows a recurrent selection strategy that allows full progeny testing prior to random mating. It is noteworthy that this strategy is based on a cycle that requires multiple years to complete. Other methods of recurrent selection have been devised, some that require an additional generation of pollen control,

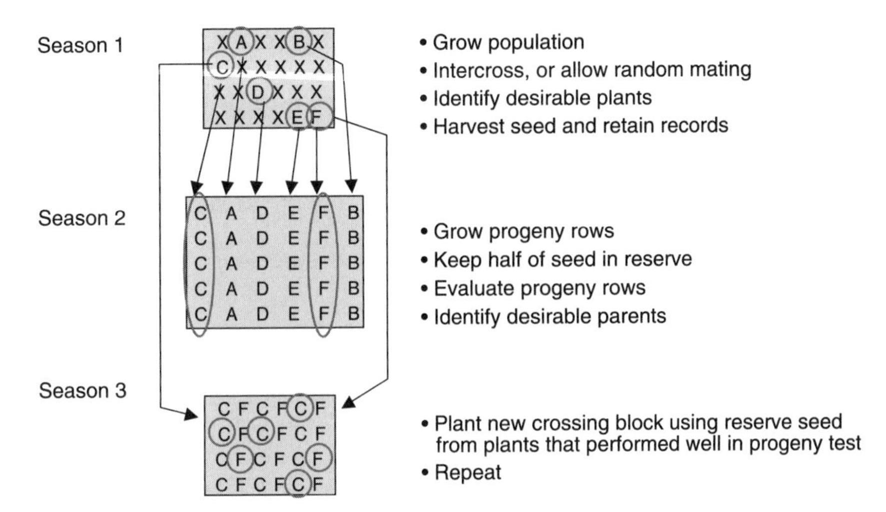

Figure 3.13. An example of a recurrent selection strategy with progeny testing. Many variations on this type of strategy have been devised.

and others where some level of controlled mating and selection can be performed every year. Important considerations in the selection of a recurrent selection strategy are the heritability of the trait, and the time at which it is expressed. Traits with low heritability, such as grain yield, are more responsive to progeny testing. It is very difficult to predict crop yield from a single plant, so it is far more accurate to test yield in a whole row of progeny. Traits with high heritability can be selected on a single-plant basis, and traits that are expressed prior to pollination can be selected more effectively by eliminating those that do not express the desired phenotype prior to pollination. Recurrent selection is often used to develop base populations from which other forms of selection or crop improvement can be made.

3.4.3.3. Synthetic Varieties. Most modern agricultural practices require plant varieties that are predictable and uniform. Even though mass and recurrent selection are practiced to improve base populations, these populations may be too variable for modern production practices, or they may be difficult to maintain in a state that will perform as predicted. A mass-selected population may continue to improve with time, but it might also be inadvertently selected into a state that could theoretically cause damage or liability. Consider an alfalfa variety that has been selected to remain in the vegetative state for an extended period of time. This trait might be a desirable characteristic if the crop is used for repeated cutting for green forage. But someone needs to produce the seed to grow that variety and may unintentionally reselect the variety to flower early and produce copious amounts of seed.

A synthetic variety is an early random mating population derived from a mixture of a group of "reproducible components" (Fig. 3.14). The components can be inbred lines, clones, or hybrids. For example, in perennial forages, synthetics are initiated from a small set of parental lines with proven merit in progeny tests. Because they are perennial, these parents can be maintained indefinitely, and can be propagated vegetatively in small quantities. It is also advantageous that these species produce large quantities of small seed. In other species, a synthetic variety may be initiated from inbred lines. Equal quantities of intercrossed seed from each founding line (Syn-0) is harvested and used to plant

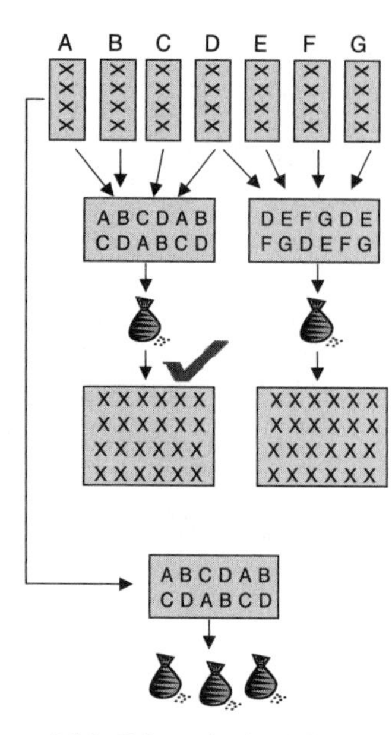

Figure 3.14. Schematic simplification of the development of a synthetic plant variety in an outcrossing species. The Syn-1 generation is produced by random mating of reproducible components (inbred lines or clones). If it is found to be desirable as a new plant variety, it can be reproduced and sold by repeating the identical crossing block. This type of breeding method is most practical in a perennial forage species. If adequate seed cannot be produced in Syn-1 generation, the Syn-2 generation (harvested from Syn-1) may be used instead.

a progeny generation called Syn-1. Seed from Syn-1 may be sold as a variety or used to produce a next generation called Syn-2. Generally, having fewer intercrossing generations is more desirable, but the number of generations will be determined primarily by limitations of the species and requirement for seed. In order to maintain uniformity and vigor, synthetic varieties must be reconstituted regularly.

3.4.3.4. Hybrid Varieties. A hybrid variety is a special type of synthetic variety that is defined as the first or second generation derived from crosses among inbred lines. Historically, many hybrid varieties were composed of double crosses (e.g., [A × B] × [C × D]). However, most modern hybrids are now produced from seed that results directly from the mating of two inbred lines (Fig. 3.15). Seed from such a mating is expected to be highly heterozygous, and highly homogeneous: two attractive traits for most crops. The primary advantage of a hybrid variety is that it can provide a performance advantage resulting from heterotic effects at many heterozygous loci, yet it is highly uniform and predictable. A disadvantage is that the uniformity lasts for only one generation, so seed must always be purchased from a hybrid seed production facility. This is an advantage if you are the hybrid seed producer, or if you are a plant breeding company that needs to control the distribution of your variety. Farmers are willing to pay the added cost in many cases because the hybrid varieties have uniformly good properties that lead to assured production.

Maize is a crop that is normally cross-pollinated, but it can easily be self-pollinated in order to derive inbred lines. For this reason, and because maize shows a large amount of hybrid vigor, most modern maize production is based on hybrid varieties. Because of the

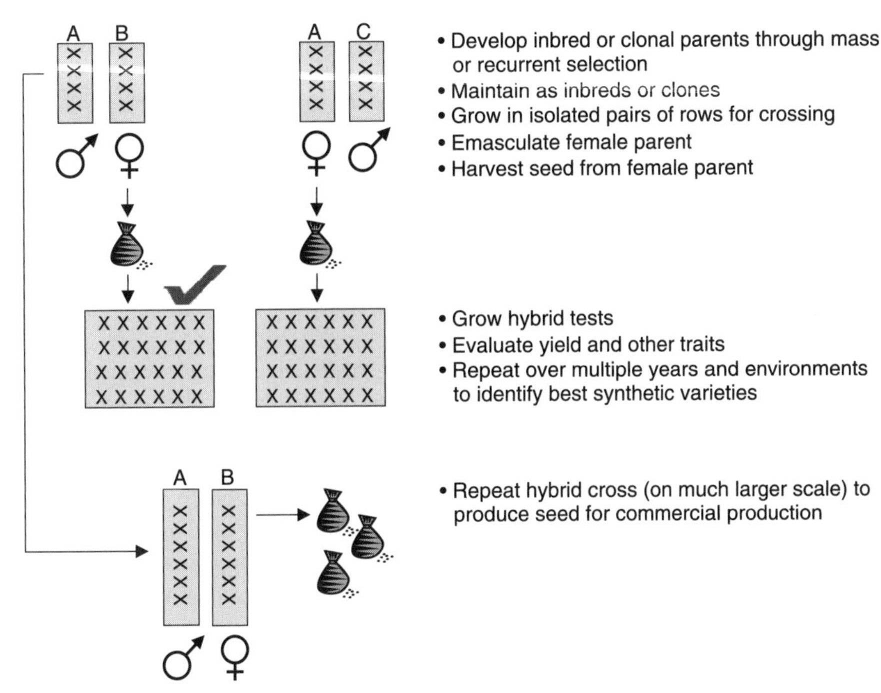

Figure 3.15. Schematic simplification of the development of a hybrid plant variety. In corn, the parents (i.e., A, B, and C) are inbred lines that have been derived through other breeding methods. In other crops, the parents may be clonally propagated. Parents are grown in adjacent rows for crossing, and the female parent is emasculated so that it will not self-pollinate. Seed harvested from the female parent is tested in performance trials. If a hybrid variety is successful, the cross is repeated on a large scale for commercial production.

high value of the maize industry in many developed countries, the development of hybrid varieties, and subsequent hybrid seed production, is dominated by industry.

Hybrid variety development involves extensive testing of many different hybrid crosses that are developed by intercrossing inbred lines in many different combinations. Inbred line development is not random. Inbred lines are generally developed in two streams from two different genetic backgrounds that are known to produce good hybrids when crossed with each other. Lines in each stream are frequently test-crossed with each other using elaborate schemes that can help guide inbred line selection in both streams. Inbred lines, and the populations from which they are derived, are carefully guarded secrets of every commercial maize breeding company.

Hybrid maize seed production involves planting alternate rows of two different inbred lines, and removing the male parts (tassels) of the line from which you intend to harvest hybrid seed (the "female" parent). The removal of tassels must be done carefully before they emerge and shed pollen. Detassling cornfields is a common summer job in rural areas where maize seed is produced. It is hot and tiring work, and not advisable for anyone with pollen allergies.

Hybrid varieties can theoretically be produced in any crop species, but for some species it is not practical or commercially viable. In many cross-pollinated species, inbred lines are difficult or impossible to produce, and so synthetic varieties are used. In self-pollinated varieties, hybrids may not show as much advantage as they do in cross-pollinated varieties, because these species naturally tolerate inbreeding. Furthermore, it is more difficult to manually enforce an adequate number of hybrid matings. However, hybrid varieties are frequently used in high-value horticultural crops that produce a large amount of seed from a single mating. Hybrids have also been used in some self-pollinated crops in which mechanisms of male fertility can be used to ensure cross-fertilization. Some of these crops include sugarbeet and sunflower.

3.4.4. Clonally Propagated Species

Some crop plants are propagated naturally and/or artificially through vegetative propagation rather than through sexually produced seeds. Globally, the most important example is potato, but other examples include banana, strawberry, yam, sweet potato, and many tree crops. Although the crossing behavior of clonal crops is not relevant to propagation, it is still important in the breeding strategy. Most clonally propagated crops are cross-pollinated, so breeding methods are most similar to those used in cross-pollinating seed crops. However, the ability to maintain an "immortal" genotype makes selection of a population less important, and selection of individual plants becomes far more relevant. The selection of tree crops presents special challenges because of the long juvenile period, so many fruit tree varieties have been identified by careful observation of hybrids from serendipitous crosses that may have taken place many years ago.

3.5. BREEDING ENHANCEMENTS

This section provides a brief description of several of the most important techniques that can be used to enhance the success of a breeding program. Many additional techniques are discussed in other literature. Perhaps the most important modern breeding enhancement is plant transformation: the ability to transform plants with DNA that originates from different species. This topic is discussed in other chapters, but it is interesting to note that the way in which genetic transformation can be incorporated into a breeding program bears many resemblances to the use of mutation breeding, discussed in Section 3.5.3. Furthermore, marker-assisted selection (Section 3.5.2) is often used as a follow-up to genetic transformation in order to recombine a transformation event into new breeding populations.

3.5.1. Doubled Haploidy

The derivation of pure lines (Fig. 3.3) is one of the most important steps in breeding self-pollinated varieties. In the SSD method (Fig. 3.10), this step could be considered "wasted time" if there were a shortcut to produce pure lines. This shortcut exists, and it is called *doubled haploidy*.

The principle behind doubled haploidy is that every plant species produces haploid gametes during meiosis. Haploid gametes are found in the female (egg) and in the male (pollen) tissues. By forcing these gametes to double the chromosomes in their nuclei, we can immediately produce a cell type that is both diploid and homozygous. There are several techniques by which this phenomenon can be induced. The most common technique is through artificial culture of the male gamete (microspore) or the tissue containing those gametes (anthers). By culturing those tissues on a growth medium, we cause the haploid cells to undergo mitotic divisions. At some stage, a natural doubling may take place when mitotic nuclei fail to divide into separate cells. This can also be induced by the addition of a chemical called *colchicine* that interferes with normal cell division. The culture can then be forced to regenerate into a normal diploid plant (see Chapter 5).

Techniques to produce doubled haploids have been developed in many crop species, and the technique is used routinely in species such as wheat and barley. While the cost of producing doubled haploids is often greater than that of producing SSD lines, the ability to accelerate the breeding program and get to market faster with a new variety is often worth the added cost.

3.5.2. Marker-Assisted Selection

Earlier it was mentioned that individual genes contributing to complex plant traits can sometimes be discovered through their association with genetic markers. This procedure, called *quantitative trait locus* (QTL) *analysis*, provides the foundation for a more efficient type of genetic selection called marker-assisted selection (MAS) (Fig. 3.16). Rather than selecting traits, which are the outcome of many genes, MAS is based on selecting specific alleles at marker loci that are known to be linked to the genes that cause the desired trait. The theoretical advantages of MAS are that it (1) avoids errors caused by environmental variance; (2) can be applied at a juvenile stage before a trait is expressed; (3) can be applied on a single plant, whereas phenotypic selection of some traits might require seed

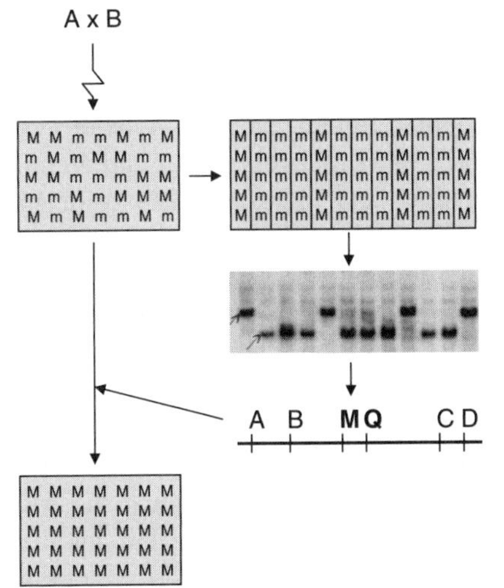

Figure 3.16. A simplified strategy for marker-assisted selection (MAS). Here, a significant association between a QTL (Q) and a molecular marker allele (M) is identified in an experimental population. This information is applied in future populations in order to select Q indirectly through its linkage to M.

or tissue from many plants to be effective; and (4) may be less expensive than phenotypic selection. Although MAS does not replace the requirement for parent selection, sexual recombination, and breeding strategies, it can significantly increase the efficiency by which superior genotypes are selected. For this reason, MAS is considered to be an important modern enhancement of traditional plant breeding.

The theoretical advantages of MAS may not always be relevant, and it is often argued that phenotypic selection is faster and cheaper than MAS for many traits. Some of the factors that can detract from the success of MAS include (1) some breeding facilities lack the technical facilities and expertise to apply MAS, (2) incomplete linkage between a marker and a target QTL may reduce the effectiveness of MAS, (3) the marker must be polymorphic on the parents, and (4) MAS is effective only if the alleles being selected are important relative to other alleles in the population. This last factor is the key to the success or failure of every MAS application. It may seem like an obvious statement, but MAS relies on the ability to predict the value of alleles. The quality of those predictions rests on many factors, but a key factor is the behavior of an allele in the presence of other alleles and other physical environments where it has not yet been tested. For example, a breeder might identify that allele A_1 at locus A has a positive effect on yield. But this prediction would be made in a limited set of environments, and with a limited set of germplasm. A breeder who crossed a parent containing allele A_1 with a new parent containing allele A_4, and selected for A_1 using a linked marker, might never discover that allele A_4 is actually better than allele A_1, or perhaps that allele A_1 causes plants to be susceptible to a disease that was not present when A_1 was first characterized. For these reasons, MAS should never be applied independently from phenotypic selection, and most successful applications of MAS have been as an enhancement to phenotypic selection rather than as a replacement.

3.5.3. Mutation Breeding

Mutations are genetic modifications that occur in the DNA of plants, producing new alleles that are different from the alleles that the plant inherited from its parents. Mutations can be small and localized, or they can cause major structural rearrangements of entire chromosomes. Localized mutations include base substitutions and small insertions or deletions. Because most amino acids are coded by two or more different codons, many base substitutions are "silent," detectable only through DNA sequence analysis. Most mutations that occur in noncoding DNA are also silent, although they can sometimes affect gene expression or chromosome structure. Mutations that cause the transcription of a different amino acid are more likely to cause phenotypic change, most likely through their influence on protein folding or their alteration of an active site in an enzyme.

Fundamentally, the success of all plant breeding depends on mutations that have occurred at some point in the evolution of a species. However, the great majority of random mutations are deleterious, so breeders rely on a relatively small number of mutations that have been presorted through natural selection because they provide some type of selective advantage in at least one environment. Beneficial mutations that arise naturally are very rare, and most probably go unnoticed. However, it is possible to artificially induce mutations at frequencies that are much higher than the natural rate. This can be done through radiation (usually applied to seeds prior to planting) or through chemical induction.

The artificial induction of new mutations has been employed when natural variation for a trait is not available. However, it has the following disadvantages, and for these reasons, it is considered by most breeders to be a technique of last resort:

1. Most importantly, mutations are almost always deleterious, and it is highly unlikely to find a beneficial mutation in a specific gene that affects a specific trait. Therefore, any mutation breeding strategy must be capable of examining large numbers of mutated progeny.
2. Mutations are often not noticeable in a first generation because they are in a heterozygous state. Therefore, breeders must usually look at the offspring of a mutated population.
3. Mutations are usually induced simultaneously in many different genes. Therefore, even if a line is found with a desirable mutation, the same line probably caries many other undesirable mutations that must be bred out.
4. Finally, there is the question of crop safety. It is possible that a new mutation may have unpredictable effects on nontarget traits. This possibility can never be completely ignored, and for this reason, artificially induced mutations are considered by some regulatory systems (including Canada's) to be equivalent to artificial genetic transformation events.

3.5.4. Apomixis

Apomixis is a genetic phenomenon whereby seed is produced without pollination. There are several types of apomixis, but most types result in the production of seed that is identical in genotype to that of the parent plant. Dandelion is a notorious weed that exhibits apomixis; thus all of the seed from a single plant is likely to be identical. Very few cultivated crops exhibit apomixis; these include Kentucky bluegrass (the other part of your lawn), and some lesser-known tropical grasses. However, many research initiatives have attempted, or are attempting, to introduce apomixis into other crops such as maize in order to take advantage of the perpetual hybrid vigor that could be obtained if this were successful. One might speculate that the amount of commercial interest in this endeavor is low, since this would theoretically allow agricultural producers to save their own seed in a crop that might otherwise require the continual purchase of hybrid seed.

3.6. CONCLUSIONS

Plant breeders—who are part scientists, part artisans, part entrepreneurs, part extension workers, and part economists—are a special breed in themselves. Breeders have adopted (and often instigated) many of the genetic discoveries made in the previous century, and have developed highly scientific approaches to plant variety development. Yet these approaches still leave ample room for personal philosophy, artistic license, and all of the practical challenges of balancing objectives with reality. Over the next century, breeding will incorporate new discoveries and new technologies, but it will almost certainly continue to rely on the principles of sexual recombination and selection. Breeders, whether they are part of industry, government organizations, or universities, will continue to be an essential part of every bioeconomy, and the designation "plant breeder" will continue to describe an interesting, challenging, and rewarding career.

LIFE BOX 3.1. GURDEV SINGH KHUSH

Gurdev Singh Khush, Former Head of Plant Breeding, Genetics and Biotechnology, International Rice Research Institute, Philippines, and Adjunct Professor, University of California Davis; Winner of the Japan Prize (1987), World Food Prize (1996) and Wolf Prize (2000); Member of National Academy of Sciences and Royal Society of London

Gurdev Khush

I was born in a farming family in Punjab, India in 1935. As I was growing up I took part in various farming operations and developed an interest in plants. Farm yields were extremely low and poverty was rampant in farming communities. My father was the first person in our village of about 5000 to graduate from high school. He inculcated in me the value of education. I chose to major in plant breeding as an undergraduate at the Government Agricultural College (now Punjab Agricultural University) in Ludhiana and graduated in 1955. Facilities for higher education in India at that time were very limited and I decided to study abroad. I borrowed some money and proceeded to England where I worked in a factory for a year and a half. I returned the borrowed money and saved enough for travel to the USA. I enrolled at the University of California, Davis in 1957 for a doctorate in Plant Genetics. I had the good fortune to work under the supervision of a world renowned biologist Professor G. Ledyard Stebbins. After completing my Ph.D. in 1960, I joined the group of another equally outstanding geneticist, Professor Charles M. Rick, as a post-doctoral associate and worked on cytogenetics of tomatoes for seven years. My solid background in plant genetics proved to be extremely useful in my future career as a plant breeder. In 1966 I was offered the position of a Plant Breeder at the International Rice Research Institute (IRRI) and I moved to the Philipines in August 1967.

The 1950s and 1960s were decades of despair with regard to the world's ability to cope with food-population balance, particularly in the tropics. The cultivated-land frontier was closing in most Asian countries, while population growth rates were accelerating because of rapidly declining mortality rates resulting from modern medicine and health care. IRRI was established to address the problem of stagnant rice yields, the main cause of poverty and hunger in Asia. Conventional rice varieties were tall and lodging susceptible. When nitrogenous fertilizer was applied those varieties grew even taller, lodged badly, and yields were actually reduced. A breakthrough occurred in doubling the yield potential of rice through reducing the plant stature by introduction of a dwarfing gene. The first short-statured rice variety IR8 was lodging resistant and highly responsive to nitrogenous fertilizer. It had double the yield potential of conventional varieties. However, it had poor grain quality and was susceptible to diseases and insects. The major focus of my research was to develop improved germplasm with high yield, shorter growth duration, superior grain quality and disease- and insect-resistance. I developed numerous breeding lines with the above traits. IR 36 was the first variety with all the desirable traits. It had high yield potential, short vegetative growth duration, excellent grain quality, multiple resistance to major diseases and insects and tolerance to adverse soil conditions such as iron toxicity and zinc deficiency. It was grown on 11 million hectares of rice land during the 1980s. No other variety of rice or any other crop had been as widely planted before. Thirty-four varieties were released under IR designation (IR8–IR74). Seeds of improved breeding lines were shared with national program scientists at their request and through international nurseries. Thus, seed materials were sent to 87 countries irrespective of geographic location or ideology. These materials were evaluated for adaptation to local growing conditions. Some were released as varieties and others were used as parents in local breeding programs. Thus, 328 IR breeding lines have been released as 643 varieties in 75 countries. It is estimated that 60% of the world rice area is now planted to IRRI-bred varieties or their progenies. Large scale adoption of these varieties has led to major increases in rice production. Average rice yield has doubled from 2 to 4 tons per hectare. Rice production increased from 257 million tons in 1966 to 615 million tons in 2005, an increase of 140 percent. The price of rice is 40% lower now than in the mid 1960s. This has helped poor rice consumers who spend 50% of their income on food grains. Thus, these IRRI-bred varieties have had a significant impact on food security and poverty alleviation and fostered economic development particularly in Asia, where 90% of the world's rice is grown.

I was fortunate to have had the opportunity to lead one of the largest and most successful plant breeding programs at IRRI. I had a team of motivated plant breeders, plant pathologists, entomologists, and cereal chemists supported by a dedicated Filipino staff. We had a large collection of germplasm, liberal financial support, modern laboratories and adequate field space. The opportunity to work with scientists in rice growing countries was another reason for our success. In addition to conventional hybridization and selection procedures, my team employed other breeding approaches such as ideotype breeding, hybrid breeding, wide hybridization, rapid generation advance, molecular marker assisted selection (MAS), and genetic engineering. I had the opportunity of working with numerous trainees from rice growing countries that came to IRRI for a degree (MSc and PhD) and non degree training. Upon returning to their countries they became our valued collaborators. Several of our trainees are now holding positions of leadership in their respective countries. This had a multiplying effect and all the rice growing countries are now using crop development methodologies and germplasm initially developed at IRRI.

The science of plant breeding is now at a crossroads. Breakthroughs in cellular and molecular biology have added new tools to the breeder's toolbox. MAS has increased the efficiency of selection and reduced the time taken for varietal development. Genetic engineering has permitted the introduction of genes into crop varieties from unrelated sources across incompatibility barriers. In 2006, 102 million hectares were planted to transgenic crops in 22 counties. The science of genomics is likely to improve the efficiency of plant breeding further. The entire genome of rice has been sequenced and efforts are underway to determine the functions of an estimated 40,000 rice genes through functional genomics. Similar efforts are underway in many other crops. Once useful genes for crop improvement are identified, it will be possible to move these genes into elite germplasm through conventional or biotechnological approaches.

It is important that plant breeders have a good background in biotechnology and that they work with specialists in the field. The marriage between the ancient profession of plant breeding and the new field of biotechnology will be good for future advances in crop improvement.

LIFE BOX 3.2. P. STEPHEN BAENZIGER

P. Stephen Baenziger, Eugene W. Price Distinguished Professor, University of Nebraska

Stephen Baenziger with Dr. Sanjaya Rajaram of ICARDA and CIMMYT looking at an in situ collection of wild barley in Syria, near the origin of barley.

"Give us this day our daily bread." Although I am not particularly religious, those words have always moved me. When I was in high school, I thought of becoming a human nutritionist so that I could work on world hunger. The U.S. Senate had a subcommittee led by Sen. McGovern on hunger in America that catalogued the dismal state of the poor and Paul Ehrlich published "The Population Bomb" highlighting, quite incorrectly, that massive famines were set to occur in the 1970s. In college, I was a biochemistry major which was the pre-med major, a group of students whom I never really enjoyed being with because they seemed more interested in their grades than the knowledge (getting into medical school was very competitive), so I gravitated to plant biology, a field that the pre-meds did not know existed. The professors in plant biology were spectacular (Winslow Briggs, Lawrence Bogorad) and I decided that, as a nutritionist, I would better define a problem, but not really solve its root causes. Food would still be limiting. Hence, I decided to work on the production side to ensure that there was ample food for those who needed it. At this time, the Green Revolution in wheat, led by Norman (Norm) Borlaug of CIMMYT, and in rice, led by Henry (Hank) Beachell, then Gurdev Khush of IRRI, had greatly increased the food supply and the predicted famines never occurred. In graduate school, David Glover, who was working on breeding high lysine maize (now referred to as quality protein maize) offered me an assistantship and sealed my fate to become a plant breeder. It was also the last time that I worked on maize. My first job was to develop small grains (wheat and barley) germplasm with improved disease resistance and tolerance to acid soils (note I only audited one plant pathology course in graduate school and never took a soils course) for the USDA. Probably the most interesting aspect of this position, in addition to the excellent scientists that nurtured me, was that the position had been vacant for 4 years and most of the germplasm was transferred or gone. Hence, we needed to rebuild the program from scratch. In winter wheat and barley breeding, it takes 12 years to release a new cultivar and usually at least 8 years to release good germplasm. It was quite clear that time was working against us, so we began a doubled haploid program in hopes we could rapidly inbreed lines and shorten the time to release. Though I have never had sufficient funds to use doubled haploids except for very special genetic studies, this approach is now very common in well funded commercial breeding programs. Working on germplasm improvement also showed me that despite the massive genetic resources available to wheat and barley, germplasm can be limiting so transformation studies are very important in crop improvement.

After working with the USDA and a short period with Monsanto, I became the small grains (winter wheat, barley, and triticale) breeder at the University of Nebraska. The collaborative USDA-University of Nebraska wheat breeding effort under the stewardship of John Schmidt, Virgil Johnson, Rosalind Morris, and Paul Mattern had been one of the most successful breeding programs in the United States. At one time 96% of the wheat grown in Nebraska, 40% of the hard winter wheat grown in the U.S., and 20% of the wheat grown in the U.S. came from their program. Here I learned that breeding can have an impact. I also learned that each crop has special tools that can be used to approach specific scientific questions. While maize had excellent molecular markers, wheat initially had few. However, wheat had chromosome substitution lines (developed at Nebraska by Rosalind Morris) where we could study single chromosome effects across the diverse environments of the Great Plains. In this work, we found that chromosome 3A would increase or decrease grain yield by 15% in the two backgrounds that Rosalind Morris developed. We then used cytological tools to break up these chromosomes by recombination and coupled them with molecular markers to study this chromosome in great detail. In this way, we developed the populations and the phenotypic data while waiting for the molecular marker technology to catch up. It took Rosalind most of her professional career to develop the substitution lines, and after 20 years we are still studying various aspects of this chromosome because grain yield is still the most important trait in plant breeding. These studies involve huge numbers of lines and the randomized complete block designs were inadequate with the highly variable conditions under which wheat is grown. Working with statisticians we implemented various statistical methods (nearest neighbor, incomplete block designs) to remove spatial variation in the fields, and to improve our phenotypic estimates. Large experiments require these statistical approaches wherever fields lack uniformity. If a breeder must be knowledgeable in a number of scientific disciplines, and if breeding is built upon the work of previous breeders, perhaps my program has benefited as much or more than most breeding efforts. However, I hope that curiosity and constant questioning of how to measure and understand the traits that breeders work with has been my contribution. That, my cultivars, and my students will be my legacy.

REFERENCES

Allard RW (1999): *Principles of Plant Breeding*. 2nd ed. Wiley, New York.

Barton NH, Keightley PD (2002): Understanding quantitative genetic variation. *Nat Rev Genet* **3**:11–21.

Donald CM (1968): The breeding of crop ideotypes. *Euphytica* **17**:385–403.

Duvick DN, Cassman KG (1999): Post–Green Revolution trends in yield potential of temperate maize in the North-Central United States. *Crop Sci* **39**:1622–1630.

Fehr WR, Hadley HH (1980): *Hybridization of Crop Plants*. American Society of Agronomy and Crop Science Society of America, Madison, WI.

Frey KJ, Holland JB (1999): Nine cycles of recurrent selection for increased groat oil content in oat. *Crop Sci* **39**:1636–1641.

Hoisington D, Khairallah M, Reeves T, Ribaut J-M, Skovmand B, Taba S, Warburton M (1999): Plant genetic resources: What can they contribute toward increased crop productivity? *Proc Natl Acad Sci USA* **96**:5937–5943.

Orr HA (2005): The genetic theory of adaptation: A brief history. *Nat Rev Genet* **6**:119–127.

Paterson AH (1998): *Molecular Dissection of Complex Traits*. CRC Press, Boca Raton, FL.

Wrike G, Weber WE (1986) *Quantitative Genetics and Selection in Plant Breeding*. Walter de Gruyter, Berlin.

Wright S (1982): The shifting balance theory and macroevolution. *Annu Rev Genet* **16**:1–20.

CHAPTER 4

Plant Development and Physiology

GLENDA E. GILLASPY

Department of Biochemistry, Virginia Tech, Blacksbury, Virginia

4.0. CHAPTER SUMMARY AND OBJECTIVES

4.0.1. Summary

From fertilization to seed to maturity, plants are genetically programmed to grow, develop, and reproduce. Agriculture is greatly dependent on seed production (yield), and yield depends on how plants cope with their environment and other organisms. Since they cannot move around, plants are adept at responding to their environment. They develop and respond primarily by altering their biochemistry, especially in response to plant hormones. In addition, understanding how to manipulate plant development in vitro is necessary for the successful engineering of transgenic crop plants.

4.0.2. Discussion Questions

1. Describe the general morphological features of a plant.
2. How is plant fertilization different from animal fertilization?
3. How does the study of mutant plants shed light on gene function?
4. Many genes involved in embryo development also have functions during later stages of development. It has been difficult to clarify these later-stage roles. Why?
5. How do GA and ABA physiology affect germinating seeds?
6. What is an apical meristem? Name one gene involved in shoot meristem identity, and describe the role it plays during development.
7. What is etiolation?
8. How do the PHY proteins function as light receptors?
9. How do the quiescent zone and root cap structures and properties differ?
10. How do guard cells participate in photosynthesis and respiration?
11. How do the adaxial and abaxial surfaces of the leaf differ?

Plant Biotechnology and Genetics: Principles, Techniques, and Applications, Edited by C. Neal Stewart, Jr.

12. You have isolated a gene whose expression is confined solely to the developing leaf primordia, and have obtained a loss-of-function mutant for this gene. Speculate as to what phenotype might result in this mutant, and explain the basis for your speculation.
13. What is the difference between a daylength-neutral and a long-day plant?
14. Describe the ABC model of flower development and speculate as to what phenotype would result if a C function gene were overexpressed in all whorls.
15. In snapdragon, *Floricula* mutants contain shoots with the characteristics of an inflorescence meristem in place of a floal meristem. Hypothesize what the wild-type function of the *Floricula* gene is, and speculate as to why investigators are interested in overexpressing this gene in Aspen trees.
16. Describe the major effects of plant hormones on growth and development. Also, describe how each hormone is percieved by the plant cell and how the signal is transduced throughout the cell.

4.1. PLANT ANATOMY AND MORPHOLOGY

Before considering the developmental and physiological processes that can impact plant biotechnology, one should have some basic knowledge of plant anatomy and morphology. This section is designed to provide a closer look at internal structures and cells within the plant.

Most plants are composed of the *shoots*, or aboveground tissues, and *roots*, the below-ground tissues (Fig. 4.1). The shoot apex consists of the topmost tissues of a seedling or plant and contains the *shoot apical meristem* (SAM) and the developing leaves or leaf primordia. The SAM is a dome-shaped region of dividing cells at the tip of the stem (Fig. 4.1).

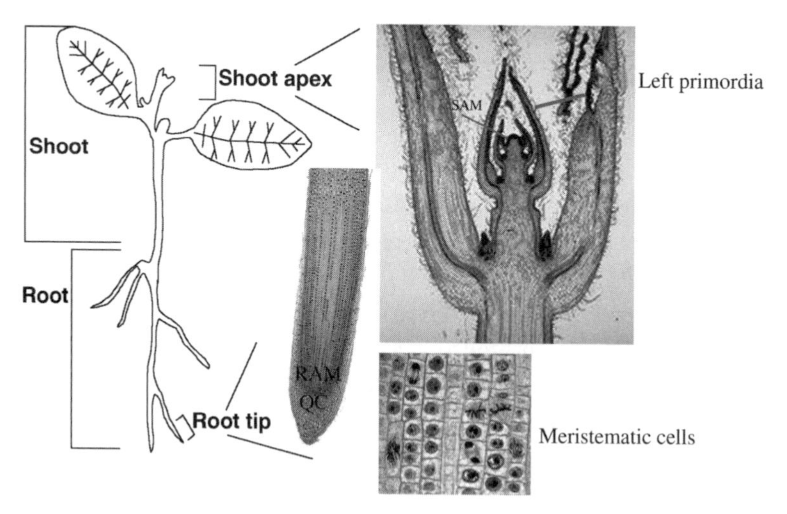

Figure 4.1. Plant anatomy and morphology. The seedling shoot and root systems are indicated, as are the shoot and root apical meristems, tissues that direct the major growth and differentiation of plants. Active cell division within the meristem is shown in the last panel; note the presence of two nuclei in some cells. Modified with permission from Dr. Dale Bentham's Website (http://biology.nebrwesleyan.edu/benham/plants/index.html).

The SAM is the control center of the plant and directs the development of all aboveground differentiated tissues such as the stems, leaves, thorns, flowers, and fruits. Cells within meristems undergo cell division quickly, and are usually smaller because they have smaller vacuoles than differentiated plant cells (Fig. 4.1).

The root also contains a similar control center, the *root apical meristem* (RAM) that functions in generating new root cells within the root tip (Fig. 4.1). A section through the root shows that roots are often full of starch granules that can be visualized by staining with potassium iodide, which turns starch a blue-brown color. One can also see the meristematic zone at the root tip, the root cap, a protective covering, and the ordered files of cells resulting from the root initial cells within the root apical meristem. One may also be able to view the quiescent center (QC), so called because cells are "sleeping" or slow to undergo cell division.

Axillary buds are the third type of meristems that give rise to new tissues. Axillary buds may be found on stems, and under the right conditions can give rise to new shoot apical meristems.

Plant cells within shoots and roots are organized into specialized tissues that enable the organism to carry out necessary functions. The tissue systems of plants are the dermal, vascular, and ground tissue systems. The *dermal* system is composed of the epidermal, or outermost, cell layer, which covers the entire plant. The *vascular* tissue system is composed of the xylem, phloem, and other conducting cells that transport water and nutrients. This tissue is present in most plant tissues, but can be arranged differently within each organ. The *ground tissue* is composed of the cells in between the epidermis and the vascular tissue.

There are many different specialized plant organs. In addition to the shoot and root apical meristems, most angiosperms contain stems, leaves, lateral roots, and reproductive tissues such as flowers and their component tissues (anthers, filaments, pollen, etc.). Each of these tissues can impact the development and physiology of the plant, and as such must be considered when manipulating gene expression in transgenic plants. Specific considerations for each of these tissues will be discussed as we chart the development and physiology of an average plant in the succeeding sections.

4.2. EMBRYOGENESIS AND SEED GERMINATION

4.2.1. Gametogenesis

The lifecycle of flowering plants alternates between a haploid organism, the gametophyte, and a diploid organism, the sporophyte. Plants have male and female gametophytes, both of which are multicellular and are produced within the flower (Fig. 4.2). The mature male gametophyte, the pollen grain, has three cells: a vegetative cell and two $1N$ sperm cells. *Pollen* development (Fig. 4.3) occurs in the *anther*, which is a specialized structure of the flower, with the meiotic divisions of the microsporocytes to form a tetrad of haploid spores. The microspores are embedded in callose, and release from the tetrad requires enzymes secreted by somatic cells in the anther. Mature pollen grains have complex walls with two layers, the inner intine and the outer exine layer.

Self-incompatibility as a mechanism to limit reproduction was discussed in Chapter 2. However, fertilization also depends on the gene products that are required for normal development of the pollen and *ovules*. Scientists have identified several of these required gene products by taking a genetic approach. To identify molecules involved in either

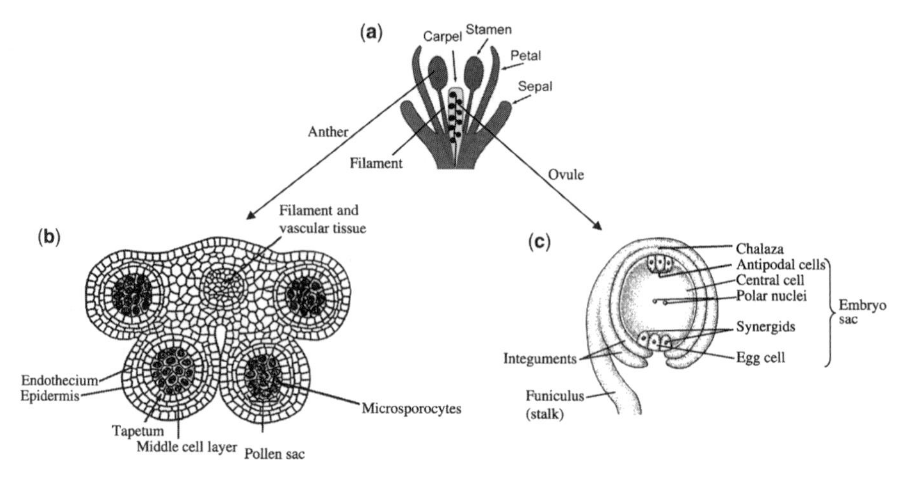

Figure 4.2. Gametogenesis. Schematics of (a) an *Arabidopsis* flower with the floral organs identified; (b) a cross section through the male organs (anther, filament) showing the site of male gamete formation; (c) the female ovule contained within the carpels of the flower showing the site of female gamete development. [Reprinted from Wilson and Yang C (2004), with permission from the Society for Reproduction and Fertility.]

gameteogenesis (the formation of gametes) or *fertilization*, geneticists have utilized mutant populations of a superb experimental model plant called *Arabidopsis thaliana*. The genome of *Arabidopsis* is fully sequenced, and many different mutant populations containing a loss of function in individual genes are available. The first mutant collections were most often composed of plants containing random, single-base-pair mutations, or T-DNA insertions. Both types of mutant collections can be screened, and mutants identified on the basis of the phenotype. For example, mutants defective in a gene required to form a female or male gamete will give rise to mature plants with low fertility. Low fertility can be somewhat easily scored in a random mutant population by looking for low seed set. In *Arabidopsis*, the seeds are produced within small, elongated fruits called *siliques*. Finding a plant with fewer siliques, or empty siliques, is an indication that a mutant has lost the function in a gene required for gamete development. One can determine which gamete has been affected by examining the appearance of both female and male gametes from the candidate mutant plant. For example, if pollen grains appear normal and germinate pollen tubes in vitro, then most likely, the defect is *not* in male gametophyte development. The scientist would then examine the appearance of the female gametophyte within the flower. Outcrosses of the candidate mutant pollen to a wild-type pistil, and the reverse outcross (candidate mutant female to wild-type male cross), can also be important in determining which gamete is defective (Wilson and Yang 2004; Boavida et al. 2005).

Using such screens and outcrosses, geneticists have isolated several genes required for pollen and ovule development. In *pop* mutants, for example, the *exine layer* of the pollen grain does not develop properly, resulting in altered hydration of pollen grains. Without normal hydration, the pollen tube guidance is not normal, and fertilization is greatly lowered. These mutants point to the idea that structural components of the pollen grain itself are important for male fertility.

Female sterile mutants have also led to the identification of genes required in female gametophytic development. ANT, BEL1, SIN1, and ATS gene products were each

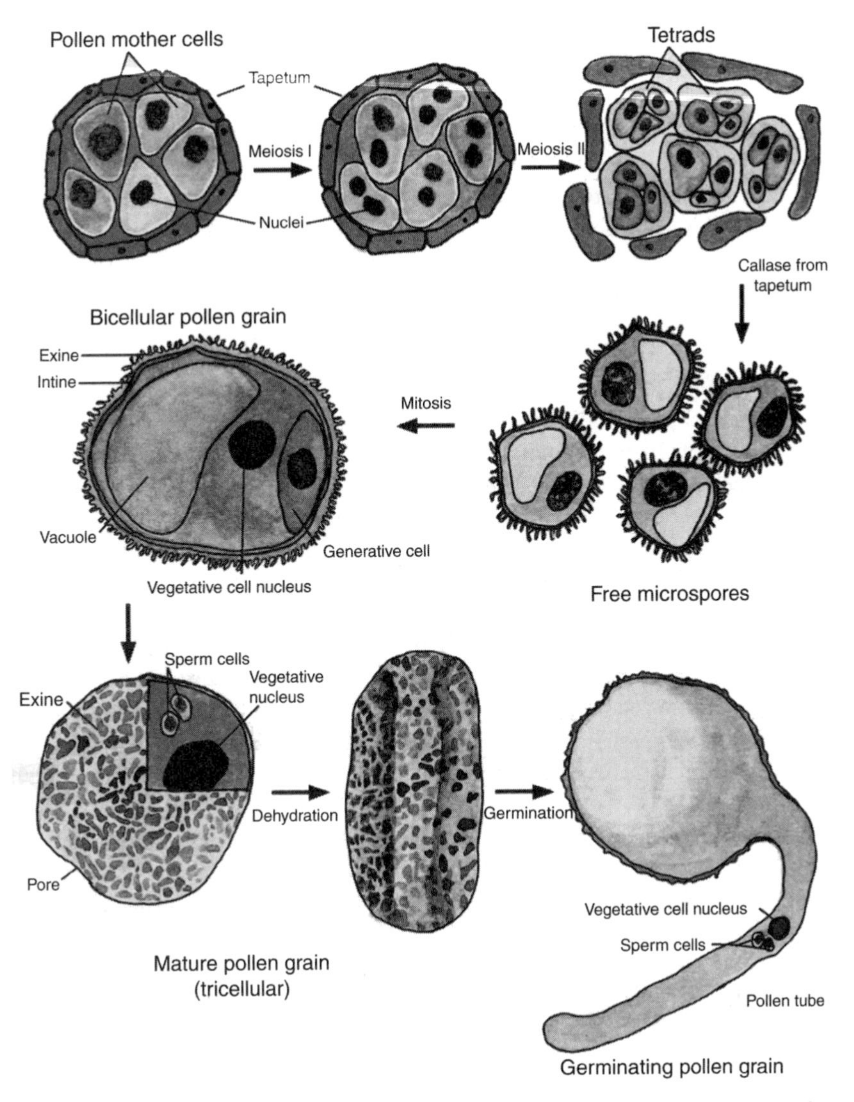

Figure 4.3. Pollen development. [Reprinted from McCormick (2004), with permission from the American Society of Plant Biologists.] See color insert.

identified in mutant screens. Each of these genes encode proteins required for ovule development. For example, the *ant* mutant cannot make the *integuments* that surround the developing egg cell; thus the ANT gene product is required for development of the integuments. The *bel1* mutant is also defective in integuments, but does develop a collar of tissue that surrounds the egg cell. Thus *bel1* mutants have an altered integument and the function of the BEL1 protein is to specify integument identity within the developing female gametophyte. The *sin1* mutant also has altered integuments that are shorter. This mutant is of special interest in that the SIN1 protein is a homolog of the DICER protein that functions in generating small, interfering RNA molecules (siRNA) that suppress *gene expression* at the posttranscriptional level. The fact that a DICER-like enzyme is required for normal ovule development strongly suggests that posttranscriptional regulation of ovule identity genes is important for maternal development.

The phenotypes of these mutants help build a model of the ovule developmental pathway. They suggest that during the process of flowering the ovule primordia initiate and then gain ovule identity. For example, primordia initiation must include ANT function, which is then followed by the action of genes that specify the integuments like BEL1. In this model, SIN1 function would follow, giving rise to the normal shape and size of the integuments. Thus, by using a combination of genetic and molecular approaches, developmental biologists can order gene function in the development of specific tissues.

4.2.2. Fertilization

The beginning of a plant's life starts with fertilization of the haploid ($1N$) egg cell within the ovule by one of the two haploid sperm nuclei carried by the pollen tube of the pollen grain (see Chapter 2 and Fig. 4.2). Development will produce a $2N$ plant embryo surrounded by maternal tissues within the carpels. Plants actually undergo a separate fertilization event that creates the $3N$ *endosperm*. The endosperm results from fusion of the other $1N$ *sperm* nuclei with the two *polar nuclei* ($2N$) within the central cell of the ovule. The resulting endosperm tissue can transfer nutrients into the developing embryo. Thus plants, like animals, have a food supply handy for the developing embryo. The triploid nature of the endosperm has been speculated to be a mechanism for controlling gene dosage or a way for maternal control of embryo development (Berger et al. 2006). An interesting phenomena called *endoreplication*, or *endoduplication*, occurs at an increased rate within the endosperm. This process involves DNA replication in the absence of cell division, resulting in a high N number within certain cells of the endosperm.

Studies on *Ephedra trifurca*, a nonflowering seed plant that is a close relative of the angiosperms, have revealed key differences in fertilization. This plant, from which Mormon tea is made, has a second fertilization event that leads to formation of a second embryo instead of endosperm development. This difference has prompted speculation that the modern endosperm of today's plants may have evolved from a second embryo like that found in *Ephedra*. We know that fertilization and development of the embryo and endosperm in angiosperms are dependent on each other; that is to say that normally the endosperm *must* develop in order for the embryo to develop. However, there is a mutant that has been identified where fertilization of the endosperm occurs in the absence of embryo fertilization and development. This mutant, called *fie* (*fertilization-independent endosperm*), suggests a connection between endosperm development and chromatin as the FIE gene product is a type of *polycomb* protein. Polycomb proteins were first discovered in *Drosophila melanogaster* and act by "locking" chromatin into accessible or nonaccessible forms that dramatically alter gene expression in the next generation. Thus, the FIE polycomb gene product may be necessary to "lock in" the appropriate chromatin pattern for the communication between the embryo and the endosperm developmental processes (Twell 2006).

4.2.3. Fruit Development

Fertilization is also important to consider in plant biotechnology as it directly impacts the process of fruit development. Fertilization is the trigger for growth of the ovary that then can develop into a fruit. The term *fruit* can be used to describe any ovary that initiates a growth program after fertilization. For example, the enlarged ovary under a decaying rose flower is called a *rose hip*, and like citrus fruits, contains high levels of vitamin C. Fruit development is a strategy thought to attract animals that will eat the fruit and disperse the seeds far from the plant. Animals and plants have coevolved, with animals trying to get the most nutrients (through digestion) from the fruit and seeds, and the plant evolving processes designed to facilitate seed dispersal in contrast to seed digestion. This coevolution may account for the incredible diversity of fruit and seed types.

Fruit development requires both fertilization and growth of the embryo within the seed; thus seed and fruit development are related. For example, in some species lopsided fruit will result when fertilization of ovules on one side of the ovary is defective. The seeds developing from fertilized ovules are thought to signal to the surrounding fruit via their production of growth hormones, such as auxin and cytokinin. There are physiological conditions, however, that will override the requirement for these seed-derived hormones. The process of fruit development in the absence of seed development is called *parthenocarpy*, which is a desirable trait for certain fresh fruit. Some commercial "seedless" varieties, like the seedless watermelon, actually have very tiny, partially developed seeds. In contrast, certain true seedless grape varieties undergo parthenocarpic fruit development in the absence of fertilization of the ovules. Studies on parthenocarpic fruit will lead to a better understanding of the processes that accompany fertilization. One useful tool will be the *fwf* (*fruit without fertilization*) mutant from *Arabidopsis*, which is a facultative parthenocarp, setting seed in a normal way when pollinated, but also forming short seedless fruit when left unpollinated. It is thought that the FWF protein acts as an inhibitor of fruit development and that this inhibition is released after fertilization. Better understanding of FWF function awaits cloning of the gene (Giovannoni 2001).

4.2.4. Embryogenesis

As described earlier, *embryogenesis* begins after the $1N$ egg cell and $1N$ sperm nuclei fuse together, forming a $2N$ embryo. Plant embryogenesis differs significantly from animal embryo development in its lack of cell migration and substantial cell specification. For example, the mature plant embryo within the seed does not contain cells specified to become flower cells or gamete-producing cells. These differentiation events will occur later in development, well after seed germination. Instead, plant embryogenesis will result in the acquisition of bilateral symmetry, an apical/basal or shoot/root axis, and the three types of tissue.

The first cell division of the plant embryo results in an asymmetric division giving rise to a small upper, terminal cell and a larger, lower basal cell (Fig. 4.4). This establishes a longitudinal or an apical/basal axis in the embryo (Weijers and Jurgens 2005). The upper cell always gives rise to the embryo proper, while the lower cell gives rise to the *suspensor* and the *hypophysis*, forming part of root meristem, root initial cells, and root cap. The suspensor is a highly specialized and terminally differentiated tissue that connects the embryo to the embryo sac and maternal ovule tissue. It functions as a conduit for nutrients and senesces after the heart stage of embryo development. This short-lived unique organ consists of only 7–10 cells total in *Arabidopsis thaliana*.

The upper cell of the two-cell embryo undergoes two more cell divisions, passing through the four- and eight-cell stages, in which a gain of embryo mass occurs. Further cell divisions result in mass of cells on top the suspensor referred to as the *globular-stage embryo* (Fig. 4.4). More cell divisions result in development of the heart-stage

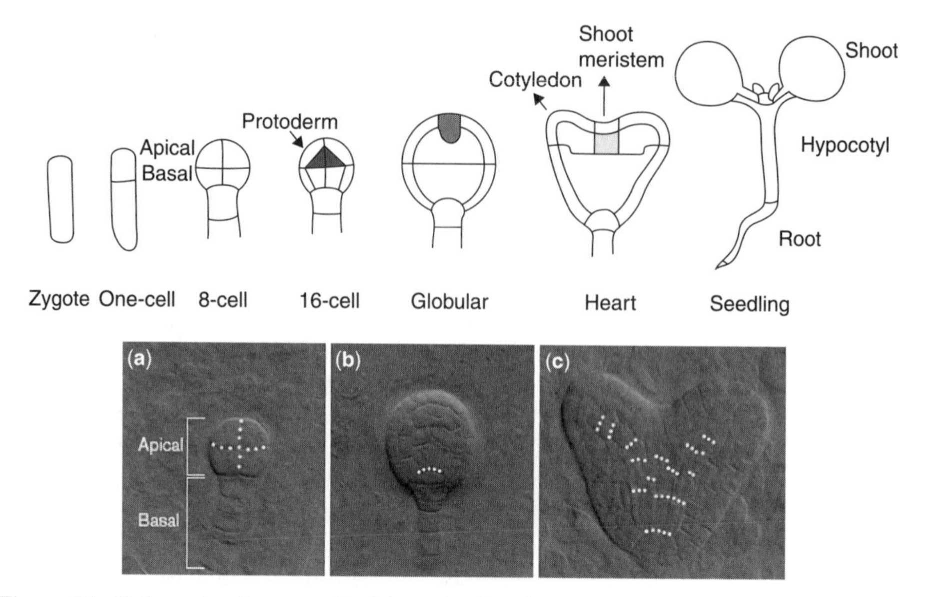

Figure 4.4. Embryo development. (a) Schematic of embryo stages. [Reprinted from Lenhard and Laux (1999), with permission by Elsevier Science Ltd.] (b) Scanning electron micrograph of *Arabidopsis* embryos in the globular and heart stages. The white lines indicate the cell division planes. [Reprinted from Costa and Dolan (2000), with permission from Elsevier Science Ltd.]

embryo, so called because of the characteristic heart shape of the embryo. This heart shape results because differentiation of cells has occurred, with some cells beginning to acquire shoot apical meristem (SAM) identity in the cleft of the heart, and two lateral domains giving rise to cells destined to form the cotyledons of the embryo. In addition, the root apical meristem (RAM) becomes specified at this stage. With the development of the SAM, RAM, and cotyledons, the embryo is now beginning a change to bilateral symmetry.

After the heart stage, organ expansion and further cell divisions result in the lengthening of the embryonic cotyledons into the "torpedo" stage (Fig. 4.4). At this point, two patterns have been established: the apical/basal patterns, which allows for shoot vs. root development; and the radial pattern, which gives rise to the three types of tissue: (1) the *protoderm* (which gives rise to the epidermis), which divides anticlinally; (2) a middle layer, the *ground meristem* (which gives rise to the cortex and endodermis); and an inner layer, the *procambium layer* (which gives rise to the vascular tissue) (Willemsen and Scheres 2004).

The last stage before the mature embryo stage is the "walking stick" stage, so called because the developing cotyledons have folded down over the SAM. To mature, the embryo must enter a dehydration phase in which metabolism pauses. In the dehydrated state the embryo within its seed coat is waiting for the appropriate environmental conditions suitable for seed *germination*. The plant hormone *abscisic acid* (ABA) is required for initiating dehydration and establishing seed dormancy. Without an ABA source or a functioning ABA signal transduction pathway, embryos can germinate "precociously" inside a fruit. Thus the study of ABA signaling pathways and the genes turned on by these pathways is directly relevant to the understanding and manipulation of seed germination.

4.2.5. Seed Germination

Germination is the process wherein the embryo imbibes water and returns to growth after dormancy. *Imbibition* is the uptake of water by the embryo within the seed. During this process, the embryonic tissues are loosened and the seed coat usually splits, allowing more water to penetrate the embryo. Once the embryonic cells are rehydrated, the metabolic processes of germination can begin.

Several common requirements are shared by very diverse types of seeds, including temperature and moisture. Some seeds have a light requirement, and some also require a cold pretreatment called *stratification*. These processes promote the increase and/or action of a plant hormone called *gibberellic acid* (GA). GA action is generally considered as antagonistic to ABA and is considered to be the dormancy-breaking hormone. One well-characterized action of GA is the induction of α-amylase production that breaks down stored starches in grain seeds. Germination can occur underground (in the dark) or above ground (in the light). Either way, the major result of germination is the expansion of the already preformed embryo (Koornneef et al. 2002).

4.2.6. Photomorphogenesis

Imbibition of a seed allows dormant cells to expand and for new cell division to occur within the embryo. The specific type of growth is influenced heavily by the presence or absence of light. Light is the most influential signal from the environment that plants perceive. When a seed germinates above ground, or in the presence of light, it immediately responds to light with an elegant and complex developmental response called *photomorphogenesis*. If a seed germinates underground or in the absence of light, it undergoes a brief and specific developmental pathway called *skotophotomorphogenesis*. The purpose of this dark developmental pathway is assumed to be the alteration of growth in the seedling that increases its chance of encountering light, a signal required for the further development of the seedling.

When germination occurs in the dark, the seedling develops into what is called an *etiolated* seedling, which is characterized by increased hypocotyl growth, an apical hook (in dicots), unexpanded cotyledons, and no chlorophyll synthesis. These adaptations to dark can allow for the elongating hypocotyl to push the SAM and cotyledons up through the soil to encounter light. The apical hook thus can protect the new SAM, and chlorophyll synthesis is not needed until light is encountered.

When the seedling encounters light, the elongation of the hypocotyl slows, the apical hook uncurls, and the *cotyledons* expand and begin to assemble functional chloroplasts containing chlorophyll. Transcription of genes encoding the chlorophyll a/b binding proteins and part of the Rubisco complex are rapidly upregulated. Thus, if a seed germinates in the presence of light, its *hypocotyl* will be much shorter than that of an etiolated seedling. The apical meristem will then give rise to the first pair of true leaves that differ in structure from the cotyledons and contain *trichomes*, or hairs.

The light receptor required for red light signal transduction is called *phytochrome*, which is composed of an open-chain tetrapyrole pigment called *phytochromobilin* and a protein dimer of 240 kDa. This pigment/protein complex allows for the perception of red light by absorption of either red or far-red light. Phytochrome is distributed throughout many different cell types in the plant, and more recent evidence suggests that it traffics from the cytosol to the nucleus in response to light, where it interacts with transcription factors such as PIF3 to influence gene expression. Many of the gene products required to construct an active photosynthesizing chloroplast are controlled by the presence of light, and thus are most likely under the control of phytochrome-mediated signal transduction pathways. Phytochrome itself is encoded by five different *Phy* genes called *PhyA* through *PhyE*.

Mutants defective in photomorphogenesis have been instrumental in identifying genes required for this process. There are two general categories of photomorphogenesis mutants: (1) *hy* and (2) *cop* and *det*. The *hy* (*hypocotyl elongated*) mutants look partially etiolated even when grown in the light, indicating that the HY gene products function in the perception of light. These screens identified some of the *Phy* genes and other positive regulators of photomorphogenesis such as HY5, a key transcription factor. In contrast, the *cop* (*constitutive photomorphogenesis*) and *det* (*deetiolated*) mutants were identified by virtue of their light-grown phenotypes when grown in the dark. Many of the *cop* mutants encode proteins that form a large complex called the *COP9 signalsome* (CNS), a nuclear complex that is similar to the 26*S* proteasome proteolytic complex that degrades ubiquitinated proteins (Rockwell et al. 2006).

The lack of etiolation in some *cop* and *det* mutants can be reversed by adding the plant steroid hormone *brassinolide* (Br), suggesting a role for Br signal transduction in photomorphogenesis. The *det2* mutant of *Arabidopsis* has sequence homology with mammalian steroid 5α-reductases. This suggests that the DET2 gene product participates in Br synthesis. Thus, light may control photomorphogenesis by downregulating Br production.

Blue light is another important stimulus for photomorphogenesis and for phototropism (growth toward light). Blue light is perceived by two types of flavin-containing proteins, *crytochrome* and *phototropin.* Both cryptochrome (CRY) and phototropin are encoded by two genes in *Arabidopsis*. CRY proteins appear to function in the nucleus, although there are indications that there may be some CRY functions in the cytoplasm as well. Evidence suggests that phytochrome and cryptochrome physically interact. CRY protein can be phosphorylated in vitro by the protein kinase activity of PHY. In addition, PHYB and CRY2 interact in plant extracts. CRY1 and CRY2 also appear to directly interact with COP1, the negative regulator of photomorphogenesis in the dark.

4.3. MERISTEMS

Plant meristems are dynamic structures whose functions are to renew themselves and to give rise to new cells with a different identity. There are three types of meristems: apical meristems including the shoot and root apical meristems (SAMs and RAMs); the lateral meristems, including the vascular and *cambial* meristems responsible for secondary growth; and the *intercalary meristems*, common to the grasses that occur at the bases of nodes. The common function of these meristems is regulation of cell division that creates new cells specified to become different cell types and renewal of the meristem itself.

4.3.1. Shoot Apical Meristem

Apical meristems are extremely important in terms of growth regulation of plants. As alluded to previously, the SAM gives rise to the aerial parts of higher plants by continuously initiating new organs. The basis of this activity is its ability to maintain a pool of pluripotent stem cells, which are the ultimate source of all tissues of the shoot. The

SAM typically consists of a dome of cells connected to two developing leaf primordia (Fig. 4.1). This area contains around 100 cells in *Arabidopsis*. The dome structure contains the least differentiated cells and consists of three different histocytological zones (Fig. 4.5). The central zone in the middle of the dome contains cells that divide infrequently, yet this is the location of the self-renewing undifferentiated stem cells. Surrounding the central zone is the peripheral zone, where the rate of cell division is higher and cells contribute to the organs of the plant, including leaves, inflorescence meristems, and floral meristems. Below the central zone is another region of rapidly dividing cells, called the "rib" meristem. Division and elongation of rib meristem cells gives rise to the stem of the plant.

The SAM also consists of different cell layers. The surface layer of cells is called the *L1 cell layer*. Cells in L1 divide only by forming anticlinal cell walls, that is, cell division is always perpendicular to the meristem surface. As a result, cells in the L1 layer and their daughter cells always remain in this layer. The *L2 cell layer*, below the L1 cells, divide the same way. The *L3* or *corpus cells*, divide in all planes, and fill the interior of the SAM.

A major issue in plant biology concerns how shoot meristems are organized and how molecular information in the SAM determines the precise placement/function of cells. More recent molecular studies indicate that the maintenance of stem cell function depends on a feedback loop involving the CLV1–3 (*Clavata*) gene products and WUS (Wuschel). In *clavata* mutants, the meristem is enlarged, due to excessive accumulation of stem cells, suggesting that CLV1–3 are required to regulate the number of stem cells in the meristem. In contrast, *wus* mutants contains a smaller meristem with differentiated cells, suggesting that WUS is a positive regulator of stem cell identity. Analysis of the interactions between these key regulators indicates that (1) the *Clv* genes repress WUS at the transcript level and (2) WUS expression is sufficient to induce meristem cell identity and the expression of the stem cell marker CLV3. As the different CLV genes encode a receptor and a ligand that binds this receptor, it appears that the CLV gene products together form a signal transduction pathway that limits the expression region of WUS. Thus the interaction between CLV and WUS maintains stem cell function and the maintenance of the meristem as a source of cells for the shoot.

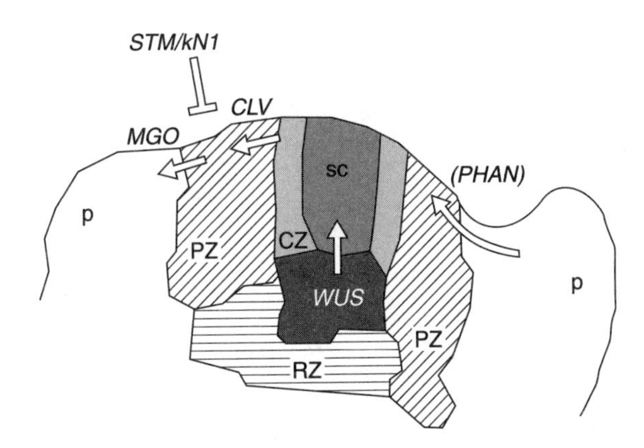

Figure 4.5. The shoot apical meristem. Schematic of a shoot apical meristem showing the central (CZ), peripheral (PZ), and rib meristem (RZ) zones. Proteins involved in meristem development are also shown. [Reprinted from Lenhard and Laux (1999), with permission from Elsevier Science Ltd.]

Other SAM regulatory genes are known to be expressed in the SAM. The *shoot meristemless* (*Stm*) homeodomain transcription factor gene is required for normal SAM function, as *Arabidopsis stm* mutants lack a functional meristem. Further, transgenic tobacco plants expressing an extra copy of the corn *KN1* (STM-related) gene develop superficial SAMs on leaves, suggesting strongly that *KN1* expression directs SAM formation. The *Mgo* genes also play a role in SAM function. The *mgo1* and *mgo2* mutants contain disorganized SAMs and fewer leaves 10 days after germination, suggesting that the SAMs of the mutants delegate fewer cells to the leaf primordia. The *Mgo* genes encode proteins similar to asymmetric cell division regulators in animal cells, suggesting a key role for the MGO proteins in meristematic cell divisions. Finally, the *Phantastica* (*Phan*) genes help specify adaxial leaf identity, and thus are involved in leaf primoridia differentiation (Traas and Bohn-Courseau 2005; Shani et al. 2006).

4.3.2. Root Apical Meristem and Root Development

Root development is illustrated in Figure 4.6. Organization of the root apical meristem (RAM) involves fewer cells than does development of the SAM. The basic organization of the SAM and RAM are similar in terms of having a central region of slowly dividing cells surrounded by cells with a higher cell division rate. Recall that specification of the RAM occurs during the embryonic heart stage. Thus at the heart stage the radial organization of tissues is in place and the RAM initials and central cells that will generate and maintain the root in the seedling are specified. The *quiescent center* (QC) is the region of slowly dividing cells within the RAM. The QC is involved in RAM activity and maintenance. In bindweed (*Convolvus arvense*) the QC cell cycle is 430 h, whereas in other cells it is about 13 h. Therefore, the QC must be viewed as the ultimate source for new cells, but not the factory that produces them.

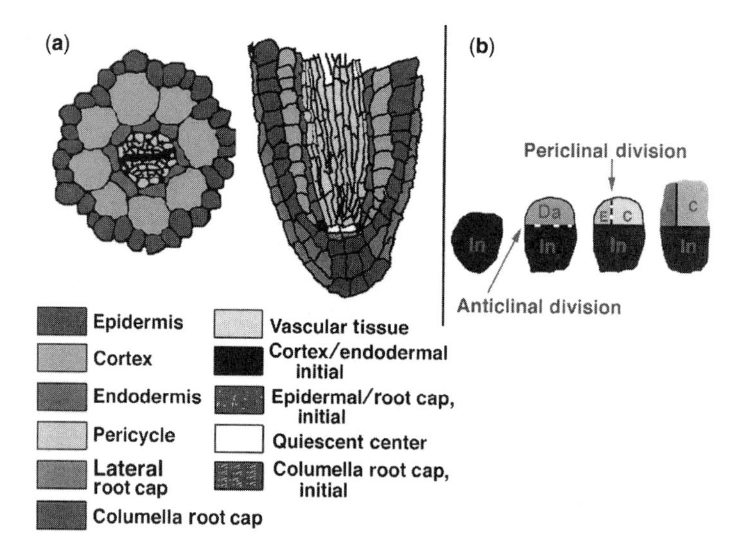

Figure 4.6. Root development. Arrangement (a) and division plane (b) of cell types within the developing root. [Reprinted from Di Laurenzio et al. (1996), with permission from Cell Press.]

Below the QC lies the *columnella root cap initial* cells, which give rise to the root cap, a protective structure. Above the QC are the epidermal initials, which will form the epidermis and lateral root cap. The cortical and endodermal initial cells give rise to the cortex and endodermis; the final layer is the vascular tissue. The initial cell that gives rise to either endoderm or cortex divides anticlinally once and then periclinally once before these identities are laid down. The portion of the root enclosed by the endodermis is often referred to as the *stele*.

Cell divisions from these initial cells follows a strict pattern of progressive differentiation resulting in an expansion (elongation zone) and a differentiation (maturation zone) to build a regular arrangement of cell files within the root body. It is not surprising that expression domains of regulatory genes are responsible for cell fate patterning in the RAM. For example, the *short root* (*Shr*) and *scarecrow* (*Scr*) genes help specify the endodermis and cortical identities of cells, respectively. SHR and SCR proteins function in a novel signaling pathway to determine radial patterning in the root. The SHR protein is translated in the stele and then moves to the adjacent cell layer, where it activates SCR transcription and initiates endodermal specification. The SCR protein is then thought to regulate the asymmetric cell division that results in the formation of cortex and endodermis.

The plant hormone *auxin*, or indole acetic acid, is required for formation of the embryonic root, lateral roots, and maintenance of the cellular organization around the initials of the seedling root. Auxin moves through the plant from the shoot, where it is synthesized, to the root using a system of influx and efflux carriers localized asymmetrically in the cells of the vascular tissues. It has been shown that the family of auxin transporters encoded by the *Pin* genes are the auxin efflux carriers and that PIN1 localization becomes progressively polarized in developing embryos. By the globular stage, PIN expression is confined to the basal portion of the embryo, and as embryogenesis proceeds, PIN becomes further localized to the developing vasculature. The effects of auxin on root patterning can be visualized in transgenic plants containing five copies of an auxin responsive gene promoter element to drive expression of the GUS (β-glucoronidase) reporter gene. When expressed in transgenic *Arabidopsis*, one can visualize auxin content by utilizing an assay that detects GUS activity. The results show that there exists an expression maxima in the root initial cells, supporting the role of auxin in root patterning. Root meristems are the focus of much research (Campilho et al. 2006; Costa and Dolan 2000).

The formation of lateral and adventitious roots also requires auxin. Lateral or secondary roots originate from the percicyle, a specific cell type contained within the stele of the root. Cells within the pericycle undergo cell division, and then further cell division and cell expansion results in the formation of a lateral root. These cells begin cell division in response to auxin and environmental cues and must establish a connection to the vascular trace of the primary root. Adventitious roots can develop from the stems of some plants when placed under inducing conditions. Tomato, for example, can develop many adventitious roots from a cut stem when placed under humid conditions.

Root hairs are another type of cell contributing to the overall root function of absorption of water and minerals. The outer, epidermal layer of the root gives rise to root hairs. Root hair formation occurs within a specific region of the root, a short distance above the region of root elongation. Root hairs are short and short-lived and develop on both primary and secondary roots. Interestingly, a root hair is a single cell that consists of a thin cell wall, a thin lining of cytoplasm that contains the nucleus, and a large vacuole-containing cell sap.

4.4. LEAF DEVELOPMENT

4.4.1. Leaf Structure

Leaves are specialized structures responsible for most of the photosynthesis that takes place in the plant, as well as functioning in respiration and transpiration. Leaves are initiated as primordia from the shoot apical meristem as described earlier. As leaf primordia are specified by gene such as the *Phan* gene, the abaxial (or top) and adaxial (bottom) surfaces develop (Fig. 4.7a). Recall that leaves differ from the cotyledons in several ways including the presence of the single-celled trichomes, or leaf hairs that function in the secretion of various compounds that can attract or repel insects (Fig. 4.7b).

A cross section of a mature leaf shows the main cell types in the leaf (Fig. 4.7c). The outer epidermal cell layers are derived from the L1 layer of the SAM in both monocots and dicots and do not contain chloroplasts. The exception to this is the *stomatal pore*, which is created from two guard cells that contain a specific number of chloroplasts, depending on the ploidy of the plant. The interior leaf cells are filled with chloroplasts that will autofluoresce when viewed under a fluorescent microscope. Dicot leaves have a distinct dorsiventrality with an upper (adaxial) layer of oblong palisade cells, and a lower (abaxial) layer of spongy mesophyll cells (both are derived from the L2 layer of

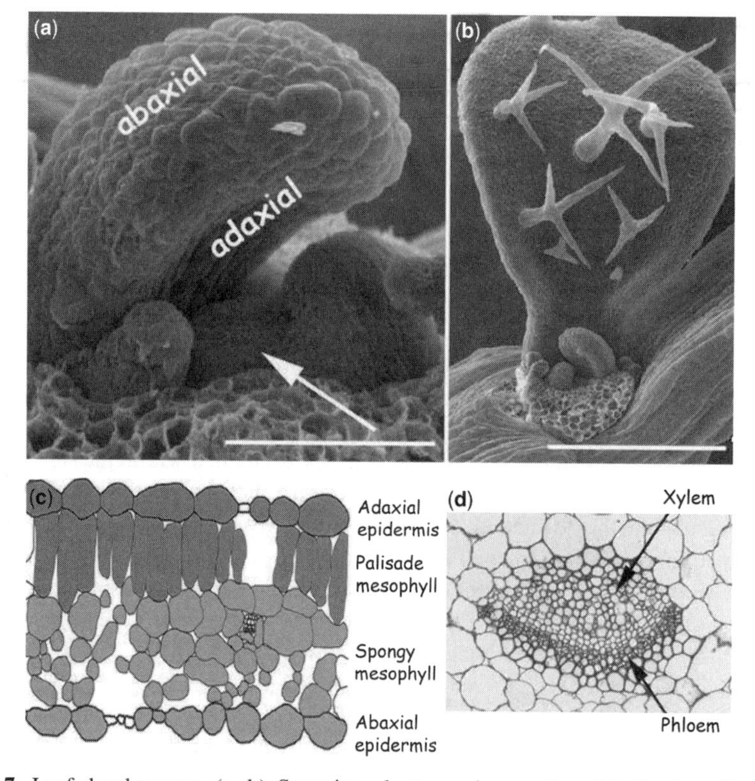

Figure 4.7. Leaf development. (a, b) Scanning electron micrographs of leaf primordia. Note the presence of trichomes in (b). (c) Schematic of leaf cross section showing the different leaf cell types. (d) Cross section through leaf vascular tissue. [Reprinted from Byrne (2006), with permission from the Public Library of Science.]

the SAM). Vascular bundles containing xylem and phloem are present in the middle or L3-derived layer. Monocot leaves vary but all contain a single photosynthetic cell type, the mesophyll, and a specialized bundle sheath surrounding the vascular tissue. There are many specializations of leaves, such as in xerophytic leaves, which are adapted to dry conditions, which contain different cell types and arrangements of these cell types (Byrne 2006).

Mature leaves are often surrounded by a waxy, cuticle layer that provides protection and prevents water loss. The epidermal cells secrete this layer and themselves provide protection to the internal tissues. Since epidermal cells do not contain chloroplasts, they are essentially colorless and facilitate the focusing of light to the active, photosynthetic mesophyll and palisade cells below. The stomatal pores present in the epidermis allow for gas exchange in photosynthesis and respiration, and are controlled by discrete signal transduction pathways that involve ABA, calcium, phosphatidic acid, and inositol-containing second messengers. These signal transduction components are thought to eventually alter ion channel activities that allow the guard cells to increase turgor, thus opening the stomatal pore, or to decrease turgor, which results in stomatal closure. Thus, in addition to its role in seed dormancy, ABA is also considered the drought-sensing hormone as its signal transduction pathway can allow for stomatal closure, an important response to drought that conserves water lost through transpiration.

It is interesting to note that most leaves contain more stomatal pores on their abaxial surface than their adaxial surface. This location places them closer to the spongy mesophyll. Indeed, the mesophyll layer within the leaf is the major site of photosynthesis in the plant, and contains two cell types in dicots: the spongy mesophyll and the palisade mesophyll cells. Both cell types are active in photosynthesis, yet have different shapes. It is thought that the oblong shape of the palisade cells helps to further focus light on the spongy mesophyll cells. The gaps around spongy mesophyll are another adaptation that accommodates the oxygen generated from photosynthesis.

4.4.2. Leaf Development Patterns

Besides photosynthesis, there are several interesting developmental considerations for leaves. Leaf primordia first arise when a small group of cells on the outer edge of the SAM gain leaf identity. These leaf primordia mature into a leaf bud utilizing a marginal meristem to form the lamina or outer edge of the leaf, and a central meristem that gives rise to the vascular tissue. Leaf buds can remain dormant in plants such as trees. Cell division within the leaf bud occurs at the base of the primordia or leaf, which means that cells are pushed up toward the tip of the growing leaf. Along with cell division, cell expansion is a critical process that produces large increases in leaf size. In general, cell expansion starts after cell division has given rise to the main structure of the leaf. Thus, the younger the leaf, the more active it is in cell division. Almost all mutants defective in the production of leaves are also affected in the SAM, containing an under- or overcommitment to leaf primordia cells. Another interesting characteristic of leaves is their placement on the plant, which is called *phyllotaxy*. Leaves are initiated in a precise pattern as the shoot meristem grows, producing either alternate, opposite, tricussate (whorled), or spiral arrangements. In many species, the number and position of leaves, or modified leaves such as the spines of a pineapple fruit, follow the Fibonacci number series (1,2,3,5,8,13, . . .). The venation pattern of leaves also varies with monocots containing parallel veination, while most dicot leaves have a reticulate pattern.

The shape of leaves is a very noticeable trait. Leaf shape is controlled by environmental and genetic programs as well as hormones. Some species such as tomato contain compound leaves, while others such as *Arabidopsis* contain simple, nonlobed leaves. Cell death within leaf primordia in plants such as philodendron produce "holes" in leaves. Some leaves such as pea also contain tendrils that function in "grasping" surrounding structures in the environment and facilitate directional growth. Corn leaves contain specialized domains called the *sheath*, *blade*, and *ligule*, which also facilitate growth by providing a way to change the position of the leaf surface, ensuring that photosynthetic tissues get maximal exposure to light.

Maize has been especially useful as a model plant to study leaf development. The *knotted* (*KN1*) gene, which is the related to the *shoot meristemless gene* (*stm*), mentioned in the discussion on shoot apical meristem development, was first identified in corn mutants that contained knots of tissue on their leaves. These *KN1* mutants were defective in the normal regulation of the *KN1* gene, which would normally be confined to the apical meristem. Instead, *KN1* mutants contain *KN1* expression in the leaves, which results in an aberrant mass or knot of tissue. The corn *KN1* gene was ectopically expressed in transgenic tomato plants to investigate the role of this homeodomain transcription factor in dicot leaf development. The results were transgenic tomato plants containing an increase in leaf complexity. Recall that most tomato species contain compound leaves with several leaflets. Ectopic expression of the corn *KN1* gene caused a large increase in the number of leaflets per leaf, suggesting that in dicots, *KN1* can alter leaf complexity specification (Fleming 2006).

4.5. FLOWER DEVELOPMENT

4.5.1. Floral Evocation

Flowers are plant's most obvious and aesthetically pleasing organ. In general, all flowers are specified in a similar manner. For flower development to occur, vegetative meristems must first undergo a transition to produce the *inflorescence meristem*. These meristems are self-renewing and also give rise to the floral meristems that produce flowers. The term *floral evocation* refers to the process of inflorescence meristem commitment. This is controlled by many factors, including plant size, whether a cold season has passed (vernalization), environmental stress, and daylength. For example, short-day plants such as cocklebur and Christmas cactus require a minimum light period (<15 h) to flower. Only one inductive period of light is needed to block flowering in many short-day plants. In contrast, long-day plants, such as *Arabidopsis*, require a longer period of light (usually 12–16 h) to flower. *Arabidopsis* is also considered to be a long-day facultative plant, as it can flower in short-day conditions but will flower much faster if placed under long-day conditions. Daylength-neutral plants, such as tomato, are not as affected by the photoperiod.

After floral evocation has taken place, a plant can be moved to noninductive conditions and still flower. Many historical studies have suggested that a hormonal factor, termed *florigen*, is produced elsewhere in the plant, such as the leaves, and then stimulates floral evocation in the meristem. Trying to determine the identity of florigen has been a focus in plant biology for years because of its importance in agriculture. Flowers are the precursor of fruit, and if flowering can be controlled, plants can be manipulated to remain in a vegetative or flowering state. Accelerated flowering could lead to a much shorter growing season, which would be advantageous for growers.

Not surprisingly, there are mutants defective in floral evocation, and their study helps us understand some of the molecular requirements for floral evocation. The *Constans* (*CO*) gene from *Arabidopsis* encodes a *zinc-finger transcription factor* whose mRNA levels rise and fall with a circadian rhythm. *CO* turns on a number of genes, including *FLOWERING LOCUS T* (*FT*), a gene known to also be involved in floral evocation. The FT protein binds to and activates other transcriptional regulators such as FD and LEAFY in the nucleus of the meristem cells. FD and LEAFY are considered to be master switches that "turn on" expression of genes needed for flowering. Thus, CO protein accumulation, controlled by the circadian rhythm, make trigger a cascade of events that results in flowering. More recent studies also indicate that increased CO protein expressed only in the leaves of transgenic plants can stimulate early flowering in *Arabidopsis*. As mentioned, LEAFY is a transcription factor involved in the switch from the inflorescence to floral meristem. *Leafy* mutants have a delay in floral meristem development and flowers are replaced by leaflike or flowerlike shoots, suggesting that the function of LEAFY is to promote floral meristem identity. Indeed, ectopic expression of LEAFY in transgenic aspen trees can speed up the flowering process in these trees, presumably by promoting floral meristem identity. Another important floral meristem mutant containing the opposite phenotype is the *terminal flower* (*tfl*) mutant. These *tfl* mutants flower early and have a determinate inflorescence which means that the inflorescence meristem is transformed into a terminal flower. Thus the function of the TFL protein is to promote inflorescence identity (Krizek and Fletcher 2005; Bernier and Perilleux 2005; Corbesier and Coupland 2006).

4.5.2. Floral Organ Identity and the ABC Model

After floral evocation and development of a floral meristem committed to the process of flowering, the individual organs present in the flower develop. A flower consists of four concentric whorls containing flower organs that in most dicots like *Arabidopsis* are arranged this way: *sepals* (Se), *petals* (P), *stamens* (St), and *carpels* (C) (Fig. 4.8). Sometimes, one of the whorls is not well developed or is repeated (like the petals in a tea rose), or sometimes one whorl is dominant so that the rest of the organs are not noticeable. On closer inspection, however, one can usually distinguish the four types of organs.

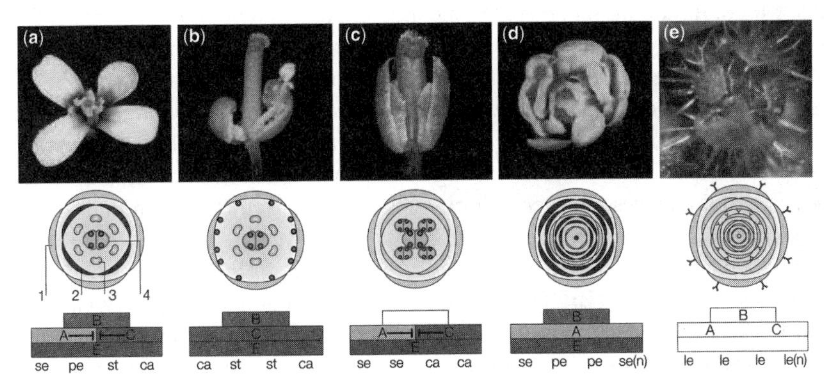

Figure 4.8. Flower development. *Arabidopsis* (a) wild-type, (b) *ap2*, (c) *pi*, (d) *ag*, and (e) *sep* flowers. Below each photo is a rendering of the ABC model as it functions in that flower. [Reprinted from Krizek and Fletcher (2005), with permission from Nature Publishing.] See color insert.

The specification of floral organ identity begins during floral evocation, for example, when the LEAFY protein acts to turn on gene expression. We have learned the most about floral organ identity from *Arabidopsis* homeotic mutants. Floral homeotic mutants were isolated that contain a transformation of one organ into another. To understand these mutants and the resulting ABC model of floral organ identity genes, one must be familiar with the normal arrangement of organs in the *Arabidopsis* flower (Fig. 4.8). This flower contains an outer whorl of four green sepals, four white petals, four to five yellow stamens, and two fused carpels. *Agamous* (*ag*) mutants are homeotic mutants that are very striking and contain an outer whorl of sepals, followed by petals, and then sepals again. Comparison of *ag* flowers to wild-type flowers indicates that *ag* mutants have lost information required to make stamens and carpels in whorls 3 and 4, and have replaced this with petals and sepals, respectively. Mutants in *ag* also contain a reiteration of this pattern resulting in an indeterminate meristem and extra rows of petals and sepals. This finding indicates that AG function is required for whorls 3 and 4 (stamen and carpel) identity. In contrast to AG, the *apetela2* (*ap2*) mutants have sepals transformed into carpels in the first whorl, and petals transformed into stamens in the second whorl, followed by stamens and carpels in the next two whorls as usual. This indicates that AP2 is required for identity of whorls 1 and 2 (sepals and petals). Finally, two different mutants with the same phenotype, the *pistillata* (*pi*) or *apetela3* (*ap3*) mutants, contain a transformation of petals to sepals in the second whorl, and of stamens to carpels in the third whorl. This indicates that the PI and AP3 proteins function in identity of whorls 3 and 4.

Together, results from these homeotic mutants suggest that three separate types of genes (denoted A, B, and C), function in floral organ identity (Fig. 4.8). The A function is controlled by the AP2 gene product and must be required for both sepals and petals in whorls 1 and 2. AP3 and PI are gene products with a B function and are required in whorls two and three to help specify petals and stamens. Finally, the C function is controlled by AG, which helps specify whorl 3 (stamens) and whorl 4 (carpels). Important to this model is the antagonism of A/C function, such that if one is lost, the other expands its function into the two whorls where it would not normally function. Another caveat is that B function must necessarily be present in combination with either A or C to specify the petals and stamens. By drawing out each mutant's observed pattern, one can see that the mutant data "fit" exactly to this model.

This elegant model can also be used to predict the phenotypes of double and triple mutants, which, for the most part, verify the model. For example, if both A and B functions are lost, this model predicts that C function will expand to all four whorls, and that carpels should be present in each whorl. The resulting double mutant is found to contain a leaflike structure in whorl 1, carpelloid leaves in whorl 2, and carpels in whorls 3 and 4, a close approximation of what the model predicts. A triple mutant that has lost A, B, and C functions is predicted to contain no floral organ identity. The observed mutant is found to contain carpelloid leaves in each whorl, which suggests that the ground state of the flower is not totally vegetative (i.e., leaflike).

A new dimension to the ABC model has recently been discovered that involves a group of four genes, called *Sepellata* (*Sep*) genes, which are required to specify each whorl in addition to the ABC genes. Loss of this E function through a quadruple mutant lacking all four genes results in whorls of carpelloid leaves, similar to the mutant lacking ABC function.

Thus, our understanding of flower development starts with CO and LEAFY transcriptional function to begin the developmental program and results in the production of

AP2, PI, AP3, AG, and SEP proteins. How do these proteins function to specify floral organs? AP3 and AG encode MADS box genes, a family of transcription factors expressed in yeast and plants that most likely function by turning on other, specific genes required to build a sepal, petal, stamen, or carpel. The ABC model predicts that expression of these genes should be confined to the specific whorls where they function. This prediction has been verified by observing the in situ mRNA expression patterns of the genes. For example, AP2 is expressed early in whorls 1 and 2.

It is important to note that several homeotic genes controlling floral development have been isolated from other plants, including *Antirrhinum* (snapdragon), supporting the importance of the ABC model in other species. For example, the *Antirrhinum deficiens* (*DefA*) gene probably functions similarly to AP3 from *Arabidopsis* (Krizek and Fletcher 2005).

4.6. HORMONE PHYSIOLOGY AND SIGNAL TRANSDUCTION

4.6.1. Seven Plant Hormones and Their Actions

Signal transduction is the cascade of events that allow a signal, usually from outside the cell, to be interpreted by the cell. Signal transduction cascades usually result in a final biological response, and often the response can be measured. Besides light and abiotic stress, the plant hormones are the major developmental and physiological signaling molecules in the plant. The seven major plant growth hormones are small molecules rather than proteins or peptides, and in some cases they are similar to certain animal cell hormones (Fig. 4.9). For example, brassinolide (Br) is a sterol, much like estrogen and testosterone, which function as sex hormones in animals. Br is critical for normal plant growth and development in plants, playing a role in stem elongation, leaf development, pollen tube growth, vascular differentiation, seed germination, photomorphogenesis, and stress responses.

Auxin, or indole 3-acetic acid, was the first plant hormone discovered and contains an indole ring much like the melatonin hormone of animals. Auxin is known to stimulate cell elongation and cell division, differentiation of vascular tissues, root initiation and lateral root development. Auxin can also mediate the bending responses to light and gravity, and within the apical bud it suppresses the growth of lateral or axillary meristems. It can delay senescence, and interfere with leaf and fruit abscission. It can induce fruit setting and delay ripening in some fruits. It can also stimulate the production of another plant hormone, *ethylene*.

Cytokinin is generally considered the second most important plant growth-regulating hormone, following auxin. Cytokinin is similar to adenine and was first discovered in 1941 as the active component in coconut milk that promoted growth of plant cells in tissue culture. Cytokinin can promote cell division and shoot growth and can delay senescence.

Abscisic acid (ABA) was first identified in a search for an abscission-promoting hormone. This is not the function of ABA, and as noted earlier, it functions in promoting dormancy and in sensing drought and other stresses. ABA is derived from mevalonic acid and carotenoids and is thus similar in structure to the developmental factor from animals called *retinoic acid*. Transport of ABA can occur in the vascular tissues. ABA stimulates closure of the stomatal pore, and can inhibit shoot growth. In seeds, it promotes dormancy and stimulates the production of seed storage proteins. It is mostly antagonistic to *gibberellic acid* (GA) and can inhibit the response of grains to GA. ABA is also involved in

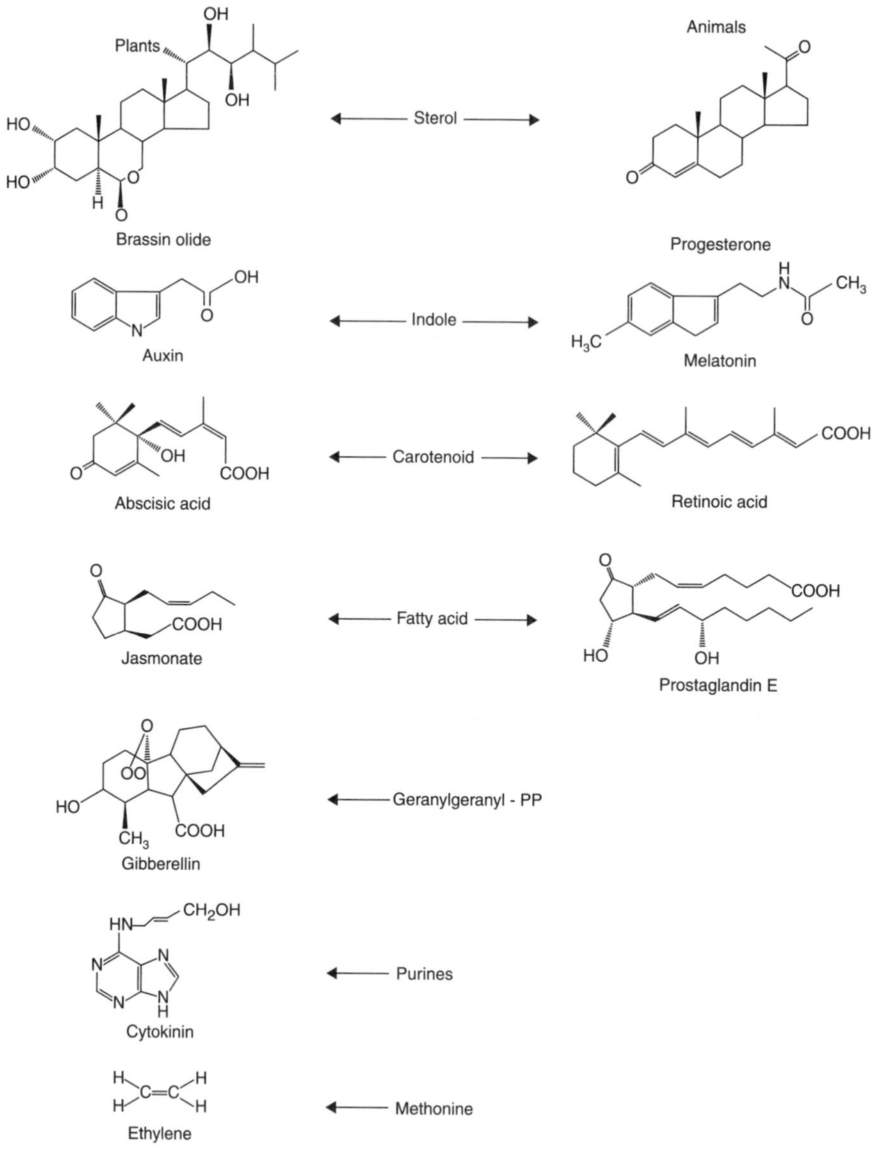

Plants

Animals

Brassin olide ← Sterol → Progesterone

Auxin ← Indole → Melatonin

Abscisic acid ← Carotenoid → Retinoic acid

Jasmonate ← Fatty acid → Prostaglandin E

Gibberellin ← Geranylgeranyl - PP

Cytokinin ← Purines

Ethylene ← Methonine

Figure 4.9. Plant hormones. Similarities between some plant and animal hormones. [Reprinted from Chow and McCourt (2006), with permission from Cold Spring Harbor Laboratory Press.]

inducing gene transcription in response to wounding, which may explain why it has a role in the pathogen defense response.

Jasmonic acid (JA) is a fatty-acid-derived plant hormone that is similar in overall structure to physiologically active small molecules from animals called *prostaglandins*. In plants, jasmonic acid is firmly associated with pathogen defense pathways. For example, it has been documented that the physical stimuli of certain insects can trigger the synthesis of jasmonic acid, which then functions to increase expression of genes involved in defending the plant, such as the *pathogenesis-related 1* (*Pr1*) gene. Microbial and viral pathogens can also trigger JA synthesis, thus the study of JA-mediated events in the plant cell are of interest to plant pathologists who wish to engineer transgenic plants that are disease-resistant.

Gibberellic acid (GA) and ethylene are two plant hormones with no similar molecular counterparts in other eukaryotic organisms. GA was first discovered from fungi that can stimulate plant cell elongation and cause significant and "leggy" growth of rice plants. GA is a series of 136 diterpene compounds that contain 19 or 20 carbons in four or five ring systems. These are named for the order in which they were discovered (GA1, GA2, etc.). The other functions of GA, as mentioned previously, are in general antagonistic to the actions of ABA. For example, ABA promotes seed dormancy, while GA is required in most cases to break seed dormancy. The actions of GA on barley germination have been well studied where it has been shown that GA promotes expression of the α-amylase genes required to break down starch in barley aleurone, an important process in the grain-malting business. GA also plays a prominent role in stimulating flower development under long days.

Ethylene, a hydrocarbon gas, is a very simple molecule that is best known for its stimulation of fruit ripening and promotion of the seedling triple response. Indeed, people of ancient cultures understood the actions of ethylene and could burn incense in a closed room to stimulate fruit ripening. The triple response of seedlings is a specific developmental program wherein an apical hook forms in the shoot, and the root becomes thicker. These adaptations may increase survival under certain conditions. In addition, ethylene can stimulate the release of dormancy, adventitious root formation, flower opening, and flower and leaf senescence.

4.6.2. Plant Hormone Signal Transduction

The first eukaryotic *signal transduction* pathways to be characterized were the peptide growth hormone pathways of animal cells. This most likely resulted from the discovery that animal oncogenes sometimes encoded altered growth factors, growth factor receptors, or other signal transduction components that regulate cell growth. A paradigm signal transduction pathway is shown in Figure 4.10 to facilitate understanding of how signal transduction works. Signals from outside a cell can be perceived, sometimes by receptors that span the plasma membrane. After stimulation of such receptors, information can be relayed by a series of small molecules or proteins to the cell nucleus, where activation of specific transcription factors can stimulate new gene expression programs. The resulting gene expression results in the production of new proteins that can function in the final biological responses to the signal.

Because plant hormones are small molecules rather than proteins, and because the plant cell wall encloses the plasma membrane, it was suggested that plant hormone signal transduction pathways would be significantly different from those of animals. While this is true in general, it is important to keep in mind that most, if not all, of the individual components of plant signal transduction pathways have similar counterparts in other eukaryotes. Plant receptors linked to plant hormone action were not discovered until the 1990s. The accelerated pace of experimentation that followed resulted in three major paradigms of plant hormone signal transduction that will no doubt be joined by other types of pathways in the future (Chow and McCourt 2006; Gibson 2004).

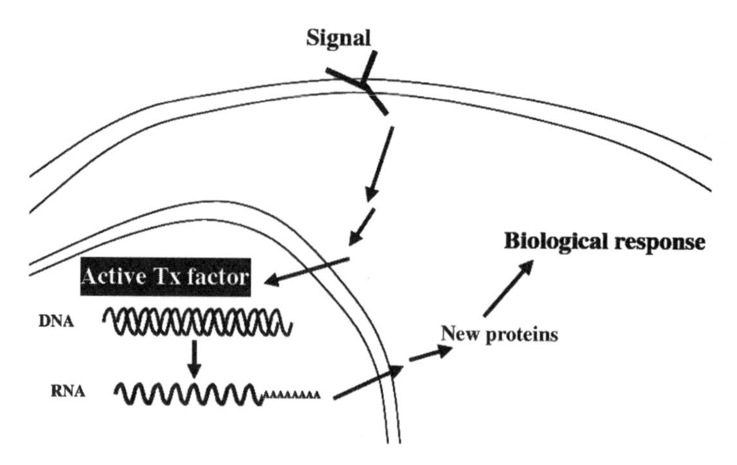

Figure 4.10. A paradigm signal transduction pathway. Signals from the outside of cells can be perceived by receptors or other proteins present at or near the plasma membrane. Once activated, these receptors can transmit signaling information (arrows) to the interior of the cell. Many signal transduction pathways converge on the stimulation of gene expression within the nucleus which results in the production of new proteins in the cytoplasm that can affect specific biological responses.

4.6.2.1. Auxin and GA Signaling. When auxin acts to promote cell division and growth, it does so mainly by increasing the expression of genes that encode required proteins for these processes. Thus, researchers have sought to understand the steps between auxin perception and the final gene expression regulation. We now know that auxin signaling involves ubiquitin-mediated protein turnover as way to control transcription of genes that allow the plant to effect a response to auxin. Molecular studies revealed the first players in auxin signaling as a group of genes encoding the IAA/AUX proteins whose expression is rapidly upregulated in response to auxin within minutes. Most of the IAA/AUX proteins are nuclear-localized and have a very short half-life. They can form heterodimers with the *auxin response factor* transcriptional regulators (ARFs), and then bind to a 6bp (6-base-pair) auxin-responsive element (ARE) present in the promoters of auxin-regulated genes. Further studies revealed that ARF:ARF homodimers were responsible for activation of gene expression in response to auxin, while ARF:AUX/IAA heterodimers blocked transcriptional activation (Quint and Gray 2006).

Genetic mutants that failed to respond to auxin in seedling growth assays identified genes that are required for some of the plant's responses to auxin. These genes, which include the *axr1* and *tir1* genes, encode proteins that function in the ubiquitin-mediated protein turnover pathway in eukaryotes. The proteasome is a large, macromolecular structure that functions to degrade proteins within the cell. Proteins destined for the 26*S* proteasome are modified by the addition of ubiquitin, itself a small protein (76 amino acids). Thus, auxin signaling requires a functioning 26*S* proteasome and enzymes necessary to add ubiquitin to target proteins.

It has been shown that the F-box protein encoded by TIR1 becomes physically associated with auxin, and thus may function as an auxin receptor. After binding to auxin, TIR1 may stimulate the proteasome to specifically degrade some of the IAA/AUX proteins. Once the IAA/AUX proteins are degraded, ARF:ARF homodimers form, bind the AREs in promoters of auxin-responsive cells, and stimulate the transcription of these genes. In this way, auxin can stimulate expression of genes required to carry out its physiological effects.

Several parts of the auxin signal transduction pathway are still not understood, and much research is underway to delineate the pathway. Interestingly, proteasomal degradation of a transcriptional repressor may be a common theme in plant hormone signal transduction pathways. Such a repressor, containing a DELLA protein domain, represses GA-regulated genes and is degraded by the proteasome after GA addition to plant cells. Thus, GA signaling may share the same general regulation in stimulating expression of genes required for the physiological responses to GA. The GA signal transduction pathway also has an identified protein, GID1, which may be the GA receptor and function as the initial step in GA perception. GID1 is a nuclear and cytosol-located protein that is homologous to the animal hormone-sensitive lipases and that binds to different GAs with saturable kinetics. This last fact is an important test that helps support the idea that a protein directly and specifically interacts with a hormone (Pimenta-Lange and Lange 2006).

4.6.2.2. Cytokinin and Ethylene Signaling. Plant cells utilize elements of the *two-component signaling pathways* in their responses to cytokinin and ethylene. The two-component systems function in microbes, yeast, and plants to convey signals between a histidine kinase receiver and a phosphorylated response receiver. These two components are joined by an intermediate in plant cells termed the *phosphorelay intermediate*. Both cytokinin and ethylene have been shown to bind to specific histidine kinases contained in the plasma membrane. This binding is thought to stimulate a phosphorylation cascade wherein the activated histidine kinase phosphorylates an intermediate protein, which then phosphorylates a specific aspartate residue on the response receiver. The response receiver must act to stimulate downstream functions that are currently uncharacterized for cytokinin signaling.

In ethylene signaling, downstream targets of the two-component signaling system have been identified, mainly through the genetic isolation of mutants altered in their responses to ethylene application. These genetic screens identified the CTR kinase that functions as a negative regulator of ethylene signaling, and the EIN and ERF proteins that function as transcriptional activators of ethylene-regulated genes. In a scenario that seems reminiscent of auxin and GA-mediated signaling, the EIN3 transcriptional activator is subjected to proteasomal degradation in the absence of ethylene. This fact implies that one critical step in ethylene perception is the increased stability of EIN3 that allows for new transcription of ethylene regulated genes.

4.6.2.3. Brassinosteroid Signal Transduction. The brassinosteriod, brassinolide (Br), is the last example of plant hormone signaling that we will consider. As was carried out for the other plant hormones, genetic mutant screens were performed to find Br-insensitive mutants. The Bri gene was identified and shown to be required for seedling responses to exogenously added Br. The BRI protein encodes a leucine-rich repeat (LRR)-containing serine/threonine protein kinase. This fact is important since these types of signaling kinases are abundant in animal cells and often serve as receptors for animal peptide hormones such as insulin. The BRI protein is predicted to span the plant cell plasma membrane, making the LRR domain accessible to the outside of the plant cell, with the kinase domain contained on the interior of the cell. This arrangement led to an integral domain-swapping experiment between the BRI protein and the XA1 protein that confers resistance to rice blast fungus. Researchers produced transgenic plants containing the outside LRR domain from BR1 and the interior kinase portion of XA1. The resulting plants could be stimulated with Br to turn on disease resistance pathways, cleverly showing that each part of these receptors is specific and can function when swapped.

Activation of the BRI receptor is thought to stimulate other protein kinases and phosphatases that help relay information to the nucleus where the BES1 transcriptional activator acts to regulate gene expression. In this way the Br signal transduction pathway is similar to other signaling pathways we have examined, with a final nuclear transcriptional output required for the final biological response to Br. It is interesting to note that there are over 170 genes in *Arabidopsis* predicted to encode LRR kinases that may function in plant signal transduction. In addition, there are several other putative receptor-like kinases that do not contain LRR domains, but also may function in signaling.

4.7. CONCLUSIONS

Even though plants do not have elaborate body plans and the number of specialized cells as organs compared with animals, developmental programs are no less elaborate. A number of crucial plant growth regulators or hormones are required for proper plant growth. We will see in the next chapter how plant biotechnologists can alter hormone type and concentration to manipulate cells in petri dishes, a requirement for plant genetic transformation.

LIFE BOX 4.1. NATASHA RAIKHEL

Natasha Raikhel, University Distinguished Professor, Ernst and Helen Leibacher Endowed Chair, University of California Riverside

Natasha Raikhel

I originated from and grew up in the Soviet Union. I immigrated with my husband and first born son to Athens, Georgia in 1978 (my second son was born in Athens, Georgia) with a personal fortune of only $25. I remember feeling somewhat lost and wondered how I could and would ever make the language, scientific and social transitions required of me. I did not realize at the time that I was lucky in many ways and that fortune had favored me.

I knew only one American scientist when I first arrived, but I encountered many helpful people that were critical to my survival. I also entered a social context within academia that differed in several important ways from the system I left behind. The American academic system is characterized by greater diversity and openness of thought and a healthy atmosphere of competition that drives one to take intellectual risks and achieve more. At its best, this environment also leads to a constant renewal of possibility, a wealth of new ideas and a rich milieu of thoughtful exchange that fosters both collective and individual progress. In America, I found a place where prestige and intellectual and economic rewards were all reasonable potential goals. Although I did not find the streets paved with gold, I actually found the far greater treasure of opportunity.

What I achieved was also due to timing. I am a product of this age of molecular biology and its corollary age of rapidly expanding knowledge bases and burgeoning information systems that our technological growth has made possible. This lucky moment in history has allowed all of us the privilege to be pioneers of new and fascinating frontiers. When I came to this country, molecular approaches in plant biology were just beginning. I did not have to catch up, because I learned along with many people who were also just beginners in molecular biology. So, once again, I was lucky with good timing.

I am a cell biologist working with plants. I am fascinated by plants: we live on this planet because of plants and I want to unlock secrets of plant cell biology. In my laboratory we are using a model plant, *Arabidopsis*. This plant has a small genome and has been sequenced with many of its proteins identified. It is, therefore, a very convenient model organism for studying processes that are important in all plants including crop plants. My group has worked on the trafficking of molecules through the cell's vesicles and vacuoles and we are interested in the synthesis of the cell wall in plants. A cell contains compartments called organelles. Compartments in cells are necessary to isolate and secure a large number of molecules that play an individual role(s) in various functions of the cell. In order for cells to function properly, molecules have to be produced and delivered to their proper destinations within the cell. Because plants are immobile and cannot run, they have to be very versatile in their ability to respond to environmental stresses and survive. Therefore, plant cells have evolved a highly complex organization of functions to sustain life. The failure of any of these functions could poison other dynamic processes occurring within the intracellular environment and actually cause the destruction of the entire cell. Alternatively, improving the success rate of sending novel proteins and carbohydrates to desired parts of the cell can result in the improved nutritional value of crops and increased biomass production.

We live in an era of unprecedented biological discovery. Technologies to sequence entire genetic codes have yielded a wealth of data that require a focused interdisciplinary approach to assimilate and exploit this information. Once we understand the functions of all gene products (proteins), how they interact and how pathways in the cell interact, we can really start to answer questions about how cells function and how the whole plant works. We call this new science "systems biology." The essence of systems biology is to model organisms and predict how various pathways in the organism interact when one pathway is affected. This requires the infusion of plant biology with disciplines such as mathematics, statistics, informatics, chemistry and engineering.

It is very important that the new generation of plant biologists have multidisciplinary experience and training. I think that the community of *Arabidopsis* researchers will make the systems approach work because they are exemplary forward thinkers, effective trainers and extremely open in sharing knowledge and tools. I hope that many talented young students are drawn to plant biology. Our field allows young people to reach for the stars and grow to the best of their potential.

I have tried, as I built a career as an American scientist, to foster and mentor those who will carry our field on into the future, to be persistent in the pursuit of worthy goals and to change and learn new things when this is necessary. Although the research in my own group is extremely important to me, I have realized that I have experience that enables me to do more for the scientific community. Lately, my career has shifted somewhat from building a personal reputation towards accepting the responsibility of leadership within our field. But leadership does not occur in isolation. We all lead and follow within a group, hopefully as a team. In his essay, *Tradition and Individual Talent*, the poet T. S. Eliot says that no artist has his complete meaning alone. I would expand that thought to include today's scientist, who also cannot have his or her complete meaning alone. It is the American context, which at its best, celebrates diversity, the acceptance of new ideas and the ever present possibility to start again and create a wonderful life.

LIFE BOX 4.2. DEBORAH DELMER

Deborah Delmer, Professor Emeritus UC Davis; Rockefeller Foundation (retired); Winner of the ACS Anselme Payen Award; Member of the National Academy of Sciences

Deborah Delmer

I must confess that there was something rather haphazard in the path I took to become a serious scientist. A major positive influence was my father—a small-town country doctor in Indiana, who had a passion for his work that certainly impressed me. Ours was a family in which Mom and my brother had a very close alliance, while the same was true for me and my father. And so I suppose it was natural that he hoped very much that I would follow in his path—which meant enrolling at Indiana University and pursuing a career in medicine. I personally also found this attractive but my boyfriend had other ideas—that I should study rather to be a nurse—a career that should be more suitable for what he hoped would be my main calling in life—his wife and mother to his children. As it turned out, I pleased neither of them. During my first week at Indiana University, I hoped to still my own confusion by talks with faculty at orientation day. I started in alphabetical order and, within half an hour, had signed up for anthropology as a major. But then I wandered on to "B" and there was this handsome young Professor who had a crowd of students enthralled by his passionate advocacy of the field of bacteriology. I joined the crowd and asked him,

"Could I major in this as a Pre-Med?" "Yes" "Isn't it dangerous to work with bacteria?" "No, it's FUN!!!!." And I was hooked. Major A changed to Major B and I never looked back. For some freaky reason relating to the fact that I was an honors major, I had the Chairman of the English Department as my advisor freshman year, and he urged me to go for a B.A. degree instead of a B.Sc.—and this turned out also to be quite lucky—in addition to science, I took honors courses in creative writing, advanced English literature, and also many extra courses in Russian and spent a summer in Russia back when the cold war was really cold. It's true that I lived my life with a secret fear all my life that I never was a strong in math and chemistry as my other colleagues. Yet I really hate the specialization we impose on our science majors now—and have no regrets at having such an enriching undergraduate experience.

I loved Bacteriology—I think as much for the terrific faculty as for the discipline—and, to my father's disappointment, I decided that graduate school was a more appealing choice than medical school. Escaping the boyfriend meant going away as far as possible from Indiana for graduate school—and so I chose Marine Microbiology at the Scripps Institution of Oceanography in California—but soon found that I got seasick easily. Again, on a random choice, I moved sideways to the new Biology Department on the new campus of UC San Diego. Again, fate played a role, and I was given a rotation with Carlos Miller who was on sabbatical at UCSD. Carlos had a key role in the discovery of the plant hormone cytokinin, and was a lovely gentle fellow who had much patience with this student who had never studied botany because of all those English and Russian courses. But he convinced me to stay with plants, and I ended up doing my thesis work characterizing the pathway for tryptophan biosynthesis in plants. By then I was married to a graduate student in astrophysics who was offered a great post-doctoral opportunity at the University of Colorado. Wanting to stay with plants, I arranged my own post-doc at Colorado with an up-and-coming young fellow named Peter Albersheim who was just beginning his groundbreaking work on the structure of the plant cell wall. It was this focus that was to set me on my own career path that focused for the rest of my academic career on the study of plant cell walls.

Although Peter had concentrated on cell wall structure, I felt more inclined to enzymology and decided to tackle a major unanswered question that occupied me for the rest of my career—the mechanism of biosynthesis of the world's most abundant organic compound—cellulose. By now I was a young faculty member at Michigan State University, and I chose the cotton fiber as a model system because it was a veritable factory for cellulose. We struggled without success trying to identify an enzyme system that could make cellulose—but here is a lesson for the young. While still maintaining our focus on the key issues, I also knew that one has to show productivity—and so I initiated some other projects that were "easier" to succeed with—the first demonstration of the role of lipid intermediates in plant glycoprotein synthesis, the pore size of plant cell walls, insights into the biosynthesis of callose, and a rather comprehensive characterization of cotton fiber development.

Again, fate intervened when for personal reasons, I relocated to The Hebrew University in Jerusalem. There we continued to focus on cellulose biosynthesis with two of my "favorite" projects—the finding that sucrose synthase—a key enzyme in synthesis of the precursor to cellulose, UDP-glc, had a plasma membrane-associated as well as the well-known soluble form, and the discovery that cells adapted to growth on an inhibitor of cellulose synthesis could survive with almost no cellulose in their walls—the latter showing just how adaptable plants can be when challenged. But the enzyme cellulose synthase still remained elusive. And here we can learn another lesson—don't be afraid to collaborate and delve into new areas of science. In order to get more comfortable working in molecular biology, I arranged a sabbatical with Dave Stalker at Calgene, Inc. in Davis California. Dave's group was interested in getting more good cotton fiber-specific promoters and we were interested to try to identify the gene for cellulose synthase—so we combined forces using our own cotton fiber cDNA library—Dave got his promoters and together we identified for the first time two sequences that encoded proteins that had all the domains expected for a cellulose synthase (plus a few more interesting domains!) and was highly-expressed just at the time fibers underwent a 100-fold increase in cellulose synthesis as they initiated secondary wall synthesis. Discovery of these genes allowed the *Arabidopsis* gurus to find similar genes and show that when disrupted they did indeed lead to loss of ability to synthesize cellulose synthesis. From there, the field now has been joined by a healthy number of young new faces, and new discoveries about the process seem to emerge every month. We too found that *Arabidopsis* had many advantages and used it to advantage once I relocated my lab to UC Davis. Yet the power of being able to combine my old skills in biochemistry with the new skills in molecular biology I think has proved to be a very important aspect of my work.

Finally, my dad—who loved medicine because it combined good science with helping people—would be proud of me at last. At age 60, I made the unusual choice to retire from academia and work for the Rockefeller Foundation where I spent 5 years developing a portfolio of grants that built capacity in biotechnology in the developing world—especially sub-Saharan Africa—and supported projects that aimed to demonstrate that biotechnology can offer solutions to at least some problems that breeders find intractable. I've enjoyed this new challenge immensely—and now have another new one—retirement! But I continue to consult on issues of international agriculture and, in a twist, have also found my knowledge of cellulose synthesis again valuable to those working on biofuels. So it's been an interesting life—and it's not over yet!

REFERENCES

Berger F, Grini PE, Schnittger A (2006): Endosperm: An integrator of seed growth and development. *Curr Opin Plant Biol* **9**:664–670.

Bernier G, Perilleux C (2005): A physiological overview of the genetics of flowering time control. *Plant Biotechnol J* **3**:3–16.

Boavida LC, Becker JD, Feijo JA (2005): The making of gametes in higher plants. *Int J Dev Biol* **49**:595–614.

Byrne ME (2006): Shoot meristem function and leaf polarity: The role of class III HD-ZIP genes. *PLoS Genet* **2**:e89.

Campilho A, Garcia B, Toorn HV, Wijk HV, Scheres B (2006): Time-lapse analysis of stem-cell divisions in the *Arabidopsis thaliana* root meristem. *Plant J* **48**:619–627.

Chow B, McCourt P (2006): Plant hormone receptors: Perception is everything. *Genes Dev* **20**:1998–2008.

Corbesier L, Coupland G (2006): The quest for florigen: A review of recent progress. *J Exp Bot* **57**:3395–3403.

Costa S, Dolan L (2000): Development of the root pole and cell patterning in Arabidopsis roots. *Curr Opin Genet Dev* **10**:405–409.

Di Laurenzio L, Wysocka-Diller J, Malamy JE, Pysh L, Helariutta Y, Freshour G, Hahn MG, Feldmann KA, Benfey PN (1996): The SCARECROW gene regulates an asymmetric cell division that is essential for generating the radial organization of the Arabidopsis root. *Cell* **86**:423–433.

Fleming AJ (2006): Leaf initiation: The integration of growth and cell division. *Plant Mol Biol* **60**:905–914.

Gibson SI (2004): Sugar and phytohormone response pathways: Navigating a signalling network. *J Exp Bot* **55**:253–264.

Giovannoni J (2001): Molecular biology of fruit maturation and ripening. *Annu Rev Plant Physiol Plant Mol Biol* **52**:725–749.

Koornneef M, Bentsink L, Hilhorst H (2002): Seed dormancy and germination. *Curr Opin Plant Biol* **5**:33–36.

Krizek BA, Fletcher JC (2005): Molecular mechanisms of flower development: An armchair guide. *Nat Rev Genet* **6**:688–698.

Lenhard M, Laux T (1999): Shoot meristem formation and maintenance. *Curr Opin Plant Biol* **2**:44–50.

McCormick S (2004): Control of male gametophyte development. *Plant Cell* **16**:5142–5153.

Pimenta-Lange MJ, Lange T (2006): Gibberellin biosynthesis and the regulation of plant development. *Plant Biol (Stuttg)* **8**:281–290.

Quint M, Gray WM (2006): Auxin signaling. *Curr Opin Plant Biol* **9**:448–453.

Rockwell NC, Su YS, Lagarias JC (2006): Phytochrome structure and signaling mechanisms. *Annu Rev Plant Biol* **57**:837–858.

Shani E, Yanai O, Ori N (2006): The role of hormones in shoot apical meristem function. *Curr Opin Plant Biol* **9**:484–489.

Traas J, Bohn-Courseau I (2005): Cell proliferation patterns at the shoot apical meristem. *Curr Opin Plant Biol* **8**:587–592.

Twell D (2006): A blossoming romance: Gamete interactions in flowering plants. *Nat Cell Biol* **8**:14–16.

Weijers D, Jurgens G (2005): Auxin and embryo axis formation: The ends in sight? *Curr Opin Plant Biol* **8**:32–37.

Willemsen V, Scheres B (2004): Mechanisms of pattern formation in plant embryogenesis. *Annu Rev Genet* **38**:587–614.

Wilson ZA, Yang C (2004): Plant gametogenesis: Conservation and contrasts in development. *Reproduction* **128**:483–492.

CHAPTER 5

Tissue Culture: The Manipulation of Plant Development

VINITHA CARDOZA

BASF Plant Science LLC, Durham, North Carolina

5.0. CHAPTER SUMMARY AND OBJECTIVES

5.0.1. Summary

Unique in biology, plant cells are totipotent; whole plants can be regenerated from single nonsexual cells. As a necessary precursor to most plant transformation systems, there must be methods established to manipulate plant tissues and cells in sterile media: tissue culture. From tissues taken from plants, media components and hormones can be manipulated to recover organs or induce somatic embryos. Tissue culture is not only a necessary enabling technology for transgenic plant production but is also used for in vitro propagation of valuable plants.

5.0.2. Discussion Questions

1. Differentiate between organogenesis and somatic embryogenesis.
2. Name plant growth hormones used to manipulate tissues in vitro.
3. How can you develop virus-free plants?
4. What is callus? What are the uses of callus in tissue culture methods?
5. What are protoplasts, and what are their uses?
6. How can haploid plants be produced using tissue culture? Why is this useful?

5.1. INTRODUCTION

Plant tissue culture is the in vitro (literally "under glass") manipulation of plant cells and tissues, which is a keystone in the foundation of plant biotechnology. It is useful for plant propagation and the study of plant hormones, and is generally required to manipulate and regenerate transgenic plants. Whole plants can be regenerated in vitro using tissues,

Plant Biotechnology and Genetics: Principles, Techniques, and Applications, Edited by C. Neal Stewart, Jr.

cells, or a single cell to form whole plants by culturing them on a nutrient medium in a sterile environment. Elite varieties can be clonally propagated, endangered plants can be conserved, virus-free plants can be produced by meristem culture, germplasm can be conserved, secondary metabolites can be produced by cell culture. Besides this, tissue culture serves as an indispensable tool for transgenic plant production. For nearly any transformation system, an efficient regeneration protocol is imperative. This can be attributed to *totipotency* of plant cells and manipulation of the growth medium and hormones. Plant cells are unique in the sense that every cell has the potency to form whole new plantlike stem cells (stem cell production in mammals is located in time and space, and most mammalian cells cannot be converted to stem cells). However, having an understanding of each plant species and *explant* (donor tissue that is placed in culture) is essential to the development of an efficient regeneration system. The physiological stage of the explant plays a very important role in its response to tissue culture. For example, young explants generally respond better than do older ones.

This chapter examines the history and uses of plant tissue culture and shows how it is integral to plant biotechnology, and presents the basic principles of media and hormones used in plant tissue culture, various culture types, and regeneration systems. Some people consider tissue culture as more of an art than science since the researcher must develop an eye for differentiating between good and bad (useful and nonuseful) cultures, which has often proved to be the difference between success and failure in plant biotechnology.

5.2. HISTORY

The history of plant tissue culture dates back at least to 1902, when Gottlieb Haberlandt, a German botanist, proposed that single plant cells could be cultured in vitro. He tried to culture leaf mesophyll cells, but did not have much success. Roger J. Gautheret, a French scientist, had encouraging results with culturing cambial tissues of carrot in 1934. The first plant growth hormone indoleacetic acid (IAA) was discovered in the mid-1930s by F. Kogl and his coworkers. In 1934 Professor Philip White successfully cultured tomato roots. In 1939 Gautheret successfully cultured carrot tissue. Both Gautheret and White were able to maintain the cultures for about 6 years by subculturing them on fresh media. These experiments demonstrated that cultures could be not only be initiated but also maintained over a long period of time. Later in 1955 Carlos Miller and Folke Skoog published their discovery of the hormone kinetin, a cytokinin. Recall from the previous chapter that cytokinin is an important class of plant growth regulators. In 1962, Toshio Murashige and Skoog published the composition a plant tissue culture medium known as *MS* (named for the first letters of their last names) *medium*, which now is the most widely used medium for tissue culture. Murashige was a doctoral student in Professor Skoog's lab, and they developed the now-famous MS medium working with tobacco tissue cultures. The formulation of MS medium took place while they were trying to discover new hormones from tobacco leaf extracts, which, when added to tissue cultures, enabled better growth. In a sense, their experiments could be deemed failures since they did not discover a new hormone. Nonetheless, they came up with a seemingly ideal medium for most plant tissue culture work that is used in practically every plant biotechnology laboratory around the world. This major breakthrough in the field of plant tissue culture has enabled nearly all the other breakthroughs cataloged in this book. MS medium seems to be ideal for many cultures since it has all the nutrients that plants require for growth and contains them in the proper relative ratios. The medium has high macronutrients, sufficient micronutrients, and iron in the slowly available chelated form. The success of tobacco culture using MS medium laid the foundation for future tissue culture work, and this has now become the medium of choice for most tissue culture work. MS medium has been improved on in the past 45 years, but the article by Murashige and Skoog (1962) remains one of the most highly cited papers in plant biology.

5.3. MEDIA AND CULTURE CONDITIONS

5.3.1. Basal Media

The success of tissue culture lies in the composition of the growth medium, hormones, and culture conditions such as temperature, pH, light, and humidity. The growth medium is a composition of essential minerals and vitamins that are necessary for a plant's growth and development; everything, including sugar, which the plant needs to thrive—all must be in sterile or *axenic* conditions. The minerals consist of macronutrients such as nitrogen, potassium, phosphorus, calcium, magnesium, and sulfur, and micronutrients such as iron, manganese, zinc, boron, copper, molybdenum, and cobalt. Iron is seldom added directly to the medium, it is chelated with EDTA (ethylenediaminetetraacetic acid) so that it is more stable in culture and can be absorbed by plants over a wide pH range. Note that EDTA is used in many foods as a preservative. If iron is not chelated with EDTA, it forms a precipitate, especially in alkaline pH. Vitamins are necessary for the healthy growth of plant cultures. The vitamins used are thiamine (vitamin B_1), pyridoxine (B_6), nicotinic acid (niacin), and thiamine. Other vitamins such as biotin, folic acid, ascorbic acid (vitamin C), and vitamin E (tocopherol) are sometimes added to media formulations. Myoinositol, a sugar alcohol, is added to most plant culture media to improve the growth of cultures. In addition, plants require an external carbon source—sugar—since cultures grown in vitro rarely photosynthesize sufficiently to support the tissues' carbon needs. Sometimes cultures are grown in the dark and do not photosynthesize at all. The most commonly used carbon source is sucrose. Other sources used are glucose, maltose, and sorbitol. The pH of the medium is important since it influences the uptake of various components of the medium as well as regulating a wide range of biochemical reactions occurring in plant tissue cultures (Owen et al. 1991). Most media are adjusted to a pH of 5.2–5.8. The acidic pH does not seem to negatively affect plant tissues but delays the growth of many potential contaminants. However, a higher pH is required for certain cultures. Cultures can be grown in either liquid or solid medium (Fig. 5.1). The medium is most often solidified as it provides a support system for the explants and is easier to handle. *Explant* is the term denoting the starting plant parts used in tissue culture. Solidification is done using agar derived from seaweed or agar substitutes such as Gelrite™ or Phytagel™ commercially available as a variety of gellan gums. These are much clearer than agar. Other than this membrane rafts or filter paper (Fig. 5.1) are also used for support on liquid medium.

A plethora of media formulations are available for plant tissue culture other than MS (Murashige and Skoog 1962) are also used (Gamborg et al. 1968; Schenk and Hildebrandt 1972; White 1963; Linsmaier and Skoog 1965; McCown and Lloyd 1981). McCown's woody plant medium (WPM) has been widely used for tree tissue culture.

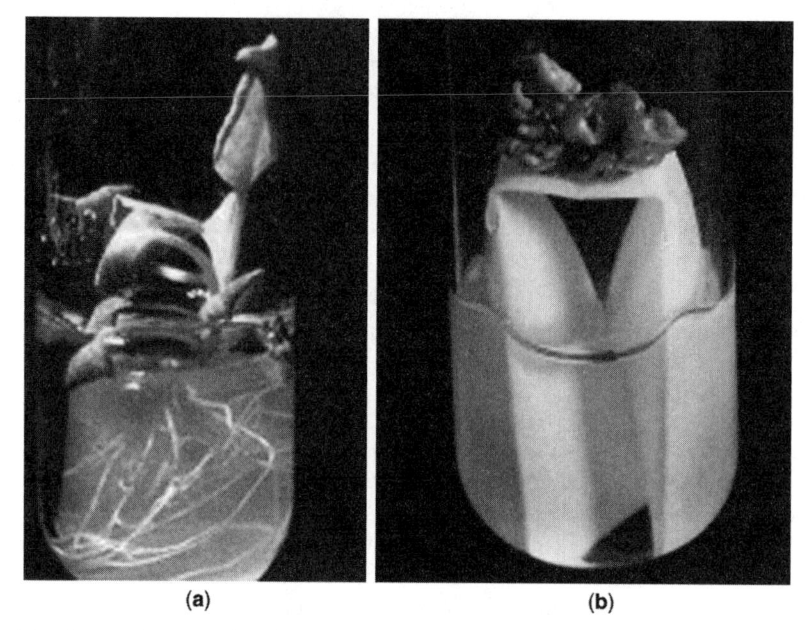

Figure 5.1. Tissue cultures in liquid (a) and solid (b) culture media. See filter paper bridge in liquid medium.

Nitsch and Nitsch (1969) developed an another culture. Knudson's medium (Knudson 1946) was developed for orchid tissue culture and is also used for fern tissue culture. With so many choices in media formulations, one might wonder about how to choose a medium to culture the species of interest. The choice of medium is typically determined empirically for optimal response of the plant species; explants used for culture and plant taxonomy are good starting points. For example, nearly all tissue cultures of plants in the Solanaceae (the nightshade family) use MS media. Recall that MS media was developed using tobacco, a member of this plant family. Many times a mix-and-match scheme of macro- and micronutrients from one medium and vitamins from another has also been successful. The composition of nutrients varies from medium to medium. For example, MS medium has higher macronutrients than does WPM, which is suitable for most plant species, but woody plants respond better in WPM than MS medium. It is important to select the right medium for culture according to how the plant empirically responds in tissue culture.

5.3.2. Growth Regulators

The basal medium (e.g., MS) is designed to keep plant tissues alive and thriving. Plant growth regulators or hormones are needed to manipulate the developmental program of tissues—say, to make callus tissue proliferate, or produce roots from shoots. Growth regulators are the items most often manipulated as experimental factors to enhance tissue culture conditions. The most important growth regulators for tissue culture are auxins, cytokinins, and gibberellins. Both natural and synthetic auxins and cytokinins are used in tissue culture. Auxins promote cell growth and root growth. The most commonly used auxins are IAA (indoleacetic acid), IBA (indolebutyric acid), NAA (naphthaleneacetic acid), and 2,4-D (2,4-dichlorophenoxyacetic acid). Cytokinins promote cell division and shoot growth. An auxinlike compound TDZ (thidiazuron) has increased success rate of plant regeneration in many species. The most commonly used cytokinins are BAP (benzylaminopurine), zeatin, and kinetin. In addition to auxins and cytokinins, other hormones such as abscisic acid (Augustine and D'Souza 1997; Cardoza and D'Souza 2002) and jasmonic acid (Blázquez et al. 2004) have also been used in plant cell culture. Other adjuvants (additional components that enhance growth) that have known to have a positive effect on morphogenesis are polyamines such as spermidine, spermine, and putrescene (Cardoza and D'Souza 2002; El Hadrami and D'Auzac 1992; Potdar et al. 1999). By manipulating the amount and combination of growth hormones, regeneration of whole plants from small tissues is possible (Fig. 5.2).

Another critical aspect in plant tissue cultures is the management of the gaseous hormone ethylene. When plants are grown in vitro in closed culture vessels, there is a buildup of ethylene, which is typically detrimental to the cultures. The addition of ethylene biosynthetic inhibitors such as silver nitrate (Giridhar et al. 2001), AVG (aminoethoxyvinylglycine), and silver thiosulphate (Reis et al. 2003) have been shown to increase the formation of shoots.

Tissues are transferred to fresh media periodically—every week to monthly, depending on the species and experiment. Without subculturing, tissues will deplete the media and often crowd each other, competing for decreasing resources.

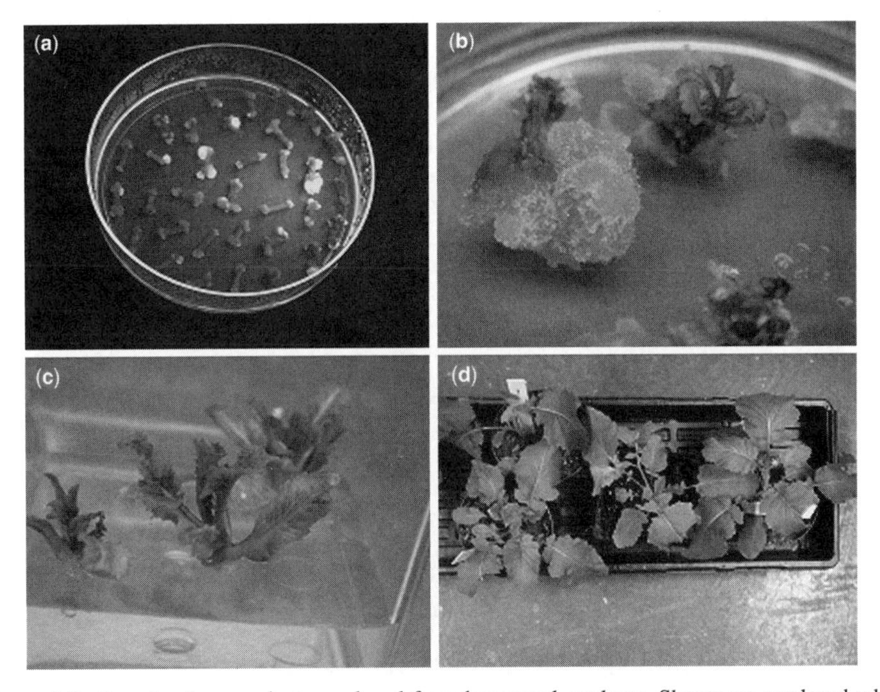

Figure 5.2. *Brassica juncea* plants produced from hypocotyl explants. Shoots are produced when a combination of auxin and cytokinin is used. (a) Callus from hypocotyl explants (note the green fluorescent protein fluorescent sectors on some of the calli); (b) shoots from callus; (c) shoots elongating; (d) whole plantlets transferred to soil. See color insert.

5.4. STERILE TECHNIQUE

5.4.1. Clean Equipment

Successful tissue culture requires the maintenance of a sterile environment. All tissue culture work is done in a laminar flow hood. The laminar flow hood filters air with a dust filter and a high-efficiency particulate air (HEPA) filter (Fig. 5.3). It is important to keep the hood clean, which can be done by wiping it with 70% alcohol. The instruments used should also be dipped in 70% ethanol and sterilized using flame or glass beads. Hands should be disinfected with ethanol before handling cultures in order to avoid contamination. It is imperitive to maintain axenic conditions throughout the life of cultures: from explant to the production of whole plants. Entire experiments have been lost because of an episode of fungal or bacterial contamination at any stage of culture (see Fig. 5.4). Especially problematic are fungal contaminants that are propagated by spores that might blow into a hood from an environmental source. Therefore, it is important to work away from the unsterile edge of a laminar flow hood. Culture rooms or chambers must be maintained as clean as possible to control any airborne contaminants.

5.4.2. Surface Sterilization of Explants

Plant tissues inherently have various bacteria and fungi on their surfaces. It is important that the explant be devoid of any surface contaminants prior to tissue culture since contaminants can grow in the culture medium, rendering the culture nonsterile. In addition, they compete with the plant tissue for nutrition, thus depriving the plant tissue of nutrients. Bacteria and especially fungi can rapidly overtake plant tissues and kill them (Fig. 5.4). The surface sterilants chosen for an experiment typically depend on the type of explant and also plant species. Explants are commonly surface-sterilized using sodium hypochlorite (household bleach), ethanol, and fungicides when using field-grown tissues. The time of sterilization is dependent on the type of tissue; for example, leaf tissue will require a shorter sterilization time than will seeds with a tough seed coat. Wetting agents such as Tween added to the

Figure 5.3. Researcher working in the laminar flow hood.

Figure 5.4. Fungal contamination has taken over the whole explant.

sterilant can improve surface contact with the tissue. Although surface contamination can be eliminated by sterilization, it is very difficult to remove contaminants that are present inside the explant that may show up at a later stage in culture. This internal contamination can be controlled to a certain extent by frequent transfer to fresh medium or by the use of a low concentration of antibiotics in the medium. Overexposing tissues to decontaminating chemicals can also kill tissues, so there is a balancing act between sterilizing explants and killing the explants themselves.

5.5. CULTURE CONDITIONS AND VESSELS

Cultures are grown in walk-in growth rooms (Fig. 5.5) or growth chambers. Humidity, light, and temperature have to be controlled for proper growth of cultures. A 16-h light

Figure 5.5. A walk-in growth room where cultures are grown.

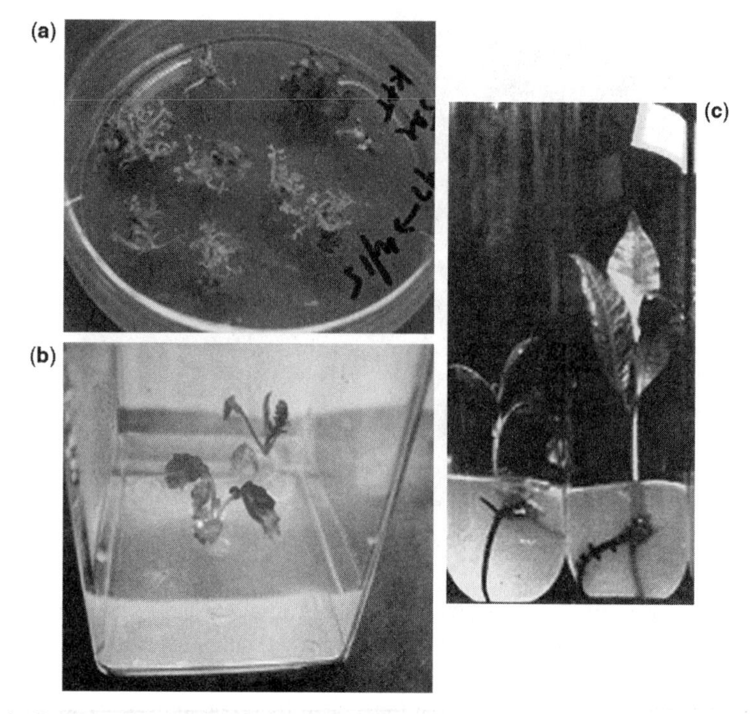

Figure 5.6. Cultures grown in different kinds of vessels: (a) petri plate; (b) Magenta GA7 box; (c) test tube.

photoperiod is optimal for tissue cultures, and a temperature of 22–25°C is used in most laboratories. A light intensity of 25–50 μmol m^{-2} s^{-1} is typical for tissue cultures and is supplied by cool white fluorescent lamps. A relative humidity of 50–60% is maintained in the growth chambers. Some cultures are also incubated in the dark. Cultures can be grown in various kinds of vessels such as petri plates, test tubes, "Magenta boxes," bottles, and flasks (Fig. 5.6).

5.6. CULTURE TYPES AND THEIR USES

5.6.1. Callus Culture

Callus is an unorganized mass of cells that develops when cells are wounded and is very useful for many in vitro cultures. Callus is developed when the explant is cultured on media conducive to undifferentiated cell production—usually the absence of *organogenesis* (organ production) can lead to callus proliferation. Stated another way, callus production often leads to organogenesis, but once callus begins to form organs, callus production is halted. Auxins and cytokinins both aid in the formation of most callus cells. Callus can be continuously proliferated using plant growth hormones or then directed to form organs (Fig. 5.7) or somatic embryos (Fig. 5.8). Callus cultures can be transferred to a new medium for organogenesis or embryogenesis or maintained as callus in culture. Although callus has been induced for various reasons, one important

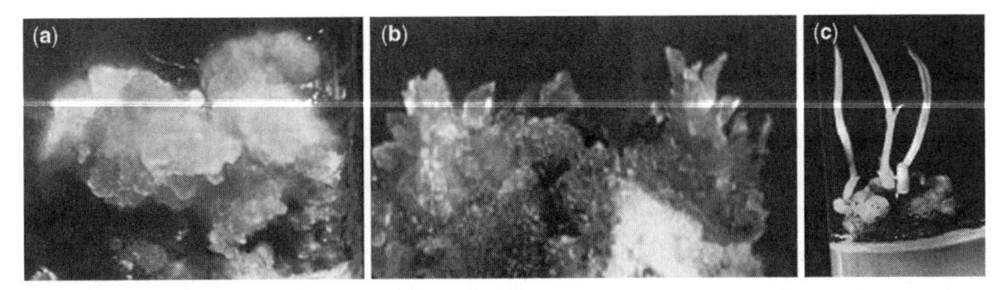

Figure 5.7. (a) Callus tissue; (b,c) shoots arising from callus (an example of organogenesis). See color insert.

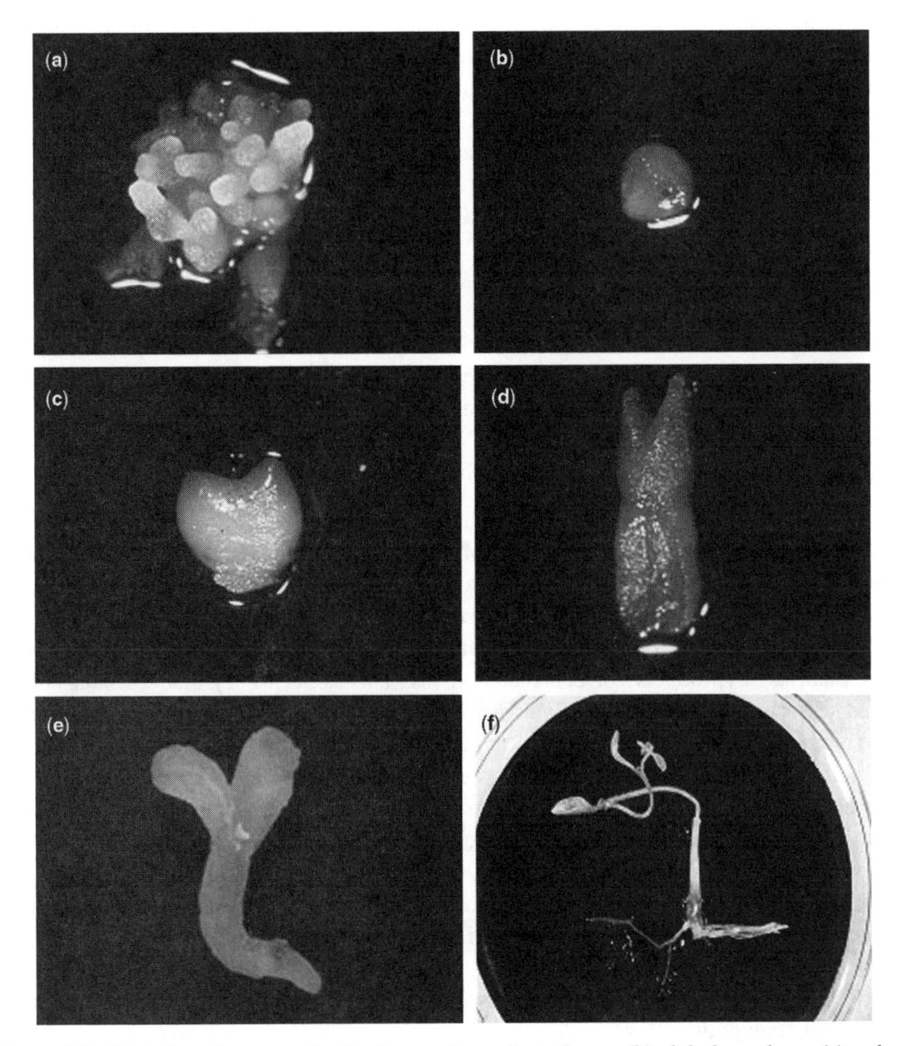

Figure 5.8. Somatic embryogenesis: (a) cluster of somatic embryos; (b) globular embryo; (c) embryo becomes heart-shaped as it grows; (d) torpedo-shaped embryo is the next developmental stage; (e) the embryo forms cotyledons as it begins to mature; (f) germinating embryo (courtesy of Wayne Parrott and Benjamin Martin).

application of callus is to induce *somaclonal* variation through which desired mutants can be selected.

5.6.1.1. Somaclonal Variation. Plant cells undergo varying degrees of cytological and genetic changes during in vitro growth. Some of the changes are derived from preexisting aberrant cells in the explants used for culture. Others represent transient physiological and developmental disturbances caused by culture environments. Still others are a result of epigenetic changes, which can be relatively stable but are not transmitted to the progeny. Some variations are a result of specific genetic change or mutation and are transmitted to the progeny. Such genetically controlled variability is known as *somaclonal variation*. Somaclonal variation serves as both a boon and a bane in tissue culture. It may hamper clonal propagation, but at the same time generate desirable somaclonal variants that can be selected for the development of novel cell lines. Induced somaclonal genetic variability of callus can give rise to genetically variable plantlets regenerated from callus and are of immense importance in the development and selection of various stress tolerant cell lines. Salt-tolerant (Ochatt et al. 1999), heavy-metal-tolerant (Chakravarty and Srivastava 1997), disease-resistant (Jones 1990), and herbicide-resistant (Smith and Chaleff 1990) cell lines have been selected via somaclonal mutations using callus tissue.

5.6.2. Cell Suspension Culture

Lose friable callus can be broken down to small pieces and grown in a liquid medium to form cell suspension cultures. Cell suspensions can be maintained as batch cultures grown in flasks for long periods of time. Somatic embryos have been initiated from cell suspension cultures (Augustine and D'Souza 1997). Cell cultures have also been employed for the production of valuable secondary metabolites.

5.6.2.1. Production of Secondary Metabolites and Recombinant Proteins Using Cell Culture. Plant cell cultures can be useful for the production of secondary metabolites and recombinant proteins. *Secondary metabolites* are chemical compounds that are not required by the plant for normal growth and development but are produced in the plant as "byproducts" of cell metabolisms. That is not to say that secondary metabolites serve no function to the plant; many do. Some are used for defense mechanism or for reproductive purposes such as color or smell. Some important secondary metabolites present in plants are flavonoids, alkaloids, steroids, tannins, and terpenes. Secondary metabolites have been produced using cell cultures in many plant species and have been reviewed by Rao and Ravishankar (2002). The process can be scaled up and automated using bioreactors for commercial production. Many strategies such as biotransformation, cell permeabilization, elicitation, and immobilization have been used to make cell suspension cultures more efficient in the production of secondary metabolites. Secondary metabolite production can be increased by metabolic engineering, in which enzymes in the pathway of a specific compound can be overexpressed together, thereby increasing the production of a specific compound.

Transgenic plant cell cultures are gaining popularity in the large-scale production of recombinant proteins, thus making them integral parts of molecular farming. What makes molecular farming economically attractive is that production costs can potentially be much lower than those of traditional pharmaceutical production. Plant cell cultures are also advantageous for molecular farming because of high level of containment that they offer relative to whole, field-grown plants and the possibility of commercially producing recombinant proteins. Tobacco suspension culture is the most popular system so far; however, pharmaceutical proteins have been produced in soybean (Smith et al. 2002), tomato (Kwon et al. 2003), and rice (Shin et al. 2003) cells. So far, more than 20 pharmaceutical compounds have been produced in cell suspension cultures, which include antibodies, interleukins, erythropoietin, human granulocyte–macrophage colony-stimulating factor (hGM-CSF), and hepatitis B antigen (Shadwick and Doran 2005).

5.6.3. Anther/Microspore Culture

The culture of anthers or isolated micropsores to produce haploid plants is known as *anther culture* or *microspore culture*. Microspore culture has developed into a powerful tool in plant breeding. Embryos can be produced via a callus phase or be a direct recapitulation of the developmental stages characteristic of zygotic embryos (Palmer and Keller 1997) (Fig. 5.9). It has been known that late uninucleate to early binucleate microspores are the best explants for embryogenesis. In this case, the somatic embryos (explained in Section 5.7.2, below) develop into haploid plants. Doubled haploids can then be produced by chromosome-doubling techniques. Thus microspore culture enables the production of homozygous (at every locus) plants in a relatively short period as compared to conventional breeding techniques. These homozygous plants are useful tools in plant breeding and genetic studies. In addition, haploid embryos are used in mutant isolation, gene transfer, studies of storage product biochemistry, and physiological aspects of embryo maturation (Palmer and Keller 1997).

5.6.4. Protoplast Culture

Protoplasts contain all the components of a plant cell except for the cell wall. Using protoplasts, it is possible to regenerate whole plants from single cells and also develop somatic

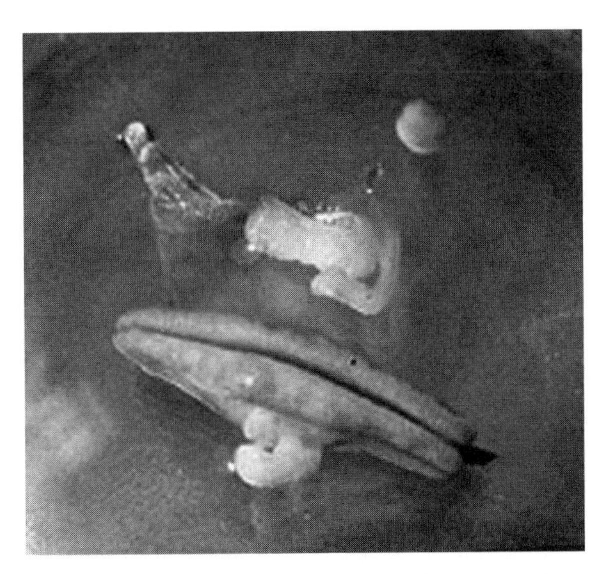

Figure 5.9. Somatic embryos regenerated from an anther.

hybrids as described below. Cell walls can be removed from explant tissue mechanically or enzymatically; the latter is used most often. Enzymatic cell wall degradation was pioneered by Cocking (1960). Ever since then, protoplast production has been applied to various crop and tree species. Plant cell walls consist of cellulose, hemicellulose, and pectin, with lesser amounts of protein and lipid (Dodds and Roberts 1995). Hence a mixture of enzymes is necessary for degrading the cell wall. The enzymes that are commonly used are cellulase and pectinase. Following enzyme treatment protoplasts are purified from cellular debris by filtering using a mesh and then flotation on either sucrose or ficoll. They are cultured in a high-osmoticum medium to avoid bursting. Protoplasts are cultured either on liquid or solid medium. Protoplasts embedded in an alginate matrix and then cultured on solid medium have better success rates of regeneration. The alginate provides cellular protection against mechanical stress and gradients in environmental conditions during the critical first few days of protoplast culture.

5.6.4.1. Somatic Hybridization. Protoplast fusion and somatic hybridization techniques provide the opportunity for bypassing reproductive isolation barriers, thus facilitating gene flow between species. Fusion of protoplasts is accomplished by the use of PEG [poly(ethylene glycol)]. Protoplast fusion has helped in the development of somatic hybrids or *cybrids* (cytoplasmic hybrids). Protoplasts offer the possibility of efficient and direct gene transfer to plant cells. DNA uptake has been found to be easier in protoplasts than into intact plant cells. Although protoplasts seem to be a very attractive means for plant regeneration and gene transfer, they are very vulnerable to handling. One has to be very careful when manipulating protoplasts. They have to be cultured on a medium with a high osmoticum such as sucrose or mannitol; otherwise the protoplasts will burst open, which is why plant regeneration from protoplasts has proven to be difficult. Therefore, protoplasts are now used in cell culture studies mostly to study localization of proteins and transient transgene assays.

5.6.5. Embryo Culture

Embryo culture is a technique in which isolated embryos from immature ovules or seeds are cultured in vitro. This technique has been employed as a useful tool for direct regeneration in species where seeds are dormant, recalcitrant, or abort at early stages of development. Embryo culture also finds use in the production of interspecific hybrids between inviable crosses, whose seeds are traditionally condemned and discarded because of their inability to germinate. In plant breeding programs, embryo culture goes hand in hand with in vitro control of pollination and fertilization to ensure hybrid production. Besides this, immature embryos can be used to produce embryogenic callus and somatic embryos (Ainsley and Aryan 1998) or direct somatic embryos (Cardoza and D'Souza 2000).

5.6.6. Meristem Culture

In addition to being used as a tool for plant propagation, tissue culture is a tool for the production of pathogen-free plants. Using apical meristem tips, it is possible to produce disease-free plants. This technique is referred to as *meristem culture*, *meristem tip culture*, or *shoot tip culture*, depending on the actual explant that is used. Although it is possible to produce bacterium- or fungus-free plants, this method has more commonly been used in the elimination of viruses in many species (Kartha and Gamborg

1975; Brown et al. 1988; Ayabe and Sumi 2001). Apical meristems in plants are suitable explants for the production of virus-free plants since the infected plant's meristems typically harbor titers that are either nearly or totally virus-free. Meristem culture in combination with thermotherapy has resulted in successful production of virus-free plants when meristem culture alone is not successful (Kartha 1986; Manganaris et al. 2003; Wang et al. 2006).

5.7. REGENERATION METHODS OF PLANTS IN CULTURE

In plant biotechnology, tissue culture is most important for the regeneration of transgenic plants from single transformed cells. It is safe to say that without tissue culture there would be no transgenic plants (although this situation is slowly changing—nonetheless tissue culture is required to regenerate intact plants in most species).

5.7.1. Organogenesis

Organogenesis is the formation of organs: either shoot or root. Organogenesis in vitro depends on the balance of auxin and cytokinin and the ability of the tissue to respond to phytohormones during culture. Organogenesis takes place in three phases. In the first phase the cells become competent; next, they dedifferentiate. In the third phase, morphogenesis proceeds independently of the exogenous phytohormone (Sugiyama 1999). Organogenesis in vitro can be of two types: direct and indirect.

5.7.1.1. Indirect Organogenesis. Formation of organs indirectly via a callus phase is termed *indirect organogenesis*. Induction of plants using this technique does not ensure clonal fidelity, but it could be an ideal system for selecting somaclonal variants of desired characters and also for mass multiplication. Induction of plants via a callus phase has been used for the production of transgenic plants in which (1) the callus is transformed and plants regenerated or (2) the initial explant is transformed and callus and then shoots are developed from the explant.

5.7.1.2. Direct Organogenesis. The production of direct buds or shoots from a tissue with no intervening callus stage is termed *direct organogenesis* (Fig. 5.10). Plants have been propagated by direct organogenesis for improved multiplication rates, production of transgenic plants, and—most importantly—for clonal propagation. Typically, indirect organogenesis is more important for transgenic plant production.

5.7.1.2.1. Axillary Bud Induction/Multiple-bud Initiation. This technique is the most common means of micropropagation since it ensures the production of uniform planting material without genetic variation. Axillary shoots are formed directly from preformed meristems at nodes (Fig. 5.11), and chances of the organized shoot meristem undergoing mutation are relatively low. This technique is often referred to as *multiple-bud induction*. Many economically important plants have been propagated using this method. Multiple-bud initiation has been successful in crop plants but in only a few tree species such as *Millingtonia hortensis* (Hegde and D'Souza 1995) and *Fagus sylvatica* (Chalupa 1996).

Figure 5.10. Direct organogenesis (shoot and root formation from leaf explant) of *Curculigo orchioides.*

Multiple-bud initiation still remains a challenge in many tree species since many tree species are recalcitrant in tissue culture.

5.7.2. Somatic Embryogenesis

Somatic embryogenesis is a nonsexual developmental process that produces a bipolar embryo with a closed vascular system from somatic tissues of a plant. Somatic embryogenesis has become one of the most powerful techniques in plant tissue culture for mass clonal propagation. Somatic embryogenesis may occur directly or via a callus phase. Direct somatic embryogenesis is preferred for clonal propagation as there is less chance of introducing variation via somaclonal mutation. Indirect somatic embryogenesis is sometimes used in the selection of desired somaclonal variants and for the production of transgenic plants. Large-scale production of somatic embryos using bioreactors and synthetic seeds from somatic embryos has been successful. Somatic embryos can be cryopreserved as synthetic seeds and germinated whenever necessary. One advantage of somatic embryogenesis is that somatic embryos can be directly germinated into viable plants without organogensis; thus it mimics the natural germination process.

Figure 5.11. (a) Multiple-bud initiation from cotyledonary nodes; (b,c) multiple buds of *Medicago truncatula* and cashew, respectively, elongate to form shoots; (d) rooting of elongated shoots of cashew.

5.7.2.1. Synthetic Seeds. Encapsulated somatic embryos are known as *synthetic seeds*. Somatic embryos are typically encapsulated in an alginate matrix, which serves as an artificial seed coat. The encapsulated somatic embryos can be germinated ex vitro or in vitro to form plantlets. Synthetic seeds have multiple advantages—they are easy to handle, they can potentially be stored for a long time, and there is potential for scaleup and low cost of production. The prospects for automation of the whole production process is another advantage because the commercial application of somatic embryogenesis requires high-volume production. Synthetic seeds can be stored at 4°C for shorter periods or cryopreserved in liquid nitrogen for long-term storage (Fang et al. 2004). Production of synthetic seeds and germination of these seeds to plantlets has been accomplished in sandalwood, coffee, bamboo, and many other plant species.

5.8. ROOTING OF SHOOTS

Efficient rooting of in vitro–grown shoots is a prerequisite for the success of micropropagation. The success of acclimatization of a plantlet greatly depends on root system production. Rooting of trees and woody species is difficult as compared to herbaceous species. Rooting of shoots is achieved in vitro or ex vitro. *Ex vitro* (out of glass) rooting reduces the cost of production significantly. Ex vitro rooting is carried out by pretreating the shoots with phenols or auxins and then directly planting them in soil under high-humidity conditions. With this method, acclimation of the rooted shoots can be carried out simultaneously. In vitro rooting consists of rooting the plants in axenic conditions. Despite the cost factor, in vitro rooting is still a very common practice in many plant species because of its several advantages. Tissue culture conditions facilitate administration of auxins and other compounds, avoid microbial degradation of applied compounds, allow addition of inorganic nutrients and carbohydrates, and enable experiments with small, simple explants. Several factors are known to affect rooting. The most important factor is the action of endogenous and exogenous auxins. In many cases a pulse treatment with auxins for a short period has also been sufficient for root induction.

Phenolic compounds are known to have a stimulatory effect on rooting. Among the phenolic compounds, phloroglucinol, known as a root promoter, has a positive effect on rooting (Hegedus and Phan 1987). Catechol, a strong reducing agent, has been reported to regulate IAA oxidation and thus affect rooting in plant tissue culture (Hackett 1970).

5.9. ACCLIMATION

Once plants are generated by tissue culture, they have to be transferred to the greenhouse or field. This requires that the plants be hardened-off before transfer to the field. During this acclimation process, plants are first transferred to a growth chamber or greenhouse and covered by domes to minimize the loss of water. Tissue culture conditions are at approximately 100% humidity, whereas relative humidity outside the vessels is typically much lower. In addition, the plants must be "weaned" off the rich media so they can grow as normal plants in soil. Once the plants are acclimatied under greenhouse conditions, they are ready for transfer to the field. Acclimation is a very important step in tissue culture since it is possible to lose plants if they are not properly hardened-off.

5.10. CONCLUSIONS

Plant tissue culture is an essential tool in plant biotechnology that has enabled mass clonal propagation, production of secondary metabolites, preservation of germplasm, and production of virus-free plants. Moreover, it serves as an indispensable tool for regenerating transgenic plants. All this has been possible by manipulating plant tissues and various kinds of media developed by plant tissue culturists and by the use of plant hormones. It has been one of the very exciting discoveries for plant biologists and will continue to be most useful in the coming years.

ACKNOWLEDGMENTS

The author would like to thank Dr. Leo D'Souza from the Laboratory of Applied Biology, St Aloysius College, Mangalore, India for suggestions on the manuscript and for kindly providing most of the illustrations for this chapter. Many thanks to Dr. Wayne Parrott and Benjamin Martin from the University of Georgia for providing the somatic embryogenesis photos.

LIFE BOX 5.1. MARTHA S. WRIGHT

Martha S. Wright, Research Scientist (retired), Syngenta and Monsanto

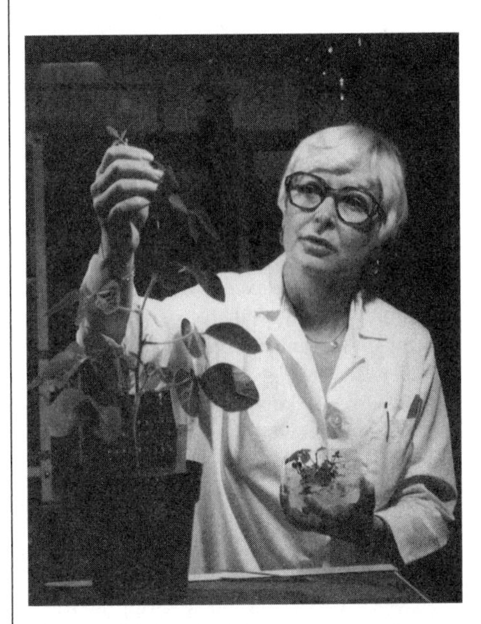

Martha Wright with a regenerated soybean plant (1981).

My love of science emerged in high school when I entered the Kansas City Science Fair in 1956. For my project, I disassembled an animal from each of the phyla and put their skeletons back together for comparison. The project didn't win anything, but my mom was glad that I wasn't boiling lizards on her stove anymore. At Lindenwood College, now Lindenwood University, in St. Charles, Missouri, I originally majored in business because my father said I'd always be able to get a job as a secretary. Remember, this was 1958 and I grew up in Kansas. In my sophomore year, I was fortunate to have an advisor in the business department who noted that I was bored. After a discussion, she urged me to sign up for an advanced biology course. Ultimately, I graduated in 1962 from Lindenwood with a major in biology and minors in chemistry, classics and business.

During my last 2 years of college, I worked summers in hospital laboratories. My first permanent job after graduating was in the Agricultural Division of Monsanto in St. Louis. I was hired because I had worked with radioactivity while in college. One of my biology professors had worked on the Manhattan Project. My first assignment at Monsanto was to work on an insecticide. For the next 15 years, I worked on a series of projects, some having to do with animals and some with plants. I especially enjoyed my early work with Roundup®. We were able to determine the mode of action through a series of experiments using *Lemna* as a model system. In the early 1970's, Ernie Jaworski, my supervisor at Monsanto at that time, went to Saskatoon on a sabbatical in Olaf Gamborg's laboratory. When he returned he handed me some cell cultures and they became mine, and that was the beginning of my true career.

In the early days of field crop cell culture, the "holy grail" was soybean and it was thought to be impossible to regenerate from cell culture. An understanding of the way plant hormones act at different stages was probably the single most important factor to aid soybean regeneration. I was fortunate to work with Michael Carnes as we unraveled the hormone profiles of several field crops, soybean, maize, alfalfa, etc. By the mid 1980's, we had published 3 methods for regenerating soybeans from cell culture. Concurrently, molecular biologists were having some success with plant cell transformation.

Eventually, several crops were in the race to the field. I left Monsanto in 1987 and started working at then Ciba-Geigy, now Syngenta, in North Carolina. Here the emphasis was clearly on corn to support our seed company. Monocots were proving to be especially difficult to transform and it seemed that if the regeneration system worked, it didn't mesh with the transformation procedure, and vice-versa. In 1991, we published the recovery of fertile transformed maize plants using the Biolistics® gun.

During the rest of my career, I either worked directly with, or supervised work with, soybean, corn, vegetables, cotton, rice, etc. I learned from every one of the wonderful, intelligent, dedicated people with whom I worked throughout my career and I am eternally grateful to them for being part of my work and my life. In 2001, I retired from Syngenta. I was happy, healthy and satisfied with my career. Now I'm giving back in various volunteer capacities.

I feel our work broke the mystique of plant regeneration from cell culture, and ultimately allowed the transformation of recalcitrant crops. Transformation, in most cases, depends upon being able to work at the cellular level, thus without the ability to regenerate from cell culture, recovery of fertile transformants is not possible.

Today, it is critical that we continue the hunt for beneficial crop genes while we look for other crops where we can make a valuable difference. Of equal importance is to educate the public on the safety of enhanced crops. We must convince the non-scientists that we too have children and do not want harm to come to them or anyone, now or in the future. This stance also requires responsibility on our part. Enhanced crops mean more people get to eat and more people are healthy and can devote their energies to improving the world. That's the goal. And that's always been my goal in agricultural research.

LIFE BOX 5.2. GLENN BURTON COLLINS

Glenn Burton Collins, Professor Emeritus, Department of Plant and Soil Sciences, University of Kentucky

Glenn Collins

My interest in science began during my growing-up years living and working on a farm. There I developed a fascination with the diversity of the plants and animals in our fields and streams. I became even more fascinated and intrigued with living organisms as I began to take science classes and started to understand how living organisms functioned and adapted to their diverse environments. The defining moment which led me down my specific educational and career pathway was in the summer between my sophomore and junior years of college when I got a job working for a plant breeder. I worked for the same breeder during the remainder of my baccalaureate degree program and I subsequently did my master's degree under his direction. I then pursued my Ph.D. degree in Genetics with a minor in Plant Physiology at North Carolina State University. Back in those days, we did not typically take postdoctoral appointments unless we had problems getting an offer of a permanent position.

I have been in an academic appointment at the University of Kentucky since completing my Ph.D. degree in 1966. Training in the field of genetics and plant breeding/cytogenetics was a wonderful platform for being positioned to participate in and contribute to the advancements in plant biotechnology. I headed up a team that developed and released ten new cultivars and eleven germplasm lines during the fourteen years that I was in my faculty position as the plant breeder. At the same time, my program made major contributions to crop improvement by developing alternate strategies for crop improvement that included improved plant tissue culture systems; producing haploids and doubled haploids from microspores in cultured anthers; and in generating new interspecific hybrid combinations in plants using in vitro embryo rescue and protoplast fusion. I moved into a more basic-science-oriented faculty position in 1980 that was defined as plant somatic cell genetics. This position change was well timed for the vast opportunities which were made available by recombinant DNA tools and genetic engineering approaches for putting foreign genes into plants using in vitro cultured explants in an aseptic tissue culture environment. We had already developed efficient totipotent in vitro systems for several plant species including for several *Trifolium* species in the legume family. We generated transgenic soybeans in the late 1980s and to date we have introduced genes for disease resistance, herbicide tolerance and for biochemical trait modifications into a number of plant species.

In addition to these cited examples of contributions to the shaping of plant biotechnology, I give a lot of credit to people who have been in or associated with my program and provided major advancements to the field of biotechnology both while in my laboratory and then in their own career positions. These include seventeen doctoral students; twelve M.S. students; twenty-five postdoctoral fellows; twenty-four visiting scientists; plus the staff in my laboratory and my many collaborators.

Another very significant contribution which we have made to the advancement of plant biotechnology has been the training of a very large number B.S. degree recipients through our interdisciplinary program in Agricultural Biotechnology, which was initiated in 1988 as a research oriented baccalaureate degree. A majority of these graduates have gone into doctoral and professional degree programs with a substantial number of them in biotechnology careers. Many other graduates have accepted positions in the field of biotechnology with private companies or in University and government laboratories.

I have a difficult time feeling precise and inclusive when I think about trying to predict future advancements in plant biotechnology. The reason is because the advancements are so rapid, numerous, and diverse as we utilize functional genomics and other approaches to identify genes and the traits that they control in plants, that predicting the myriad of applications is mind-boggling. The knowledge-base that will be generated will certainly provide the opportunity to improve crops for their current traditional uses and also to engineer plants for new uses in health, nutrition, energy and environmental applications.

REFERENCES

Ainsley PJ, Aryan AP (1998): Efficient plant regeneration system for immature embryos of triticale (*x Triricosecale Wittmack*). *Plant Growth Reg* **24**:23–30.

Augustine AC, D'Souza L (1997): Somatic embryogenesis in *Gentum ula* Brongn. (*Gentum edule*) (Willd) Blume. *Plant Cell Rep* **16**:354–357.

Ayabe M, Sumi S (2001): A novel and efficient tissue culture method—"stem-disc dome culture"—for producing virus-free garlic (*Allium sativum* L). *Plant Cell Rep* **20**:503–507.

Blázquez S, Piqueras A, Serna MD, Casas JL, Fernández JA (2004): Somatic embryogenesis in saffron: Optimization through temporary immersion and polyamine metabolism. *Acta Hort* **650**:269–276.

Brown GR, Kwiatkowski S, Martin MW, Thomas PE (1988): Eradication of PVS from potato clones through excision of meristems from *in vitro* heat-treated shoot tips. *Am Pot J* **65**:633–638.

Cardoza V, D'Souza L (2002): Induction, development and germination of somatic embryos from nucellar tissues of cashew (*Anacardium occidentale* L). *Sci Hort* **93**:367–372.

Cardoza V, D'Souza L (2000): Direct somatic embryogenesis from immature zygotic embryos in cashew (*Anacardium occidentale* L). *Phytomorphology* **50**(2):201–204.

Chakravarty B, Srivastava S (1997): Effects of genotype and explant during *in vitro* response to cadmium stress and variation in protein and proline contents in linseed. *Ann Bot* **79**:487–491.

Chalupa V (1996): *Fagus sylvatica* L. (European Beech). In *Biotechnology in Agriculture and Forestry*, Vol 35 (Trees IV), Bajaj YPS (ed). Springer-Verlag, Berlin Heidelberg, New York, pp 138–152.

Cocking EC (1960): A method for the isolation of plant protoplasts and vacuoles. *Nature* 187–927.

Dodds JH, Roberts LW (1995): Isolation and culture of protoplasts. In *Experiments in Plant Tissue Culture*, 3rd ed, Dodds JH, Roberts LW (eds). Cambridge Univ Press, Cambridge, UK, pp 167–182.

El Hadrami I, D'Auzac J (1992): Effects of polyamine biosynthetic inhibitors on somatic embryogenesis and cellular polyamines in *Hevea brasiliensis*. *J Plant Physiol* **140**:33–36.

Fang J, Wetten A, Hadley P (2004): Cryopreservation of cocoa (*Theobroma cacao* L.) somatic embryos for long-term germplasm storage. *Plant Sci* **166**:669–675.

Gamborg OL, Miller RA, Ojima K (1968): Nutrient requirements of suspension cultures of soybean root cells. *Exp Cell Res* **50**:151–158.

Giridhar P, Reddy BO, Ravishankar GA (2001): Silver nitrate influences *in vitro* shoot multiplication and root formation in *Vanilla planifolia* Andr. *Curr Sci* **81**:1166–1170.

Hackett WP (1970): The influence of auxin, catechol and methanolic tissue extracts on root initiation in aseptically cultured shoot apices of juvenile and adult forms of *Hedera helix*. *J Am Soc Hort Sci* **95**:398–402.

Hegde S, D'Souza L (1995): *In vitro* propagation of an ornamental tree *Millingtonia hortensis* L.f. *Gartenbauwissenschaft* **60**:258–261.

Hegedus P, Phan CT (1987): Activities of five enzymes of the phenolic metabolism on rooted and acclimatised vitreous plants in relation with phenolic treatments. *Acta Hort* **221**:211–216.

Jones PW (1990): *In vitro* selection for disease resistance. In *Plant Cell Line Selection*, Dix PJ (ed). VCH, Weinheim, pp 113–150.

Kartha KK (1986): Production and indexing of disease free plants. In *Plant Tissue Culture and Its Agricultural Application*, Withers LA, Alderson PG (eds). Butterworths, London, pp 219–238.

Kartha KK, Gamborg OL (1975): Elimination of cassava mosaic disease by meristem culture. *Phytopathology* **65**:826–828.

Knudson L (1946): A new nutrient solution for germination of orchid seed. *Am Orch Soc Bull* **15**:214–217.

Kwon TH, Kim YS, Lee JH, Yang MS (2003): Production and secretion of biologically active human granulocyte-macrophage colony stimulating factor in transgenic tomato cell suspension cultures. *Biotechnol Lett* **25**:1571–1574.

Linsmaier EM, Skoog F (1965): Organic growth factor requirements of tobacco tissue cultures. *Physiol Plant* **18**:100–127.

Manganaris GA, Economou AS, Boubourakas IN, Katis NI (2003): Elimination of PPV and PNRSV through thermotherapy and meristem-tip culture in nectarine. *Plant Cell Rep* **22**:195–200.

McCown BH, Lloyd G (1981): Woody plant medium (WPM)—a mineral nutrient formulation for microculture for woody plant species. *Hort Sci* **16**:453.

Murashige T, Skoog F (1962): A revised medium for rapid growth and bioassay with tobacco tissue cultures. *Physiol Plant* **15**:473–497.

Nishihara M, Seki M, Kyo M, Irifune K, Morikawa H (1995): Transgenic haploid plants of *Nicotiana rustica* produced by bombardment mediated transformation of pollen. *Trans Res* **4**:341–348.

Nitsch JP, Nitsch C (1969): Haploid plants from pollen grains. *Science* **163**:85–87.

Ochatt SJ, Marconi PL, Radice S, Arnozis PA, Caso OH (1999): *In vitro* recurrent selection of potato: Production and characterization of salt tolerant cell lines and plants. *Plant Cell Tiss Org Cult* **55**:1–8.

Owen HR, Wengaerd D, Miller AR (1991): Culture medium pH is influenced by basal medium, carbohydrate source, gelling agent, activated charcoal, and medium storage method. *Plant Cell Rep* **10**:583–586.

Palmer CE, Keller WA (1997): Pollen embryos. In *Pollen Biotechnology for Crop Production and Improvement*, Shivanna KR, Sawhney VK (eds). Cambridge Univ Press, Cambridge, UK, pp 392–422.

Potdar UA, Khan BM, Rawal SK (1999): Genotype dependent somatic embryogenesis in sunflower (*Helianthes annuus* L.). In *Plant Tissue Culture and Biotechnology*, Kavi Kishor, PB (ed). Univ Press, Hyderabad, India, pp 64–70.

Rao SR, Ravishankar GA (2002): Plant cell cultures: Chemical factories of secondary metabolites. *Biotechnol Adv* **20**:101–153.

Reis LB, Paiva Neto VB, Todedo Picoli EA, Costa MGC, Rego MM, Carvalho CR, Finger FL, Otoni WC (2003): Axillary bud development of passionfruit as affected by ethylene precursor and inhibitors. *In Vitro Cell Dev Biol-Plant* **39**:618–622.

Schenk RU, Hildebrandt AC (1972): Medium and techniques for induction and growth of monocotyledonous and dicotyledonous plant cell cultures. *Can J Bot* **50**:199–204.

Shadwick FS, Doran PM (2005): Foreign protein expression using plant cell suspension and hairy root cultures. In *Molecular Farming: Plant-Made Pharmaceuticals and Technical Proteins*, Fischer R, Schillberg S (eds). Wiley, New York, pp 13–26.

Shin YJ, Hong SY, Kwon TH, Jang YS, Yang MS (2003): High level of expression of recombinant human granulocyte-macrophage colony stimulating factor in transgenic rice cell suspension cultures. *Biotechnol Bioeng* **82**:778–783.

Smith WA, Chaleff RS (1990): Herbicide resistance. In *Plant Cell Line Selection*, Dix PJ (ed). VCH, Weinheim, pp 151–166.

Smith ML, Mason HS, Shuler ML (2002): Hepatitis B surface antigen (HbsAg) expression in plant cell culture: Kinetics of antigen accumulation in batch culture and its intracellular form. *Biotechnol Bioeng* **80**:812–822.

Sugiyama M (1999): Organogenesis *in vitro*. *Curr Opin Plant Biol* **2**:61–64.

Wang L, Wang G, Hong N, Tang R, Deng X, Zhang H (2006): Effect of thermotherapy on elimination of apple stem grooving virus and apple chlorotic leaf spot virus for *In vitro*-cultured pear shoot tips. *Hort Sci* **41**:729–732.

White PR (1963): *The Cultivation of Animal and Plant Cells*, 2nd ed. Ronald Press Company, New York.

CHAPTER 6

Molecular Genetics of Gene Expression

MARIA GALLO

Agronomy Department, Cancer/Genetics Research Complex, University of Florida, Gainesville, Florida

ALISON K. FLYNN

Veterinary Medical Center, University of Florida, Gainesville, Florida

6.0. CHAPTER SUMMARY AND OBJECTIVES

6.0.1. Summary

Along the information pipeline from DNA (a gene) to the production of a protein, there are many steps where gene expression can be controlled. In eukaryotes, such as plants, transcriptional control is considered the major form of gene regulation. Because of its importance, transcriptional regulation has been the best studied and probably the most manipulated. However, it is becoming clearer each day that posttranscriptional mechanisms of gene regulation are critical because levels of transcription are not always well correlated with functional protein levels. Additionally, as the area of proteomics advances, and as we move from genetically engineering plants to improve their performance or enhance their utility in a traditional agricultural setting, to using plants as biofactories to produce proteins, posttranslational regulation will gain in importance.

6.0.2. Discussion Questions

1. What are the differences between DNA and RNA?
2. Describe the main parts of a gene and their functions.
3. How important are *cis*-regulatory elements and *trans*-acting factors in gene regulation?
4. What are the control points that can regulate gene expression?

6.1. THE GENE

6.1.1. DNA Coding for a Protein via the Gene

From Chapter 2, we saw that there are several definitions of a gene. In this chapter, *gene* means a specific segment of DNA, including its regulatory regions, that code for a protein.

Plant Biotechnology and Genetics: Principles, Techniques and Applications, Edited by C. Neal Stewart, Jr.

In this chapter, we describe the *central dogma* of genetics, which involves information flow from DNA to RNA via transcription in the nucleus, followed by RNA transport into the cytoplasm, where it is translated into protein. Let's first look at DNA. What exactly is DNA? DNA, or *deoxyribonucleic acid*, is simply a chemical, a double-stranded, helical polynucleotide, to be specific. However, in the proper biological context, this chemical determines such traits as the color of a petunia petal, the scent of a citrus blossom, the sweetness of a corn kernel, the strength of a cotton fiber, and the yield of a wheat head in the face of biotic and abiotic stress. The majority of a plant's DNA is found within the nucleus of each cell. Specific segments of the nuclear DNA, called *genes*, contain all the information required for the cell to make proteins (polypeptides) that are responsible for traits. Each protein-coding gene codes for a particular polypeptide, which is composed of a unique linear arrangement of amino acids as determined by the gene sequence.

6.1.2. DNA as a Polynucleotide

Before describing how the DNA of a gene can lead to the production of a protein (gene expression), the chemical structure of DNA must be understood. DNA is composed of two strands of *deoxyribonucleotides* [sugar (*deoxyribose*) + phosphate + a nitrogenous *base*—(either adenine (A), guanine (G) (both are *purines*), cytosine (C), or thymine (T) (both are *pyrimidines*)] (Fig. 6.1). The two strands have a right-handed (clockwise) helical shape, the so-called double helix (Watson and Crick's model), with the sugars and phosphates forming the backbone (or outside), and the bases located in the center of the molecule (Fig. 6.2). It is important to note here that the phosphates of the DNA backbone are negatively charged, and this will allow proteins that have positively charged domains to bind to the DNA. The importance of such DNA–protein binding will be discussed later in this chapter in terms of controlling gene expression. The deoxyribonucleotides of each strand are paired through specific hydrogen bonding of their respective bases: A on one strand always pairs with T on the other via two hydrogen bonds, and G on one strand always pairs with C on the other via three hydrogen bonds. This hydrogen bonding keeps the two strands together. Knowing the sequence of only one of the strands will provide all the information required to make the other strand through this specific or complementary base-pairing mechanism. It is also sufficient information for scientists to deduce the sequence of the second strand. It is important to note that the strands have directionality, each has a $5'$ end and a $3'$ end, and when the DNA strands pair, they are said to be antiparallel (Fig. 6.1). Since the bases are what distinguish the nucleotides from one another, a gene sequence conventionally is written by listing the linear sequence of the bases of one strand (the *coding* strand; see below) starting from the $5'$ end and proceeding to the $3'$ end.

6.2. DNA PACKAGING INTO EUKARYOTIC CHROMOSOMES

In a cell, the DNA described above is not "naked," but in association with proteins that together are packaged as chromosomes that can fit within the nucleus. Specifically, eukaryotic chromosomes are composed of DNA (2 nm in diameter) in association with histone and nonhistone proteins to form a nucleoprotein structure called *chromatin* (200 nm in

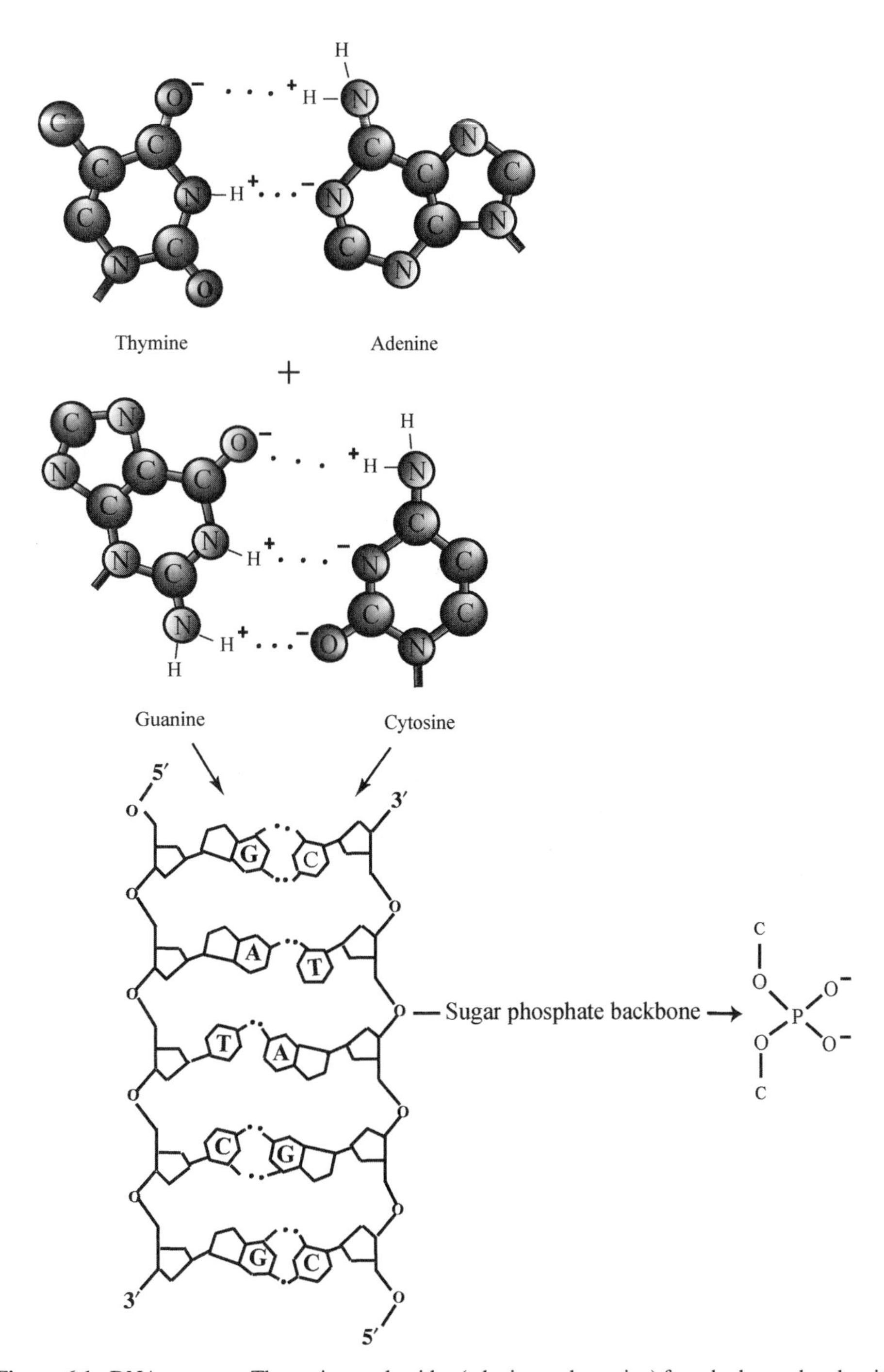

Figure 6.1. DNA structure. The purine nucleotides (adenine and guanine) form hydrogen bonds with the pyrimidine nucleotides (thymine and cytosine, respectively). Nucleotides are strung together by a sugar phosphate backbone that has an antiparallel orientation (5′–3′) to the complementary base pairs.

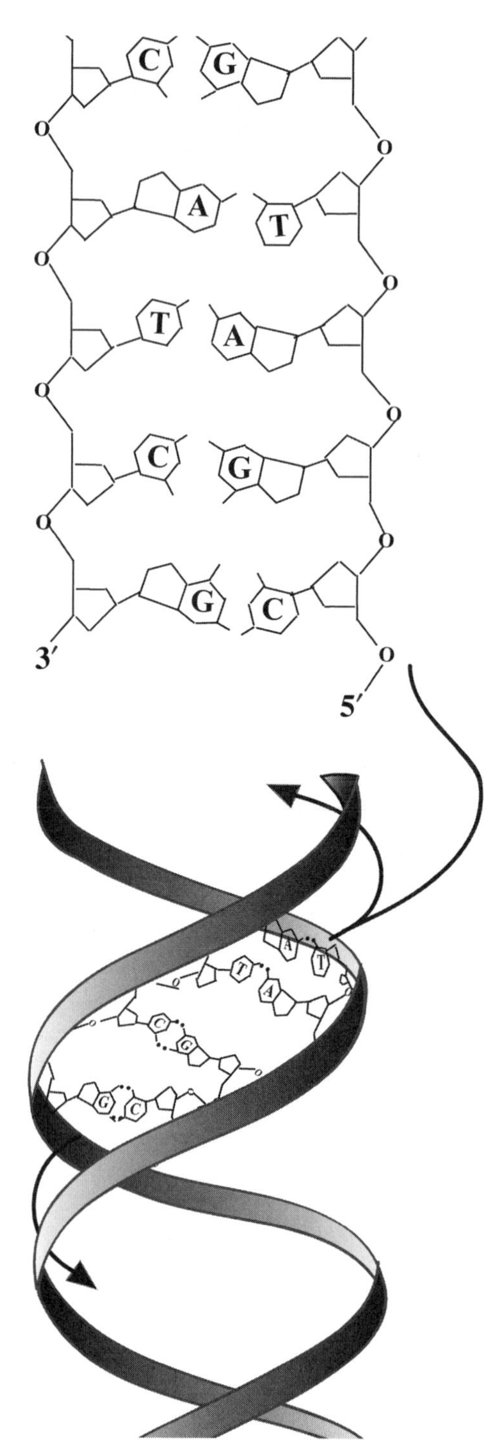

Figure 6.2. The two antiparallel chains of nucleotides strung together by a negatively charged sugar phosphate backbone form a right-handed double helix with the base pairs in the center of the helix.

diameter). Initially, the histones produce a complex with the DNA to form the first structural unit of chromatin called the *nucleosome*. The nucleosome consists of DNA wrapped twice around a core of eight histones to form a 10-nm fiber (Fig. 6.3). This fiber is further folded to result in a chromatin fiber. Chromatin is dynamic in terms of its ability to coil and uncoil

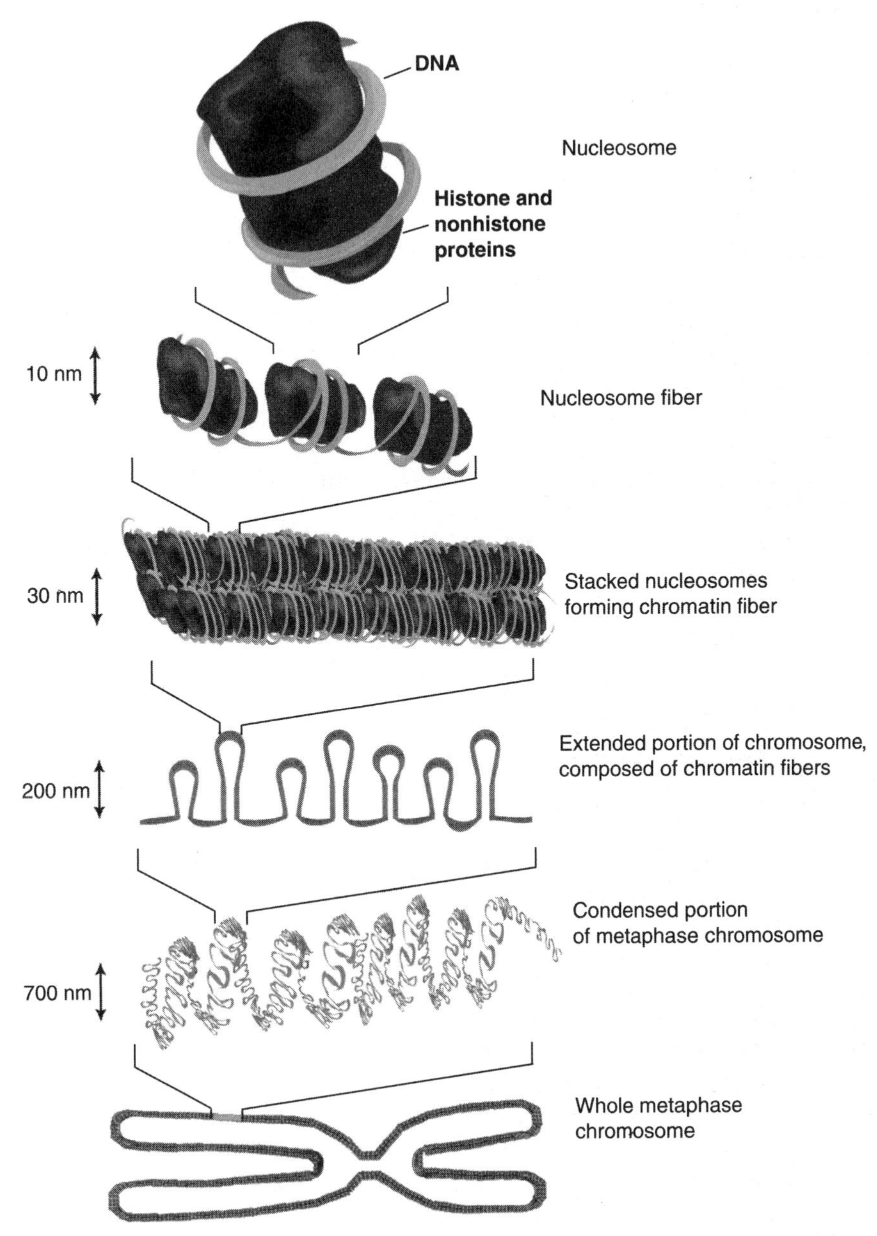

Figure 6.3. Chromatin structure. The different levels of chromatin structure are shown. The basic building block consists of the nucleosome–DNA wrapped around a core of histone and nonhistone proteins. Nucleosomes are strung together by strands of DNA, which are densely packed to create chromatin. Tight winding of chromatin creates the metaphase chromosome.

during the lifecycle of a cell. Chromatin is in its most condensed or coiled form during mitosis, when it forms a metaphase chromosome (700 nm in diameter). Regulation of gene expression, as detailed below, involves nucleosome uncoiling and this change in DNA conformation is termed *chromatin remodeling*. So, chromatin not only is necessary for packaging DNA to conveniently fit within the nucleus of a cell but also plays an important role in gene expression.

6.3. TRANSCRIPTION

6.3.1. Transcription of DNA to Produce Messenger RNA (mRNA)

How does the information contained in a protein-coding gene on a chromosome within the nucleus lead to the formation of a polypeptide in the cytoplasm? The key is that the DNA of a gene does not directly participate in the synthesis of a polypeptide. The gene's information or message is faithfully carried by another molecule out of the nucleus and into the cytoplasm. The first step in this information flow from DNA to polypeptide is to synthesize this "messenger" from the gene in a process called *transcription*. The transcribed messenger molecule, also referred to as a *transcript*, is another polynucleotide aptly named *messenger ribonucleic acid* (mRNA). Like DNA, mRNA is composed of nucleotides that are assembled in a $5' \rightarrow 3'$ direction; however, mRNA is made up of *ribonucleotides*, because its sugar is a *ribose*. Messenger RNA also differs from DNA in that it is a single-stranded molecule and, in place of T, it has another base, uracil (U), which can form a complementary base pair with A. Only one DNA strand of a gene is used as a "template" during transcription to create the mRNA. The order or linear sequence of bases in this DNA *template* strand ($3' \rightarrow 5'$) determines the sequence of the mRNA ($5' \rightarrow 3'$) because transcription works through complementary base pairing. Consequently, the mRNA made is a complement of the DNA template strand of the gene and an exact copy of the other DNA strand of the gene (the *coding* strand) except for having a U where a T would be located (Fig. 6.4).

Transcription is carried out by the enzyme *RNA polymerase II* (RNAP II) in eukaryotes such as plants. RNAP II does not act alone. Its binding and activity are controlled by both DNA sequences located within the gene (*cis-regulatory region*) and by proteins (*trans-acting factors*) called *transcription factors*, which can be general in helping transcribe many genes or specific to one or a few genes. Gaining a better understanding of the roles that *cis*-regulatory regions and transcription factors play in gene regulation is an active area of current research. The *general transcription factors* (GTFs) are necessary for RNAP II to transcribe DNA. The specific transcription factors affect the efficiency or the rate of RNAP II transcription for specific genes.

The *cis*-regulatory region controlling transcription by RNAP II, called the *promoter*, is located at a gene's 5′ end (using the coding strand as a reference). The promoter is composed of a core promoter plus other promoter elements that help define when and where a gene is transcribed. The core promoter element is where RNAP II and the GTFs bind to begin transcription. The *transcription start site* (the gene location where the first ribonucleotide of the RNA being synthesized will base-pair) is designated as the +1 site (i.e., the first base in the transcript), so the gene promoter is therefore located upstream of (or before) the +1 site and its nucleotides are given negative numbers, whereas all nucleotides after the +1 site are positive sequential numbers (Fig. 6.5). As will be detailed below, the actual

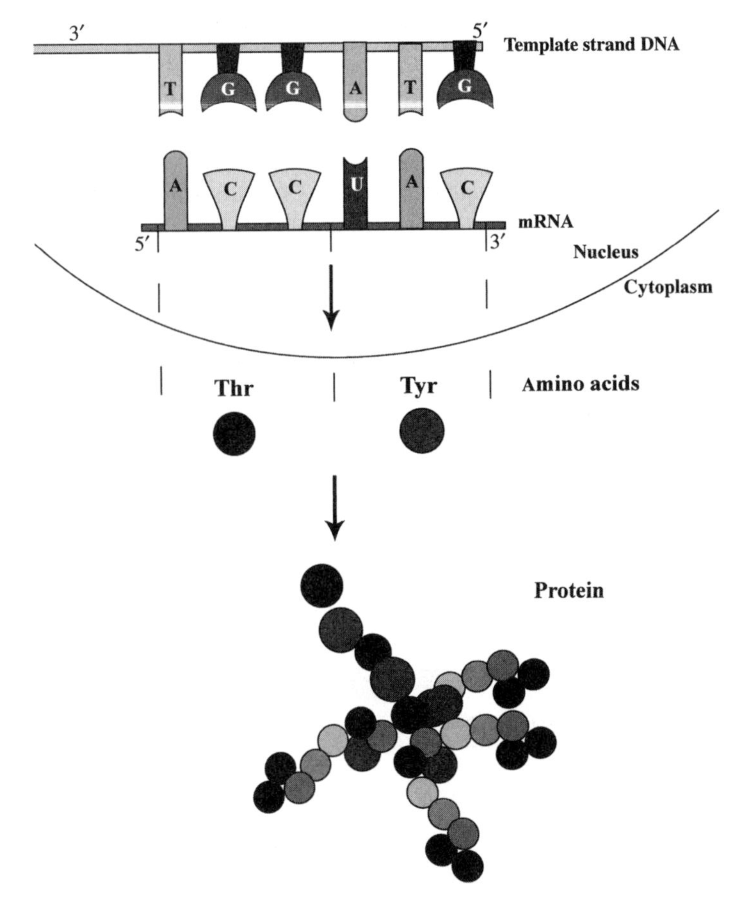

Figure 6.4. The central dogma: DNA is transcribed to RNA in the cell nucleus. RNA is translated to protein in the cell cytoplasm. See color insert.

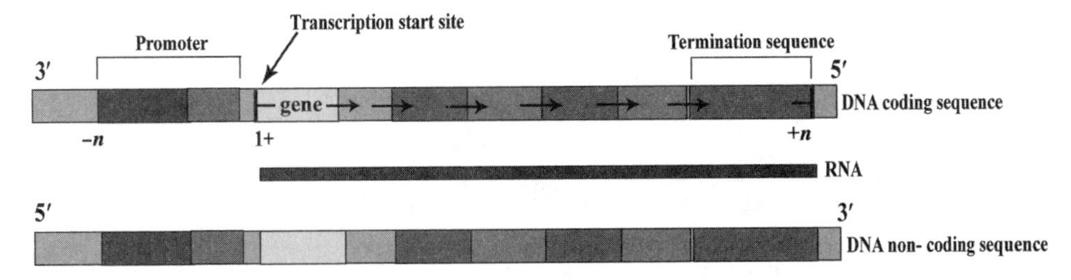

Figure 6.5. The structure of the promoter in relation to the gene and termination sequence. The transcription start site is designated by 1^+, and DNA sequences that are downstream or toward the 3′ end of the DNA strand are described by a negative number. In contrast, nucleotide sequences upstream (toward the 5′ end of the DNA strand) are described with a positive number.

protein-coding portion of the gene will begin with an ATG sequence (AUG in the mRNA), but the +1 site is generally well upstream or in front of that sequence. Therefore, the portion of the gene from the +1 site up until the ATG sequence is termed the *5′ untranslated region* (5′UTR; this sequence is located in the gene and in the transcribed mRNA, but does not get read for translation). Similarly, at the end of a gene, there is also a portion that is transcribed into mRNA, but is not translated, and that is termed the *3′ untranslated region* (3′UTR).

A core promoter element found in most eukaryotic genes consists of a *consensus sequence* (the bases most often found at certain positions that have been conserved throughout evolution) located at approximately −25 to −30 called the *TATA box* or the *Goldberg–Hogness box* (Goldberg 1979). It is called TATA because the bases T and A are prominent. Initially, RNAP II and the GTFs are bound to the core promoter element in an inactive state called the *preinitiation complex* (PIC). Then 11–15 base pairs of the gene around the transcription start site break their bonds, thereby changing the DNA conformation into an open complex, and the template strand of the promoter becomes located in the active site of RNAP II to initiate transcription at a basal level (Fig. 6.6).

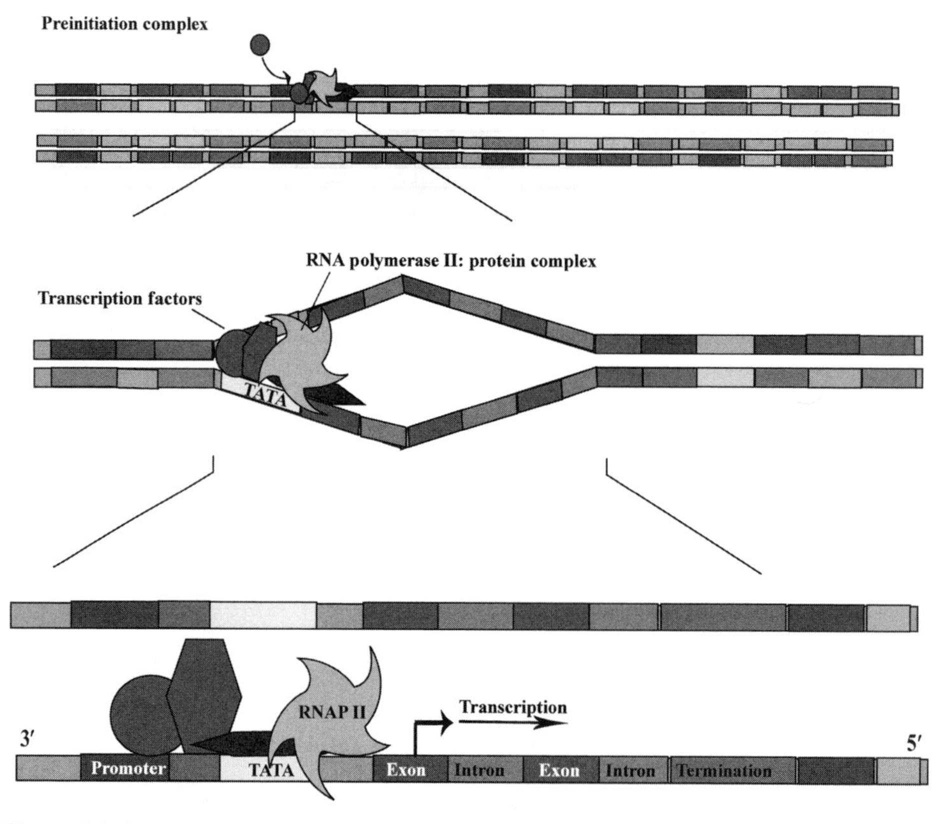

Figure 6.6. Overview of the early steps of transcription. A preinitiation complex is formed by a complex of transcription factors and RNA polymerase II (RNAP II). Association of the preinitiation complex with the start sequence (TATA) of the coding strand of DNA causes a conformation change and hydrogen bond breakage. This causes the DNA strands to separate so that transcription can proceed. See color insert.

Promoter elements that are not required for transcription initiation, but influence the level, rate, timing or tissue specificity of transcription are the *CAAT box* (CCAAT), and gene-specific response elements. The CAAT box is generally located at sites −70 to −80. The gene-specific response elements vary in their sequence and location within the promoter. A third type of *cis*-regulatory element is an *enhancer*, the location of which varies from gene to gene. Unlike a promoter element, an enhancer can function even at long distances (>1 kb) upstream or downstream of the transcriptional start site (Khoury and Gruss 1983), and its orientation can be inverted without losing its function. The CAAT box, gene-specific response elements, and enhancers carry out their functions by binding specific transcription factors (Fig. 6.7).

6.3.2. Transcription Factors

Transcription factors are regulatory proteins that bind to DNA and to other regulatory proteins to effect gene expression, as described above. Thus, there are transcription factor genes whose expression affects the regulation of other genes. In order to carry out their functions, they generally have specific portions, or domains. There are two main domains in transcription factors, a *DNA binding domain* and a *trans-acting domain*. The *DNA binding domain* does just that; it allows the transcription factor to bind directly to a DNA *cis*-regulatory element. DNA binding domains are characterized by specific structures or motifs. For example, some DNA binding domains have a helix–turn–helix motif, a zinc-finger motif, or a leucine zipper motif. The *trans-acting domain* of a transcription factor allows it to bind to RNAP II or to other transcription factors, thus allowing protein–protein interactions. So, with two such domains, a transcription factor

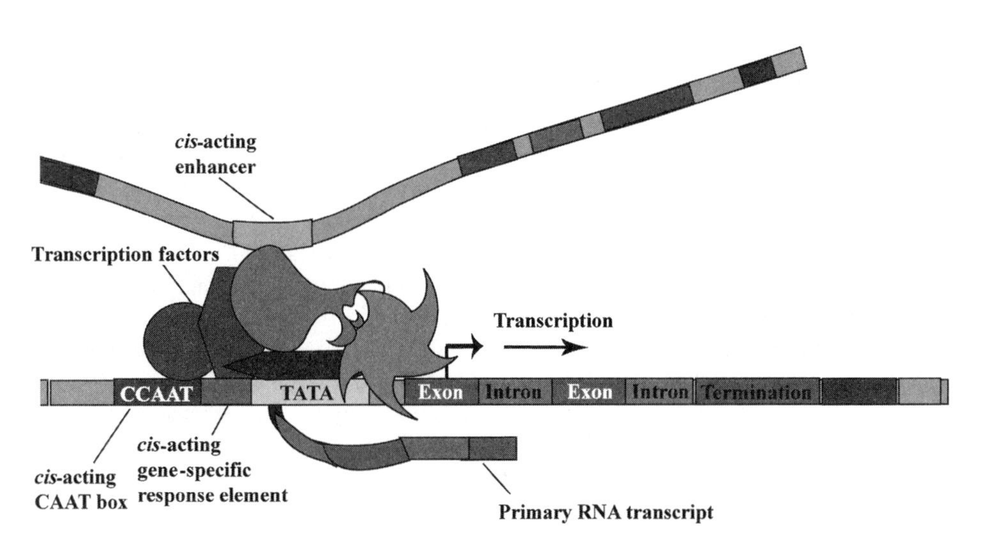

Figure 6.7. Regulation of transcription. The *cis*-acting elements are segments of DNA that regulate transcription; these segments may be adjacent to the gene such as the promoter (CAAT box) and the *cis*-acting gene-specific response elements, or they may be distant to the gene such as enhancers. The *trans*-acting elements are transcription factors and other regulatory proteins that may associate with the promoter, other proteins, or both. See color insert.

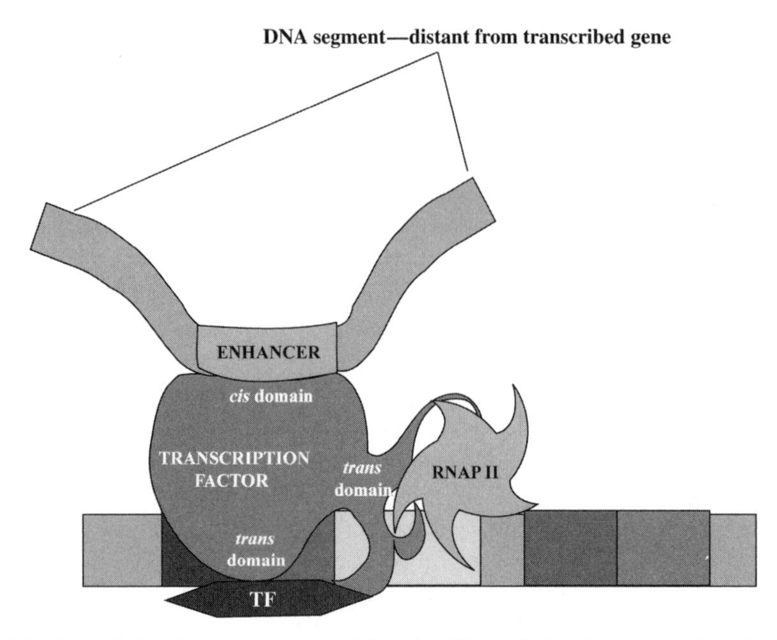

Figure 6.8. Transcription factors structure and function. Transcription factors may have domains that bind *cis*-acting elements such as enhancers, and domains that also bind *trans*-acting elements such as RNA polymerase (RNAP II) and other transcription factors. See color insert.

can simultaneously bind DNA and other transcription factors or RNAP II to regulate gene transcription (Fig. 6.8).

6.3.3. Coordinated Regulation of Gene Expression

Eukaryotes can coordinately express subsets of many different genes in response to particular biotic and abiotic signals because those genes will contain common *cis*-regulatory or response elements in their promoters or enhancers that allow them to recognize the same signals. These elements have a consensus sequence that can bind specific transcription factors allowing for transcription of those genes. A gene may also contain several different response elements allowing it to be expressed following a number of stimuli. For example, the CBF transcription factors of *Arabidopsis* (Gilmour et al. 1998) can bind to the cold- and dehydration-responsive *cis*-regulatory element called CRT/DRE (C-repeat/dehydration-responsive element) (Baker et al. 1994; Yamaguchi-Shinozaki and Shinozaki 1994) that is found in the promoters of many cold- and dehydration-responsive genes of *Arabidopsis*. So, following cold or water-stress stimuli, those genes containing the CRT/DRE responsive element will be transcribed and provide *Arabidopsis* with increased tolerance to freezing, as well as drought.

6.3.4. Chromatin as an Important Regulator of Transcription

DNA wrapped around histones and coiled to produce chromatin is not accessible for transcription. It is not physically possible for RNAP II to make contact with the DNA for transcription. Chromatin remodeling, as mentioned earlier, is required to allow the appropriate regions of a gene to bind transcription factors and RNAP II for transcription. This remodeling "opens up" the DNA to make it accessible to RNAP II and transcription factors. Following remodeling, the promoter region no longer contains histones, thereby making the *cis*-regulatory elements free to bind to the necessary transcription factors and RNAP II to begin transcription. Chromatin remodeling is done by various multiprotein complexes that have ATP-ase activity (use energy) to bind directly to particular regions of the DNA to move the nucleosomes to a new position to expose the DNA for transcription (Vignali et al. 2000).

Chromatin structure also can be changed through the covalent addition of acetyl groups (CH_3CO) to the histones of the nucleosome. When the acetyl groups are added to the histone tails, they are no longer positively charged and consequently the negatively charged DNA can disengage from them. The acetyl groups are added by enzymes called *histone acetyltransferases* (HATs). It is known that certain transcription factors have acetyltransferase activity or can recruit these enzymes to the DNA, thereby altering chromatin structure and allowing for transcription. Chromatin structure can be restored by histone deacetylase complexes (HDACs) that remove the acetyl groups from these histones. A good example of such regulation of gene expression can be found in the control of flowering in *Arabidopsis*. If the gene *flowering locus C* (*FLC*) is expressed, flowering does not occur. However, if the *flowering locus D* (*FLD*) gene is active, it produces a deacetylase that removes acetyls from histones around *FLC*. Consequently, transcription of *FLC* cannot occur owing to the restoration of chromatin structure, and silencing of *FLC* allows flowering (Fig. 6.9) (He et al. 2003).

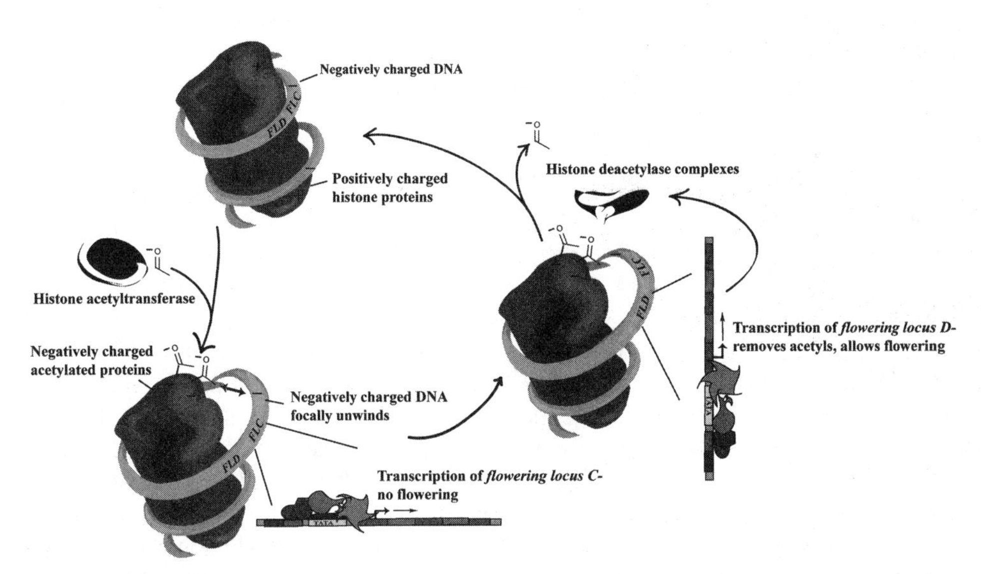

Figure 6.9. Control of transcription by chromatin remodeling. Genetic regulation of flowering in *Arabidopsis*. If the gene *flowering locus C* (*FLC*) is expressed, flowering does not occur. However, if the *flowering locus D* (*FLD*) gene is active, it produces a deacetylase that removes acetyls from histones around *FLC*. Consequently, transcription of *FLC* cannot occur because of the restoration of chromatin structure, and silencing of *FLC* allows flowering.

6.3.5. Regulation of Gene Expression by DNA Methylation

DNA methylation (—CH_3 groups attached to DNA of the promoter or coding region) is a major factor in regulating gene expression. There appears to be an inverse relationship between percent methylation and the degree of expression. *Hypo*methylation is associated with higher gene expression, whereas *hyper*methylation is associated with greater gene silencing. The most common methylated base in eukaryotic genomic DNA is 5-methylcytosine (m^5C).

Plants generally have higher levels of DNA methylation than do mammals. Also, in plants methylation occurs mainly in transposable elements and other repeat sequences. If a transposon is methylated, it will be inactive and not hop around the genome, but it can be activated if the methylation is removed. However, as in mammals, methylation of the cytosine on both strands of the CpG dinucleotide (linear sequence of cytosine followed by a guanine separated by a phosphate, to be distinguished from a cytosine base-paired to a guanine) is common in plants and is carried out by DNA methyltransferases such as MET1 in *Arabidopsis*. This enzyme is responsible for maintenance of global genomic methylation. Plants mutant for MET1 have significantly lower levels of methylation and show late flowering phenotypes (Kankel et al. 2003). Also, transgenes that are genetically engineered into plants and become highly methylated are not expressed. However, if these plants have a defective MET1, these transgenes will no longer be silenced. Plants also have methylation of CpNpG trinucleotides ("N" can be any of the four DNA bases) and asymmetric CpNpN trinucleotide sites that are performed by specific enzymes that are unique to plants such as *chromomethylases* (CMT) and *domain-rearranged methylases* (DRM). The CMTs appear to be involved in maintaining methylation of sites that are heavily methylated to keep them silenced. The DRMs function in RNA-directed DNA methylation by somehow recognizing *small interference RNA* (siRNA—these RNAs are usually 20–25 nucleotides long and inhibit expression of specific genes) and then methylating the appropriate DNA sequences. Additionally, it has been shown that chromatin-remodeling factors, as described above, can be necessary for maintaining methylation.

6.3.6. Processing to Produce Mature mRNA

Controlling transcription is one of the most important ways to alter gene expression for biotechnology applications. Many of the mechanisms that plants possess to regulate the transcription of DNA to mRNA have been introduced above. Promoters, transcription factors, chromatin remodeling, and DNA methylation are all crucial for transcriptional control. However, transcription is only the first step in gene regulation. The mRNA that is made through transcription is not mature and is termed a *pre-mRNA* or a *heterogeneous nuclear RNA* (hnRNA). Before a gene transcript is transported out of the nucleus and into the cytoplasm where it will ultimately be translated into protein, it must be processed in several ways: *5′ capping, 3′ polyadenylation (polyA tail)*, and *splicing* out of *introns* and putting together of *exons* (Fig. 6.10). The first processing step, occurring when approximately 20–30 ribonucleotides of the hnRNA have been made, is the addition of a 7-methylguanosine to the 5′ end of the transcript. This cap structure may play a role in mRNA stability by physically protecting the mRNA from 5′ → 3′ *exonucleases*, types of *RNAses*, once it is in the cytoplasm. Most eukaryotic gene protein-coding regions are interrupted by non-protein-coding sequences (introns) that are removed from the hnRNA, so they are not found in mature mRNA. The hnRNA has consensus sequences at the

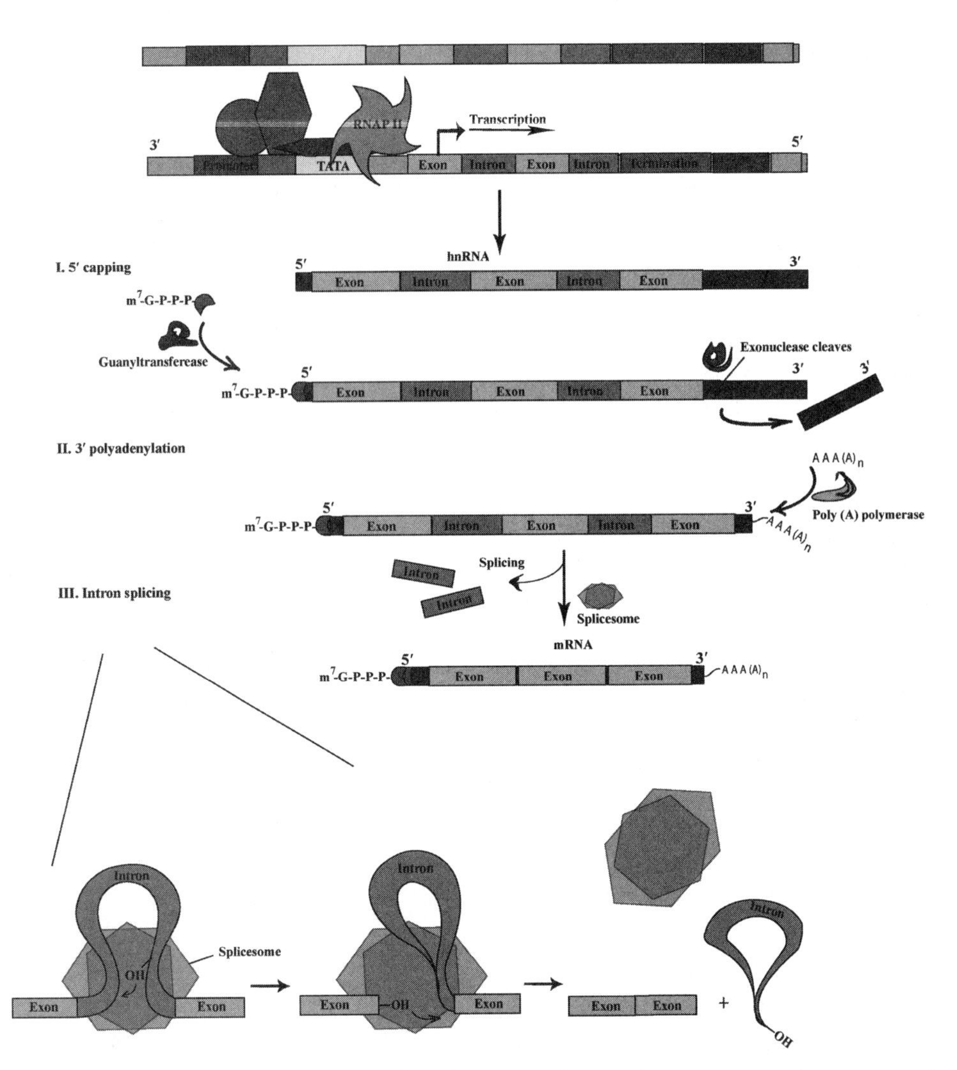

Figure 6.10. Overview of mRNA processing. Three steps of processing must occur prior to export of the mature mRNA out of the nucleus: 5′ capping, 3′ polyadenylation, and intron splicing.

exon–intron junctions (marked by "/") that are required for proper splicing out of the introns. The 5′ exon–intron junction consensus sequence is AG/GURAGU, and the 3′ exon–intron junction consensus sequence is YAG/RNNN (Y, pyrimidine; R, purine; N, either purine or pyrimidine). Also, about 100 nucleotides upstream of the 3′ exon–intron junction there is a branchpoint conserved sequence. A *splicesome* composed of small nuclear RNAs (snRNAs—these RNAs are 100–300 bases in length) and various proteins forms over the intron and helps in the splicing process. Most mRNAs contain a polyadenylated 3′ end consisting of 200 A residues. This polyA tail acts as a protective buffer against RNAses that could digest the mRNA from the 3′ end, and thus stabilizes the molecule. Approximately 10–30 bp upstream of this polyA tail is the invariant hexamer sequence AAUAAA (Fig. 6.10).

6.4. TRANSLATION

How does the information in the mRNA result in the synthesis of a polypeptide? Multiple cellular players are involved in the synthesis of a polypeptide. First, the structure of a polypeptide needs to be understood. Polypeptides are made up of a linear sequence of amino acids. There are 20 common types of amino acids (Table 6.1), and to form a polypeptide, amino acids are joined together in a chain by peptide bonds (Fig. 6.11). Proteins can be composed of either a single polypeptide or multiple polypeptide chains that are the same or different in amino acid sequence (Fig. 6.11).

Once the mRNA is transported out of the nucleus, it must be properly "read" or *translated* by *ribosomes* in order to produce a polypeptide. But how many nucleotides of the mRNA are needed to code for one amino acid? Three consecutive nucleotides, called a *codon*, are required to be read to specify one amino acid. This code is *nonoverlapping*, meaning that once a triplet is read, the cellular machinery reads the next three nucleotides and so on in linear fashion. Therefore, within a given *reading frame* (there are three possible reading frames; see Fig. 6.12), a nucleotide cannot be present in more than one codon. Since there are four nucleotide possibilities (A, G, C, or U) at each of the three codon positions, there are $4 \times 4 \times 4 = 64$ different combinations or codons (Table 6.2). A codon is written in the $5' \rightarrow 3'$ direction as it would be read on the mRNA molecule. Since there are more codons than amino acids, some codons actually specify the same amino acid, and so the code is considered to be degenerate in that regard. Three codons (UAA, UAG, and UGA) do not code for any amino acid. These are *stop codons*, and when any one is read, it signals the cellular machinery to stop translation.

TABLE 6.1. The 20 Amino Acids Commonly Found in Proteins

Amino Acid	Three-Letter Abbreviation	One-Letter Abbreviation
Alanine	Ala	A
Arginine	Arg	R
Asparagine	Asn	N
Aspartate	Asp	D
Cysteine	Cys	C
Glutamine	Gln	Q
Glutamate	Glu	E
Glycine	Gly	G
Histidine	His	H
Isoleucine	Ile	I
Leucine	Leu	L
Lysine	Lys	K
Methionine	Met	M
Phenylalanine	Phe	F
Proline	Pro	P
Serine	Ser	S
Threonine	Thr	T
Tryptophan	Try	W
Tyrosine	Tyr	Y
Valine	Val	V

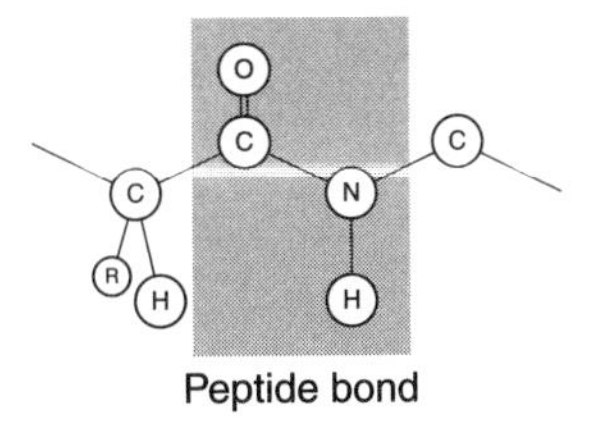

Single polypeptide chain

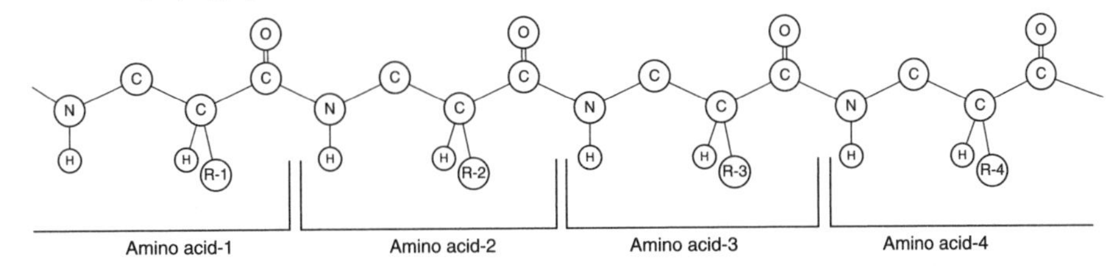

Two identical polypeptide chains

Phe-Val-Asn-Gln-His-Leu-Cys-Gly-Ser-His-Gly-Val
S
S
Phe-Val-Asn-Gln-His-Leu-Cys-Gly-Ser-His-Gly-Val

Two unique polypeptide chains

Phe-Val-Asn-Gln-His-Leu-Cys-Gly-Ser-His-Gly-Val
S — S
Gly-Ile-Val-Glu-Cys-Cys-Ala-Ser-Val-Cys-Ser-Leu

Figure 6.11. Polypeptide structure. The building block of a polypeptide is the peptide bond formed between amino acids. Peptide bonds connect amino acids to create a polypeptide chain. Proteins are formed through the association of individual polypeptide chains that may be identical to each other or unique in sequence.

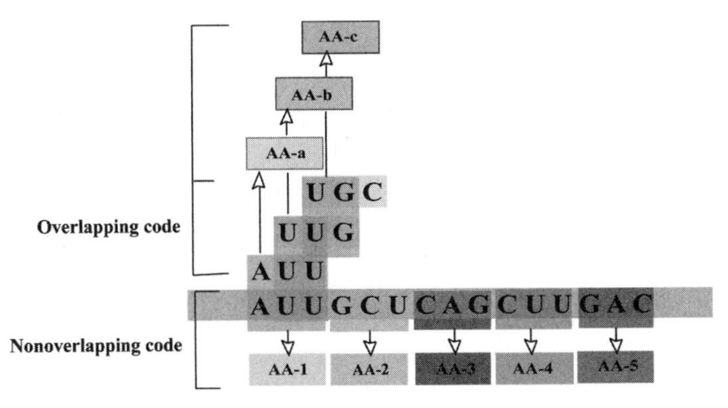

Figure 6.12. The genetic code gives rise to either overlapping or nonoverlapping reading sequences. A codon consists of three consecutive nucleotides that code for an amino acid. The nucleotides in a codon may give rise to multiple amino acids depending on the reading frame.

TABLE 6.2. The Genetic Code—mRNA Codons and Amino Acids Encoded[a]

	Second Base								
First Base	U		C		A		G		Second Base
U	UUU	Phe	UCU	Ser	UAU	Tyr	UGU	Cys	U
	UUC		UCC		UAC		UGC		C
	UUA	Leu	UCA		UAA	Stop	UGA	Stop	A
	UUG		UCG		UAG		UGG	Trp	G
C	CUU	Leu	CCU	Pro	CAU	His	CGU	Arg	U
	CUC		CCC		CAC		CGC		C
	CUA		CCA		CAA	Gln	CGA		A
	CUG		CCG		CAG		CGG		G
A	AUU	Ile	ACU	Thr	AAU	Asn	AGU	Ser	U
	AUC		ACC		AAC		AGC		C
	AUA		ACA		AAA	Lys	AGA	Arg	A
	AUG	Met	ACG		AAG		AGG		G
G	GUU	Val	GCU	Ala	GAU	Asp	GGU	Gly	U
	GUC		GCC		GAC		GGC		C
	GUA		GCA		GAA	Glu	GGA		A
	GUG		GCG		GAG		GGG		G

[a]The codons are written in the $5' \rightarrow 3'$ direction.

Since an mRNA is a long molecule containing many nucleotides, where does the translational machinery start looking to begin reading each codon? The first codon read is called the *initiation* (or *start*) *codon*, and it is usually AUG that codes for methionine [earlier we mentioned that the protein-coding portion of the gene (DNA) began with ATG]. In eukaryotes, the initiation codon is surrounded by a consensus sequence termed the *Kozak sequence* (ACCAUGG) (Kozak 1986, 1987), which indicates to the translational machinery to begin translation with this codon. If this sequence is not present, this codon will be missed and the cellular machinery will continue to scan down the mRNA until it finds a suitable initiation codon, if present. As mentioned above, three different reading frames are possible. The start codon defines what the correct reading frame will be for any particular gene. As you will see later, this is an important consideration for biotechnology.

6.4.1. Initiation of Translation

Translation of the mRNA is done in connection with organelles called *ribosomes* and another type of RNA termed *transfer RNA* (tRNA). In eukaryotes, ribosomes are complex and composed of two subunits, one large and the other small. The large subunit contains three types of ribosomal RNAs (rRNAs) (28*S* rRNA, 5*S* rRNA, and 5.8*S* rRNA), along with 49 proteins. The small subunit contains the 18*S* rRNA, and 33 proteins. A ribosome will bind to the 5′ end of the mRNA and move down toward the 3′ end as translation proceeds. Specifically, starting at the 5′ cap of the mRNA, the small subunit of the ribosome along with initiation factors will bind and move down the mRNA until it encounters the proper initiation codon. Then the correct amino acid (the initiation codon codes for methionine; therefore methionine is always the first amino acid in the initial polypeptide) is brought to it via a tRNA molecule and combines with additional factors to form an initiation complex. The tRNA molecule is said to be "charged" when it carries an amino

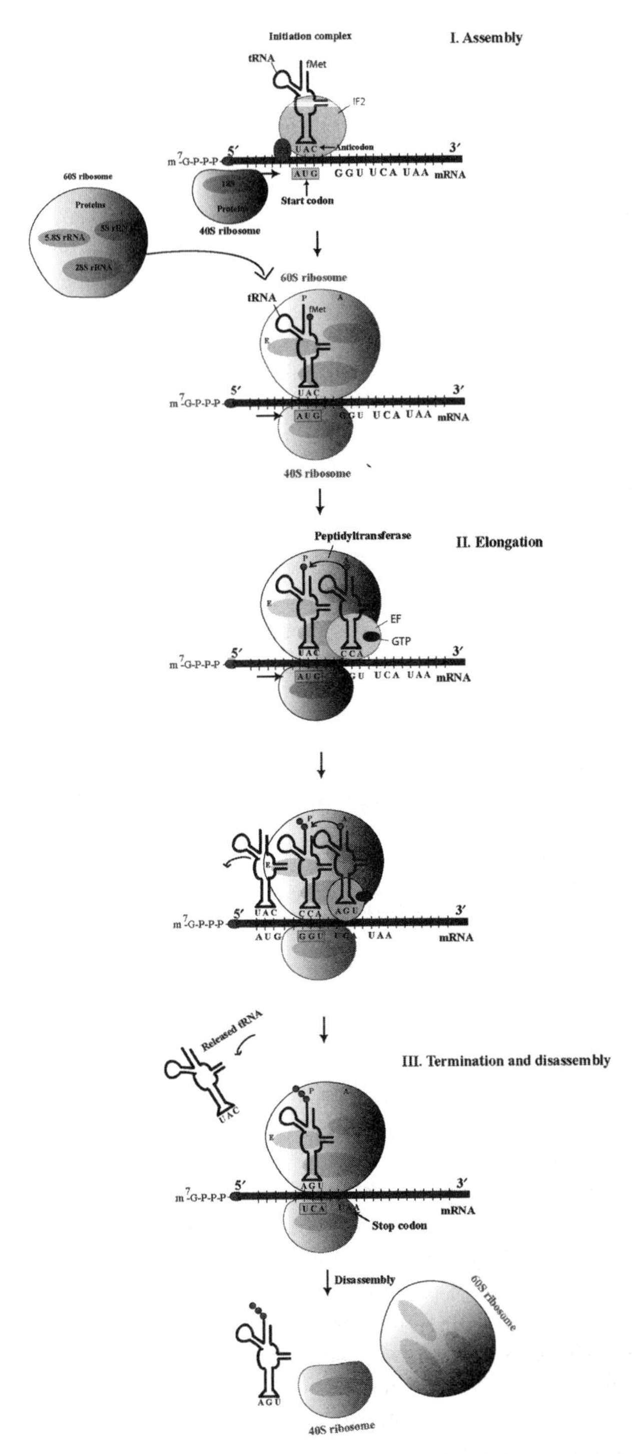

Figure 6.13. Overview of translation showing the structure of tRNA, 60*S* and 40*S* ribosomal subunits. The three steps of translation are shown: ribosome assembly, elongation of the polypeptide chain, and termination. See color insert.

acid. The charged tRNA molecule recognizes the codon through complementary base pairing with a region of it called an anticodon (Fig. 6.13).

6.4.2. Translation Elongation

Now polypeptide synthesis takes place with amino acids joining together as successive codons are read in the elongation phase of translation. Before elongation can occur, the large ribosomal subunit joins to create a complete ribosome. The ribosome now has three sites that can accommodate a tRNA molecule: a peptidyl (P), an aminoacyl (A), and an exit (E) site. The initiator tRNA occupies the P site of the ribosome, which is positioned over the initiator AUG codon and is adjacent to the A site, which at this stage is available and is over the next codon to be read. Then the appropriately charged tRNA for this next codon in the A site enters it, and its anticodon pairs with the codon. A peptide bond then forms between the amino acids that are attached to the tRNAs in the P and A sites. Now the initiator amino acid is released from its tRNA and the ribosome moves down the mRNA or translocates to position the growing polypeptide in the P site and free the A site, which once again positions over the next codon to be translated. The initiator tRNA that no longer is charged is in the E site and it is then free to leave the ribosome and become charged again. This elongation cycle is repeated until the entire polypeptide chain is made.

6.4.3. Translation Termination

Polypeptide synthesis is over when the ribosome encounters a stop codon in its A site. Since no tRNAs can base pair with these stop codons, proteins called "release factors" bind to the ribosome instead. These release factors allow the polypetide chain to be released from the P site as well as the mRNA to no longer bind to the ribosome. The ribosome also splits into its two subunits.

6.5. PROTEIN POSTRANSLATIONAL MODIFICATION

Following translation, polypeptides can be modified in a number of ways before they are fully functional, and in fact, different organisms modify proteins in different ways that can have biological significance. The initiator amino acid, methionine, can be changed or removed. More amino acids can be added, or the polypeptide can be "trimmed" by removing amino acids. Also, amino acids can be modified by the addition of carbohydrate sidechains, phosphates, methyl groups, or conjugated with metals. These modifications can significantly alter the function of proteins, and subsequently control cellular function. For example, phosphorylation is an important mechanism for controlling intracellular signaling. In order to be a functional protein, polypeptides also must be appropriately folded into a three-dimensional conformation, which can occur either spontaneously or under the direction of molecular "chaperones." As mentioned earlier, some proteins are composed of a single polypeptide, whereas others are multimeric, composed of one or more additional polypeptides that form the complete protein. Posttranslational modifications can fundamentally alter gene expression by changing protein function, allowing the cell to rapidly respond to variable internal and external stimuli. Understanding how to control

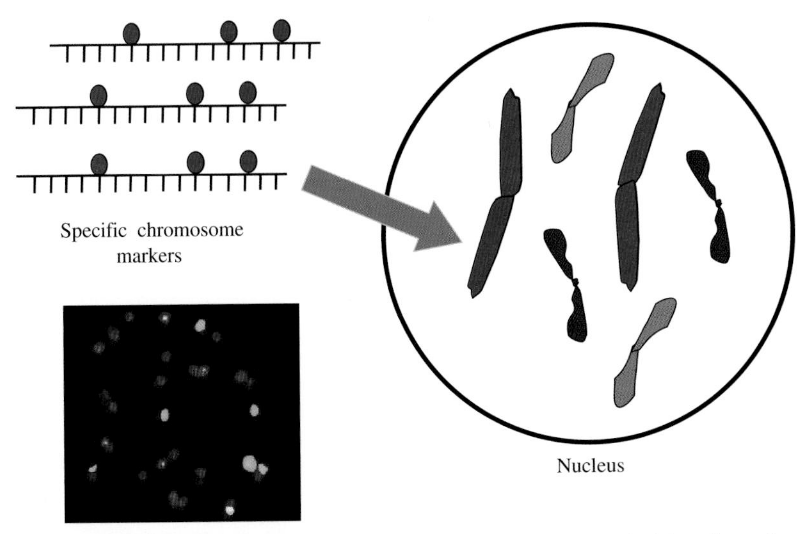

Figure 2.10. Fluorescent in situ hybridization (FISH) shows the physical location of a specific transgene or DNA. The inset (bottom left; courtesy of Chris Pires) shows *Brassica napus* mitotic metaphase chromosomes stained blue with two different centromere probes (red and green).

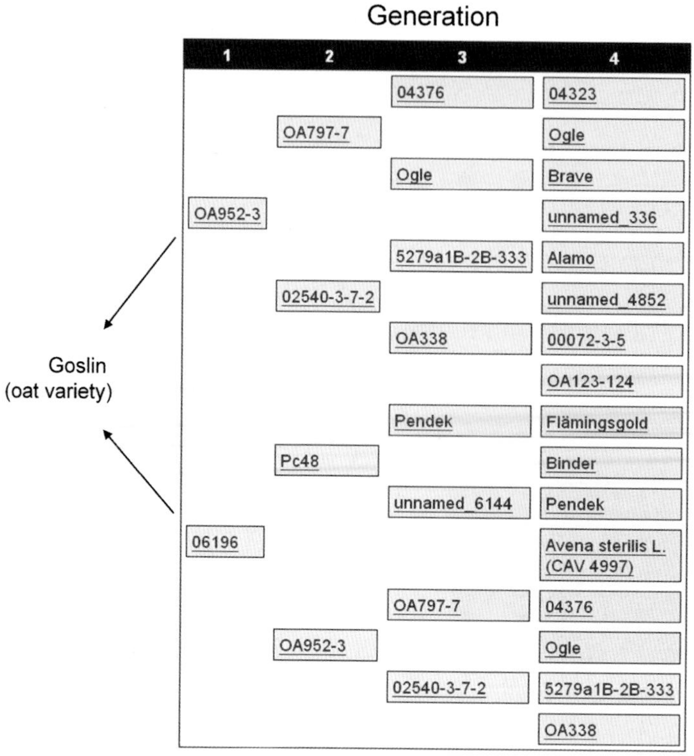

Figure 3.6. The pedigree of an oat variety named "Goslin." The parents of the cross from which Goslin was selected are shown on the left in column 1, grandparents and great-grandparents are shown in columns 2 and 3, and so on. Lines identified by numbers (e.g., OA952-3) were probably elite breeding lines that did not become varieties. This pedigree tree was drawn using an online database (http://avena.agr.gc.ca) that records pedigrees of historical oat varieties for many generations.

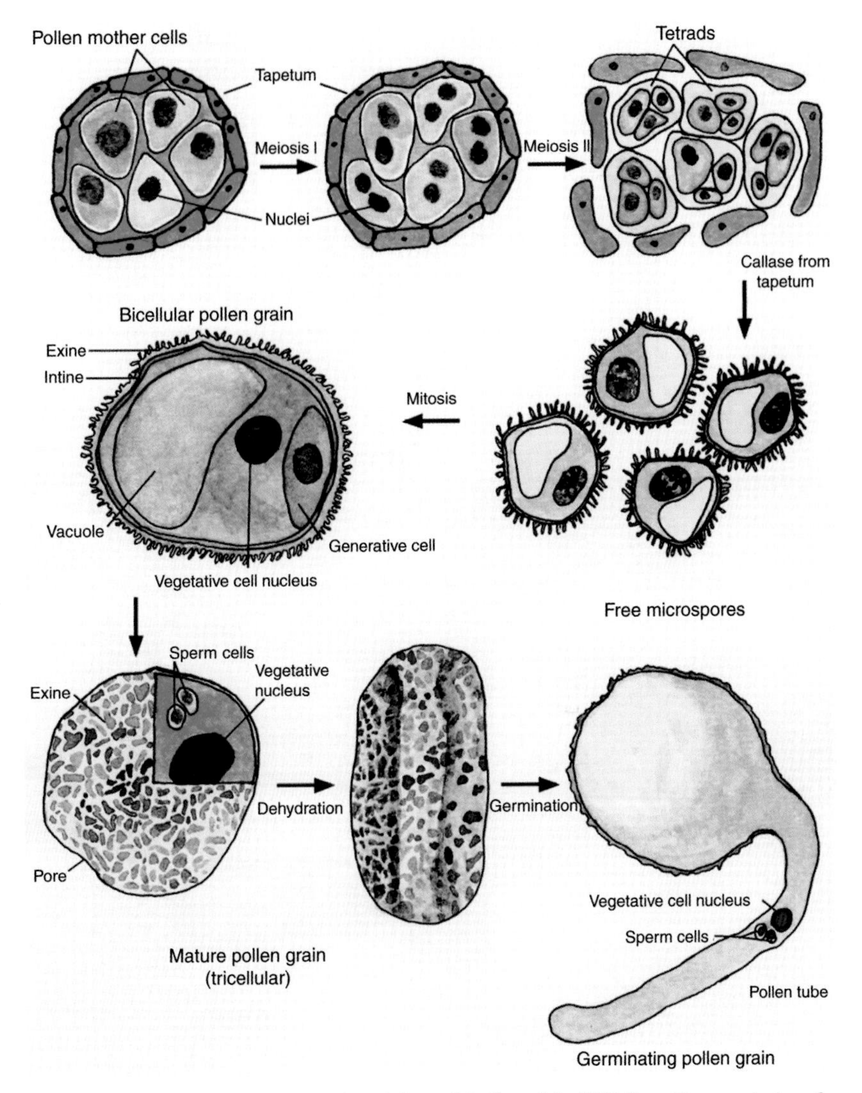

Figure 4.3. Pollen development. [Reprinted from McCormick (2004), with permission from the American Society of Plant Biologists.]

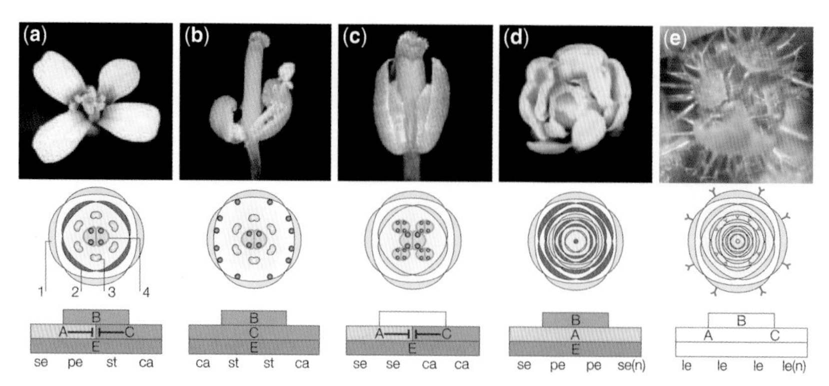

Figure 4.8. Flower development. *Arabidopsis* (a) wild-type, (b) *ap2*, (c) *pi*, (d) *ag*, and (e) *sep* flowers. Below each photo is a rendering of the ABC model as it functions in that flower. [Reprinted from Krizek and Fletcher (2005), with permission from Nature Publishing.]

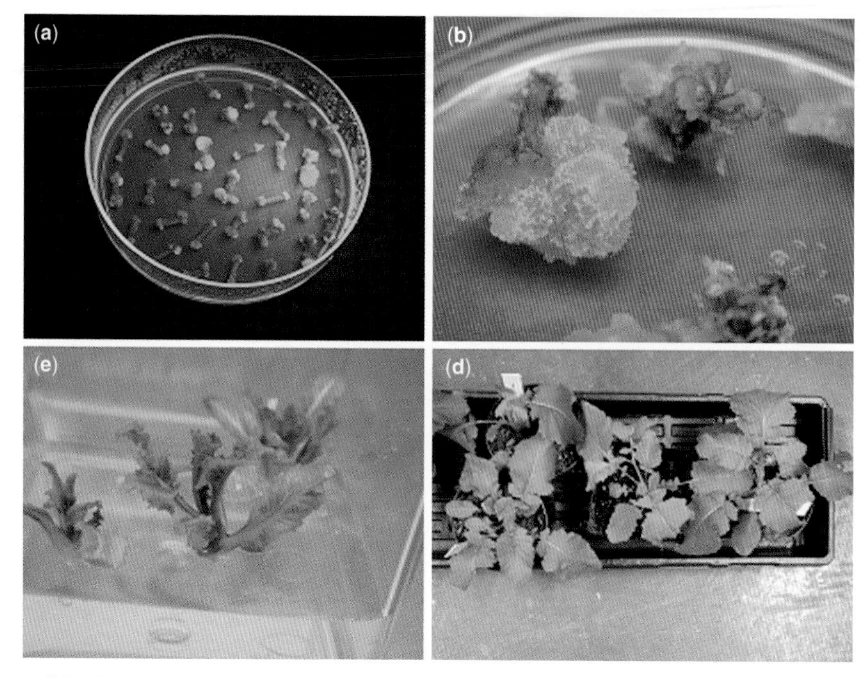

Figure 5.2. *Brassica juncea* plants produced from hypocotyl explants. Shoots are produced when a combination of auxin and cytokinin is used. (a) Callus from hypocotyl explants (note the green fluorescent protein fluorescent sectors on some of the calli); (b) shoots from callus; (c) shoots elongating; (b) whole plantlets transferred to soil.

Figure 5.7. (a) Callus tissue; (b,c) shoots arising from callus (an example of organogenesis).

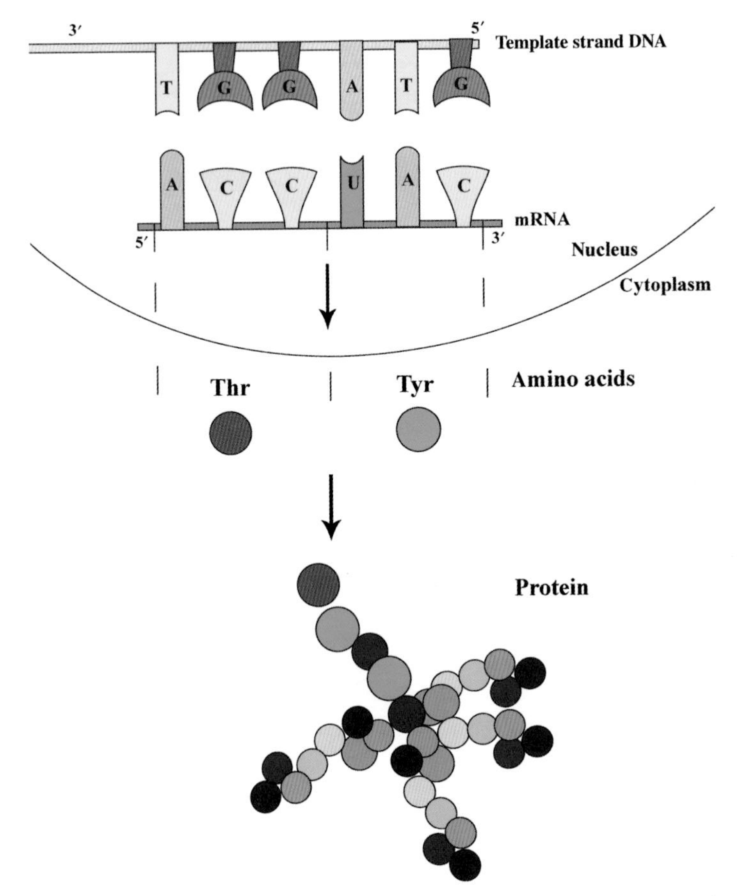

Figure 6.4. The central dogma: DNA is transcribed to RNA in the cell nucleus. RNA is translated to protein in the cell cytoplasm.

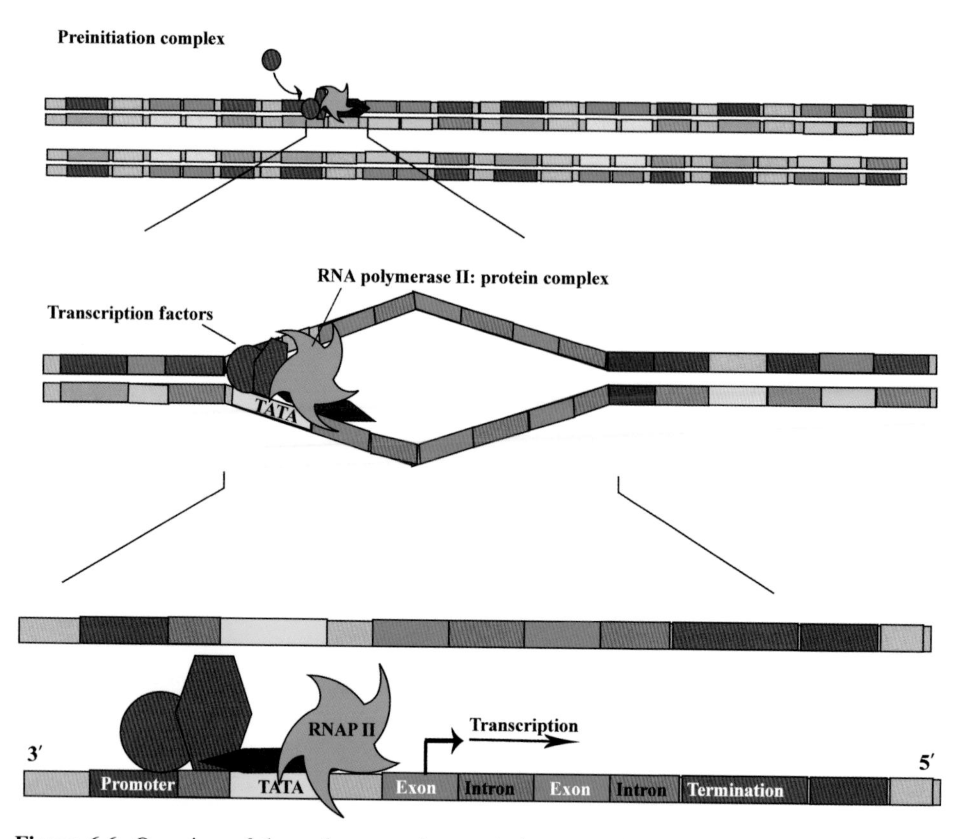

Figure 6.6. Overview of the early steps of transcription. A preinitiation complex is formed by a complex of transcription factors and RNA polymerase II (RNAP II). Association of the preinitiation complex with the start sequence (TATA) of the coding strand of DNA causes a conformation change and hydrogen bond breakage. This causes the DNA strands to separate so that transcription can proceed.

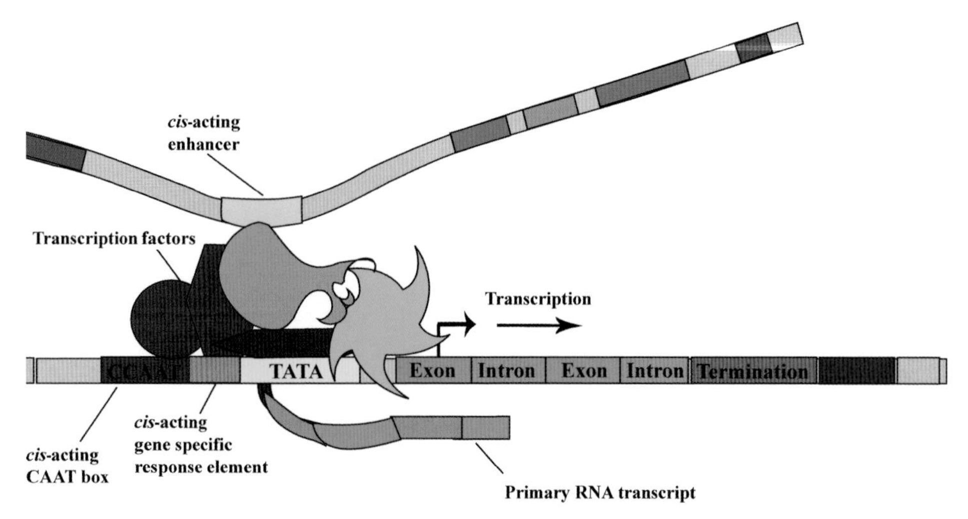

Figure 6.7. Regulation of transcription. The *cis*-acting elements are segments of DNA that regulate transcription; these segments may be adjacent to the gene such as the promoter (CAAT box) and the *cis*-acting gene-specific response elements, or they may be distant to the gene such as enhancers. The *trans*-acting elements are transcription factors and other regulatory proteins that may associate with the promoter, other proteins, or both.

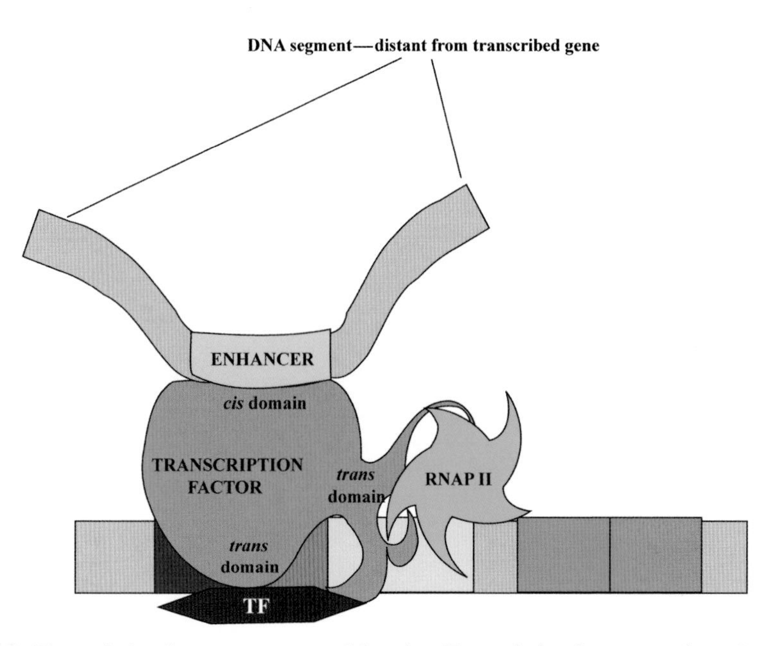

Figure 6.8. Transcription factors structure and function. Transcription factors may have domains that bind *cis*-acting elements such as enhancers, and domains that also bind *trans*-acting elements such as RNA polymerase (RNAP II) and other transcription factors.

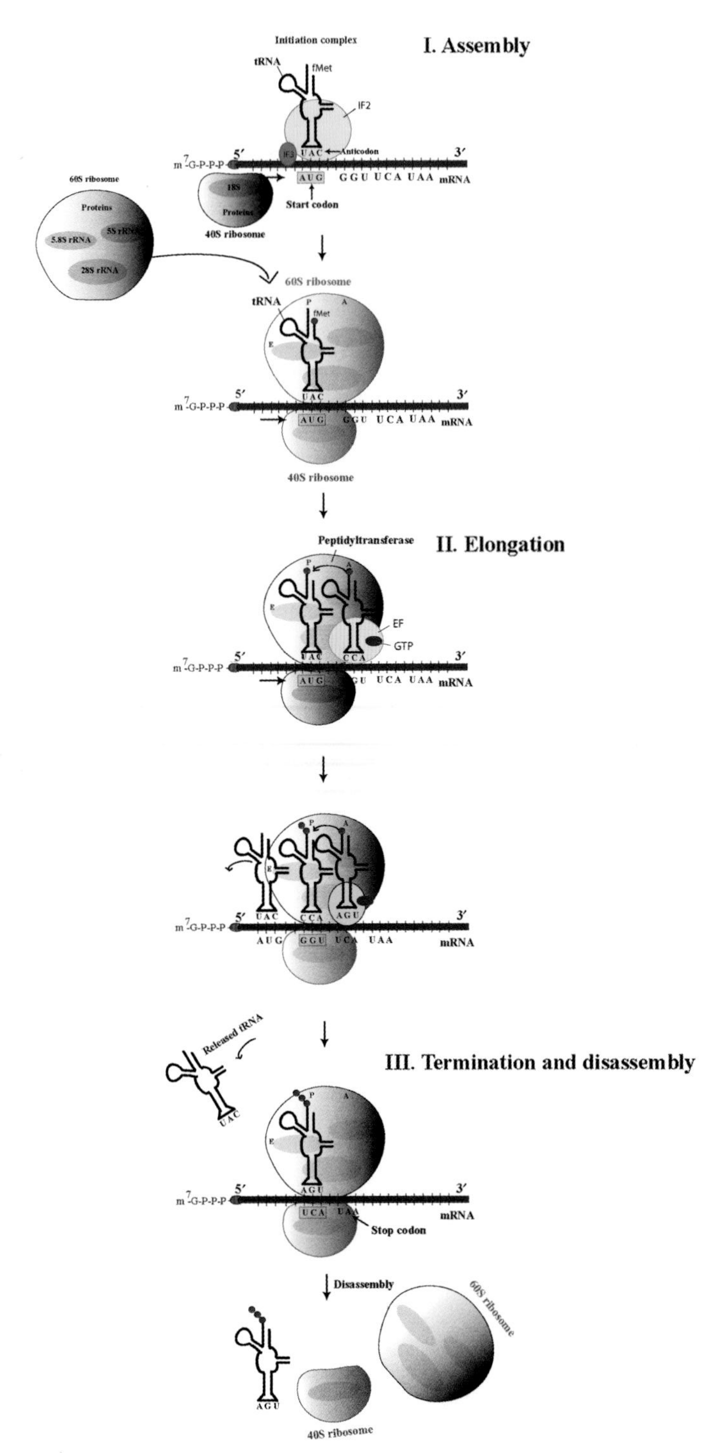

Figure 6.13. Overview of translation showing the structure of tRNA, 60*S* and 40*S* ribosomal subunits. The three steps of translation are shown: ribosome assembly, elongation of the polypeptide chain, and termination.

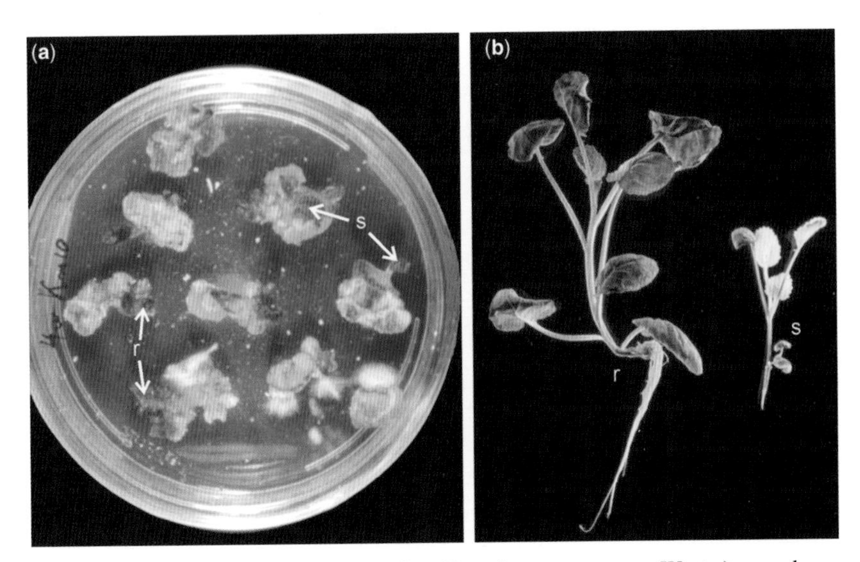

Figure 9.3. Selection of transgenic canola (*Brassica napus* cv Westar) on kanamycin-containing tissue culture media (reproduced with permission from Pierre Charest, PhD thesis, Biology Department, Carleton University, 1988). Stem explants were first infected with an *Agrobacterium tumefaciens* strain harboring a transformation vector with a chimeric *nptII* gene designed to confer kanamycin resistance on transformed plant tissue. (a) After cocultivation of plant tissue and *Agrobacterium* to allow transformation to occur, the plant tissues were transferred to tissue culture media containing kanamycin for growth of callus tissue and shoot differentiation. Much of the nontransformed tissues turned white and stopped growing because they were sensitive to the antibiotic(s). Transformed tissues remained green and continued to grow and differentiate because they were resistant to kanamycin (r). (b) Transgenic shoots that differentiated in the presence of kanamycin were excised from the callus and transferred to media for the regeneration of roots (r). Escapes that were not truly kanamycin-resistant were unable to regenerate roots in the presence of the antibiotic(s).

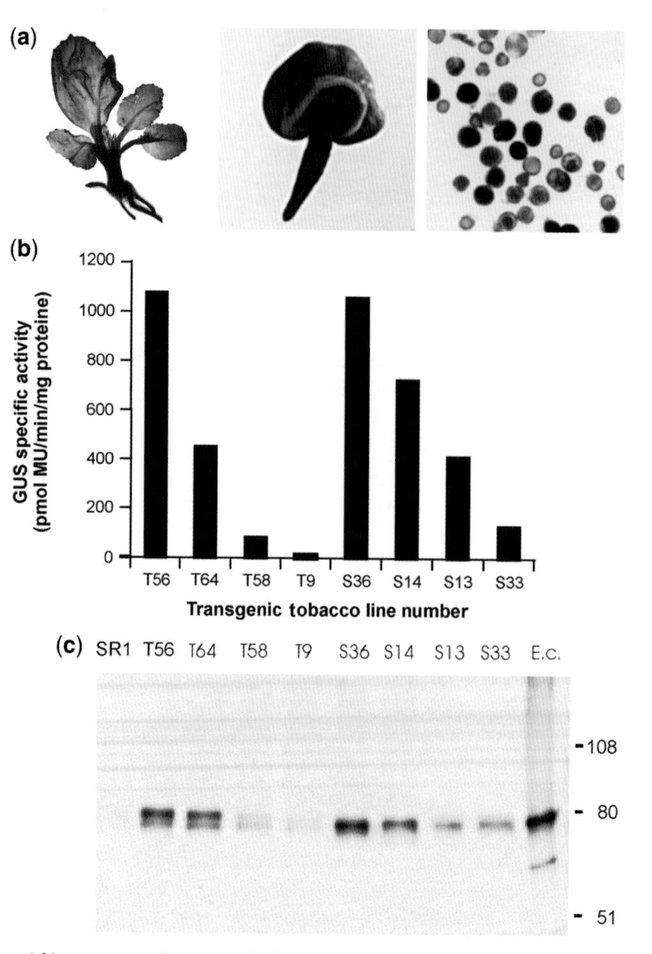

Figure 9.4. The *uidA* gene, coding for GUS, as an example of a reporter gene that has been extensively used in plants. (a) Histochemical staining for GUS activity using the substrate 4-methyl umbelliferyl glucuronide (MUG) allows detection of gene activity in specific tissues of transgenic plants. Shown are the staining of cauliflower plantlets in which constitutive expression of GUS is conferred by a strong constitutive promoter, tCUP (photo courtesy of Dan Brown, Agriculture and Agri-Food Canada, London, Ontario, Canada); excised embryos from transgenic canola seeds in which seed-specific expression is conferred by the napin promoter; and transgenic canola pollen in which cell-specific expression is conferred by the pollen-specific promoter (Bnm1 promoter). Note here that pollen cells are segregating as transformed and nontransformed cells indicated by the presence and absence of staining. (b) Measurement of GUS enzyme specific activity using the substrate 5-bromo-4-chloro-3-indolyl glucuronide (X-gluc). Each separate transgenic line of tobacco differs in the level of gene expression due to variation in the influences on the inserted genes from the genetic elements and chromatin environment at the different sites of insertion. These are often called *position effects* (also see Fig. 9.5). To compare differences among genes and elements introduced into transgenic plants, analyses must account for a large number of transgenic lines to reduce the influence of position effects. Reporter genes provide a valuable means for gathering large amounts of data. Here, a comparison of the promoter strengths of the 35*S* (plant lines with the *S* designation) and tCUP (plant lines with the *T* designation) constitutive promoters is inferred by comparing the activities of the reporter gene. (c) To ensure that the reporter gene reflects transcriptional activity, RNase protection assays are used to measure the relative amounts of GUS mRNA accumulating in the transgenic lines. This assay involves the formation of stable RNA duplexes with a radiolabeled antisense RNA probe followed by RNase digestion of the single-stranded RNA molecules so that the protected double-stranded RNA can be separated by gel electrophoresis and quantified.

Figure 9.8. Fusion of a reporter and selectable marker gene to create a bifunctional gene: (a) GUS:NPTII fusion reporter system for plants that incorporates the *nptII* gene for kanamycin selection and the GUS reporter gene in a single module; (b) transformed tobacco shoots selected on kanamycin; (c) shoots with roots regenerated on kanamycin; (d) a transgenic seedling after two generations showing retention of GUS gene activity indicated by the histochemical staining with the GUS substrate X-Gluc (provided by Raju Datla, Plant Biotechnology Institute, National Research Council of Canada, Saskatoon, Canada).

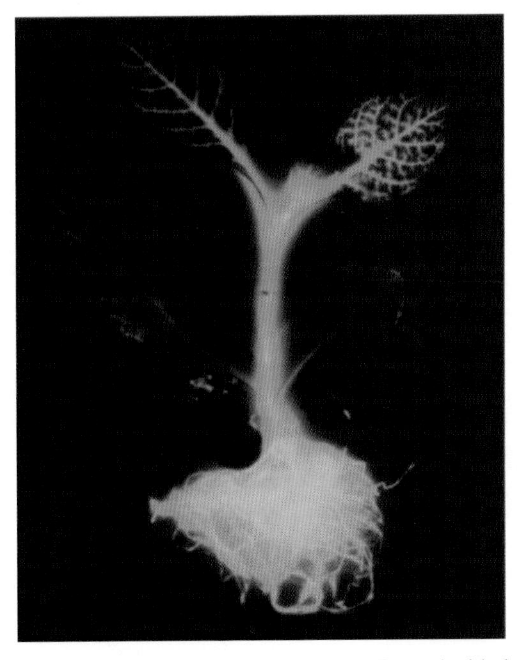

Figure 9.9. Luminescence detected in transgenic tobacco transformed with the firefly luciferase gene driven by the 35*S* promoter and watered with a solution of luciferin, the luciferase substrate. [Reprinted with permission from Ow et al. (1986), copyright 1986, AAAS.]

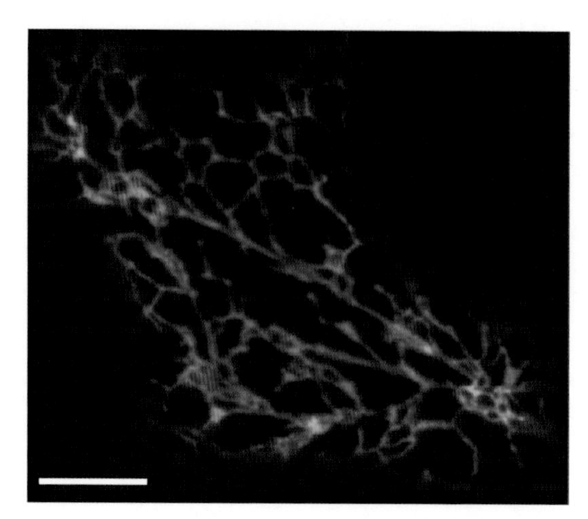

Figure 9.10. Confocal laser-scanning microscopy of leaf mesophyll cells transiently expressing peptides fused to green fluorescent protein (green image) and yellow fluorescent protein (red image). Green fluorescent protein is fused to the HDEL tetrapeptide (spGFP-HDEL) to achieve ER retention and thus reveals the cortical ER network in leaf cells. The proximity of the Golgi to the ER network is revealed by the yellow fluorescent protein fused to a Golgi glycosylation enzyme (ST-YFP). (Bar = 10:m.) [Reprinted from Brandizzi et al. (2004), with permission.]

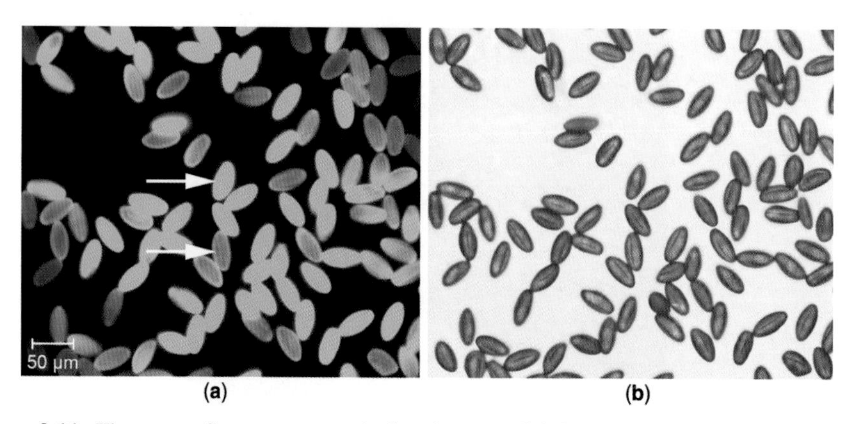

Figure 9.11. The green fluorescent protein has been useful for marking whole plants using a 35*S*-GFP construct and plant parts such as pollen using a GFP under the control of a pollen-specific promoter (LAT59) from tomato: (a) 867 ms, 200× under blue light; (b) 1.7 ms, 200× under white light. The arrows in (a) show GFP fluorescence of pollen cells. (Photos courtesy of H. S. Moon and Neal Stewart.)

Figure 9.12. Novel fluorescent proteins whose genes were recently cloned from corals and expressed in tobacco (a) and *Arabidopsis* (b) plants.

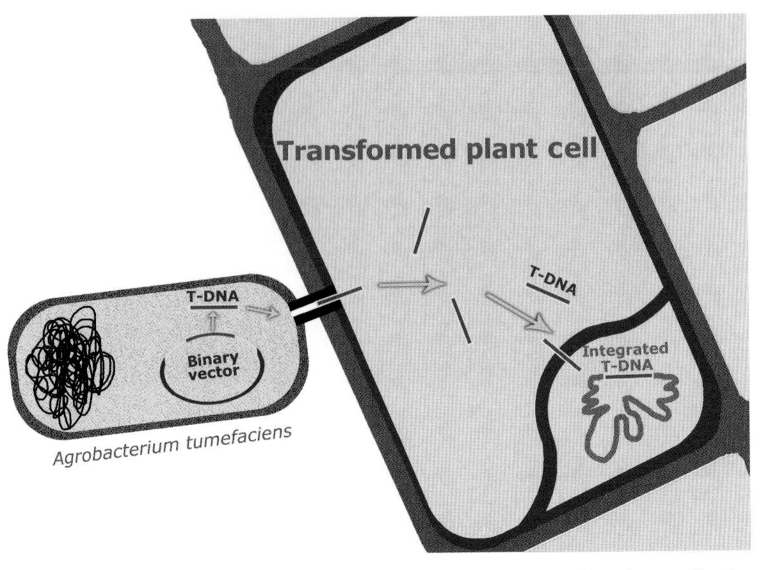

Figure 10.4. Schematic of *Agrobacterium*-mediated transformation of a plant cell, showing production of the T strand from the binary vector, transport through the bacterial pillus, and integration into plant chromosomal DNA.

Figure 10.5. Agroinfiltrated *Nicotiana benthamiana* plants showing high levels of GFP expression. The aerial parts of the tobacco plant were submerged in an *Agrobacterium* suspension, and the plant was then placed under vacuum for infiltration. (Image provided by John Lindbo, Department of Plant Pathology, University of California-Davis.)

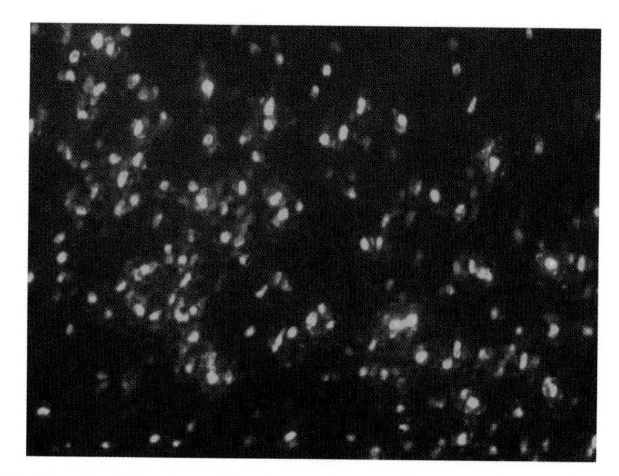

Figure 10.9. Particle bombardment-mediated transient GFP expression in lima bean cotyledonary tissues. This target tissue is flat, nonpigmented, and ideally suited for tracking GFP expression in individual transiently transformed cells.

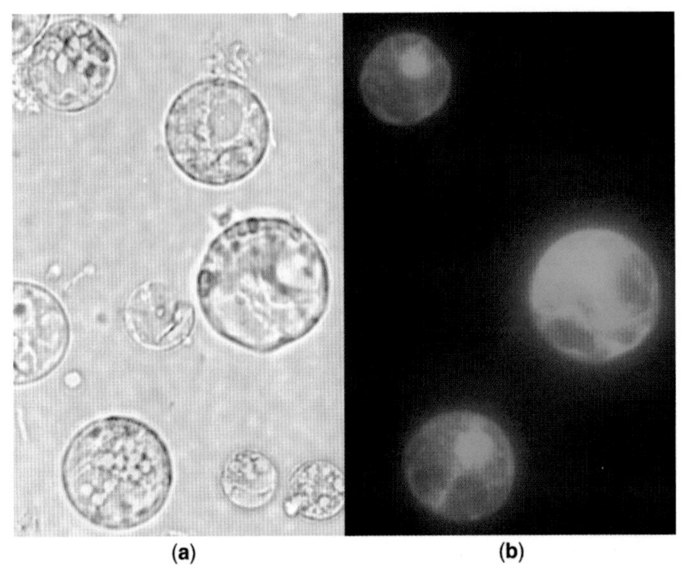

Figure 10.10. Maize protoplasts, electroporated with a *gfp* gene, showing brightfield (a) and with GFP filters (b). (Illustrations provided by Pei-Chi Lin and JC Jang, Department of Horticulture and Crop Science, OARDC/The Ohio State University.)

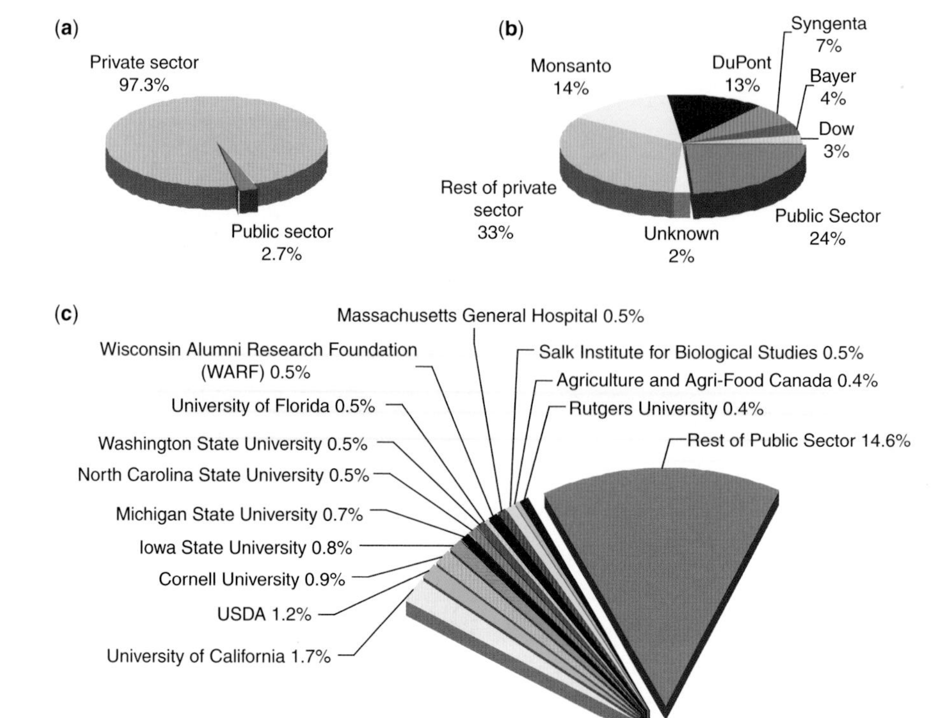

Figure 14.2. Distribution of assignment of US patents from 1982 to 2001 to private and public institutions: (a) all patents; (b) agricultural biotechnology patents; (c) public-sector agricultural biotechnology patents [adapted from Graff et al. (2003); used with permission].

posttranslational regulation is becoming increasingly valuable as we engineer some plants to be protein production factories, accumulating high levels of desirable, functional proteins for numerous applications ranging from industrial to medical.

LIFE BOX 6.1. MAARTEN CHRISPEELS

Maarten Chrispeels, Professor of Biology and Director, Center for Molecular Agriculture, University of California San Diego; Member, National Academy of Science

Maarten Chrispeels giving a public lecture on molecular agriculture (2002).

I was born in a small Flemish town not far from Brussels, Belgium, and after an uneventful youth and a solid classical education with three foreign languages and six years of Latin, I enrolled in the College of Agriculture in Ghent. I wanted to become a biochemist. In the fall of 1960, after graduation at the age of 22, I found myself on the Mauretania, sailing for America with a Fulbright travel fellowship and a fellowship from the University of Illinois to start graduate studies in the Department of Agronomy. My Ph.D. research and postdoctoral work (with Joe Varner) were in plant cell biology. A couple of papers in *Plant Physiology* landed me a job as assistant professor of biology at the then newly founded University of California San Diego (UCSD). Upon arriving there I switched from studying α-amylase secretion by barley aleurone cells to the biosynthesis and secretion of hydroxyproline-rich glycoproteins, which had just been discovered but were little studied. We discovered that these proteins move from the ER to the Golgi apparatus where glycosylation of hydroxyproline residues takes place. After a sabbatical leave in England I switched to study the synthesis and intracellular transport of proteases to the protein storage vacuoles (PSVs) in cotyledons during germination. We made use of antibodies—quite a novelty at the time—to demonstrate by immuno-electrton microscopy that the protease that digests storage proteins is in the ER before it arrives in the PSVs. It then occurred to me that if I wanted to study protein transport to PSVs I should be looking at developing seeds and not at germinating seeds, because seed development is characterized by massive protein synthesis and transport to vacuoles. So, I switched again and started working on the synthesis, posttranslational modification and transport of storage glycoproteins and lectins in developing bean seeds. About that time others invented gene cloning and plant transformation and soon we had bean storage protein and lectin genes and were expressing them in tobacco seeds to identify vacuolar targeting domains.

Wishing to expand my horizons a bit more we purified bean PSV membranes (tonoplasts really) and cloned the gene for the most abundant tonoplast protein. This turned out to be a protein with six membrane-spanning domains. What could its function be? Not until we obtained a homolog from *Arabidopsis* was Christophe Maurel in the lab able to show that this new family of proteins constituted the plant aquaporins. Aquaporins had been described the year before by Peter Agre who later received the Nobel Prize for this work. From then on we dropped the vacuolar targeting work and worked on aquaporins, as they were capturing the imagination of many plant physiologists. Somewhere along the line I had time to do other things. My former mentor Joe Varner was always heavily involved in "service to the profession" and I also accepted to become first Associate Editor and later Editor in chief of *Plant Physiology*, an excellent journal then in need of a physical and intellectual facelift. I believe that I contributed to this facelift and my successor Natasha Raikhel took the journal to new heights.

When, in 1978, the USDA created its first Competitive Grants Program, I called my friend Joe Key who had just been named the Director and volunteered to come to Washington DC on short notice to put together a panel to evaluate grants in the area of "Genetic mechanisms for crop improvement". He took me up on my offer and a few weeks later I was working in DC having received a leave of absence from UCSD. While on sabbatical leave in Canberra, Australia I became involved in a biotech project. We had isolated the cDNA for α-amylase inhibitor from common bean, and with my friend T. J. Higgins we expressed this gene in developing pea seeds. Larry Murdock from Purdue University showed that the larvae of the pea bruchid, which normally burrow into dry pea seeds, starved to death on these transgenic pea seeds, presumably because the bruchid digestive amylases are inhibited by the bean inhibitor. At this time I also realized that there was no good textbook to help university teachers who wished to teach courses in plant biology with an applied or biotechnology flavor, and I started work on the first edition of a text that in its second edition was called *Plants, Genes and Crop Biotechnology*. David Sadava and I put together a completely integrated textbook that had elements of plant physiology and biochemistry, human nutrition, plant breeding, human population changes and world food production, soils and plant nutrition, and biotechnology applications.

By the year 1997, 30 years after my arrival in San Diego, plant biology had grown from just three faculty members to about 15 in three different institutions—UCSD, The Salk Institute and The Scripps Research Institute—and we founded the San Diego Center for Molecular Agriculture—a virtual center with a grandiose name—whose purpose it would be to simply enrich our own intellectual lives. Creating "a community of scholars" as we fondly call academia, is actually quite difficult and requires effort and commitment from all parties.

LIFE BOX 6.2. TONY CONNER

Tony Conner, Senior Scientist, New Zealand Institute for Crop & Food Research; Professorial Fellow, Bio-Protection & Ecology Division, Lincoln University, New Zealand

Tony Conner with a transgenic potato plant.

Towards designer plants. The first transgenic plants were developed in 1983 while I was studying toward my PhD in plant genetics at the University of California, Davis. At the time, my research involved somatic cell selection in *Nicotiana plumbaginifolia* as a model system. Upon graduation it was an obvious step to move toward developing transformation systems for crop plants. I was very fortunate to be offered a position back in my home country of New Zealand to establish a research programme in applying the emerging tools in plant biotechnology to crop improvement.

It was an exceptionally exciting time to be involved in plant science. My research initially focused on potatoes, asparagus, and a few other vegetable and arable crops. In those early days it was rewarding to be associated with the first examples of *Agrobacterium*-mediated transformation of monocotyledonous plants (asparagus) and some of the very first field tests on transgenic plants. Research advances in plant molecular biology were rapidly gaining momentum, and this was matched by the development of molecular tools for analysing genetic variation in plant populations and technologies for genetic engineering in a diverse range of plant species.

Integration of these new technologies into breeding programs of crops presented some important challenges. Often the elite material of plant breeders destined to become the future cultivars for the agricultural industries was more difficult to work with than other laboratory-based model systems. This was especially the case for developing transformation systems for gene transfer via genetic engineering.

However, public concerns about the deliberate release of transgenic crops into the agricultural environment quickly changed research agendas. Considerable effort was required to participate in the public debate on the merits and biosafety of transgenic crops and absorbed much of my time for about a decade. During this time my research efforts were directed more to investigating the environmental impacts and food safety of transgenic crops.

More recently my research focus changed to refining vectors systems for gene transfer to plants. This work has been motivated by the need to eliminate components of vectors that regulatorys [regulatory systems] find less acceptable.

This eventually led to our development of intragenic vector systems, which involve identifying functional equivalents of vector components from plant genomes and using these DNA sequences to assemble vectors for plant transformation.

Gene transfer using intragenic vectors allows the well-defined genetic improvement of plants without the introduction of foreign DNA. Biologically, the resulting plants are not transgenic, although the tools of molecular biology and plant transformation have been used in their development. The genetic make-up of the resulting plants is equivalent to a minor rearrangement of the endogenous DNA sequences within the species. This is very similar to "micro-translocations" that can occur naturally in plant genomes or as a consequence of deliberate mutation breeding. For the transfer of genes from within the gene pools of crops, intragenic vectors may help to alleviate some of the public concerns over the deployment of GM crops in agriculture, especially ethical issues associated with the transfer of DNA sequences across wide taxonomic boundaries. Nowadays, my research is moving toward functional genomics of potato to better understand how important traits are controlled by specific genes and their alleles. I envisage this will lead to valuable sources of gene sequences for transfer to existing elite potato cultivars via intragenic vectors.

Early in my career I never considered it would be possible, in my lifetime, for science to generate the full genome sequence of a higher organism. Yet, within the next 5–10 years the annotated sequence, at least for the gene-rich regions, of the genomes of all major crops will be known. This will provide unprecedented opportunities for mining the germplasm collections of plant breeders for novel alleles that represent variant versions of genes with altered functions. The resulting novel DNA sequences can then be used for highly targeted genetic changes in crop plants by transformation of elite crop cultivars.

The next few decades are going to be exceptionally exciting for plant genetics as research moves toward the targeted design and development of genetically enhanced plants for sustainable production of high quality and healthy food. My career has been an exciting and fulfilling journey so far. But I often think: "what if I was thirty years younger?" What a tremendous career opportunity modern plant genetics would offer.

REFERENCES

Baker SS, Wilhelm KS, Thomashow MF (1994): The 5′ region of *Arabidopsis thaliana* cor15a has *cis*-acting elements that confer cold-, drought- and ABA-regulated gene expression. *Plant Physiol Biochem* **30**:123–128.

Gilmour SJ, Zarka DG, Stockinger EJ, Salazar MP, Houghton JM, Thomashow MF (1998): Low temperature regulation of the *Arabidopsis* CBF family of AP2 transcriptional activators as an early step in cold-induced *COR* gene expression. *Plant J* **16**:433–442.

Goldberg ML (1979): *Sequence Analysis of Drosophila Histone Genes*. PhD thesis, Stanford Univ, CA.

He Y, Michaels SD, Amasino RM (2003): Regulation of flowering time by histone *Arabidopsis*. *Science* **302**:1751–1754.

Kankel MW, Ramsey DE, Stokes TL, Flowers SK, Haag JR, Jeddeloh JA, Riddle NC, Verbsky ML, Richards EJ (2003): *Arabidopsis MET1* cytosine methyltransferase mutants. *Genetics* **163**:1109–1122.

Khoury G, Gruss P (1983): Enhancer elements. *Cell* **33**:313–314.

Kozak M (1986): Point mutations define a sequence flanking the AUG initiator codon that modulates translation by eukaryotic ribosomes. *Cell* **44**:283–292.

Kozak M (1987): At least six nucleotides preceding the AUG initiator codon enhance translation in mammalian cells. *J Mol Biol* **196**:947–950.

Vignali M, Hassan AH, Neely KE, Workman JL (2000): ATP-dependent chromatin-remodeling complexes. *Mol Cell Biol* **6**:1899–1910.

Yamaguchi-Shinozaki K., Shinozaki K (1994): A novel *cis*-acting element in an *Arabidopsis* gene is involved in responsiveness to drought, low-temperature, or high-salt stress. *Plant Cell* **6**:251–264.

CHAPTER 7

Recombinant DNA, Vector Design, and Construction

MARK D. CURTIS

Institute of Plant Biology, University of Zurich, Switzerland

7.0. CHAPTER SUMMARY AND OBJECTIVES

7.0.1. Summary

Genomics, biotechnology, and biology in general have been enabled by methodologies to manipulate DNA in a test tube (a very tiny test tube). Restriction enzymes are used as molecular scissors, and ligases are used as molecular "glue." The polymerase chain reaction has become invaluable in amplifying and cloning DNA. In addition, recombination systems have been developed as alternatives to restriction enzymes as cloning tools. All these methods are useful in creating plasmids containing chimeric DNA constructs that will be transformed into plants.

7.0.2. Discussion Questions

1. What basic elements should be included in the design and construction of an efficient ubiquitous and constitutive plant gene expression vector?
2. Discuss the advantages and disadvantages of recombination cloning technologies versus traditional restriction digestion and ligation technology.
3. Describe a novel strategy to generate a T-DNA vector that allows the expression of several genes from a single position in the genome.
4. Discuss the advantages and disadvantages of using plastid vectors for plant transformation and gene expression.
5. Describe ways in which transgene technology could be made more acceptable to the public.

Plant Biotechnology and Genetics: Principles, Techniques, and Applications, Edited by C. Neal Stewart, Jr.

7.1. DNA MODIFICATION

Recombinant DNA technology relies on the ability to manipulate DNA using nucleic acid–modifying enzymes. The isolation of these enzymes followed shortly after James Watson and Francis Crick's description of the double helical structure of DNA in 1953. Recall that DNA is made up of two twisting complementary strands, comprising alternating units of deoxyribose sugar and phosphates that run in opposite directions. Attached to each deoxyribose sugar is a nitrogen-rich base. The bases, adenine (A), thymine (T), guanine (G), and cytosine (C), on opposite strands are held together by hydrogen bonds to form base pairs (bp); A with T and G with C. The complementary nature of the strands means that each strand provides a template for the synthesis of the other (Fig. 7.1).

In 1955, Arthur Kornberg and colleagues isolated DNA polymerase I, an enzyme capable of using this template to synthesize DNA in vitro in the presence of the four bases, in the form of deoxribonucleoside triphosphates (dNTPs). Although this was the first enzyme to be discovered that had the required polymerase activities, the primary enzyme involved in DNA replication is DNA polymerase III.

While DNA polymerases can replicate a second strand of DNA, they cannot join the ends of DNA together. The discovery of circular DNA molecules (plasmids, discussed later) suggested that such an enzyme must exist. In 1966, Bernard Weiss and Charles Richardson isolated DNA *ligase*, an enzyme that allowed DNA to be "glued" together, catalyzing the formation of a phosphodiester bond (Fig. 7.2).

Soon after this discovery, investigations into bacterial resistance that "restricted" viral growth revealed that *endonucleases* within the cells could destroy invading foreign DNA molecules. Among the first "restriction enzymes" to be purified were *Eco*RI from *Escherichia coli*, and *Hin*dIII from *Haemophilus influenzae*. Restriction enzymes went on to become one of the most useful tools available to molecular biologists and deserve special consideration.

Restriction enzymes (restriction endonucleases) are produced by a wide variety of prokaryotes. These enzymes identify specific nucleotide sequences in DNA of 4–8 bp, usually palindromes, and cleave specific phosphodiester bonds in each strand of the DNA. The methylation of these specific nucleotide sequences in the host DNA protects the cell from attack by its own restriction enzymes. There are many different site-specific restriction enzymes. These are named after the bacterial species and strain of origin. The restriction endonuclease *Eco*RI, for example, was the first restriction endonuclease identified from the bacterium *Escherichia coli*, strain *R*Y13 (other examples are shown in Table 7.1).

Such enzymes recognize a specific double-stranded DNA sequence and cleave the strands to produce either a 5′ overhang, a 3′ overhang, or blunt ends (Fig. 7.3).

DNA fragments that contain single-stranded overhangs ("sticky ends") are the easiest to join together. Two DNA molecules, with compatible single-stranded overhangs, can hybridize to bring the 5′ phosphate and 3′ hydroxyl residues together, allowing DNA ligase to catalyze the formation of phosphodiester bonds (recall Fig. 7.2). In this way, two DNA molecules from different sources can be combined to produce an artificial or "recombinant" DNA molecule (Fig. 7.4). All of biotechnology hinges on recombinant DNA—combining DNA from various sources to do something new. Using two restriction enzymes with different recognition sequences, one can combine two DNA molecules in a predetermined orientation (Fig. 7.5).

The first recombinant DNA molecule was created in Paul Berg's lab in 1972. This pioneering work formed the basis of the recombinant DNA revolution; however, it was not

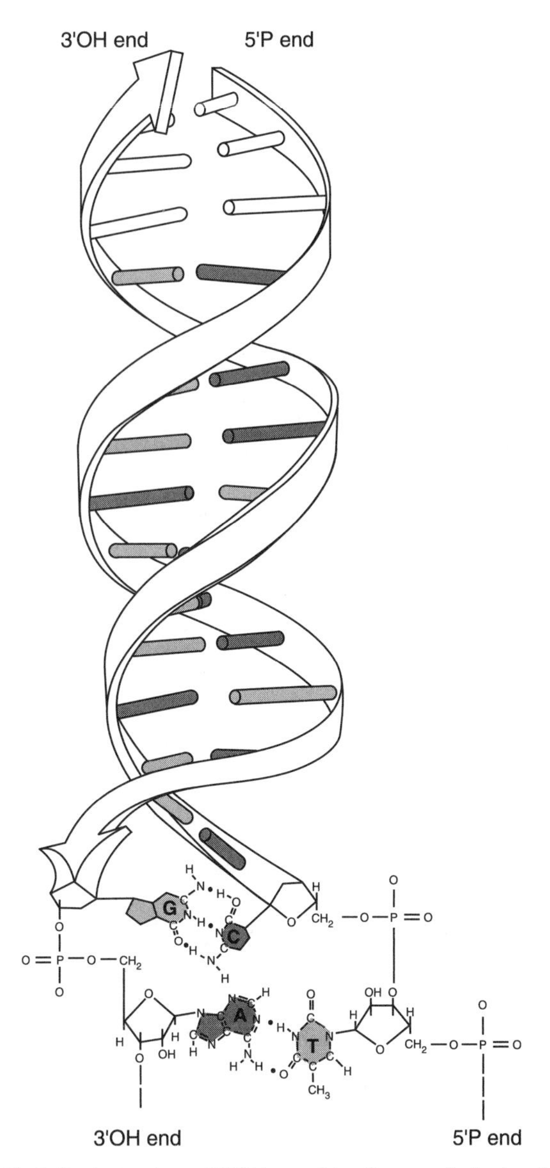

Figure 7.1. The double helical structure of DNA provides the genetic instructions for the development of an organism. The specific sequence arrangement of the bases G, A, T, and C encode regulatory features such as the promoter and terminator sequences of genes, and the triplet code determines the amino acid sequence of proteins. In plants, as with all eukaryotes, most of the DNA is packed into chromosomes and located in the cell nucleus, while in bacteria the DNA is found directly in the cytoplasm and is most often circular.

until a year later in 1973 that Stanley Cohen and Herbert Boyer created the first genetically modified organism using these approaches. Combining Cohen's expertise in plasmids and Boyer's expertise in restriction enzymes, a strand of DNA was cut and pasted into a plasmid and maintained and replicated in the bacterium *E. coli*. The transfer of such recombinant

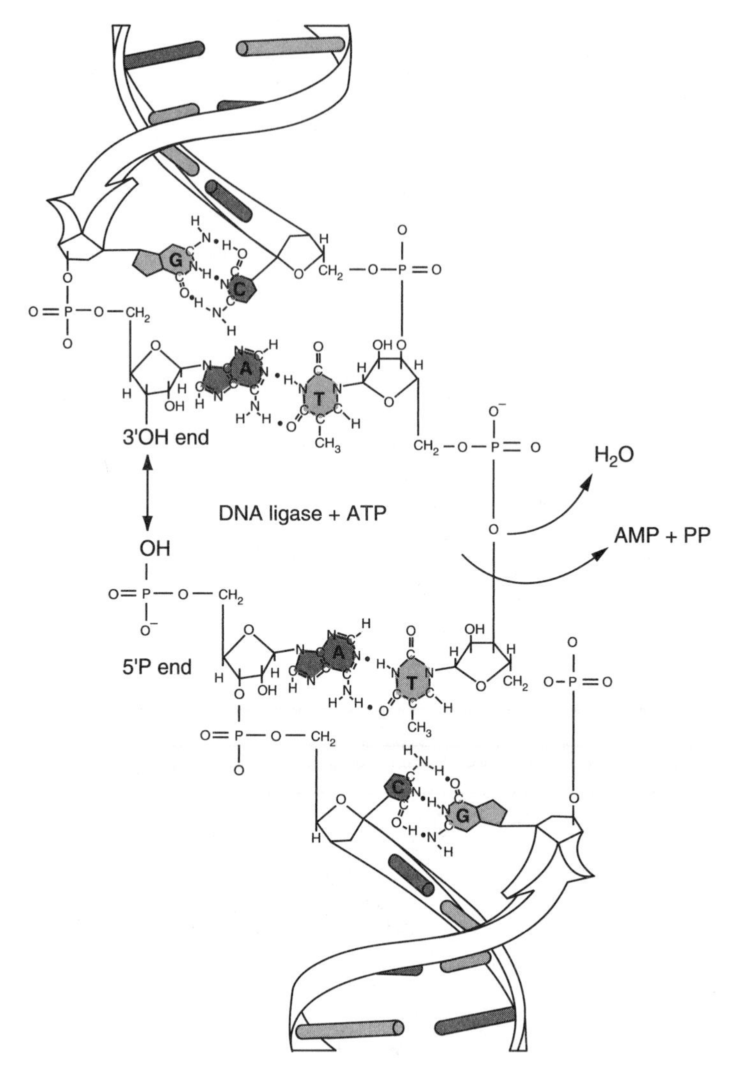

Figure 7.2. The joining of two linear DNA fragments, catalyzed by DNA ligase, creating phosphodiester bonds between the 3′ hydroxyl of one nucleotide and the 5′ phosphate of another.

DNA molecules to a host cell for amplification is achieved in a process known as *transformation*. Observations in the late 1920s, by Fred Griffith and later by Oswald Avery in the early 1940s, suggested that bacteria could undergo rare natural transformation events. The frequency of these events increased when bacterial cells were treated with cold calcium chloride, which enhanced their competence, prior to a brief heatshock treatment at 42°C. Alternative electroporation approaches are now commonly used for transformation. These yield higher transformation frequencies and allow bacterial artificial chromosomes (BACs), too large for conventional transformation, to be taken up successfully by bacterial cells (Sheng et al. 1995). This general procedure formed the basis of clonal propagation, or amplification, of DNA and initiated the development of DNA cloning vectors.

TABLE 7.1. Restriction Endonucleases

Enzyme	Source	Recognition sequence	Cut		Ends
*Eco*RI	*Escherichia coli* RY13	GAATTC	G	AATTC	5′overhangs
		CTTAAC	CTTAA	G	
*Bam*HI	*Bacillus amyloliquefaciens* H	GGATCC	G	GATCC	5′overhangs
		CCTAGG	CCTAG	G	
*Hind*III	*Haemophilus inflenzae* Rd	AAGCTT	A	AGCTT	5′overhangs
		TTCGAA	TTCGA	A	
*Kpn*I	*Klebsiella pneumoniae*	GGTACC	GGTAC	C	3′overhangs
		CCATGG	C	CATGG	
*Not*I	*Nocardia otitidis*	GCGGCCGC	GC	CGCCGG	5′overhangs
		CGCCGGCG	GGCCGC	CG	
*Pst*I	*Providencia stuartii*	CTGCAG	CTGCA	G	3′overhangs
		GACGTC	G	ACGTC	
*Sma*I	*Serratia marcescens*	CCCGGG	CCC	GGG	Blunt ends
		GGGCCC	GGG	CCC	
*Sac*I	*Streptomyces achromogenes*	GAGCTC	GAGCT	C	3′overhangs
		CTCGAG	C	TCGAG	
*Sst*I	*Streptomyces stanford*	GAGCTC	GAGCT	C	3′overhangs
		CTCGAG	C	TCGAG	
*Taq*I	*Thermophilus aquaticus*	TCGA	T	CGA	5′overhangs
		AGCT	AGC	T	
*Xba*I	*Xanthomonas campestris pv. badrii*	TCTAGA	T	CTAGA	5′overhangs
		AGATCT	AGATC	T	

7.2. DNA VECTORS

In molecular biology, a cloning vector is a DNA molecule that carries foreign DNA fragments into a host cell and allows them to replicate. Cloning vectors are frequently derived from *plasmids*, a generic term first coined by Joshua Lederberg in 1952, to describe any extrachromosomal hereditary determinant. Plasmids, found in bacteria but not in plants and other "higher" organisms, are convenient vectors used to manipulate DNA for genetic engineering. Plasmids were discovered in bacteria as double-stranded, covalently closed circular, extrachromosomal DNA molecules. They have evolved mechanisms to maintain a stable copy number in their host, to ensure that copies are shared between daughter cells and to encode *genes* that provide a selective advantage to their host.

DNA replication determines the plasmid copy number and this is rigorously controlled and closely coordinated with the *cell cycle*. The process of DNA replication is initiated at distinct sites known as *origins of replication* (ori) and proceeds in both directions along the DNA. In simple organisms, such as *E. coli*, there is only one origin (oriC); however, more complex organisms, with larger genomes, require many origins to ensure complete DNA synthesis prior to cell division. Origins are usually defined by a segment of DNA, comprising several hundred base pairs, which binds DNA polymerase and other proteins required to initiate DNA synthesis. The plasmid DNA must also replicate in its host organism to ensure that each daughter cell receives a copy of the plasmid. The regulation of this replication determines the number of plasmid copies contained within each cell. Control of plasmid

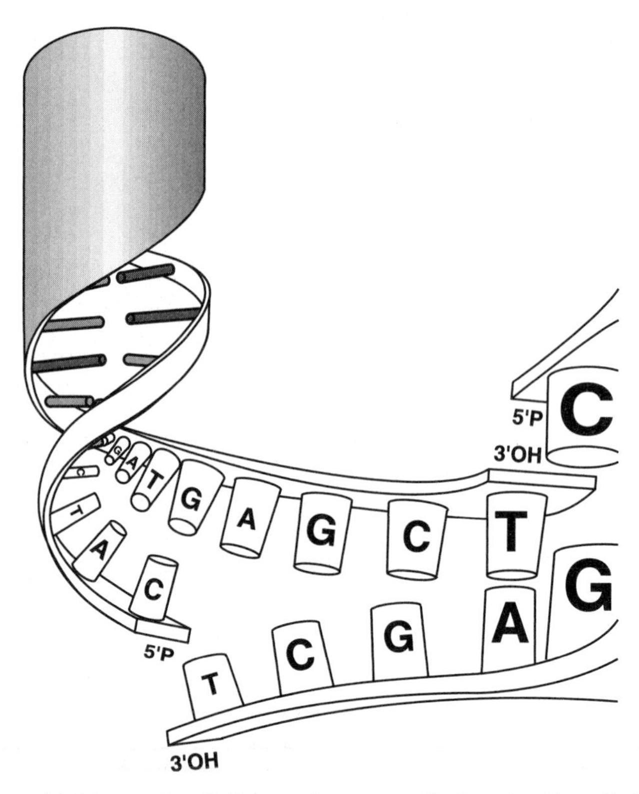

Figure 7.3. The restriction enzyme *Sac*I recognizes a specific 6-nucleotide palindromic sequence wherever it occurs in the DNA and cleaves the DNA asymmetrically at specific phosphodiester bonds to produce 3′ overhangs or "sticky ends."

replication is either "relaxed" or "stringent," a characteristic determined by the origin of replication. Plasmids with stringently controlled replication have low copy number, replicating alongside the host's chromosome, once per cell cycle, while plasmids with relaxed replication control have high copy number, replicating throughout the host's cell cycle, resulting in many hundreds of copies per cell. Whether replication is relaxed or stringent, the rate of plasmid DNA synthesis is controlled to maintain harmony with the host's replication. In general, relaxed plasmid replication is controlled by the supply of an RNA molecule, known as *RNA II*, which is required to prime (or start) DNA synthesis [for a review, see Eguchi et al. (1991)]. The supply of RNA II is regulated by another RNA molecule, RNA I, which is complementary to the RNA II molecule. When these two molecules hybridize, with the help of a protein known as the *Rop protein*, the priming of DNA synthesis is prevented. Therefore, plasmid replication is inhibited when RNA II is in short supply. Stringently controlled plasmid replication uses a different mechanism. Here plasmid copy number is regulated by the supply of the plasmid-encoded RepA protein, a *cis*-acting protein, which negatively regulates its own transcription and positively regulates the origin of replication [for a review, see Nordstrom (1990)]. Relaxed or high-copy-number plasmids are used most often as vectors to produce large quantities of cloned, recombinant DNA, while stringent or low-copy-number vectors are used to replicate massive, unstable, foreign DNA fragments such as BACs, or genes that produce lethal

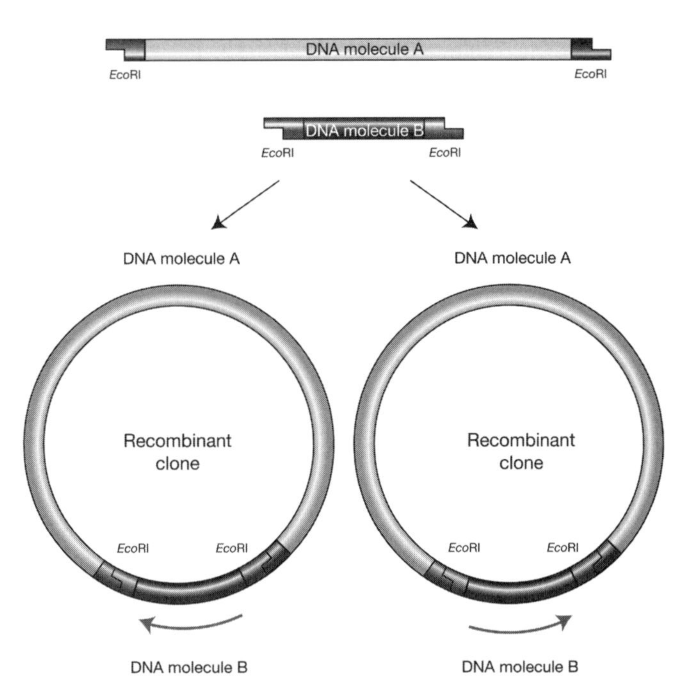

Figure 7.4. DNA fragments produced with a single *Eco*RI restriction enzyme give rise to compatible protruding termini that can anneal in either orientation, bringing together the 5′ phosphate and the 3′ hydroxyl residues on each strand. This allows DNA ligase to catalyze the formation of phosphodiester bonds, joining the two molecules together.

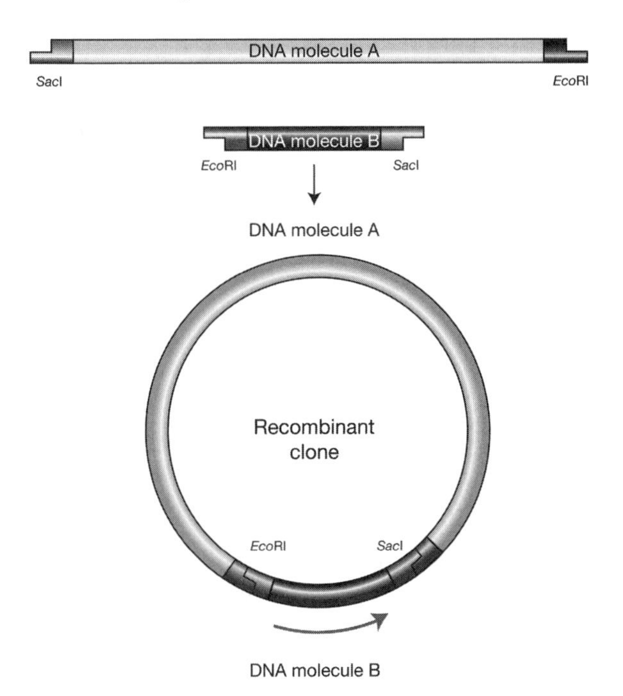

Figure 7.5. DNA fragments produced with two restriction enzymes, *Eco*RI and *Sac*I, give rise to fragments with protruding termini that can anneal in only one orientation with respect to one another, forcing the two molecules to combine in one direction only.

effects at high copy number. Unlike chromosomal DNA, plasmid DNA is dispensable to the host, so why does the host keep it? To be maintained, plasmid DNA molecules must provide their host cells with a selective advantage over their competitors. Plasmid selection is a natural phenomenon that has allowed the evolution of plasmid DNA and its maintenance in bacterial host cells. They encode genes, such as bacteriocins or antibiotics, enabling the host to kill other organisms competing for nutrients. The first bacterial plasmid identified was the fertility factor (F factor) in *E. coli*, discovered in 1946 by Joshua Lederberg and Edward Tatum. This F factor enables bacteria to donate genes to recipients by conjugation [for a review, see Clark and Warren (1979)], providing a mechanism for adaptive evolution, permitting, for example, plasmid-mediated transfer of antibiotic resistance genes or pathogenicity genes.

7.2.1. DNA Vectors for Plant Transformation

Many bacterial plant pathogens benefit from plasmid borne, pathogenicity genes, which provide them with the ability to infect or parasitize plants. One such organism, *Agrobacterium tumefaciens*, benefits from a tumor-inducing (Ti) plasmid (Fig. 7.6), which plays a central role in crown gall disease in a wide variety of plants.

The ability of *A. tumefaciens* containing a Ti plasmid to hijack a plant's protein synthesis machinery and genetically engineer the host genome, prompted the development of plasmid vectors for *Agrobacterium*-mediated plant transformation. In plant transformation vectors, the T-DNA contains only the genes intended for transfer to the nuclear genome of the

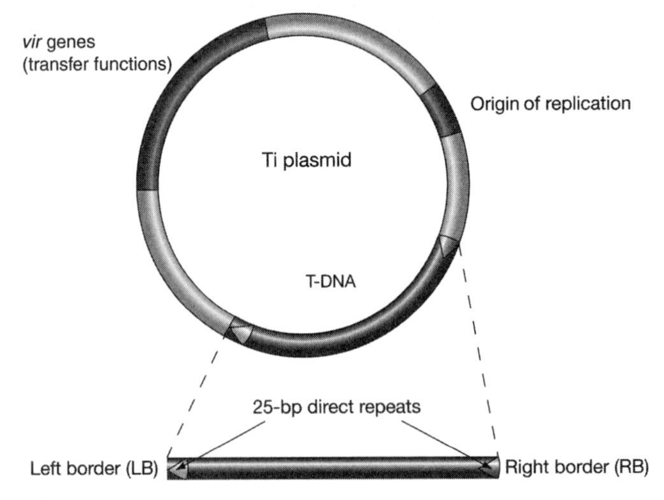

Figure 7.6. The Ti plasmid of *Agrobacterium tumefaciens* showing the origin of replication, the region encoding the virulence (*vir*) genes, and the transfer-DNA (T-DNA). The T-DNA is flanked by 25-bp direct repeats, known as the *left* and *right border* sequences (LB and RB, respectively). The *vir* genes are required for T-DNA processing and transfer to the plant cell. The T-DNA is stably integrated into the nuclear genome of the plant cell, and genes encoded within it, necessary for the biosynthesis of the plant growth hormones, cytokinin and auxin result in the formation of the characteristic tumorous growth associated with crown gall disease. The T-DNA also encodes opines (nopaline and octapine) that provide the *Agrobacterium* with an exclusive nitrogen source. This provides *Agrobacterium* carrying the Ti plasmid with a competitive advantage over *Agrobacterium* that do not.

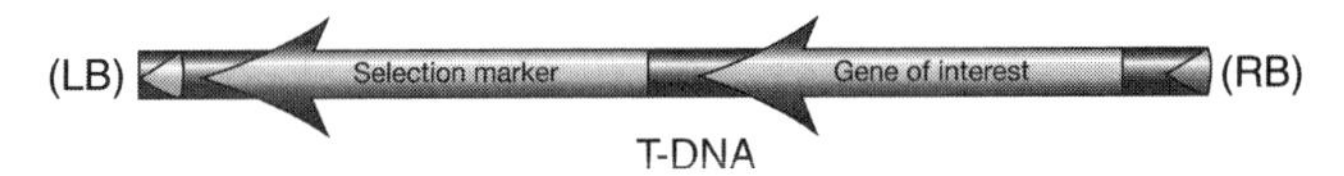

Figure 7.7. T-DNA used to genetically engineer plants frequently contains a selectable marker gene under the transcriptional control of a constitutively and ubiquitously active promoter to ensure gene expression in all tissues at all stages of development, together with the gene of interest (GOI) providing a novel phenotype for the plant.

engineered plant cell. All the phyto-oncogenes (tumor-inducing genes) have been removed (Fig. 7.7).

These plant vectors are known as *binary vectors* because they require the interaction of a second, disarmed Ti plasmid lacking a T-DNA. This second plasmid contains the *vir* region, allowing the T-DNA containing the transgenes on the binary vector to be transferred and stably integrated into the host nuclear genome (a more detailed description of plant transformation can be found in Chapter 10).

Plant binary vectors are constructed and amplified with the aid of *E. coli*, the workhorse organism in molecular biology. Once construction is completed in *E. coli*, such plasmid vectors are transferred to *A. tumefaciens*, the organism responsible for transferring genes to the nuclear genome of plant cells. These vectors therefore contain origins of replication that function in *A. tumefaciens* and *E. coli*. The pVS1 origin is derived from a *Pseudomonas* plasmid and is stably maintained in a wide variety of proteobacteria, including *Pseudomonas*, *Agrobacterium*, *Rhizobium*, and *Burkholderia*. For this reason, the pVS1 origin has been widely used to construct cloning vectors that are suitable for use in plant-associated bacteria. *A. tumefaciens* uses the repABC operon to stringently control plasmid replication and the partitioning of plasmid DNA to daughter cells. This operon is not only present on large plasmids of low copy number derived from *Agrobacterium* but is also encoded by the chromosomes of *Agrobacterium*. Unfortunately, *E. coli* does not use the repABC operon for plasmid replication, so plasmids containing only the pVS1 origin do not replicate in *E. coli*. Binary vectors designed to shuttle between *E. coli* and *A. tumefaciens* must, therefore, also contain an *E. coli*–compatible *ori*, most commonly the ColE1 origin (providing relaxed replication) (Fig. 7.8).

Since plant binary vectors provide no selective advantage to the bacteria, the vectors must be engineered to encode selectable marker genes for their propagation in *E. coli* and *A. tumefaciens* (examples of commonly used bacterial selectable marker genes are shown in Table 7.2).

A broadly active bacterial promoter must be used to transcribe the antibiotic resistance gene, so that bacteria containing the vector can survive and amplify the recombinant DNA. The same selection criteria are used for *E. coli* and *A. tumefaciens*. However, the T-DNA that is transferred to the plant cell must also contain a selectable marker, this time under the control of a broadly active plant promoter, allowing the identification and propagation of transformed plant cells (Fig. 7.8) (marker genes and the promoters that drive them are discussed in detail in Chapter 9).

7.2.2. Components for Efficient Gene Expression in Plants

The requirements for the successful introduction and efficient expression of foreign genes in plant cells have developed with our understanding of the mechanisms of plant gene

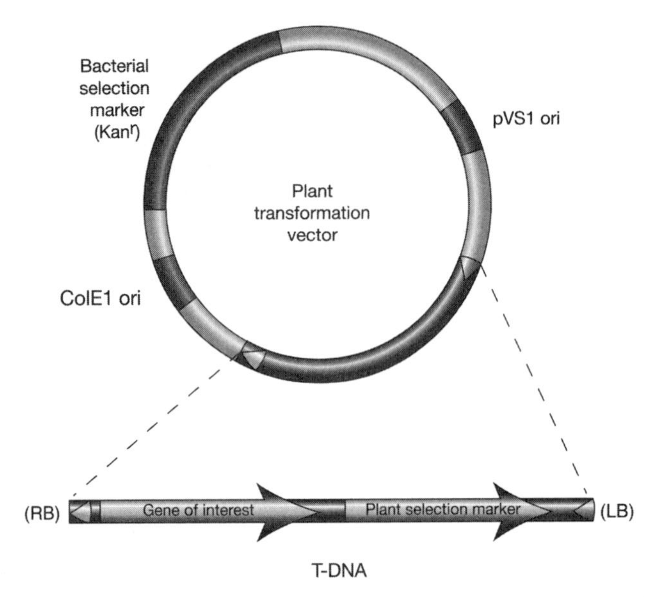

Figure 7.8. A generic plant binary vector with two origins of replication, the pVS1 ori for propagation in Agrobacterium and the ColE1 ori for propagation in *Escherichia coli*. The backbone of the vector contains an antibiotic resistance gene for bacterial selection (kanamycin resistance), and the T-DNA contains a plant selectable marker and the gene of interest (GOI).

expression and plant transformation (for more details, see Chapters 6 and 10). Failure to obtain gene expression using cistrons (gene and promoter sequences) from other species led to the first chimeric genes that used the 5′ and 3′ nopaline synthase (*nos*) regulatory sequences: the nos promoter and nos terminator. Although the nos promoter and terminator sequences are derived from the Ti plasmid of bacterial origin, they share more characteristics with eukaryotic than with prokaryotic genes. The promoter contains sequences that resemble CAAT and TATA boxes, which assist in directing RNA polymerase II (RNAP II) to initiate transcription upstream of the transcriptional start site (Fig. 7.9).

Terminator sequences contain an AATAA polyadenylation signal (which specifies transcript cleavage approximately 30 bp downstream of the signal). Soon after cleavage, multiple adenine residues are added to form a polyA tail on the 3′ end of the transcript. The polyA tail is thought to be important for mRNA stability.

TABLE 7.2. Commonly Used Bacterial Selectable Marker Genes

Antibiotic	Antibiotic Resistance Gene	Gene	Source Organism
Streptomycin/ Spectinomycin	Aminoglycoside adenyl transferase gene	*aadA*	*E. coli*
Kanamycin	Neomycin phospho transferase gene	*nptII* (*neo*)	*E. coli* Tn5
Chloramphenicol	Chloramphenicol acetyl transferase gene	*cat*	*E. coli* Tn5
Ampicillin	β-Lactamase	*bla*	*E. coli* Tn3
Tetracycline	Tetracycline/H^+antiporter	*tet*	*E. coli* Tn10

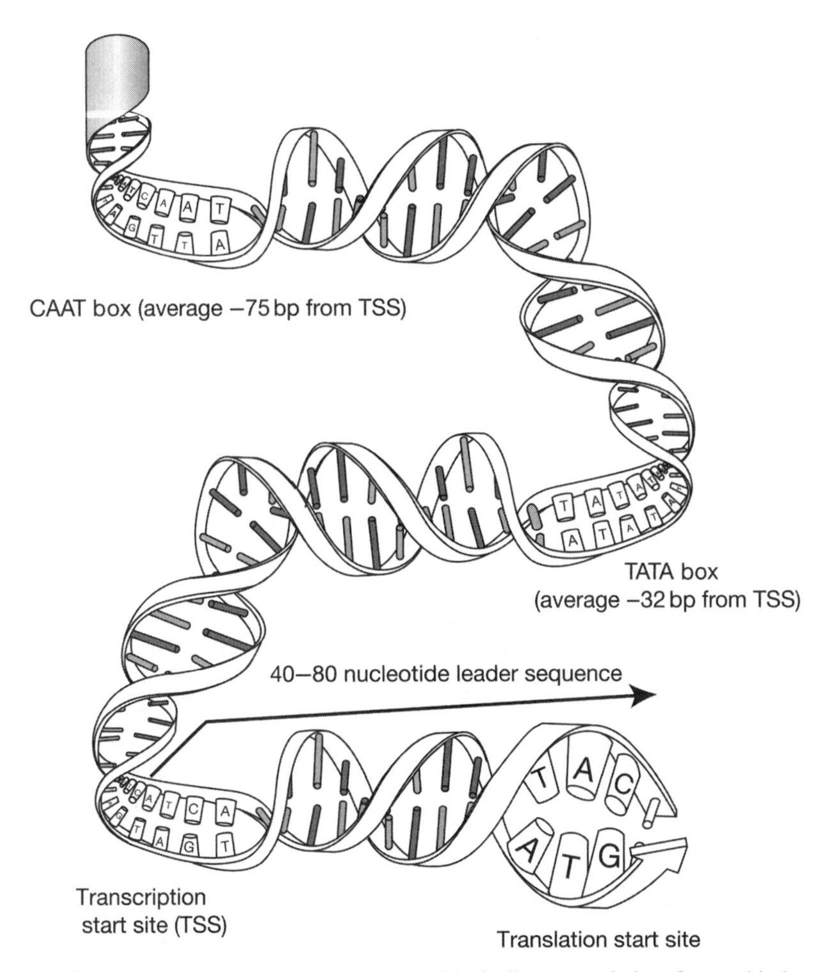

Figure 7.9. Cartoon of a generic plant promoter. Typically, transcription factors bind promoter sequences initiating the formation of the transcription complex. Components of the transcription complex bind the CAAT box and the TATA box and assist with the recruitment of RNA polymerase II, allowing the initiation of transcription. The transcription complex can cause the DNA to bend back on itself, bringing together regulatory sequences far from the site of transcription.

The efficiency of transgene expression in plants is dependent on a number of factors that affect mRNA accumulation and stability. In addition to the promoter (discussed in detail in Chapter 9), these include untranslated sequences (UTRs) both upstream (5′) and downstream (3′) of the gene, codon usage, cryptic splice sites, premature polyadenylation sites, and intron position and sequence (these important features affecting gene expression are discussed in more detail in Chapter 6). Careful consideration of these important factors should be made when designing vectors for transgene expression in plants. Once a decision has been made as to whether a transgene should be expressed ubiquitously or cell-type-specifically, inducibly or constitutively, by changing the promoter fragment used, further decisions can be made that determine whether a gene product is required at high or low levels. Often the omega sequence from the 5′ UTR of the tobacco mosaic virus (TMV) is used to enhance translation in plants. Omega contains a poly(CAA) region, which serves as a binding site for the heatshock protein, HSP101, which is required for translational enhancement. The efficiency of translation initiation is also affected by other mRNA structures, including the length of the leader—short leader sequences lead to reduced translation efficiency. Secondary structures, both upstream and downstream of the AUG start codon, can inhibit ribosome entry and again reduce translation efficiency. The consensus nucleotide sequence surrounding the AUG start codon in dicots (dicotyledonous plants) is aaA(A/C)aAUGGCu, while in monocots (monocotyledonous plants) it is c(a/c)(A/G)(A/C)cAUGGCG). The presence of upstream AUG codons are particular features of some genes that can reduce translational efficiency [for a review, see Kozak (2005)].

Foreign genes often contain nucleotide sequences that are not commonly used by plants to encode amino acids. Unusual codon usage can affect mRNA stability. For example, *Bacillus thuringiensis* (*Bt*) toxin genes are typically A/T-rich with an A or a T in the third position of codons, which occurs only rarely in plants. Extensive modification of the nucleotide sequence in the coding region of these genes can result in increased expression so that enough Bt toxin would be produced to kill target insects that fed on host plants. The plant species chosen for modification may also influence the design of the transgene construct, since the codon bias in monocot genes tends to be more stringent than it is in dicot genes.

Agrobacterium-mediated plant transformation has had a limited taxonomic host range, with most successful reports of transformation among dicots. Modifications to plant transformation protocols can, however, lead to the successful transfer of genes to plant species once thought to be beyond the host range of *Agrobacterium*, including a number of monocots, such as rice and wheat. Despite these advances, monocots are most often transformed using microparticle bombardment (Biolistics®) (for a more detailed description of microprojectile bombardment-mediated transformation, see Chapter 10). Particle bombardment does not require the use of plant binary vectors containing a T-DNA, since the DNA is physically delivered into the cell by the force of the projected particle. In early plant transformations using particle bombardment, entire plasmids were used, but more recently, only the transgene cassette (promoter, gene, and terminator sequences) has been used. This approach has reduced the transgene copy number and eliminated the insertion of unwanted vector sequences.

7.3. GREATER DEMANDS LEAD TO INNOVATION

Recombinant DNA technology has become more sophisticated as new techniques have emerged and greater demands have been made in the analysis of genes and the development of biotechnological innovations. Today it would not be unusual, in the course of analyzing a gene, to express the gene under a variety of promoters, make fusions with reporter genes (Chapter 9) for subcellular localization studies, or make fusions with a purification tag for biochemical analyses. All these types of analysis involve complex DNA manipulations so that a gene and/or its promoter can be inserted into the appropriate vector. Such manipulations have been facilitated by vectors that incorporate a series of restriction endonuclease recognition sites in a sequence known as a *polylinker* or *multiple cloning site* so that there is a convenient place in the vector to insert DNA. However, since vectors do not always contain a standardized polylinker, DNA molecules are not easily exchanged between vector types. In addition, genes and their promoters differ. Genes are rarely flanked by convenient restriction sites for cloning and often contain internal restriction

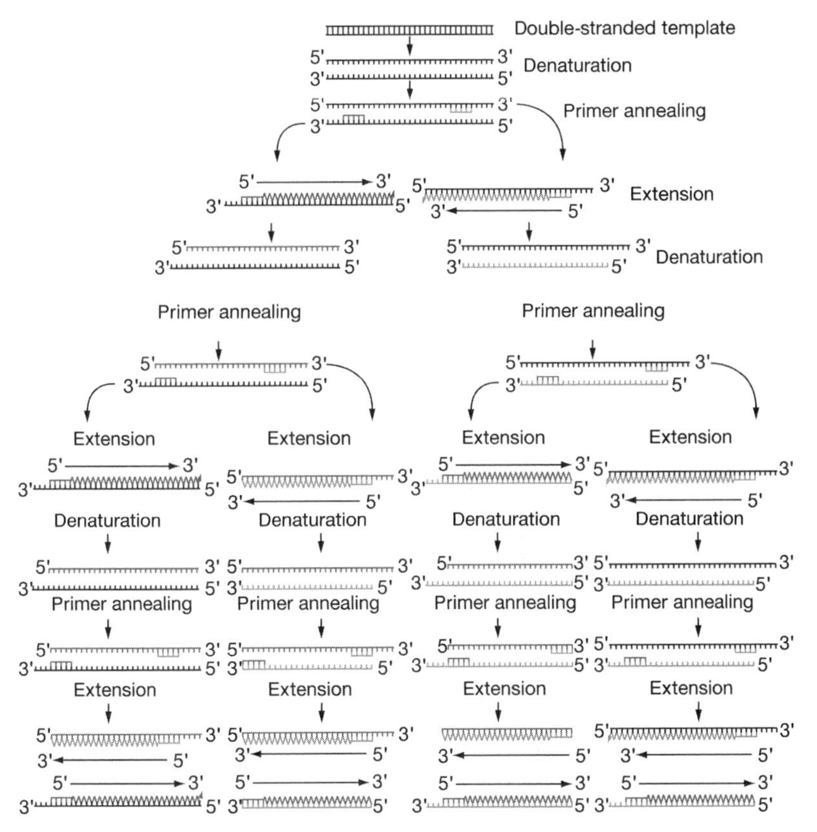

Figure 7.10. Polymerase chain reaction is a technique that allows a chosen region of DNA to be amplified in vitro by separating the double-stranded DNA template into two strands by denaturation and incubating with oligonucleotide primers and DNA polymerase to synthesize a complementary strand of each. The primers can be designed to incorporate restriction enzyme recognition sites or any other recognition sequence to facilitate the cloning of PCR fragments. Repeated cycles of denaturation, primer annealing, and extension (DNA synthesis with DNA polymerase) allow the targeted region of DNA to be amplified many thousands of times. This tool is frequently used in biotechnology, forensics, medicine, and genetic research to amplify DNA fragments.

sites that make them incompatible with some vectors. The development of the polymerase chain reaction (PCR) (Fig. 7.10) in 1985, by Kary Mullis, revolutionized the manipulation of DNA, facilitating the inclusion of restriction sites in positions flanking a gene, or its promoter, facilitating cloning as well as the removal of internal restriction sites, while maintaining the integrity of the gene. PCR amplifies specific DNA sequences in a test tube and also allows the sequences to be changed. Despite these improvements, the production of constructs is laborious, and inappropriately positioned restriction sites are still a major factor that hinders vector construction.

7.3.1. Site-Specific DNA Recombination

Several strategies have been developed to overcome the difficulties associated with conventional cloning. These have been compounded by the demands of the numerous functional genomics studies that have resulted from the availability of whole-genome sequences. These novel cloning strategies rely on *site-specific DNA recombination techniques* and significantly reduce the time and effort involved in generating recombinant DNA vectors for gene analysis and cDNA library construction (cDNA is a DNA sequence that is complementary to the coding sequence of an RNA transcript). Three systems are currently available that work efficiently for large-scale cloning projects: Gateway™ (Invitrogen), Creator™ (Clonetech), and the Univector system (Liu et al. 1998), also known as the Echo™ system (Invitrogen).

7.3.1.1. Gateway Cloning. The Gateway cloning system takes advantage of elements that evolved naturally in the life cycle of the bacteriophage *lambda* (λ). During this cycle, the bacteriophage passes from a lysogenic phase, in which the viral genome is stably incorporated into the host genome, to a lytic phase, in which the host cell ruptures (lyses) and infectious phage particles are released [for a more recent review of lambda development, see Oppenheim et al. (2005)] (Fig. 7.11).

The Gateway cloning system utilizes modified *att* recombination sites, together with an integration enzyme mix containing Integrase (Int) and Integration Host Factor (IHF) proteins (BP clonase) and an excision/integration enzyme mix containing the Int, IHF, and

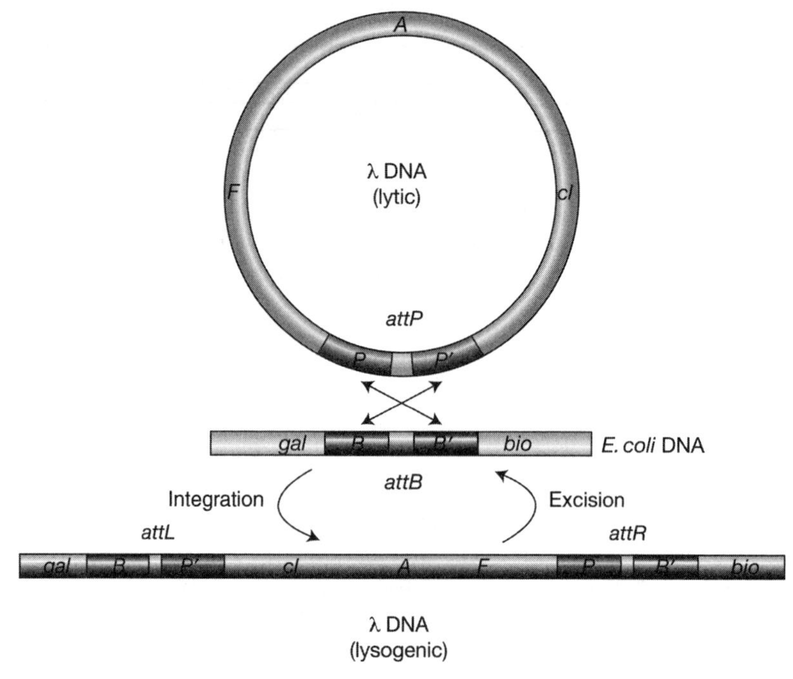

Figure 7.11. For lysogeny, the viral DNA is incorporated into the host genome by a process of recombination between common sequences, the *att* sites, in the two genomes. Bacteriophage λ contains an *attP* site (*P* for phage), and the host *E. coli* DNA contains an *attB* site (*B* for bacterium). A number of proteins are required for this recombination: λ-derived Integrase (Int) and *E. coli*–derived Integration Host Factor (IHF) allow λ to enter the lysogenic phase of its life cycle, and IHF, Int, and λ-derived Excisionase (Xis) allow λ to excise from the *E. coli* genome and enter the lytic phase of its life cycle.

Excisionase (Xis) proteins (LR clonase), derived from elements used during the bacteriophage λ life cycle. DNA fragments flanked by recombination sites can be mixed in vitro with vectors that also contain recombination sites, allowing the exchange of DNA fragments and the generation of recombinant DNA. Such an approach avoids many of the difficulties associated with conventional cloning (inconvenient restriction sites, time-consuming reactions, etc.). For Gateway cloning, the *att* sites have been modified so that the orientation of the DNA fragments can be maintained during the excision and integration process. Catalyzed by BP clonase, an *attB1* site specifically recombines with an *attP1* site to produce an *attL1* site, while an *attB2* site specifically recombines with an *attP2* site to produce an *attL2* site (Fig. 7.12). This allows PCR fragments flanked by *attB1* and *attB2* sites to be inserted into pDONR vectors containing the reciprocal *attP* sites, thereby generating "entry clones" in which the chosen DNA fragments are flanked by *attL1* and *attL2* sites.

Entry clones should be sequence-validated to provide a library of well-characterized DNA fragments for insertion into "destination vectors." Catalyzed by LR clonase, DNA fragments flanked by *attL1* and *attL2* sites are then transferred, by a second recombination

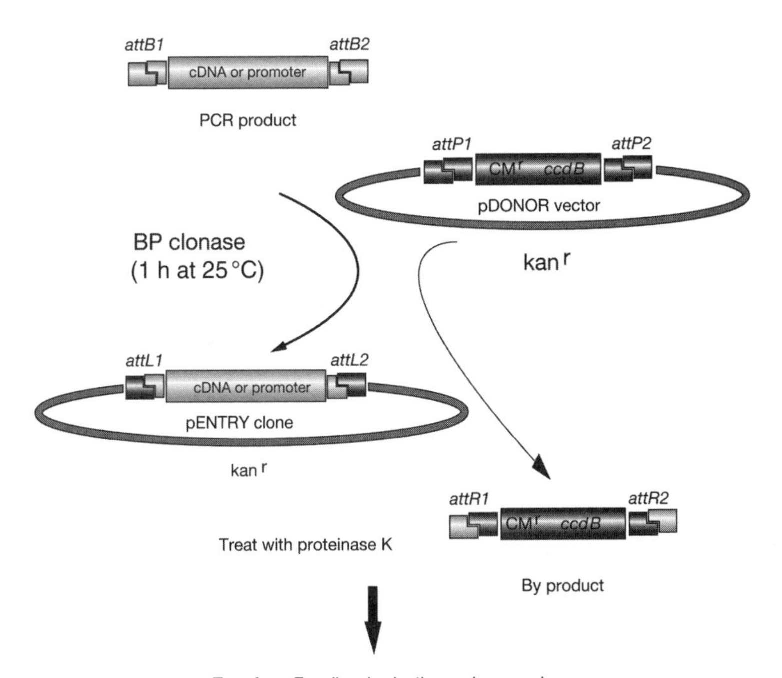

Figure 7.12. A gene or promoter is amplified by PCR using DNA target-specific primers that contain the *attB* sites (*attB1* and *attB2*) at the 5′ and 3′ ends, respectively. The purified PCR product, flanked by *attB* sites, is mixed with a pDONOR vector that contains the corresponding *attP* sites. To this DNA mix is added BP clonase enzyme (containing Int and IHF proteins). After 1 h incubation at 25°C, proteinase K is added and incubated for a further 20 min at 37°C. This mix is used to transform *E. coli* bacteria and plated on the appropriate antibiotic (in this example kanamycin) selecting transformants containing the appropriate pENTRY clone.

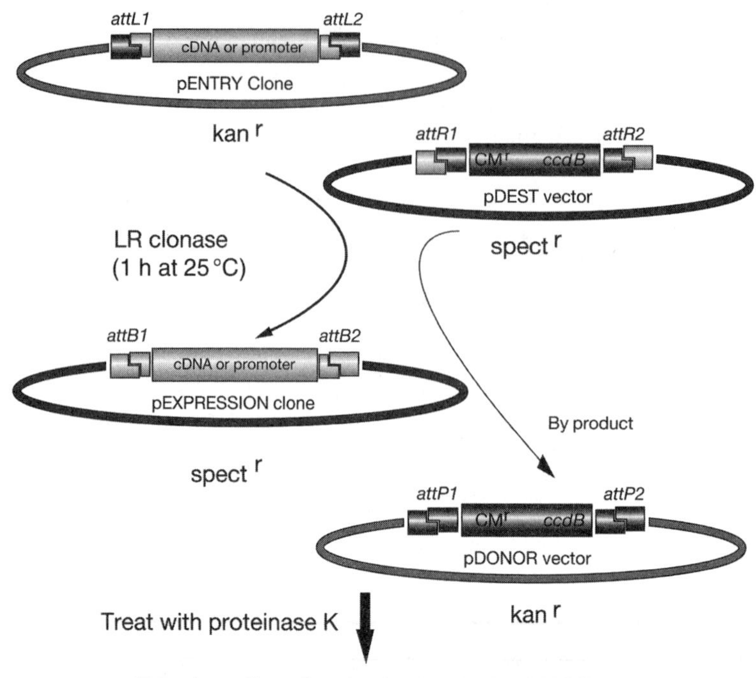

Figure 7.13. A gene or promoter contained within the pENTRY clone flanked by *attL* sites (*attL1* and *attL2*) is mixed with a pDESTINATION vector that contains the corresponding *attR* sites. To this DNA mix is added LR clonase enzyme (containing the Int, IHF, and Xis proteins). After 1 h incubation at 25°C, proteinase K is added and incubated for a further 20 min at 37°C. This mix is used to transform *E. coli* bacteria and plated on the appropriate antibiotic (in this example spectinomycin) selecting transformants containing the appropriate pEXPRESSION clone.

reaction, to pDEST vectors containing *attR1* and *attR2* sites. The resulting recombinant DNA constructs are known as "expression clones." Here, the recombination product of the *attL1* and *attR1* sites is an *attB1* site and the recombination product of the *attL2* and *attR2* sites is an *attB2* site (Fig. 7.13).

To select the correct recombination product for the BP and LR reactions, a combination of positive and negative selectable markers are employed. Positive selection is afforded by alternative antibiotic selection, while negative selection is afforded by the *ccdB* gene, the product of which inhibits the activity of DNA gyrase, leading ultimately to cell death. *E. coli* bacteria transformed with vectors containing the *ccdB* gene (i.e., pDONR or pDEST vectors) or by cointegrate intermediates, cannot grow. Only bacteria containing the desired recombinant construct that lacks the *ccdB* gene and contains the appropriate antibiotic resistance marker gene can survive (Figs. 7.12 and 7.13). The propagation of pDONR vectors and pDEST vectors is achieved using the *E. coli* strain DB3.1, which contains a mutant DNA gyrase, which is unaffected by the *ccdB* gene product.

7.3.1.2. Creator™ Cloning. The Creator cloning system is an alternative approach that allows the efficient transfer of DNA fragments from donor vectors to "creator" expression vectors. This transfer is mediated by the Cre-*loxP* site-specific recombination reaction discovered in bacteriophage P1 (Sternberg and Hamilton 1981). First, PCR products are inserted into Creator-compatible donor vectors using a proprietary enzyme called In-Fusion™, creating "master" clones (Fig. 7.14).

LoxP sites flanking the insertion site of the master clone allow cloned DNA fragments to be transferred to a single *loxP* site in an acceptor vector. This second transfer relies on the presence of Cre recombinase. The *loxP* sites contain inverted repeat sequences separated by a spacer region, within which DNA breakage and reunion take place. To select the correct recombination product, as with the Gateway system, both positive and negative selection is used. Donor vectors, or recombinants that retain the donor vector backbone, are selected against because they contain a *sacB* gene from *Bacillus subtilis* that generates a toxic metabolite in the presence of sucrose— recombinants containing the CAT gene are selected for by the presence of chloramphenicol in the medium. Using this dual-selection regime, only bacteria containing the desired recombinant construct that lacks the *sacB* gene and contains the appropriate antibiotic resistance marker gene will survive in the presence of sucrose (Fig. 7.15).

7.3.1.3. Univector (Echo™) Cloning. The Univector system was developed by Liu and coworkers (Liu et al. 1998, 2000). Like the clonetech Creator™ system, the Univector

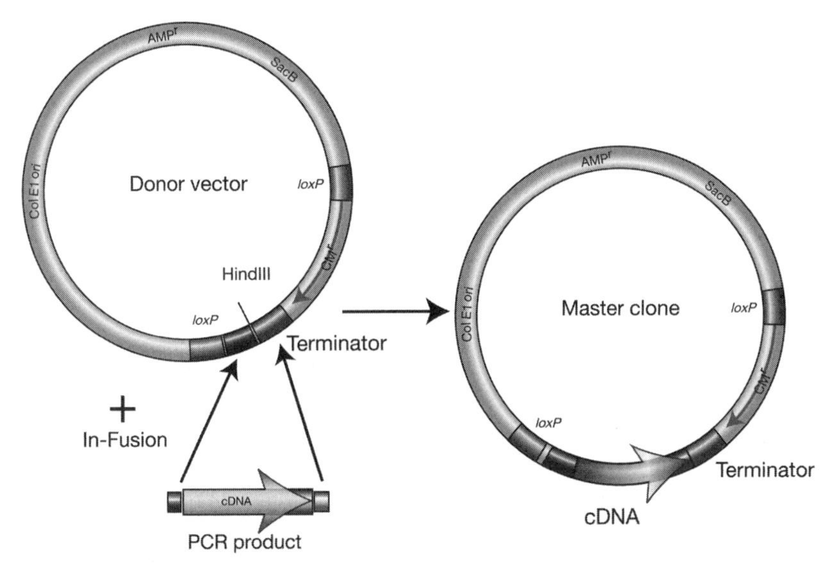

Figure 7.14. The PCR products to be cloned are generated with gene- or promoter-specific primers that contain 15-bp extensions. In this hypothetical example a cDNA is amplified. These extensions are homologous to a region of the vector that flanks any unique restriction site. The In-Fusion™ enzyme creates single-stranded regions that share homology between the vector and the PCR product, allowing the PCR product to join the specialized "donor" vector by strand displacement and In-Fusion-mediated recombination. In this hypothetical example, the cDNA is inserted upstream of a plant terminator sequence in the "donor" vector.

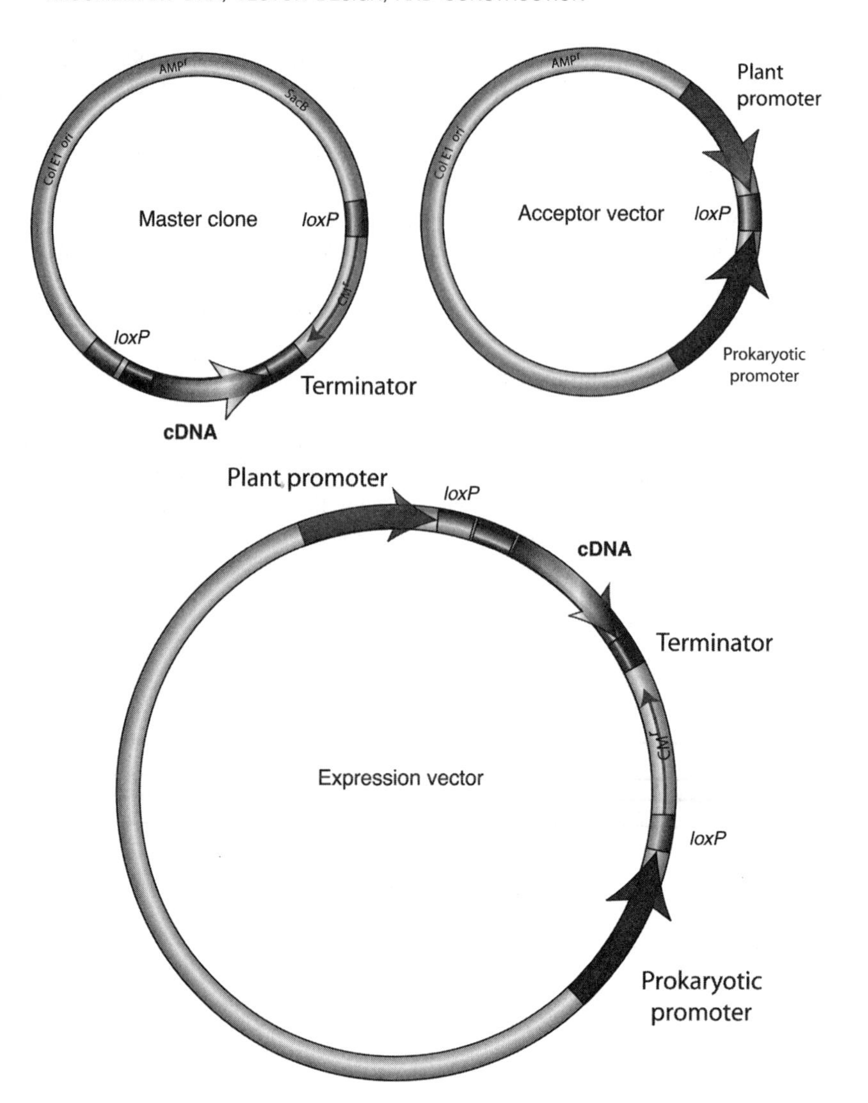

Figure 7.15. Positive selection of expression clones is provided by a chloramphenicol resistance (CAT) gene in the "acceptor" vector. This gene can be expressed only when it is transferred from the donor clone into the acceptor vector, where it is positioned adjacent to a prokaryotic promoter. In this way, only *E. coli* transformed by acceptor vectors containing an insert will survive on chloramphenicol-containing media. In this hypothetical example, a constitutive and ubiquitous plant promoter present on the acceptor vector allows expression of the cDNA only when the fragment of interest flanked by *loxP* sites is transferred to a single *loxP* site in the "acceptor" vector.

system makes use of the Cre-*loxP* site-specific recombination system to transfer DNA fragments from a reference plasmid, known as a *pUNI* plasmid, to a recipient vector, known as *pHOST*, to produce an expression vector. Both the pUNI and the pHOST vectors contain a *loxP* site that permit site-specific recombination and the production of a cointegrate (Fig. 7.16).

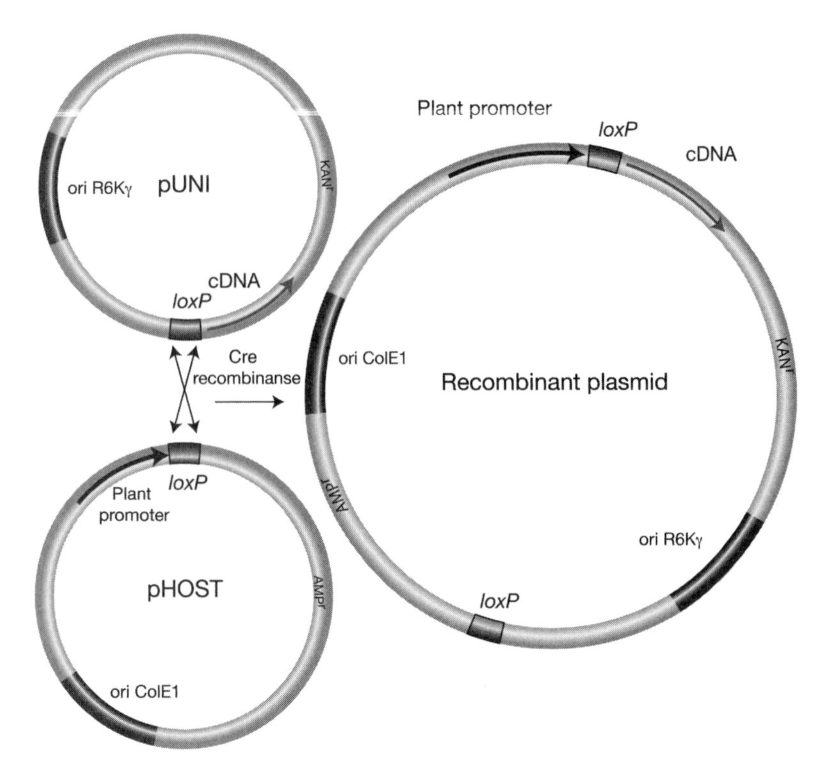

Figure 7.16. The pUNI vector contains a conditional origin of replication that functions only in *E. coli* strains that express the *pir* gene (which encodes an essential replication protein, π) and an antibiotic selectable marker. The pHOST vector contains the ColE1 origin of replication and an alternative antibiotic selectable marker. Nonrecombinant vectors are counterselected: the pUNI vector, by a host bacterial strain that does not express the *pir* gene; and the pHOST vector, by media containing the antibiotic used to select the pUNI vector. Using this dual selection system, only bacteria containing the recombined, or cointegrated, molecule survive.

7.4. VECTOR DESIGN

Recombinant DNA technology has made an enormous impact on plant biotechnology, both in the development of novel crop traits and the functional analysis of new genes and their promoters. The efficient functional analysis of DNA fragments and the effective application of the resulting discoveries to crop trait improvement are increasingly dependent on innovative vector design and construction. The design and construction of vectors has an impact on the versatility of experimental systems and influences the public acceptability of genetically modified crops.

7.4.1. Vectors for High-Throughput Functional Analysis

Obtaining nearly the entire genomic DNA sequence of the model plant organisms, rice (*Oryza satica*), a monocot, and arabidopsis (*Arabidopsis thaliana*), a dicot, have presented new challenges in the production and analysis of recombinant DNA. Large numbers of promoters and genes encoded by these genomes have been discovered, but many remain uncharacterized, providing an incentive to design and construct vectors with the capacity for high-throughput functional analysis. Traditional ligase-mediated cloning is

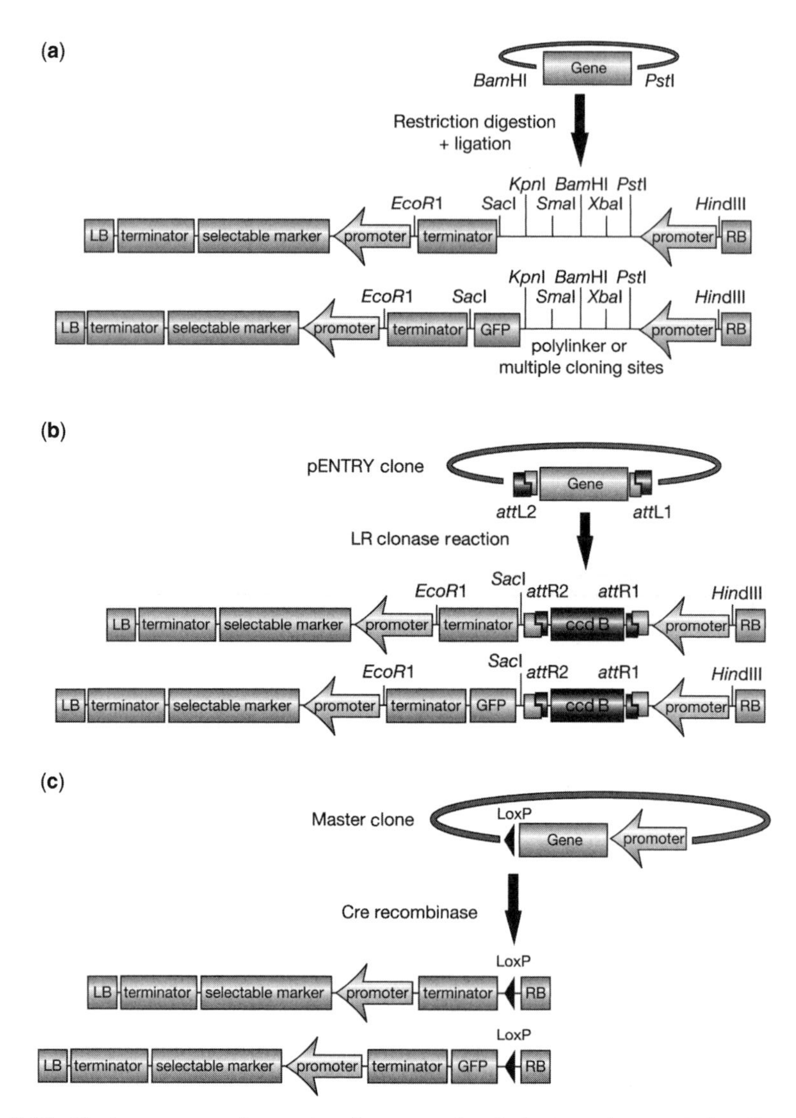

Figure 7.17. Plant gene expression vectors for conventional cloning using restriction digestion and ligation (a), Gateway^{TM} recombination cloning (b), and univector recombination cloning (c). The first vectors shown in (a), (b), and (c) are designed to allow a gene to be ectopically expressed in a plant cell. The second vectors shown for each category contain the GFP (green fluorescent protein) gene. These vectors are designed to effect protein fusions with GFP to help identify the subcellular target of a protein under investigation. Ideally, three vectors for each type are frequently made, one for each reading frame, to ensure that a perfect fusion between the GOI and the marker gene is made. The insert DNA must be in an "open" ORF configuration (described in the text) so that no stop codon is present between the GOI and the marker gene.

no longer a practical approach to facilitate the analysis of all the genes and promoters from these model organisms. Plant vectors compatible with Gateway recombination cloning and Univector recombination cloning have been generated to aid these analyses (Fig. 7.17).

Recombination-compatible collections of plant ORFs (open reading frames; a sequence encoding a polypeptide) have also been generated. Trimmed ORFs lacking 5′ or 3′ UTRs (i.e., containing protein-coding sequences only) can be shuttled rapidly and efficiently between vectors bearing compatible recombination sites. These so-called ORFeome collections have been generated so that the positions of the original translation initiation and termination codons remain intact ("closed" ORF configuration). However, since some applications to investigate gene function require the addition of *C*-terminal peptide fusions, ORFeome collections in which the stop codon is omitted ("open" ORF configuration) are also being generated. Often, the initial functional data on an ORF or gene are on the phenotype it induces when it is *ectopically* expressed (i.e., in tissues in which it is not normally expressed) under a constitutive and near-ubiquitous promoter. Gateway vectors designed for this type of analysis have been generated using the strongly active 35*S* promoter from cauliflower mosaic virus (CaMV). Some of these vectors have an additional design feature that provides stop codons adjacent to the 3′ recombination site in all three reading frames, to facilitate the expression of open as well as closed ORF configurations. Of course, not all ORFs can be misexpressed constitutively. Some cause lethal effects when expressed in this manner. In such cases, ORFs can be shuttled into vectors that are designed for conditional or inducible ectopic expression (Karimi et al. 2002; Curtis and Grossniklaus 2003; Joubes et al. 2004) or even to vector systems that allow induced expression in restricted cell types (Brand et al. 2006).

7.4.2. Vectors for RNA Interference (RNAi)

A very powerful tool that helps elucidate gene function is to reduce, or "knockdown" native gene expression in the organism using RNA interference (RNAi) (Waterhouse et al. 1998). Here, double-stranded RNA is produced by the transcription of an inverted repeated sequence of a gene. This transcript forms a hairpin–loop structure that triggers the RNAi pathway, leading to the degradation of homologous mRNAs [reviewed by Brodersen and Voinnet (2006)]. The careful construction of specialized Gateway™ destination vectors guarantees the rapid and efficient production of double-stranded RNAs (Fig. 7.18). In standard Gateway™ vectors, the *att* site modifications were designed to maintain DNA fragment orientation during the excision and integration process (Hartley et al. 2000). The arrangement of *att* sites in RNAi constructs ensures the easy insertion of two identical gene segments in opposite orientations, downstream of a constitutively active promoter (Fig. 7.18). Constitutively expressed interfering RNA can be used to silence genes throughout a plant's development, or can be expressed conditionally to provide temporal control over the onset of gene silencing.

7.4.3. Expression Vectors

The thorough analysis of gene function frequently involves expressing a GOI, not only in plants but also in multiple systems. With traditional cloning methods, independently derived expression constructs must be made. Recombination cloning technology has

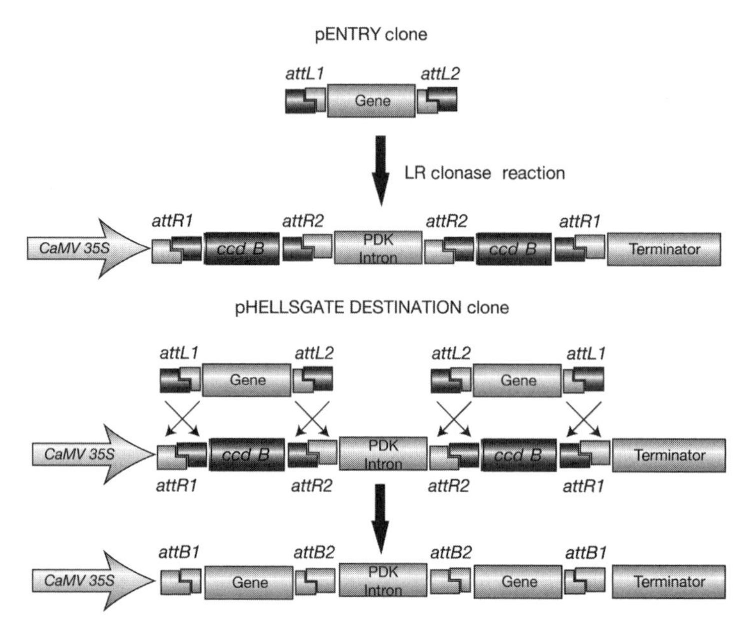

Figure 7.18. Gene silencing in plants can be achieved using inverted repeat transgene constructs that encode a hairpin RNA (hpRNA). Using Gateway™ cloning technology, the production of such inverted-repeat transgene constructs can be achieved efficiently, since DNA fragment orientation during the excision and integration process is maintained and the Gateway™ recombination cassettes are arranged in opposite orientations with respect to each other.

revolutionized gene analysis by allowing genes to be expressed from the same recombination cassette in *E. coli, Saccharomyces cerevisiae*, or baculovirus expression systems (Liu et al. 1998) (Invitrogen Carlsbad), providing easier access to tools that broaden the scope for the functional analysis of genes.

7.4.4. Vectors for Promoter Analysis

Tools that identify the spatial and temporal expression patterns of genes also provide important clues in functional genomics studies. Frequently, vectors are designed to allow the promoter or *cis* element of a GOI to be fused upstream of a reporter-coding sequence (reporter genes are discussed in Chapter 9). Such constructs are used to determine the cell type(s), organ type(s), or developmental stage in which a gene is expressed. By assembling promoter ENTRY clones in recombination-compatible vectors, researchers are compiling a library of promoters and enhancers that are universally compatible with a wide variety of vectors. The modular assembly of DNA components has recently been extended through the introduction of additional novel recombination sites (Multisite®, Gateway®, Invitrogen) with unique specificities that allow multiple DNA fragments to be assembled in a single vector (Fig. 7.19). This facilitates the simultaneous incorporation of a promoter, ORF, and epitope tag into a single plant vector derived from collections of the modular component parts.

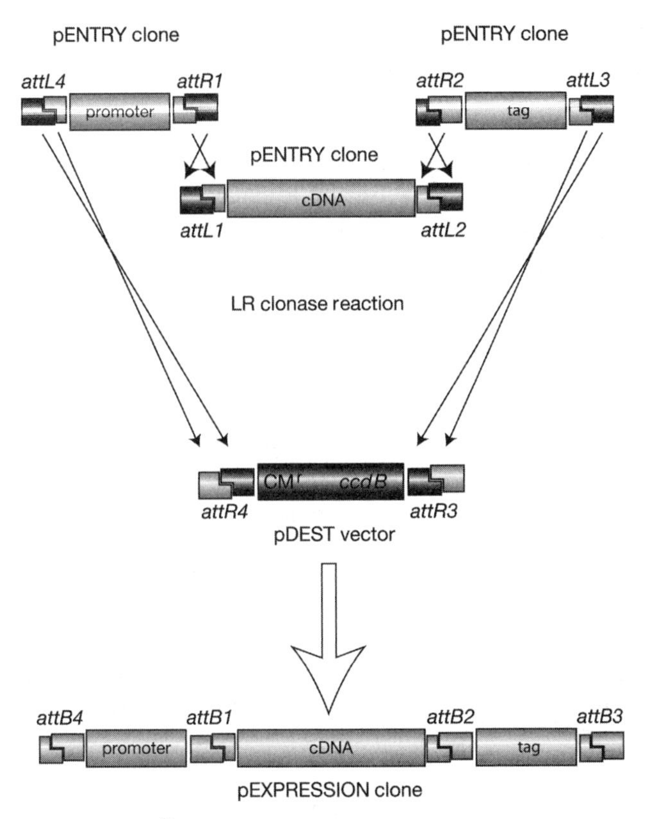

Figure 7.19. Multisite Gateway® allows several DNA fragments to be cloned within a single vector construct. More recent advances in the design of new *att* recombination sites have permitted the assembly of up to five DNA molecules within a single vector construct, but none have been designed as yet for plant transformation.

7.4.5. Vectors Derived from Plant Sequences

The revolutionary advances in recombinant DNA technology provide great opportunities to engineer new traits in crop plants that could not have been achieved through conventional breeding. Ironically, it is this ability to overcome species barriers, to widen the pool of genetic traits available for crop modification, that makes this technology extremely powerful, while at the same time being the cause of many objections to the deployment and consumption of genetically modified crop varieties. Consumer surveys have identified that public acceptance of genetically engineered organisms is linked to concerns about the origin of the genetic material used to improve crop traits. These surveys have identified that the food crops least appealing to consumers are those containing foreign genetic material derived from organisms distantly related to plants. Ironically, wild-type plant cells already contain the genetic material of three genomes, the plant nuclear genome, and two bacterially derived genomes: the chloroplast and the mitochondrial genomes, from cyanobacteria and α-proteobacteria, respectively. Some concerns could be alleviated through the careful design of the recombinant DNA vectors used to improve crop varieties. During the design stage of vector construction, measures can be taken to ensure that non-plant-derived sequences are kept to a minimum. The T-DNA of *Agrobacterium* is one source of foreign genetic material that could be eliminated using plant-derived "P-DNA" sequences (Rommens et al. 2004). These are functional analogs of *Agrobacterium*-derived T-DNAs, which have been shown to support the transfer of DNA from *Agrobacterium* to plant cells. The transfer of DNA to plant cells is a relatively rare event, and transformed cells are usually identified and regenerated with the aid of selectable markers, such as antibiotic resistance genes, again often derived from bacteria. Once these foreign selectable marker genes have served their purpose, they can be removed, since they play no further role in the expression of the transgenic trait. One method of removing such genes relies on the presence of an inducible recombination system in the plant vector, which allows excision of a marker gene positioned between recombination sites (Fig. 7.20).

Some marker genes, such as those conferring herbicide resistance, can be used to select transformants and, at the same time, provide an economically important crop improvement trait. In fact, about 75% of genetically modified crops are engineered for herbicide tolerance (Castle et al. 2004). The two most commonly used herbicide resistance genes are derived from the bacteria *Streptomyces hygroscopicus* and *Bacillus licheniformis*. Such bacterial herbicide resistance genes could be replaced by plant-derived sequences. Several plant genes that produce agronomically useful levels of herbicide resistance have now been identified.

Plant genome sequence data and advances in plant molecular biology have provided the means by which to identify and isolate plant sequences that have the potential for use in

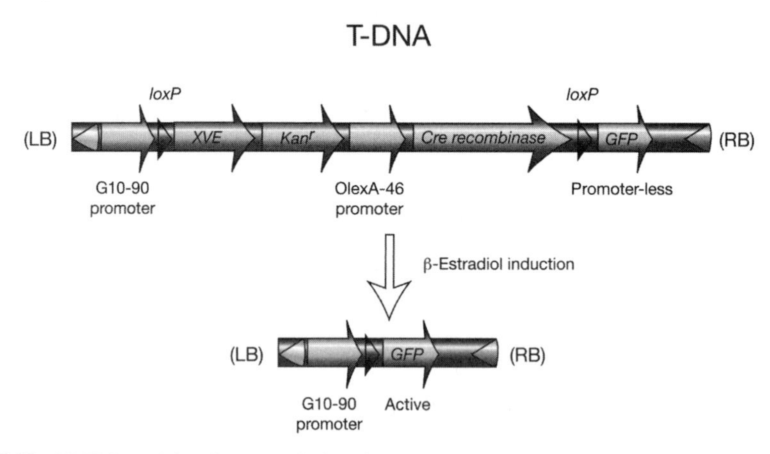

Figure 7.20. XVE is a chimeric transcription factor. It contains three functional domains, a LexA DNA binding domain (X), the VP16 activation domain (V), and the estrogen receptor binding domain (E). The G10-90 promoter drives the constitutive and ubiquitous expression of XVE in transformed plant cells. The XVE protein is then bound as a monomer in the cytosol of the cell by a chaperone protein HSP90, and the target gene is transcriptionally inactive. Application of β-estradiol causes a conformational change in E, which leads to the release of HSP90 and dimerization of the receptor. On dimerization, the receptor is activated, allowing the protein to translocate to the nucleus of the cell where it binds O*LexA* binding sites of the promoter that is placed upstream of the Cre recombinase. The VP16 activation domain activates RNA polymerase II, leading to the transcription of the Cre recombinase gene. Cre recombinase allows recombination to occur between the *LoxP* sites removing all intervening genes, including the selectable marker gene.

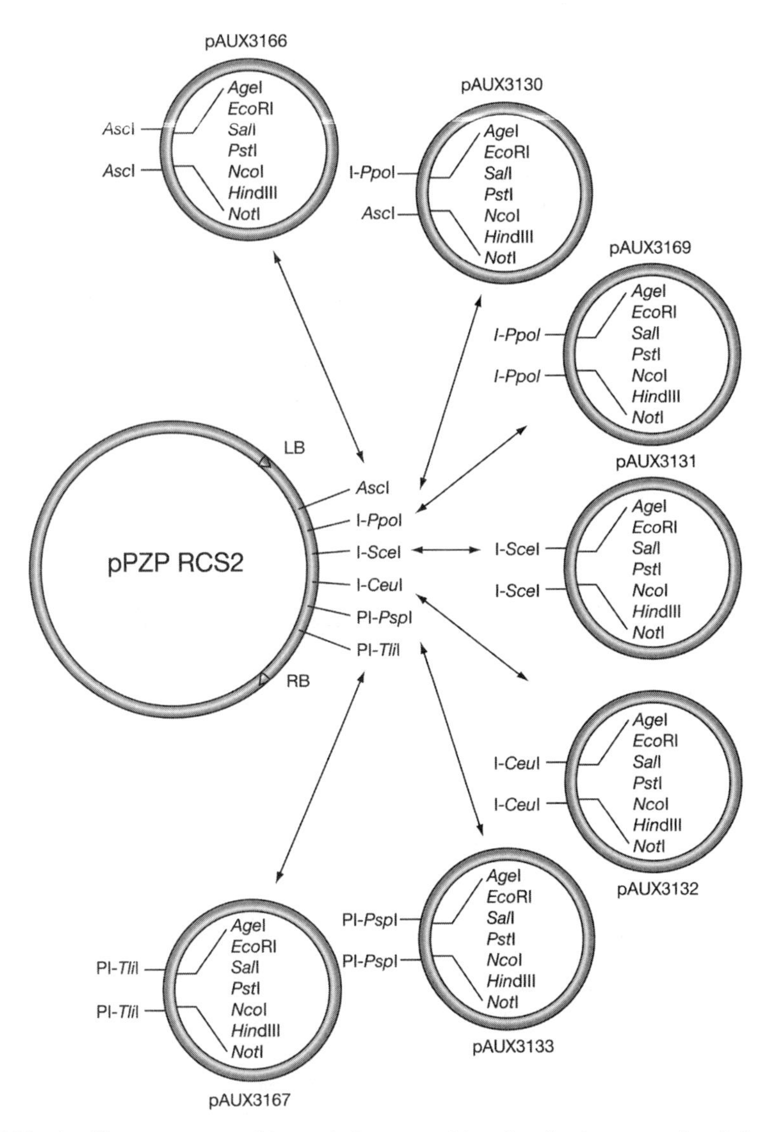

Figure 7.21. Auxillary vectors provide restriction recognition sites for the conventional cloning of an expression cassette with the desired promoter, ORF, and terminator sequences. The expression cassettes are flanked by 8-bp restriction recognition sites or homing endonuclease sites. Homing endonuclease sites are extremely rare in natural sequences and can facilitate the assembly of several expression cassettes within the single plant transformation vector pPZP RCS2.

crop improvement. Frequently, viral promoters, such as the CaMV 35*S* promoter, are used to express genes constitutively and near-ubiquitously in transgenic plants. These can be replaced by native plant promoters with similar expression profiles, such as actin or ubiquitin promoters. The use of such promoters to express transgenes both ubiquitously and constitutively may, however, cause unwanted secondary effects that might be avoided by designing and constructing vectors to deliver tissue-specific or conditional gene expression. For example, dwarfism is an agronomically important trait, which helps plants survive

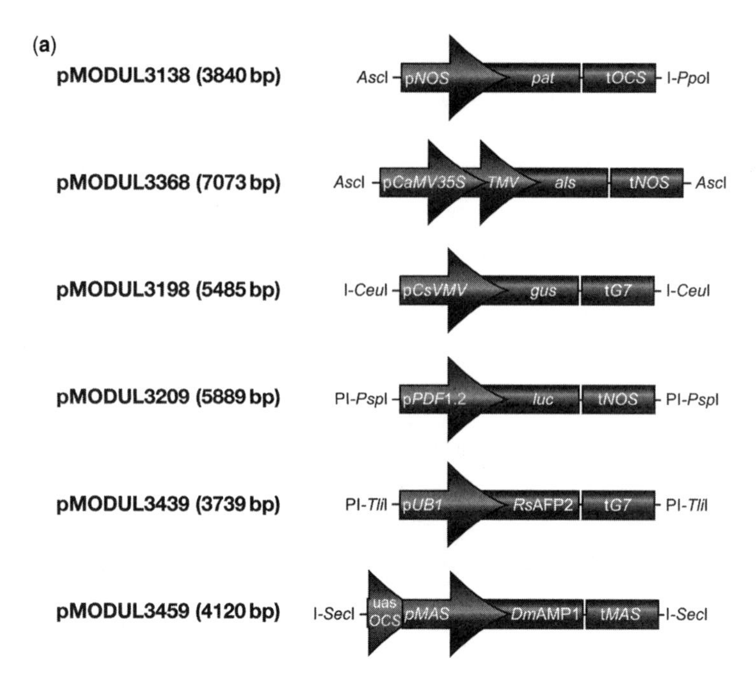

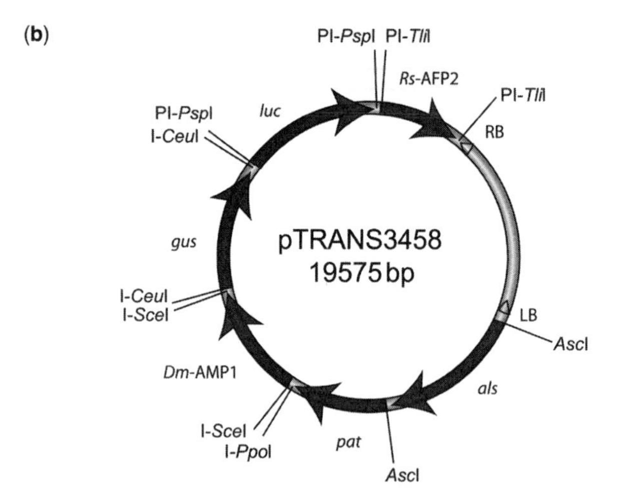

Figure 7.22. A schematic representation of six expression cassettes inserted within each of the auxillary vectors described in Figure 7.20. (a) A variety of promoters have been used to generate each expression cassette, either the nopaline synthase gene promoter (*pNOS*), the enhanced cauliflower mosaic virus RNA promoter (*pCaMV 35S*) with the untranslated leader region of tobacco mosaic virus (TMV), the promoter of cassava vein mosaic virus (*pCsVMV*), the *A. thaliana* plant defensin 1.2 promoter (*pPDF1.2*), the promoter of the GUbB1 gene of *Helianthus annuus* (*pUBI*), or the promoter of the *mannopine synthase* gene (*pMAS*) with an upstream activating sequence of the octopine synthase promoter (*uasOCS*). These promoters drive the expression of a variety of genes, including the *S. hygroscopicus* phosphinothricin acetyl transferase gene (*pat*), the coding sequence of the DmAMP1 defensin (*DmAMP1*), the acetolactate synthase coding sequence (*als*), the *E. coli* β-glucuronidase *uidA* gene (gus), the firefly luciferase gene (luc), or the *RsAFP2* defensin-coding sequence (*RsAFP2*). Each cassette also contains a terminator; the terminator from the

heavy rain and windy conditions. The dwarf plants of the so-called Green Revolution are short because they respond abnormally to the plant growth hormone, gibberellin. Attempts to generate transgenic dwarf rice, by misexpressing the *Arabidopsis gibberellin-insensitive* (*GAI*) gene, resulted in short plants that unfortunately also produced low seed set (Fu et al. 2001; Tomsett et al. 2004). Subsequent experiments have shown that this problem could be resolved, at least in *Arabidopsis*, by constructing a vector that places the *GAI* transgene under the control of an inducible promoter (Ait-ali et al. 2003).

Although many endogenous (originating from within the organism) plant promoters that can rapidly respond to the application of inducers have been identified, these often also respond to environmental factors, such as water, salt stress, temperature, illumination, wounding, or infection by pathogens. Other nonendogenous inducible systems have been developed, but these rely on DNA sequences of foreign origin [for a more recent review, see Curtis and Grossniklaus (2006)]. Since endogenous promoters can be triggered inappropriately by environmental factors, and inducers may modify native gene expression (perhaps altering the physiology and development of the plant), an alternative approach that restricts transgene activity to specific tissue types to produce the desired trait would be more profitable. In the case of *GAI* expression, a construct with a promoter that is active in vegetative tissues only (and not reproductive tissues) may result in dwarf plants that do not have reduced seed set.

7.4.6. Vectors for Multigenic Traits

The construction of vectors for crop improvement can rely on the insertion of a single gene, as is the case with the production of *Bt* toxin to protect crops against insects, or on the insertion of several genes, as is likely to be required to engineer the vast majority of agronomically important traits. Currently, multigenic traits are obtained either through sequential sexual crossing of transgenic plant lines that allows the accumulation of three or four independent transgenes in a single plant (see Chapter 3), or by the use of different transgenes held on distinct T-DNAs that are used to cotransform plants (see Chapter 10). The former approach is laborious, and the latter is technically challenging. Careful consideration of the design and construction of plant transformation vectors can resolve many of the technical difficulties, allowing polygenic traits to be expressed from a single T-DNA. One such design relies on a collection of auxiliary vectors (Fig. 7.21). Using such approaches, a construct capable of expressing up to six transgenes from a single location within the nuclear genome has been generated (Goderis et al. 2002) (Fig. 7.22).

There are many alternative approaches to "stacking" multiple genes into acceptor vectors. These make use of site-specific recombination systems and homing endonucleases that allow the sequential and indefinite delivery of expression cassettes to an acceptor vector, thereby allowing the expression of many transgenes from a single locus in the genome.

Figure 7.22 (*Continued*) *Agrobacterium tumefaciens* nopaline synthase gene (*tNOS*), the terminator of gene 7 of the *A. tumefaciens* plasmid Ti15955 (*tG7*), the terminator of the mannopine synthase gene from *A. tumefaciens* (*tMAS*), or the terminator from the *A. tumefaciens* octopine synthase gene (*tOCS*). (b) These six gene cassettes are arranged between the LB and RB of the T-DNA vector pTRANS3458.

7.5. TARGETED TRANSGENE INSERTIONS

Once a recombinant T-DNA vector has been generated, with features designed to provide stable integration and gene expression, the DNA enters the plant cell and integrates randomly within the genome. The position of integration is uncontrolled and can often result in variable levels of transgene expression. A number of factors influence the level of transgene expression in plants, including the number of transgenes inserted into the genome, local *cis*-acting elements, and RNA silencing. Nontranscribed, A/T-rich regions in eukaryotic genomes, known as *matrix attachment regions* (MARs), have been used to flank genes in T-DNA vectors (Butaye et al. 2004). These sequences have been reported to result in more reliable transgene expression shielding transgenes from RNA silencing (Mlynárová et al. 2003). However, targeting transgenes to predetermined chromosomal sites by homologous recombination would perhaps provide greater control and reduce potential positional effects. Until relatively recently, such approaches have been very inefficient in plants. Advances in the production of synthetic transcription factors [zinc-finger proteins (ZFPs)] designed to recognize specific DNA target sequences have now made it possible to increase the efficiency of targeted homologous recombination in plants, creating the potential to engineer precise deletions, insertions, or mutations within specific chromosomal regions. The production of customized ZFPs will provide a variety of precision tools to alter genomes, changing the expression of endogenous genes and transgenes in future generations of genetically engineered plants.

7.6. SAFETY FEATURES IN VECTOR DESIGN

The use of plants as bioreactors for the manufacture of polymers, antibodies, vaccines, hormones, and a variety of other therapeutic agents also presents new challenges in vector design and construction. The effective use of plants as bioreactors, involves not only the careful selection of host plants (i.e., food crop or a nonfood crop), but also innovative vector design to ensure high levels of gene expression with safety features that ensure that products will not enter the food chain. For pharmaceutical production, plants have many advantages, the most significant of which is their eukaryotic protein synthesis pathway, capable of the posttranslational modification and assembly steps required to produce active eukaryotic proteins, such as antibodies. Unfortunately, plants glycosylate proteins differently to mammals, however, recent and future advances in "humanizing" plant glycosylation pathways (for a recent review see Joshi and Lopez 2005) will make the production of "humanize" proteins feasible. A great advantage of plants is that they can be grown in huge numbers to produce very large quantities of protein, they are free of mammalian pathogens, and many plant varieties are edible, providing an easy means of administering medication. However, to reduce the risk of nonedible products entering the food chain, plant expression vectors for such products must be engineered with robust safety features. One such safety mechanism is to incorporate *intein* sequences that permit the transsplicing of proteins. This means that genes encoding the transsplicing protein fragments do not need to be located in the same genome; one can be contained in the nuclear genome and another in the chloroplast genome, for example. In this case, the nuclearly encoded protein fragment is engineered to target the chloroplast, where it is transspliced to the second protein fragment encoded by the chloroplast. This type of split-gene technology requires two types of vector construction: a T-DNA vector, for nuclear genome integration with elements for eukaryotic gene expression; and a plastid vector, with elements that allow prokaryotic gene expression. Using such vector design features, the risk of gene flow and product contamination is reduced, since chloroplasts are maternally inherited in crop species and pollen produced by these plants would contain only half of the protein-coding fragment.

Innovations in vector construction and plant transformation technology can influence the character of the resulting transgenic crop. Gene silencing and transgene containment are two important issues in plant biotechnology, the impact of which can be reduced by genetically engineering the plastid genome rather than the nuclear genome. This is because, unlike the nuclear genome, gene silencing does not occur in the plastid genome and, in most agronomically important plant species, plastids are maternally inherited, preventing pollen-mediated outcrossing. Vectors for chloroplast transformation are designed and

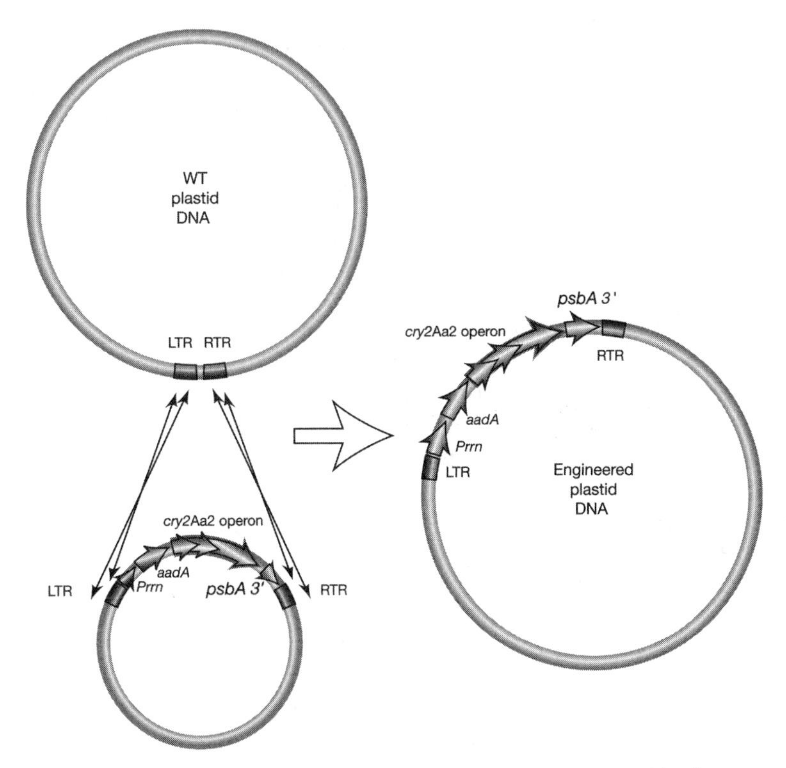

Figure 7.23. Site-specific integration is achieved by two homologous recombination events, one on either side of the DNA fragment to be integrated. During insertion, the targeted region of the vector replaces the targeted region of the plastid genome, and the vector backbone is lost. The inserted DNA fragment contains a selectable marker (here, the *aadA* gene encoding aminoglycoside 3′-adenylyltransferase, providing spectinomycin resistance) and can contain either a single gene flanked by independent 5′ and 3′ regulatory regions, including a promoter; a 5′ UTR and a 3′ UTR; or, as is the case here, multiple genes with a single promoter that regulates the expression of the operon with individual ribosome binding sites (RBS), upstream of each ORF. In this example, the *Bacillus thuringiensis* (*Bt*) *cry2*Aa2 operon is inserted in the plastid genome generating insecticidal proteins in plant cells. The 16*S* ribosomal RNA (rRNA) promoter (*Prrn*) drives the expression of the *aadA* gene and the three genes of the *cry2*Aa2 operon. The terminator is the *psbA* 3′ region of a gene encoding the photosystem II reaction center component of the tobacco chloroplast genome.

constructed so that they contain left and right plastid-targeting regions (LTR and RTR), which are 1–2 kb in size and homologous to a chosen target site (Fig. 7.23).

The design and construction of plastid vectors that allow the simultaneous expression of several genes in an operon will be particularly useful in the engineering of agronomically important traits, as described earlier. Transgene integration has been achieved at 16 independent sites distributed across the plastid genome, ensuring that the positional effects, which are often associated with DNA integration events in the nuclear genome, are eliminated. Since there are 10–100 plastid genomes per plastid and approximately 10–100 plastids per cell, as many as 10,000 transgene copies can be generated in a single cell, resulting in highly abundant transgene transcription, producing as much as 46% of the total soluble protein in a cell. As with nuclear genome transformation vectors, *loxP* sites in plastid vectors can be engineered to flank the marker gene and excised when no longer required using the Cre site-specific recombinase. Plastid transformation technology does not yet extend to major crops, but has been demonstrated in soybean, carrot, and cotton through species-specific chloroplast vectors, and plant regeneration through somatic embryogenesis.

7.7. PROSPECTS

Recombinant DNA technology, vector design, and construction form the foundations on which advances in modern plant biotechnology are built. The development of tools for the rapid amplification and manipulation of DNA sequences are essential if we are to keep pace the ever-increasing wealth of genetic information that results from the analysis of plant, animal, bacterial, and viral genomes. To exploit this information fully, functional studies must be conducted to determine the potential uses of such sequences, identifying the elements required to control gene expression and the genes required to ensure the high crop yields needed to sustain the planet's expanding population. Understanding the elements required for the efficient expression of genes in plants has already facilitated the development of new crop varieties. Novel genetic engineering approaches resulting from recombinant DNA technologies will provide the solutions to many of our future industrial, pharmaceutical, and sustainable fuel requirements. This evolving technology forms the basis of a new "Green Revolution."

LIFE BOX 7.1. WAYNE PARROTT

Wayne Parrott, Professor, Department of Crop and Soil Sciences, University of Georgia

Wayne Parrott

It was almost a given that I would have a career in agriculture—my parents were in agribusiness, and both sets of grandparents lived on a farm. Thus, upon graduating from high school, I started studying towards a degree in agronomy at the University of Kentucky.

It was not just the cultivation of plants that I found interesting—the plants themselves and their amazing diversity were just as fascinating—as was the underlying genetic basis behind all the diversity. Growing up in Central America, genetic diversity was all around me, as was access to "new and improved" varieties that breeders were releasing to farmers all the time. A key moment came while visiting my parents, who were living in Honduras at the time, and got invited to visit a banana breeding station. Relating the experience to the Agronomy Club advisor back at Kentucky, he immediately offered me a job as a student worker in his wheat-breeding program. After that point, the genetic modification of crops—rather than their actual cultivation—became my chief interest.

The next key moment came when I saw my first plant growing in a test tube—it was so fascinating, there would never be any turning back. Coincidentally, the public press was filled with reports of the first gene transfer into a plant (the "sunbean") and all the expected potential to improve agriculture as that technology developed. I was irreversibly hooked.

Following graduation, I went to the University of Wisconsin for graduate school in Plant Breeding and Genetics, where I got to indulge in studying all those aspects of plant genetics I found so fascinating. After graduation, I returned to the University of Kentucky as a postdoc. By that time, the foundation for plant tissue culture and genetic engineering technology had advanced to point it was a fruitful area of research. From there, I joined the faculty at the University of Georgia, where the technology and I have continued to grow up together.

Breakthroughs and major developments have not come continuously—they are interspersed with lots of experiments that don't work out or hypotheses that easily get disproved. Yet, when a technological hurdle is overcome, or when there is a new discovery—there is a rush and excitement that carries over to the next one. Looking back on my career, it is these moments that I most remember and collectively lead to a feeling of accomplishment.

But, research is also about people. The value of those who have served as my mentors along the way cannot be understated. I must mention Glenn Collins at the University of Kentucky, in whose lab I did an undergraduate research thesis, to whose lab I returned for a postdoc. Glenn has never stopped being my chief mentor. Then there is Richard Smith, my major professor from graduate school. I still see his work ethic and research approach in everything I do. Finally, I want to single out Roger Boerma and Joe Bouton at the University of Georgia—two established faculty members who helped me out every step of the way.

Last but not least, I have had the good fortune to have had many postdocs, graduate students, and undergraduate students whose thought-provoking questions and enthusiasm have led my research forward. Seeing them move on and progress in their own careers has been as rewarding as the research itself.

REFERENCES

Ait-ali T, Rands C, Harberd N (2003): Flexible control of plant architecture and yield via switchable expression of Arabidopsis gai. *Plant Biotechnol J* **1**:337–343.

Brand L, Hörler M, Nüesch E, Vassalli S, Barrell P, Yang W, Jefferson RA, Grossniklaus U, Curtis MD (2006): A versatile and reliable two-component system for tissue-specific gene induction in Arabidopsis. *Plant Physiol* **141**:1194–1204.

Brodersen P, Voinnet O (2006): The diversity of RNA silencing pathways in plants. *Trends Genet* **22**:268–280.

Butaye KMJ, Goderis IJWM, Wouters PFJ, Pues JM-TG, Delauré SL, Broekaert WF, Depickers A, Cammue BPA, De Bolle MFC (2004): Stable high-level transgene expression in Arabidopsis thaliana using gene silencing mutants and matrix attachment regions. *The Plant J* **39**:440–449.

Castle LA, Siehl DL, Gorton R, Patten PA, Chen YH, Bertain S, Cho HJ, Duck N, Wong J, Liu D, Lassner MW (2004): Discovery and directed evolution of a glyphosate tolerance gene. *Science* **304**:1151–1154.

Clark AJ, Warren GJ (1979): Conjugal transmission of plasmids. *Annu Rev Genet* **13**:99–125.

Curtis MD, Grossniklaus U (2003): A gateway cloning vector set for high-throughput functional analysis of genes in planta. *Plant Physiol* **133**:462–469.

Curtis MD, Grossniklaus U (2006): Conditional gene expression in plants. In *Floriculture, Ornamental and Plant Biotechnology: Advances and Topical Issues*, Vol II, Teixera da Silva JA (ed). Global Science Books London, pp 77–87.

Eguchi Y, Itoh T, Tomizawa J (1991): Antisense RNA. *Annu Rev Biochem* **60**:631–652.

Fu X, Sudhakar D, Peng J, Richards DE, Christou P, Harberd NP (2001): Expression of Arabidopsis GAI in transgenic rice represses multiple gibberellin responses. *Plant Cell* **13**:1791–1802.

Goderis IJ, De Bolle MF, Francois IE, Wouters PF, Broekaert WF, Cammue BP (2002): A set of modular plant transformation vectors allowing flexible insertion of up to six expression units. *Plant Mol Biol* **50**:17–27.

Hartley JL, Temple GF, Brasch MA (2000): DNA cloning using in vitro site-specific recombination. *Genome Res* **10**:1788–1795.

Joshi L, Lopez LC (2005): Bioprospecting in plants for engineered proteins. *Curr Opin Plant Biol* **8**:223–226.

Joubes J, De Schutter K, Verkest A, Inze D, De Veylder L (2004): Conditional, recombinase-mediated expression of genes in plant cell cultures. *Plant J* **37**:889–896.

Karimi M, Inze D, Depicker A (2002): GATEWAY vectors for Agrobacterium-mediated plant transformation. *Trends Plant Sci* **7**:193–195.

Kozak M (2005): Regulation of translation via mRNA structure in prokaryotes and eukaryotes. *Gene* **361**:13–37.

Liu Q, Li MZ, Leibham D, Cortez D, Elledge SJ (1998): The univector plasmid-fusion system, a method for rapid construction of recombinant DNA without restriction enzymes. *Curr Biol* **8**:1300–1309.

Liu Q, Li MZ, Liu D, Elledge SJ (2000): Rapid construction of recombinant DNA by the univector plasmid-fusion system. *Meth Enzymol* **328**:530–549.

Mlynárová L, Hricová A, Loonen A, Nap J-P (2003): The presence of a chromatin boundary appears to shield a transgene in tobacco from RNA silencing. *Plant Cell* **15**:2203–2217.

Nordstrom K (1990): Control of plasmid replication—how do DNA iterons set the replication frequency? *Cell* **63**:1121–1124.

Oppenheim AB, Kobiler O, Stavans J, Court DL, Adhya S (2005): Switches in bacteriophage lambda development. *Annu Rev Genet* **39**:409–429.

Rommens CM, Humara JM, Ye J, Yan H, Richael C, Zhang L, Perry R, Swords K (2004): Crop improvement through modification of the plant's own genome. *Plant Physiol* **135**:421–431.

Sheng Y, Mancino V, Birren B (1995): Transformation of *Escherichia coli* with large DNA molecules by electroporation. *Nucleic Acids Res* **23**:1990–1996.

Sternberg N, Hamilton D (1981): Bacteriophage P1 site-specific recombination. I. Recombination between loxP sites. *J Mol Biol* **150**:467–486.

Tomsett B, Tregova A, Garoosi A, Caddick M (2004): Ethanol-inducible gene expression: First step towards a new green revolution? *Trends Plant Sci* **9**:159–161.

Waterhouse PM, Graham MW, Wang MB (1998): Virus resistance and gene silencing in plants can be induced by simultaneous expression of sense and antisense RNA. *Proc Natl Acad Sci USA* **95**:13959–13964.

CHAPTER 8

Genes and Traits of Interest for Transgenic Plants

KENNETH L. KORTH

Department of Plant Pathology, University of Arkansas, Fayetteville, Arkansas

8.0. CHAPTER SUMMARY AND OBJECTIVES

8.0.1. Summary

The whole purpose of biotechnology is to manipulate the genome of important plants, typically by adding a few genes at a time. Traits can be manipulated by inserting DNA originating from any organism with that trait of interest into the target plant. Thus far in crop biotechnology, much work has been accomplished in conferring traits to plants such as the ability to survive herbicide treatment, insect resistance, disease resistance, and stress tolerance. However, there is growing interest in producing drugs and industrial proteins in plants as well as enhancing the nutrition of plant products.

8.0.2. Discussion Questions

1. What are the differences between "input" and "output" traits? Considering the environmental and biological factors that limit production in a farmer's field, what are some new input traits that might be good candidates for improvement using biotechnology?
2. Consider the possibility that you are employed by an agricultural biotechnology company, and they ask you to find a bacterial gene for resistance to a specific herbicide. The herbicide has been manufactured by the company for many years. Using a strategy similar to that used to find glyphosate resistance, where might you start to look for a bacterium resistant to that herbicide?
3. Other than the products discussed in this chapter, what other sorts of genes or strategies might be useful in engineering transgenic plants resistant to insects or pathogens?
4. Golden Rice producing provitamin A has the potential to help many impoverished people who might benefit from eating it. Although application of this technology

Plant Biotechnology and Genetics: Principles, Techniques, and Applications, Edited by C. Neal Stewart, Jr.

is supported by many people and organizations, there are also some who oppose the technology. Considering their possible motivations and potential biases, discuss some of the reasons that groups have come out in favor or in opposition to Golden Rice.

5. What are the potential benefits of producing pharmaceutical proteins in plants? What are some of the disadvantages or potential dangers?
6. Animal genes can be inserted into plants and expressed. Would you be opposed to eating foods from plants expressing proteins encoded by animal genes? By human genes? Discuss the reasons for your answers.

8.1. INTRODUCTION

As discussed in Chapter 6, the specific order of the nucleotide bases of DNA determines the function that a given sequence encodes. However, those four DNA bases are contained in a repetitious sugar–phosphate backbone that is essentially identical in DNA from any source. Because of this similarity of DNA structure in all organisms, there are no chemical limits on DNA from any organism being transferable to another, and this has allowed the development of transgenic plants carrying genes from many different sources, including microbes, insects, and animals, including humans. Essentially, sources for transgenes are as deep as our genomic knowledge in all of biology.

Many important traits in agriculture, such as crop yield, are often controlled by the action of multiple genes working together. However, other useful traits can be controlled by just a single gene. Because it has been easier to identify single-gene traits and produce transgenic plants with a limited number of introduced genes, most transgenic plants being grown today originated via the transfer of just one or a few foreign genes. In this chapter some of the most common genes and traits that have been engineered into transgenic crops will be discussed, and we will also take a look into the future to some potential applications of transgenic plants that could benefit consumers by providing improved foods and products.

8.2. IDENTIFYING GENES OF INTEREST VIA GENOMIC STUDIES

Advances in technologies used to determine DNA sequence and mRNA accumulation have allowed detailed inquiry into the impressive quantities of information contained in the genome of an organism. *Genomics* is a broadly defined term, but it generally refers to a strategy of using high-throughput, large-scale molecular techniques to analyze DNA sequence or gene expression patterns.

Deciphering and interpreting the vast information of a genome sequence are the focus of great efforts, and it is hoped that this information will lead to development of new tools for crop improvement. In most crop species, this is a difficult task. For example, the soybean genome consists of around 1.1 billion base pairs (bp) of DNA, whereas the maize genome is considerably larger, at approximately 2.4 billion bp. For comparison, the size of the human genome is slightly over 3 billion bp. These billions of base pairs of sequence are filled with many regions that are highly repetitive, and many others that do not seem to encode for any protein products. Identifying the important regions of plant DNA and those that contribute to useful traits for farmers can require a combination of traditional breeding techniques, high-tech molecular analyses, genetic studies, and newly developed computational strategies. The financial and intellectual commitments made toward completion of deciphering the human genome were instrumental in leading to development of new technologies for large-scale analysis of genes and proteins. Those technological developments continue today, and are being applied to analysis of every class of organism—including important crop plants.

Although all plant families and species have their specific traits that make them unique, there are many genes that are conserved across species. In fact, there are many genes with conserved functions across plants and animals. By determining the function of a given gene in one species, it might allow us take a reasonable guess about the function of the corresponding, or *homologous*, gene in another species. For this reason, some plants that are viewed as models attract a lot of attention. For example, the species *Arabidopsis thaliana* is a small, fast-growing member of the mustard family, and has a relatively small genome confined to just five chromosomes. For these reasons, it serves as a good model for studies of plant development and response to the environment. The *Arabidopsis* genome was the first plant to be fully sequenced, and its genome of approximately 120 million bp was reported in 2000. Having the complete genome of a plant, even one of no value as a crop such as *Arabidopsis*, has proved very valuable in determining the function of individual genes. As genomic DNA sequence information from crop plants continues to increase, the similarities and differences among gene structures and presence in different plant species is becoming clearer. It is hoped that by comparing the structures of these different genomes, the gene regions that are important for valuable traits can be identified.

From a technical perspective, improved methods have made it increasingly feasible to determine DNA sequences of an organism. However, although knowledge of the genomic sequence of a species is a valuable tool, it does not necessarily tell us about the function of genes or how they contribute to phenotype. It can be particularly difficult to associate specific genes with valuable traits, especially when the gene might have a minor, but important, effect on a trait. Therefore, genomic approaches to understanding gene functions or patterns of gene expression are being widely applied. Gene expression studies are typically aimed at indicating presence of a particular mRNA transcript. For most genes, their ultimate function is dependent on the presence of the mRNA transcript whose nucleotide sequence information can be translated into amino acid sequence.

Expression of many genes is regulated at the level of mRNA accumulation and can be associated with their ultimate function in the plant. For example, many genes thought to be involved in plant defense against pathogens will have greatly increased amounts of their encoded mRNAs during infection by a pathogen. To study this phenomenon, scientists often take the approach of inoculating a plant with a pathogen, and then measuring mRNA transcript levels. If a given gene is *upregulated* at the level of mRNA accumulation, then this gene is a good candidate for one involved in defense responses. By measuring large numbers of transcripts under certain sets of environmental conditions, profiles of gene expression begin to emerge and gene sets involved in plant defenses (or other traits) can be identified.

A common technique for measuring mRNA transcript accumulation of large numbers of genes is a DNA *microarray* (Alba et al. 2004). This technique takes advantage of the ability of two nucleotide segments with complementary sequences to bind together, or *hybridize*. If one of the sequences is somehow tagged with a label that can be measured, then the amount of binding can be quantified. In a DNA microarray, specific sequences are typically bound to a substrate such as a glass slide on a small scale. Differing technologies allow for the binding of hundreds of thousands of individual sequences onto specific locations within areas as small as $1\,cm^2$. Generally, DNA sequences from a given species are spotted onto a microarray, and then hybridized with labeled copies of mRNA (usually in the form of cDNA) from a specific tissue or after some treatment, such as pathogen inoculation. If a given mRNA is present at high levels in a treatment, then a high degree of binding to its corresponding DNA sequence on the array will be detected. The level of binding of transcript sequences is usually compared with levels in some untreated control tissue. This general approach, known as *comparative gene expression*, allows one to observe the transcript profiles of tens of thousands of genes in a single experiment.

For species where genomic DNA sequence information is not as available, or where DNA microarrays are not developed, the strategy of using *expressed sequence tags* (ESTs) can also provide information on mRNA profiles. In this strategy, mRNA is collected from the tissue of interest and then converted via reverse transcription into cDNA. Individual clones from the collection of cDNAs, known as a "library," are then partially sequenced and the information is compiled in a database. The presence of a given EST in a database then reflects the presence of its corresponding mRNA transcript in the original tissue. By determining how often an mRNA occurs in a given tissue, and by comparing its abundance after other treatments or in other tissues, a profile of when that particular transcript is present can sometimes emerge. This technique was first developed to study human gene expression, but it is now widely applied in many types of organisms, including many crop plants.

Ultimately, the protein products of most genes, or the metabolites that those proteins produce, are the things that will function leading to a particular plant trait. It is therefore useful to analyze the endproducts of gene expression. In fact, the accumulation of a given RNA transcript measured in most gene expression studies does not always correlate with the level or activity of the protein it encodes. This can be due to many factors, such as regulation of RNA stability, protein translation rates, or posttranslational regulation of protein stability or enzyme activity. As with genomic studies, the identification of an individual protein from among tens of thousands can be a technical challenge. *Proteomic* approaches use different techniques to examine the large mixture of proteins present in a given tissue or after some treatment. This usually involves separating individual proteins on the basis of some physical characteristics such as protein size or charge. After the proteins are separated from one another their amino acid sequence can be identified using techniques such as mass spectrometry. If the proteomic data are accompanied by a wealth of DNA sequence or gene expression data, these data can be even more valuable, as the amino acid sequences can be correlated with specific gene sequences in that plant. Likewise, *metabolomics* is the large-scale analysis of chemical compounds that accumulate and contribute to the characters of a plant. These metabolites can be important not only for plant defense and physiology but also in nutrition and food production; therefore they are valuable contributors to a number of traits in crop plants that are of interest to farmers and consumers.

Through genomic, proteomic, and metabolomic (*omics*) approaches, scientists have attempted to take a large-scale, or *systems biology*, view of the events occurring at the cellular level in an organism. The technologies developed and used in these methods generate huge amounts of data. Trying to make sense of these data is a considerable challenge in itself, and this has given rise to a discipline called *bioinformatics*, which applies computational and mathematical methods to help scientists understand biological data (Rhee et al. 2006).

As the amount of genomic detail for crop plants continues to rapidly expand and be understood, it will provide more candidate genes as tools for biotechnological applications. Uses for this knowledge could come in the form of transgenes to be transferred between species, or as tools for plant breeders who utilize DNA marker-assisted selection in crop improvement. The amount of information contained within a single plant species' genome is immense, and the potential that it holds for genetic improvement is therefore also large. Understanding and applying that potential is the challenge for scientists trying to identify genes that can contribute to traits of value to growers and consumers.

8.3. TRAITS FOR IMPROVED CROP PRODUCTION

The growth of healthy plants that yield quality products requires farmers to deal with ever-changing environmental conditions and pests. Transgenic approaches to helping farmers with these challenges are being broadly used today, while additional products are in the developmental pipeline. Plants with improved tolerance to high temperatures, saline conditions, and drought are likely to find their way into production in the future. The most common uses of transgenic plants in agriculture today are engineered resistance to herbicides, insects, and pathogens. In doing so, transgenic plants are addressing some of the oldest problems in crop production.

8.3.1. Herbicide Resistance

The first transgenic application to be widely adopted in agriculture was *resistance to herbicides*. Weeds are generally regarded to be the most serious problem for farmers and result in reduced yields because they compete with crop plants for water, light, and nutrients. Chemical herbicides are widely used by many farmers because they are cost-effective and efficient at killing weeds. Most effective herbicides for agricultural production must be somewhat selective, meaning that they should kill the target weeds but not the crop plant. Using single-gene traits in transgenic plants can provide a very specific way to protect the crop plant from the effects of a given herbicide.

Herbicides generally work by targeting metabolic steps that are vital for plant survival. For example, *glyphosate* kills plants by inhibiting the production of certain amino acids that the plant requires for survival. Glyphosate is the active ingredient in the herbicide RoundUp™. Thus, crops such as soybean and corn that have been engineered to be resistant to glyphosate were given the name "RoundUp Ready."

Glyphosate works by binding to and inhibiting the enzyme 5-enolpyruvylshikimate-3-phosphate synthase (EPSPS), which is active in the shikimate pathway leading to the synthesis of chorismate-derived metabolites, including the aromatic amino acids (tyrosine, phenylalanine, and tryptophan) (Fig. 8.1).

To make plants resistant to glyphosate, a form of the EPSPS enzyme that is functional in plants but is not affected by the herbicide was used. In addition to being present in plants, the EPSPS protein can also be found in bacteria. So scientists at Monsanto, the inventors of RoundUp, looked for and identified a form of EPSPS from a soil bacterium that was not sensitive to treatment with glyphosate. The initial steps in this process were relatively straightforward, as they simply plated soil bacteria on media containing glyphosate to identify strains that were resistant to the chemical. The EPSPS gene from the bacterium was then isolated and transferred into plants where its expression was regulated by putting it

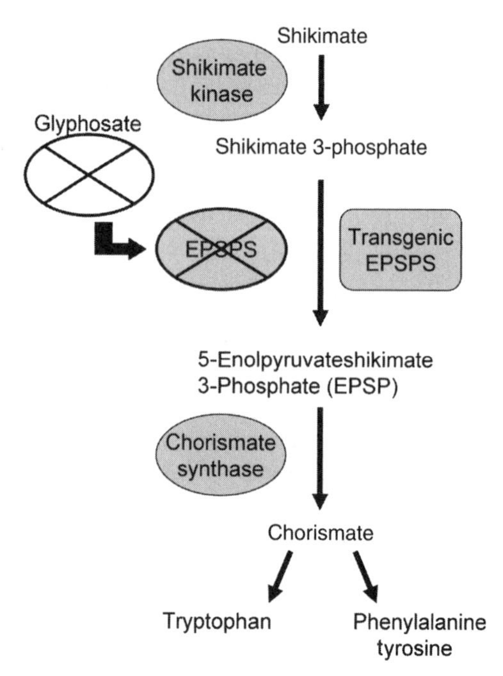

Figure 8.1. Resistance to glyphosate in RoundUp Ready™ plants is engineered by expressing a form of the 5-enolpyruvylshikimate-3-phosphate (EPSP) synthase (EPSPS) enzyme that is resistant to the herbicide. In the absence of this transgenic enzyme, glyphosate inhibits the plant EPSPS and ultimately blocks the synthesis of chorismate, the branchpoint precursor to the essential aromatic amino acids: tryptophan, phenylalanine, and tyrosine. The transgenic EPSPS is unaffected by glyphosate, and can carry out the synthesis of EPSP leading to chorismate production.

downstream of a strong promoter, the cauliflower mosaic virus 35*S* promoter, which drives gene expression throughout the plant (see the next chapter) (Shah et al. 1986). Because *Agrobacterium*-mediated transformation methods are not very efficient in soybean, the particle bombardment method was used to make the initial transgenic event. This event was then used to transfer the glyphosate-resistant bacterial EPSPS gene to many other commercially grown soybean varieties using traditional breeding techniques.

The normal plant version of EPSPS is encoded by DNA in the nuclear genome. Following translation of the mRNA sequence to amino acid sequence in the cytoplasm, EPSPS is transported into the chloroplast, where the shikimate pathway is active. To ensure that the bacterial form of EPSPS would make its way into the chloroplast after the protein was synthesized, a short DNA sequence encoding a chloroplast transit peptide was fused to the 5′ end of the bacterial EPSPS open reading frame. This transit peptide sequence fused at the amino terminus of the bacterial EPSPS serves as an intracellular signal for proper protein localization. The transit peptide sequence originated from a gene encoding a protein normally found in the chloroplasts that carries out carbon fixation, ribulose-1,5-bisphosphate carboxylase/oxygenase (Rubisco). Once the bacterial EPSPS gets into the chloroplast, it can function in place of the plant enzyme during the biosynthesis of aromatic amino acids.

RoundUp Ready soybeans were one of the first transgenic crops to be approved and used on a large scale. Once they were commercialized, they gained rapid acceptance by farmers and are now the most popular transgenic plant in the world. Glyphosate has several features that make it an attractive herbicide for growers. The compound is readily taken up and transported throughout the treated plant, traits that make it especially effective as an herbicide. Because glyphosate is rapidly degraded by soil microorganisms, it does not persist long in the environment after application. This is a benefit from both an environmental standpoint and a crop management perspective, because farmers can plant any crop in a sprayed field relatively soon after herbicide application. Because it is so effective at selectively killing weeds and not the herbicide-resistant crop plant, more farmers using glyphosate have adopted "no-till(age)" or "low-till(age)" methods, resulting in less soil erosion and lower fuel costs because they take fewer trips through a field. Furthermore, because animals do not make aromatic amino acids, they do not possess the shikimate pathway that is the target of glyphosate and so the herbicide has low toxicity in animals. In 1996, the first year they were commercially available, RoundUp Ready soybeans made up about 2% of the total soybeans grown in the United States. By 2000, that amount had risen to 54%, and in 2005 it was up to 87% (US National Agriculture Statistics Service; http://www.nass.usda.gov/). Now, glyphosate resistance has been engineered into a large number of crops that are grown globally, including Latin America and Asia. Predictably, adoption of glyphosate-resistant crops has resulted in a vast increase in the amount of this herbicide applied worldwide; however, there has been a decrease in the use of other herbicides, especially on soybean. This increase has also been encouraged by glyphosate coming off patent in 2001. Now glyphosate is sold as a generic by many companies as well in RoundUp formulations by Monsanto. The large amounts of glyphosate that are now applied to crops have led to concerns that glyphosate-resistant weed biotypes will be selected for and propagate in agricultural fields. Furthermore, farmers are required to pay a significant technology fee to Monsanto for the right to grow RoundUp Ready crops.

Glyphosate resistance is conferred through the expression of an active target enzyme, EPSPS, which is not affected by the herbicide. An alternative strategy to engineer herbicide resistance is to express a protein that will inactivate the herbicide if it is sprayed onto plants. This is the approach used in resistance against the herbicide glufosinate, the active ingredient in the product Liberty™, generating a trait in crop plants often called "LibertyLink." *Glufosinate* kills plants by inhibiting the plant enzyme glutamine synthetase (GS), which is responsible for synthesis of the amino acid glutamine. As part of the chemical reaction that produces glutamine, GS utilizes excess plant nitrogen in the form of ammonium that is incorporated into the amino acid. When GS is inhibited in glufosinate-treated plants, ammonium concentrations inside the plant rise to toxic levels (Fig. 8.2).

The glufosinate compound is naturally produced in some *Streptomyces* bacteria. In addition to having phytotoxic activity, glufosinate also servers as an antibiotic because it is toxic to some other bacteria. Bacterial strains that are resistant to glufosinate produce an enzyme, encoded by the *bar* gene, called *phosphoinothricine acetyltransferase* (PAT) (Thompson et al. 1987). The *bar* gene was isolated from a strain of *Streptomyces hygroscopicus*, which degrades glufosinate, and has been transferred into several crop plants. The LibertyLink trait is currently widely used in transgenic corn, canola, and cotton varieties.

Similar to the strategy in making LibertyLink crops, resistance to the herbicide bromoxynil (Buctril™) was engineered by expressing the protein of a bacterial gene that will inactivate the herbicide. Bromoxynil kills plants by inhibiting function of photosystem II, a crucial component of photosynthesis. Buctril-resistant cotton is already widely grown in the United States, and other crops resistant to this herbicide, such as tobacco and potato, are nearing final stages of commercialization.

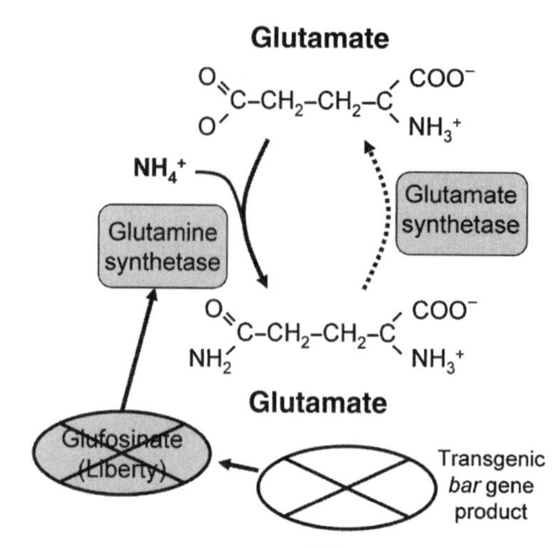

Figure 8.2. Resistance to glufosinate in LibertyLink™ plants is engineered by expressing an enzyme that directly targets and inactivates the herbicide. Glufosinate kills plants by inhibiting glutamine synthetase. This enzyme is responsible for production of the amino acid glutamine in a reaction that can sequester excess nitrogen by incorporating ammonia (NH_4^+). If this enzyme is inactivated by glufosinate, excess ammonia accumulates and the plant is killed. An enzyme encoded by the bacterial *bar* gene in transgenic plants inactivates glufosinate.

8.3.2. Insect Resistance

Insect damage to crops poses a problem for farmers worldwide. In spite of the great amounts of money and effort spent on attempts to control insect pests, staggering losses to insects are still incurred before and after harvest. In an effort to control these pests, synthetic chemical insecticides are widely used where mechanized agriculture is practiced, but insects nevertheless continue to represent a significant hindrance for food production. In much of the world, insect damage proceeds unchecked by chemical pesticides, and growers and consumers suffer significant losses in both yield and quality.

A number of proteins with negative effects on insects have been tested as potential weapons for use in engineering insect-resistant transgenic crops. Genes for several proteins have been expressed in transgenic plants and were shown to inhibit insect growth or cause higher insect death rates. These include genes for protease inhibitors, which interfere with insect digestion; lectins, which kill insects by binding to specific glycosylated proteins; and chitinases, enzymes that degrade chitin found in the cuticle of some insects. Although each of these genes has been shown to have some negative impact when consumed by insects and may have some utility in insect control, none have been as effective or widely adopted as genes encoding endotoxins from the bacterium *Bacillus thuringiensis* (Bt). The natural insecticidal activity of Bt endotoxin proteins represents an attractive alternative to synthetic chemical pesticides, which often have nonselective toxic effects on beneficial insects, birds, fish, and mammals. The transgenic plant produces its own insecticidal protein that is delivered only to insects that dare eat the plant. Rather than using the entire bacterium to kill insects, only a single-plant-encoded transgene product is used.

A bacterial species producing Bt toxins was first isolated and described over 100 years ago. A microbiologist named Ernst Berliner formally named the species *Bacillus thuringiensis* in 1915. His work followed and confirmed the discovery in 1902 of a bacterial disease affecting silkworms (*Bombyx mori*) in Japan. Obviously, infection by Bt is detrimental for silkworm production. However, it was later noted that Bt had toxic effects on caterpillar larvae of most Lepidoptera species (moths and butterflies), which gives the Bt species great potential as a tool for protecting crop plants. In later years, additional strains of Bt were identified that are toxic to Coleoptera (beetles), Diptera (flies and mosquitoes), and even nematodes. The specificity of insecticidal activity of Bt on a particular insect species is determined by the form(s) of the *cry* gene(s) carried by the bacterium. Only certain species of insects are controlled by particular endotoxins.

The *cry* genes encoding the toxic proteins in Bt take their name from the *cry*stal inclusions formed inside the bacterium when it enters into its spore-forming stage. These crystals often contain more than one specific type of *cry* gene product. Before they become toxic, the *cry*-encoded Bt proteins exist as protoxins and must be activated inside the insect digestive tract. Once they are ingested by a susceptible insect, the crystals break down in the alkaline environment of the insect midgut, generally dissolving at pH $\geq$ 8.0. At that point, the termini of the Bt protoxin proteins are cleaved by specific proteases inside the gut, yielding the toxic protein. The active protein will then bind to specific protein receptors on the insect microvillar membrane of the midgut (Fig. 8.3). In most cases, when Bt proteins are expressed in transgenic plants the entire coding region of the protoxin is not transferred to the plant. Rather, a shortened version of the gene will typically be expressed because levels of Bt protein accumulation are higher using this strategy (Barton et al. 1987; Fischoff et al. 1987).

After binding to a receptor, the active Bt toxin will enter the insect cell membrane, where multiple copies of the protein will oligomerize and form pores. This results in ion leakage through the membrane, which causes membrane collapse from osmotic lysis. Once the membranes on the epithelia of the gut cells are disrupted, the insects effectively starve and die. In the case of a true *B. thuringiensis* infection, bacterial cells would form spores during the latter stages of infection and insect collapse, thereby readying themselves for subsequent infections of other insects. In transgenic plants, susceptible insects usually stop feeding within a few hours after feeding on the plants, and die a short time later.

It is generally the presence or absence of specific forms of midgut receptors that determines whether a particular insect species is susceptible to a given Bt protein (Hofmann et al.

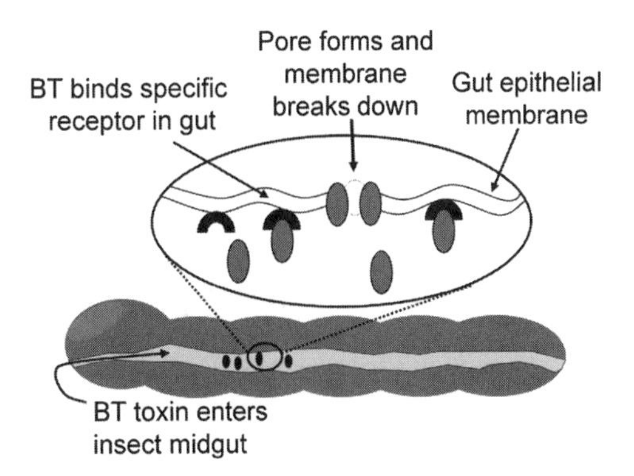

Figure 8.3. The Bt toxin binds to very specific receptors on the epithelial membrane of the insect gut. The toxin then forms channels in the membrane that leads to ion leakage and ultimately, death of the insect. This mode of action explains the specificity of Bt (from the presence of the necessary receptors) and also shows why the toxin needs to be eaten by the insect to function.

1988). For example, the most widely deployed *cry* genes in transgenic plants are members of the *cry*1A gene family, which are toxic to a broad range of Lepidoptera pests. However, this form of Bt has relatively little effect on Coleoptera species because the insects lack the specific receptors that recognize Cry1A proteins. Likewise, some beetle species, such as the Colorado potato beetle (*Leptinotarsa decemlineata*), are targeted by the Cry3A Bt toxin, whereas most lepidopterans are unaffected. Therefore, specific *cry* genes have been expressed in transgenic crops to tailor varieties to control specific pests and not affect non-target species. For example, several variations of *cry*1A genes have been transferred to corn to control European corn borer (*Ostrinia nubilali*), a lepidopteran pest that feeds on the insides of corn stems; whereas *cry*3Bb1 expression has been used in corn varieties to control western corn rootworm (*Diabrotica virgifera)* larvae, a coleopteran species that feeds primarily on roots. By using this strategy, varieties resistant to a particular insect pest can be effective in growing regions where particular pests are problematic.

Because of the steps necessary to activate them and their target sites in the digestive tract, the Cry toxins are not effective as contact insecticides. Rather, insects are killed only when the toxins are ingested. This means that most nontarget and beneficial insects are not affected in fields of Bt crops. Furthermore, most insect and noninsect species lack the specific membrane receptors for Bt and often have digestive conditions that degrade the Bt toxin if it is consumed; therefore Bt is essentially nontoxic for most arthropods, animals, and birds (and humans). In fact, Bt sprays (the intact microbes) are considered to be so safe that certified organic food production in the United States allows for the direct application of Bt crystalline spores on plants immediately prior to harvest as a control for insects. Organic growers use Bt in this form as a valuable tool for insect control. One disadvantage of this approach in comparison to transgenic Bt production in plants is that Bt applied externally to plant surfaces does not penetrate the plant tissue and is not very stable, since it breaks down with time and exposure to ultraviolet light. Even so, because organic producers sometimes depend on application of Bt as a management tool, they are especially concerned about the possibility of Bt-resistant insect populations developing because of the growing and widespread application of engineered Bt crops.

As with herbicide-resistant crops, adoption of Bt transgenic crops has also been extensive. Damage by insects can be a severe problem in cotton, and this crop is heavily treated with synthetic chemical pesticides in many production schemes. In 2005 transgenic cotton represented almost 80% of the total of that crop grown in the United States, and it is widely grown in other parts of the world, including China. Transgenic corn is now grown on well over 50% of all the acreage in the United States. In the case of both cotton and corn, traits of herbicide and insect resistance are often combined in the same plant lines as "stacked" traits.

8.3.3. Pathogen Resistance

Plant pathogens such as viruses, fungi, and bacteria are a severe and constant threat to agricultural crop production. Multiple transgenic approaches have been used to attempt plant disease control, although relatively few of these have yet made their way into the field of production.

The most effective way to control pathogens in a field setting is to use plants that are resistant to the problem pathogen. Resistance to a particular pathogen can often be conferred by a single plant gene (an *R* gene), the product of which is active in recognition of the presence or activity of a single virulence factor from the pathogen (encoded by an *Avr* gene). In plant pathogen systems, this relationship is known as a *gene-for-gene interaction* (Fig. 8.4). Plant breeders have historically taken advantage of this system, although it can sometimes take many years to identify a plant line with the desired resistance and to breed that trait into useful cultivars. Another disadvantage to the breeding approach is that unwanted or undesirable genes may sometimes be linked to the *R* gene, and it can be difficult to separate them from the *R* gene using traditional breeding methods. Finally, useful *R* genes are sometimes not easy to transfer because of barriers in crossing different species. Therefore, the ability to clone and transfer a single *R* gene from one plant variety or species to another represents an encouraging option to adapt and speed up the process.

A promising approach at engineering resistance is seen in the application of a specific resistance gene to ward off a bacterial disease in rice (Ronald 1997). Bacterial blight is a destructive disease of domesticated rice (*Oryza sativa*) in Africa and Asia, caused by the pathogen *Xanthomonas oryzae* pathovar *oryzae*. Scientists looking for alternative sources of resistance to bacterial blight identified a wild relative of rice, *O. longistaminata*, native to Mali, which is resistant to the pathogen but has very low quality and yield in terms of grain production. Through careful genetic and molecular studies, an *R* gene called *Xa21* was isolated from the wild species. This gene has been introduced into domesticated rice using particle bombardment and it confers strong resistance against strains of *X. oryzae* carrying the *Avr* gene recognized by *Xa21*. Through efforts of scientists scattered across the globe, the *Xa21* gene has been incorporated into several rice varieties of agricultural importance. The use of transgenic rice as a food crop is still controversial, and its adoption has been slow compared to crops like soybean, corn, and cotton. So although transgenic lines of blight-resistant rice are poised for application, they are currently not widely grown for food production. At least one-third of the world's population, including many developing countries, depends on rice as the major source of calories they consume.

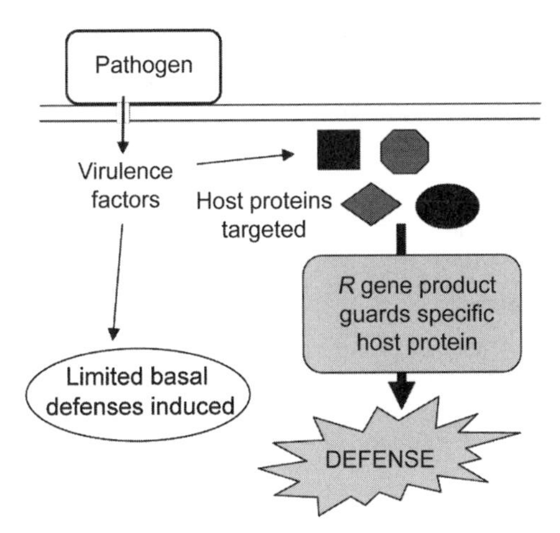

Figure 8.4. Resistance to specific strains of plant pathogens can be conferred by the protein product of a single resistance (*R*) gene. Most plant *R* genes function by recognizing the activity or presence of a specific virulence factor from the pathogen. In addition to the ability to induce basal defenses, these pathogen "effectors" are also active in attacking various host proteins. The protein products of *R* genes guard against pathogens via surveillance of specific targeted host proteins. When these *R*-gene-mediated defenses are triggered, the plant responds with a hypersensitive response and rapid activation of defense gene expression.

Therefore, development of disease-resistant rice could potentially make a major impact on alleviating hunger.

It has been known for decades that a previous inoculation with a virus can often protect a plant from subsequent infections by closely related viruses. This form of immunization of the plant has been known as *cross-protection* and has been employed with active viruses in limited cases. Crop plants can be intentionally inoculated with mild strains of a virus in the hope that this will protect the plant against future viral outbreaks. Much like vaccination with live viruses in humans, this strategy does have certain risks. In the case of inoculating with mild strains of a plant virus, there is a chance that the mild strain will present a drag on yield or that a virulent strain will emerge from the population and cause severe disease. With the advent of genetic engineering in plants, it became possible to express just a portion of plant viruses within the host. It turns out that this approach can likewise lead to resistance to closely related viruses.

Most plant viruses are relatively simple in terms of their genetic makeup, consisting of just a few genes carried by either an RNA or DNA genome encased in a protein coat. By expressing a portion of the viral genome constitutively in plants, a system of specific targeting of incoming, similar RNA sequences can be activated in a potential host plant. This *RNA silencing* system is active in many organisms, including humans, and might have evolved partially as a surveillance–protection system against invading viruses.

A great success story using RNA-mediated virus resistance has developed in the production of papaya in Hawaii (Gonsalves 1998) (see Life Box 8.1). Virtually the entire production of this crop in Hawaii was threatened in the mid-1990s by the spread of the papaya ringspot virus (PRSV). Infection with the virus was so common, and the effects on yield were so severe by the late-1990s that many fields had already been abandoned. By expressing the coat protein gene of a mild strain of PRSV in papaya (Fig. 8.5), transgenic plants

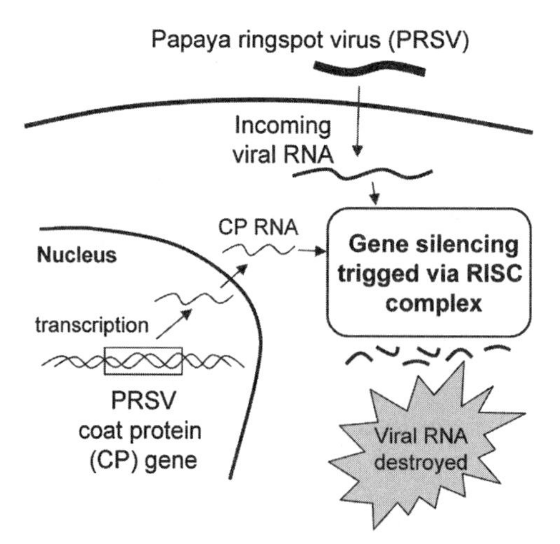

Figure 8.5. Transgenic resistance to papaya ringspot virus (PRSV) is possible because of the process of RNA-mediated gene silencing. To make virus-resistant plants, a portion of the coat protein (CP) gene of PRSV was transferred to and expressed in transgenic papaya plants. Following transcription, the RNA triggers targeted, sequence-specific degradation of similar RNA sequences, such as that found on incoming PRSV viral RNA. The initial degradation of RNA is carried out by an enzyme called *DICER*, and the process is mediated by an enzymatic structure called the *RNA-induced silencing complex* (RISC). Ultimately, this can lead to RNA cleavage, as well as blockage of transcription or translation of the target gene.

were made resistant to incoming pathogenic viruses (Fitch et al. 1992). Varieties of transgenic papaya were first introduced commercially in Hawaii in 1998, and so far, the transgenic lines have remained virus-resistant over the years. Just as in other transgenic crops, after the initial transgenic transformation in a single variety, the gene of interest was transferred to other desirable commercial varieties using standard breeding techniques.

A similar approach has been used successfully to control cucumber mosaic virus (CMV) in transgenic squash production. A particularly exciting application of RNA-mediated virus resistance might be viable in the control of the feathery mottle virus in sweet potato in Africa. Sweet potato serves as a staple crop in some countries, such as Kenya, and viral diseases can be especially severe there and in developing countries. Transgenic varieties resistant to this virus have been developed and might be an effective tool in managing production and increasing yields.

8.4. TRAITS FOR IMPROVED PRODUCTS AND FOOD QUALITY

In the early years of commercialization of plant biotechnology, efforts and products focused on traits that aid in the growing of crop plants, such as resistance to herbicides or insects—these are called *input traits*. It is likely that many future applications of plant biotechnology will also target *output traits*, centered on improved plant-based products that will find their way to consumers.

8.4.1. Nutritional Improvements

Humans depend on plants as food for survival. In addition to the calories that they provide, plants produce nutrients, vitamins, and essential amino acids that we require. Much more so than animals, plants have an incredible capacity for producing a variety of complex chemical compounds. Through methods in biotechnology, efforts are being made to take advantage of this capacity for chemical synthesis to improve or alter the nutritional values of plants.

One of the best known examples of nutritional improvement of a food crop has been the development of *Golden Rice*, a transgenic plant that produces high levels of β-carotene or provitamin A in the grain (Ye et al. 2000). Over one-third of the world's population depends on rice as a major component of their diet. Although rice can be a good source of calories, it is not high in protein or vitamins. Although dietary vitamin deficiencies are uncommon today in industrialized countries, they can still be a serious problem in developing countries in parts of southern Asia and sub-Saharan Africa, where rice is a staple and there is a lack of a diverse diet including meat, fruits, and vegetables. Vitamin A deficiency is especially serious, and the World Health Organization estimates that as many as 4 million children suffer from a severe deficiency. Humans depend on dietary sources of vitamin A, and deficiency of this vitamin is the leading preventable cause of blindness in children and significantly increases the mortality rate due to illnesses such as measles and malaria. Providing vitamin A supplements as capsules to children and new mothers is one approach to solving this problem, but to be effective, supplements need to be administered several times per year, which can present logistical challenges in many areas. An alternative strategy is to provide provitamin A in the form of β-carotene in rice.

Carotenoids are a subset of compounds within a large and variable class of plant metabolites called *terpenoids* or *isoprenoids*. This class of compounds is all based on a five-carbon building block, which can be assembled into multimers to form complex molecules. Many familiar plant scents and flavors, such as mint and pine resin, are based on terpenoids. The five-carbon precursor to terpenoids can be produced via two independent pathways, in either the cytoplasm or in plastids. Carotenoids are 40-carbon compounds produced from the precursor molecule via a biochemical pathway localized in plastids. The 40-carbon backbone of β-carotene is phytoene, which is assembled by combination of two 20-carbon geranylgeranyl diphosphate (GGPP) molecules by the enzyme phytoene synthase (Fig. 8.6). Double bonds are then added to phytoene through a series of desaturation steps to produce lycopene, an antioxidant compound found in most plants and that contributes to the red color of tomatoes. Finally, lycopene can be converted to β-carotene by the enzyme lycopene cyclase. Much of the understanding of how this pathway operates and could be manipulated came from the laboratories of Dr. Ingo Potrykus in Switzerland (see Life Box 8.2) and Peter Beyer in Germany. Researchers in these labs led the way in transforming rice with the necessary genes to produce carotenoids in rice grains.

Rice grains naturally produce GGPP, and so the addition of an active phytoene synthase gene expressed in rice grains under the control of a seed endosperm-specific promoter led to the production of phytoene in preliminary experiments. Transgenic plants were later produced via particle bombardment in which genes for phytoene synthase, phytoene desaturase, and a lycopene cyclase were cotransformed. These transgenic rice plants had grains with a bright yellow coloring, which was confirmed to come from the presence of

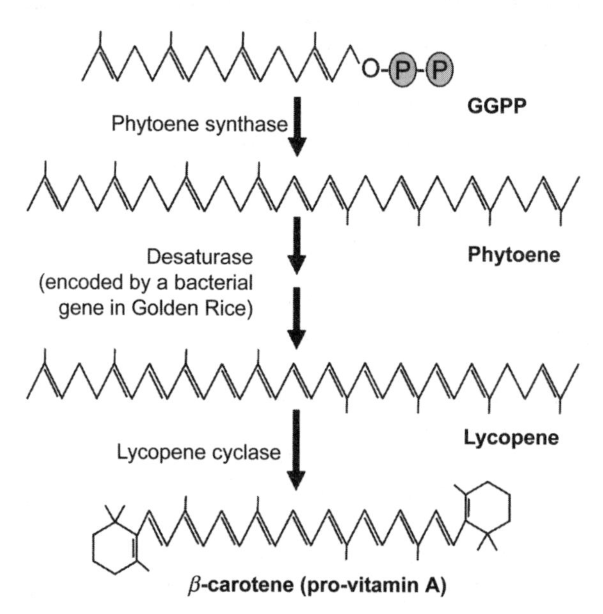

Figure 8.6. The production of β-carotene in Golden Rice was made possible by high-level, tissue-specific expression of the necessary enzymes in rice. Rice grains normally produce geranylgeranyl–diphosphate (GGPP). A gene-encoding phytoene synthase was transferred to rice from daffodil (for the original Golden Rice) or maize (in Golden Rice 2), and this led to production of phytoene in rice grains. A desaturase enzyme necessary to add double bonds to the structure was provided by transfer of a bacterial gene to rice (the two arrows at this step represent the multiple reactions that are necessary to add all double bonds). Finally, lycopene was converted in rice grains by an endogenous lycopene cyclase activity to the yellow-orange endproduct, β-carotene.

β-carotene and led to the name "Golden Rice" (Ye et al. 2000). It turned out that plants expressing just the phytoene synthase and the desaturase produced β-carotene, indicating that rice grains already contained the metabolic activity to convert lycopene to β-carotene. The gene for the desaturase originated from a bacterium, *Erwinia* spp., whereas the other genes originated from daffodil. The bacterial desaturase enzyme actually performs metabolic steps normally carried out by two separate plant enzymes. Because the daffodil gene products are normally found in plastids, they already contained sequences for a plastid transit peptide to direct newly synthesized proteins to the proper cellular location. The bacterial gene-encoding desaturase was modified by addition of a transit peptide to direct it to plastids following translation, in much the same way the bacterial EPSPS gene was modified for engineered RoundUp resistance.

Golden Rice produces carotene levels sufficient to impart a visible yellow color. One concern with these plants, however, has been that the accumulation levels of β-carotene might not be sufficient to provide enough of the compound to be of nutritional benefit. An improved version of transgenic rice referred to as *Golden Rice 2*, using a phytoene synthase gene from corn rather than daffodil, was subsequently produced that accumulated levels of carotenoids over 20 times higher than in the original Golden Rice (Paine et al. 2005). It is estimated that by eating modest amounts of Golden Rice 2, enough β-carotene can be provided to overcome vitamin A deficiency.

The large-scale dissemination of Golden Rice has been controversial (see Life Box 8.2). Advocates maintain that this rice can provide provitamin A to millions of undernourished children who need it. Rice is already widely grown and consumed in the target regions, and so packaging the technology in this form takes advantage of an existing means to distribute and administer the nutrient. Opponents of the technology counter that development of this product is a tactic used by the biotechnology industry to drive acceptance of transgenic foods worldwide. Many opponents also contend that vitamin supplements and food fortification are superior methods for fighting the problem of vitamin A deficiency. Clearly, this rice has the potential to help malnourished children, but contentious issues must be resolved before it is accepted worldwide. At the very least, development of Golden Rice demonstrates that it is possible to alter the natural abilities of plants to synthesize complex chemicals, and to enhance their nutritional value.

8.4.2. Modified Plant Oils

The fatty acids produced by plants are the source of oils used in foods, and also have applications in cosmetics, detergents, and plastics. Oilseed rape (*Brassica napus*) has been used as a plant oil source for many years. Canola is the common name for the cultivated form of this plant, and has been bred through traditional means to contain low levels of harmful glucosinolates and erucic acid. By engineering canola with a thioesterase gene that originated in the California bay tree (*Umbellularia californica*), the oils that accumulate contain much higher levels of beneficial fatty acids. The "bay leaf" thioesterase enzyme expressed in canola causes premature chain termination of growing fatty acids, and results in accumulation of 12-carbon lauric acid and 14-carbon myristic acid. The overall level of lipids is not increased in these plants, as the increase in the short-chain molecules is matched by a decrease in the amount of long-chain fatty acids such as the 18-carbon oleic and linoleic acids. These short-chain fatty acids make the canola oil much more suitable as replacement for palm and coconut oils in products such as margarine, shortenings, and confectionaries.

Soybean oil is also used in a variety of food and industrial applications. By decreasing the levels of the enzyme called Δ_{12}-desaturase in transgenic soybeans, the amount of oleic acid can be increased. To decrease levels of enzyme expression, the normal soybean *fad2* gene encoding Δ_{12}-desaturase was repressed using a technique called *gene silencing*, whereby a second copy of the gene is introduced into the plant. By overexpressing a second copy of the target gene, a response in the plant is triggered to shut down expression of both the endogenous gene and the transgene. In this case, silencing the *fad2* gene results in higher levels of *oleic acid* and corresponding lower levels of two other 18-carbon fatty acids and linoleic and linolenic acids. The only differences in the structures of these three fatty acids are the number of double bonds in the chain. As a result, high–oleic acid soybeans have low levels of saturated fats and transfats. This can alleviate the need for the hydrogenation process that is often used to make soybean oil suitable for foods like margarine, resulting in a healthier product. It also keeps the oil in a liquid form and makes it more heat-stable for cooking applications.

8.4.3. Pharmaceutical Products

Plant-manufactured pharmaceuticals (PMPs) are one of the most widely discussed applications of transgenic plants. The tremendous variety and potency of chemicals produced in plants has been long recognized, as many have powerful effects on human health and physiology (salicylic acid, cocaine, morphine, taxol, etc.). In addition to being able to produce complex metabolites, plants can also produce high levels of specific proteins when a novel transgene is introduced.

Production of human and animal *oral vaccines* in plants has been proposed as an attractive approach, especially in areas of the world where infrastructure and costs might limit storage, transfer, and administration of traditional vaccines. By including an immunogenic protein in a food, vaccination could be effected using a product that is easily grown and stored and that could be administered via consumption of the food source. For example, production of the surface antigen of the hepatitis B virus in transgenic potato has been demonstrated in clinical trials to lead to an immune response in humans consuming the potatoes. Production of proteins in transgenic bananas is also often cited as a potential source for these oral vaccines. There are several potential problems with this approach, such as the timing of administering the vaccine, dosage, and the ability of the protein to induce immunity on oral administration. Nonetheless, this strategy might have application in some specific instances for humans or in vaccination of farm animals.

Antibodies are large, complex proteins with the powerful ability to recognize and bind to specific molecular targets. Plants do not normally produce antibodies, but it has been repeatedly demonstrated that they can form functional antibodies when the encoding genes are expressed transgenically. One of the more promising approaches is the production of a specific monoclonal antibody that recognizes a cell surface protein of *Streptococcus mutans*, a bacterium that is one of the major causes of tooth decay. By binding to its surface, the antibody interferes with the bacteria's binding to tooth enamel. The planned applications for this product, produced in tobacco and called *CaroRX*, would be primarily in toothpastes and mouthwashes.

To date, the vast majority of transgenic biopharmaceuticals are produced using *E. coli*, yeast, or mammalian cell cultures. The strategy of producing pharmaceutical proteins in plants could have several advantages (Giddings et al. 2000). Transgenic plants offer the economies of scale to grow and harvest large amounts of biomass expressing the target product on relatively little land. Some applications for therapeutic proteins such as serum factors, hormones, or antibodies have traditionally relied on human or animal sources. By using plants, the risk of transferring unknown infectious agents from the donor source can be greatly reduced because plants typically do not carry animal pathogens. The idea of producing *therapeutic proteins* in crop plants is not accepted by everyone. Opponents worry that food products could be contaminated with tissue of plants intended for drug production. Companies that rely on commodities for products to which certain consumers may be sensitive have also opposed transgenic crops expressing pharmaceuticals. A prominent example was when a large beverage company opposed a pharmaceutical company who wished to grow transgenic rice near rice fields that would be used in their beverage product. Another potential hurdle is the differences in glycosylation of proteins that occur in plants and animals. The sugar moieties added to proteins can vastly affect their function and immunogenicity, and some patterns of plant glycosylation can cause unwanted allergic reactions in humans. To be used in humans, these proteins would need to be produced so that they do not elicit an immune response in the patient.

8.4.4. Biofuels

With demands for energy increasing worldwide and supplies of fossil fuels being depleted, finding alternative and renewable energy sources has become an important goal for plant scientists. Both ethanol (ethyl alcohol) and biodiesel produced using plant materials can be adapted relatively easily to existing fuel storage, movement, and uses with existing infrastructure and machinery. Applications using transgenic plants have the potential to increase the efficiency of biofuel production on several fronts.

Ethanol offers several attractive features as an energy source; it is biodegradable and renewable, and burns cleaner than do most fossil fuels. Ethanol is produced by yeast-driven fermentation of carbohydrates (sugars). In the United States corn is currently the dominant source for fermentable sugars. In this case, the complex carbohydrates of starch in corn grains are first converted to simple sugars, which the yeast can then use to produce ethanol. One suggested approach to improve ethanol production is to transgenically engineer plants to produce higher levels of the enzymes responsible for the initial steps of starch breakdown (Himmel et al. 2007). The genes encoding enzymes such as amylase, which degrades starch into simpler sugars, could possibly be expressed at high levels in corn grains or in other plants, resulting in higher percentages of readily fermentable sugars. The considerable inputs necessary for growing corn, in terms of nitrogen fertilizer, fuel, and pesticides, mean that it is likely not going to be an efficient long-term solution as a source for ethanol production. In Brazil, sugarcane is the plant source of choice for making ethanol, as the high levels of simple sugars make it superior for fermentation. In addition, sugarcane is a perennial crop that can be more easily grown with fewer inputs. The success of the Brazilian adoption of ethanol as a fuel source is widely touted as an example of how existing infrastructure and practices can be adapted for conversion to reliance on biofuels.

The use of plant material high in *cellulose* as a source for ethanol production is also being widely studied. The conversion of high-cellulose materials into fermentable sugars is an inefficient process, and so it is not currently viable as a method for biofuel production. However, plant materials such as corn stover (stalks and leaves), wood chips, or biomass crops such as perennial grasses contain energy that could potentially be converted to ethanol. Biomass crops, such as switchgrass or fast-growing trees such as willow or poplar, have advantages in that large amounts of biomass can be harvested multiple times from the same plants, and that they will grow efficiently with less need for watering and fertilizers. Although they are currently not efficient, improved methods for this cellulytic conversion of plant material to ethanol may hold some of the best promise for sustainable fuel production from plants. Transgenic approaches are being explored to produce cellulose that would be more easily converted to simple sugars by microbes for alcohol production, or in grasses and woody plants with decreased levels of *lignin* that can interfere with cellulose degradation. In addition, identification and engineering of microbes that can degrade lignin or more readily convert cellulose and sugars to ethanol are also being explored (Stephanopoulos 2007). There are a number of investigators searching for ways to modify plant feedstocks for eventual more facile cellulosic ethanol production. One idea is to encode *cellulases* and other cell-wall-degrading enzymes by the transgenic biomass crops directly.

Diesel fuel made from plant material, biodiesel, can also represent an alternative to fossil fuels. Diesel currently accounts for approximately 20% of the fuel consumed for transportation in the United States; therefore, finding a renewable replacement could have a considerable impact on the need for oil throughout the world. Biodiesel is produced from oilseed crops such as soybean and canola, through a process called *transesterification*. The properties of biodiesel are slightly different from those of petroleum-based diesel, but biodiesel can be used alone as a fuel or in a blend of the two types of fuel. Although there are currently no transgenic applications to improve biodiesel production in oilseed crops, the two major sources for biodiesel (soybean and canola) are most often grown as transgenic plants.

Because of the economic, environmental, and political concerns associated with fossil fuel consumption, the use of plants for biofuel production will almost certainly continue to increase and develop with new strategies. Genetically engineered biofuel crops will likely not be food or feed crop plants for several reasons—as noted above, food companies could be opposed to altering food crops for fuel purposes if there is a viable chance of accidental mixing of fuel and feed (Stewart 2007).

8.5. CONCLUSIONS

Clearly we are at the proverbial tip of the iceberg with regard to the numbers and types of genes identified that could be useful in plant biotechnology. Genes are currently limited by insufficient knowledge of diverse kinds of genomes and the ability to engineer in metabolic pathways. Simple solutions to problems that can be fixed with the insertion of one gene coding for one protein are myriad, but how much more important will be the ability to engineer into plants entire metabolic pathways such as was done to produce Golden Rice.

LIFE BOX 8.1. DENNIS GONSALVES

Dennis Gonsalves, Center Director, Pacific Basin Agricultural Research Center, USDA Agricultural Research Service; Recipient of the Alexander Von Humbolt Award (2002)

Dennis Gonsalves with transgenic papayas.

I was born and raised on a sugar plantation in Kohala on the island of Hawaii. My dad was a first generation Portuguese whose parents had immigrated from the Azores and from the Madeira islands. My mother was Hawaiian-Chinese with her dad emigrating from mainland China and her mom being a pure native Hawaiian. As a child and all the way through my undergraduate career, I never had ambitions to be a scientist nor even to go to graduate school. I had a key break in life when I was accepted to attend the excellent Kamehameha Schools, which had been started in the late 1800s by the Hawaiian Princess Pauahi Bishop to educate people of the Hawaiian race. I subsequently enrolled at the University of Hawaii with the intention of being an agricultural engineer so I could be back to work on the sugar plantation. However, midway through my undergraduate tenure, the program for training engineers to work on the sugar plantations was dropped and I subsequently shifted to the field of horticulture. I was just an average student. I landed a job on the island of Kauai as a technician for Dr. Eduardo Trujillo, a plant pathologist at the University of Hawaii. That one year as a technician changed my life.

Dr. Trujillo told me to look at this "new" disease of papaya which he felt was caused by a virus. I knew next to nothing about viruses, but as soon as I started work to identify the disease I knew that I wanted to be a research plant pathologist that would specialize in plant viruses. I had found my potential career niche. After working for some months as a technician, I wanted to pursue graduate work, but my grades were not good enough. I got a break when Dr. Trujillo persuaded the graduate school to accept me into a Master's program on probation. The other break or lesson also came from Dr. Trujillo who told me: "don't just be a test tube scientist, do things that will have practical applications." That philosophy would serve me well as I pursued my career, especially in biotechnology. I got my Master's degree from the University of Hawaii in 1968 under Dr. Trujillo and Ph.D. from the University of California at Davis in 1971 under Dr. Robert Shepherd, who at that time had just shown for the first time that the cauliflower mosaic virus had a DNA genome. Little did I know that it would yield the sequences for the CAMV 35*S* promoter which is widely used in biotechnology. In 1972, I took a job at the University of Florida, subsequently moved to Cornell University in 1977, and in 2002 I returned to my Hawaiian roots to work for the Agricultural Research Service of USDA in Hilo, Hawaii

I gravitated from classical virology to molecular biology and biotechnology in the mid-1980s because of the prospects for developing virus-resistant transgenic crops. The pioneering work by Roger Beachy's group provided the proof-of-concept. My lab in collaboration with others have developed commercial virus-resistant squash and papaya. However, the papaya story has garnered the most interests for several reasons. A nutshell summary of the papaya work follows.

We developed, for Hawaii, transgenic papaya that resists papaya ringspot virus (PRSV), the most widespread and damaging virus of papaya worldwide. We started developing the transgenic papaya in the mid-1980s and had obtained a resistant transgenic papaya line by 1991. Coincidentally, PRSV invaded papaya plantations in Puna on Hawaii Island in 1992 and by 1995, the papaya industry was severely affected because 95% of Hawaii's papaya was being grown in Puna. Essentially, we had a potential technology to control the virus but it had to be deregulated by APHIS and EPA, and pass consultation with FDA. We worked feverishly to test the papaya, develop data for deregulation, and get it commercialized. In 1998, we commercially released the SunUp and Rainbow papaya and essentially saved the industry from being devastated by PRSV. Nine years after commercialization, the transgenic papaya is widely grown in Hawaii and its resistance has held up well.

Aside from helping the Hawaiian papaya industry, our papaya work showed that "small" scientists can develop and commercialize a transgenic product. Basically, the work was done on a shoe-string budget and without funding from private companies. I and the team did the work because we were committed to help the papaya growers and to do it in a timely manner. If one analyzes the papaya story, one sees the ingredients for successful research and implementation because: (1) work was done proactively by anticipating the potential damage that PRSV could do in Hawaii, (2) the research was focused so we could go from concept to practicality in a timely manner, (3) the research team had a strong commitment to good science and to achieving practical results in a timely manner, (4) the clientele was brought in and consulted early, and (5) we ventured out of our fields of expertise to get the job done. This last step involved collecting data needed for deregulation, assembling the package for submission to the regulatory agencies, working on the intellectual properties of the project, and making the clientele well aware of events as the project progressed. Today, the papaya case is often used as a model on how to get the job done in timely manner and make an impact, even though your group is small and your resources rather limited.

Plant virology is in an academic heyday, in part, because the technology of developing virus resistant transgenic crops is now rather routine, and much is known about the mechanism that a governs resistance: post-transcriptional gene silencing. I expect to see continued incremental improvements on the development of effective virus-resistant transgenic crops. However, I am rather disappointed and surprised that so few transgenic virus-resistant crops (papaya, squash, and potato) have been commercialized. It is not due to lack of technology; numerous scientific reports have validated the effectiveness of virus-resistant transgenic plants with a number of plants and viruses. Yet, only transgenic squash and papaya are in commercial production today. We need to seriously ask why? Unless we effectively address this question, the huge promise that biotechnology has shown for virus-resistant crops will largely remain in the field of academia with little practical application. I suspect that the answers to this question do not lie in the technology arena, but more in the people's arena.

LIFE BOX 8.2. INGO POTRYKUS

Ingo Potrykus, Chairman, Humanitarian Golden Rice Board and Network; retired Professor in Plant Sciences, ETH Zurich

Ingo Potrykus

Rice-dependent poor societies are vitamin A-deficient because rice is their major source of calories, but does not contain any pro-vitamin A. Hundreds of millions in the developing world, therefore, do not reach the 50% level of the recommended nutrient intake (RNI) for vitamin A, required to live a healthy live. We developed "Golden Rice" to provide pro-vitamin A with the routine diet. Even with rice lines containing modest concentration of pro-vitamin A, a shift from ordinary rice to Golden Rice in the diet could save people from vitamin A malnutrition. Recent studies establish that Golden Rice, if supported by governments, could save, at minimal costs, up to 40,000 lives per year in India alone.

How did I get involved in science and genetic engineering of plants? My connection to biology dates back to my childhood and is that of an old-fashioned naturalist. Ornithology is, after 60 years, still my major hobby. My interest in molecular biology began only when I was already around 40 years old. I got fascinated by the phenomenon of totipotency of somatic plant cells. Having an engineer's mind and being concerned about the problem of food-insecurity of poor people in developing countries I could not resist of challenging that potential of totipotency for contributing to food security—and this let me into a scientific career as pioneer in the area of plant tissue culture and genetic engineering. As research group leader at the Max-Planck-Institute for Plant Genetics, Heidelberg (1974–76), the Friedrich Miescher Institute Basel (1976–86), and full professor in plant sciences at the Swiss Federal Institute of Technology (ETH) (1986–99) I had exceptional good conditions and great teams to follow the basic concept of developing genetic engineering technology for crop plants such as cereals and cassava. The task was to rescue harvests, to improve the nutritional content, and to improved exploitation of natural resources. As long as active in academia this was all "proof-of-concept" work. Only with my retirement in 1999 and the need for Golden Rice to be brought to the poor did I realized that the decisive follow-up steps of product development and deregulation are routinely ignored by academia.

The science leading to Golden Rice. By the end of the 1980s we had transformation protocols ready for rice and had already worked on insect-, pest-, and disease-resistance. The Rockefeller Foundation alerted me of the problem of micronutrient malnutrition. In 1991, I appointed a PhD student to work towards pro-vitamin A-biosynthesis in rice endosperm and Dr. Peter Beyer, an expert in terpenoid biosynthesis from the university of Freiburg, Germany joined as co-supervisor. The project was, for numerous good reasons, considered totally unfeasible and was, therefore, difficult to finance. The breakthrough came eight years later, with the concluding experiment of a Chinese post-doc. When the harvest from a co-transformation experiment involving five genes was finished, the offspring from a transgenic line harboring all genes segregated for white and yellow endosperm. This was in February 1999 and two months before my retirement.

Great recognition but no support for completion from the public domain. We presented our success to the public at my ETH-Farewell Symposium on March 31st 1999 and it was finally published in *Science* [see Ye et al. (2000)]. *Nature* refused our earlier submitted manuscript for publication because they lacked interest, but the scientific community, the media, and the public were quite interested, even became exited about this vitamin A rice. *TIME Magazine* devoted a cover story on Golden Rice July 31, 2000, and there were hundreds of articles and airings in the media. The readers of *Nature Biotechnology* voted Peter Beyer and me as the "most influential personalities in agronomic, industrial, and environmental biotechnology for the decade 1995 to 2005" [*Nature Biotechnology* **24**:291–300 (2000)] and there were numerous recognitions for the work from the scientific community and the pubic. However, nowhere in the public domain could we find support for the long and tedious process of product development for our humanitarian Golden Rice project.

The private sector helped us to continue with the humanitarian project. Only thanks to the establishment of a public-private partnership with Zeneca/Syngenta we could [could we] proceed. The basis was an agreement, in which we transferred the rights for commercial exploitation in return for support for the humanitarian project—making Golden Rice freely available to the poor in developing countries. This public-private partnership helped also to solve the next big problem: getting permission to use all intellectual property rights involved in the technology. We had been using intellectual property of 70 patents belonging to 32 patent holders! Fortunately, 58 patents were not valid in our target countries, and of the remaining 12, 6 of these belonged to our partner company and for the rest, it was not a big problem to get free licenses. Product development, de-regulation, and delivery of a GMO-product turned out to be a gigantic task, especially for two naïve university professors. We needed advice from the private sector and received help from Dr. Adrian Dubock who works for Syngenta. We were short in different areas of expertise for strategic decisions and established a Humanitarian Golden Rice Board. We needed GMO-competent partner institutions in our target countries and established a Golden Rice Network. And we needed managerial capacity and found a project manager and a network coordinator. For more details please visit www.goldenrice.org.

Lost years because of over-regulation. If it were not a GMO, Golden Rice would have been in the hands of the farmers since 2003. We have lost 6–7 years in the preparatory adoption to regulatory requirements, which all do not make any sense scientifically. An example for how irrational regulatory authorities operate could be found in our experience with a permission for small-scale field testing of Golden Rice. No ecologist around the world has been able to propose any serious risk for any environment from a rice plant containing a few micrograms of carotenoids in the endosperm, and this trait does not provide for any selective advantage in any environment. We still (Spring 2007) have not been given permission to field test Golden Rice in the field in any Asian country! Because of the requirements of the established "extreme precautionary regulation" in costs and data, we have to base all Golden Rice breeding work and variety development on one single selected transgenic event. This event selection will be completed in 2007 and from then on our partner institutions will be introgressing the transgenes from this event into a carefully chosen 30 popular southeastern Asian rice varieties. Deregulation can then be based on the single event, not on 30 different varieties. Considering the 40,000 lives Golden Rice could save in India per year, regulation is, through the delay it is causing, responsible for the death and misery of hundreds of thousands of poor people. And did regulation prevent any harm? Judging from all regulatory review and from all data from all "bio-safety research," my answer is: "most probably not."

Where is plant biotechnology going in future? The answer depends entirely upon what our society does with GMO-regulation. If this unjustified and excessive procedure is maintained, plant genetic engineering will have no future, and hundreds of millions of lives will be lost, which could be saved by applying this technology to food security problems. Plant molecular biology is, so far, an extremely successful scientific discipline, but much of its motivation and funding came from its potential application, not only in the private sector, but much so in the public sector, e.g. as contribution to the solution of humanitarian problems. With this potential being cut off, financial support for basic research will probably dry out.

I propose here my recommendations needed for humanity to maximally benefit from plant biotechnology:

- De-demonize GMO's and inform the public that these are perfectly normal plants. There is not a single crop plant which has not been extensively "genetically modified" by traditional interventions.
- Reform GMO-regulation such that it evaluates traits, not GM-technology, and takes decisions on balancing benefits versus risks. Because of the time and financial requirements of present regulation, no public institution can not afford to take a single transgenic event to the marketplace.
- Establish public funding schemes for product development and deregulation. Humanitarian problems are problems of the public sector and should not be expected to be solved by the private sector.
- Encourage establishment of public-private partnerships for the solution of humanitarian problems. The private sector has the necessary experience for solutions of practical problems.
- Establish a reward system for those in academia who sacrifice their academic career by contributing to solutions of humanitarian problems. Academia receives much of its funding because the public believes that it is helping to solve humanitarian problems.
- Change the paradigm "highest priority to biosafety". It leads to millions of deaths and there are other topics deserving higher priority such as food security and poverty alleviation.
- Prosecute those institutions who use their political and financial power to block green biotechnology in an international court. They are responsible for a crime against humanity.

REFERENCES

Alba R et al (2004): ESTs, cDNA microarrays, and gene expression profiling: Tools for dissecting plant physiology and development. *Plant J* **39**:697–714.

Barton KA, Whiteley HR, Yang N-S (1987): *Bacillus thuringiensis* δ-endotoxin expressed in transgenic *Nicotiana tabacum* provides resistance to lepidopteran insects. *Plant Physiol* **85**:1103–1109.

Fischoff DA et al. (1987): Insect tolerant transgenic tomato plants. *Bio/Technol* **5**:807–813.

Fitch MMM, Manshardt RM, Gonsalves D, Slightom JL, Sanford JC (1992): Virus resistant papaya derived from tissues bombarded with the coat protein gene of papaya ringspot virus. *Bio/Technol* **10**:1466–1472.

Giddings G, Allison G, Brooks D, Carter A (2000): Transgenic plants as factories for biopharmaceuticals. *Nat Biotechnol* **18**:1151–1155.

Gonsalves D (1998): Control of papaya ringspot virus in papaya: A case study. *Annu Rev Phytopathol* **36**:415–437.

Himmel ME, Ding SH, Johnson DK, Adney WS, Nimlos MR, Brady JW, Foust TD (2007): Biomass recalcitrance: Engineering plants and enzymes for biofuels production. *Science* **315**:804–807.

Hofmann C, Vanderbruggen H, Höfte H, Van Rie J, Jansens S, Van Mellaert H (1988): Specificity of *Bacillus thuringiensis* δ-endotoxins is correlated with the presence of high-affinity binding sites in the brush border membrane of target insect midguts. *Proc Natl Acad Sci USA* **85**:7844–7848.

Paine JA et al (2005): Improving the nutritional value of Golden Rice through increased pro-vitamin A content. *Nat Biotechnol* **4**:482–487.

Rhee SY, Dickerson J, Xu D (2006): Bioinformatics and its applications in plant biology. *Annu Rev Plant Biol* **57**:335–360.

Ronald PC (1997): Making rice disease-resistant. *Sci Am* **277**:100–105.

Shah D et al (1986): Engineering herbicide tolerance in transgenic plants. *Science* **233**:478–481.

Stephanopoulos G (2007): Challenges in engineering microbes for biofuels production. *Science* **315**:801–804.

Stewart CN Jr (2007): Biofuels and biocontainment. *Nat Biotechnol* **25**:283–284.

Thompson CJ, Movva NR, Tizard R, Crameri R, Davies JE, Lauwereys M, Botterman J (1987): Characterization of the herbicide-resistance gene bar from *Streptomyces hygroscopicus*. *EMBO J* **6**:2519–2523.

Ye X, Al-Babili S, Klöti A, Zhang J, Lucca P, Beyer P, Potrykus I (2000): Engineering provitamin A (β-carotene) biosynthetic pathway into (carotenoid-free) rice endosperm. *Science* **287**:303–305.

CHAPTER 9

Marker Genes and Promoters

BRIAN MIKI

Agriculture and Agri-Food Canada, Ottawa, Ontario, Canada

9.0. CHAPTER SUMMARY AND OBJECTIVES

9.0.1. Summary

Two essential segments of DNA are required to produce a transgenic plant that will express the trait of interest: (1) a promoter must be fused upstream of the gene of interest to control its expression—the choice of promoter is crucial in that it specifies when and where a transgene is expressed in the plant, and (2) marker genes are needed to select transgenic plants and/or monitor gene expression. Selectable markers typically confer antibiotic resistance so that transgenic cells, tissues, and plants can be selected that survive antibiotic selection. Visual marker genes often will cause a color change in the transgenic plants so that researchers can see when and where transgenes are expressed in plant tissues.

9.0.2. Discussion Questions

1. Why use marker genes?
2. What are some differences between selectable markers and scorable markers?
3. Discuss the relative merits of GUS and GFP as reporters. Does the profile of experimentation using these reporter genes overlap directly or partially?
4. What are the advantages, if any, for the use of the *manA* gene over the *nptII* gene as a selectable marker for food and feed crops, and would the use of the *manA* gene overcome public concern over the use of the *nptII* gene? Conversely, what are the disadvantages?
5. Considering the large number of selectable marker gene systems that have been developed, why are so few adopted for basic research and commercialization?
6. What experimental factors should be considered for a functional genomics study of unknown genes if the vector employs a new selectable marker gene system in the base vector?

Plant Biotechnology and Genetics: Principles, Techniques, and Applications, Edited by C. Neal Stewart, Jr.

7. Why is there a need for negative selectable markers in experimental plant science? Can you design experiments that would employ them?

9.1. INTRODUCTION

The genetic transformation of plant cells is known to occur in nature, but the technologies for the reproducible generation of transgenic plants in the laboratory are only about 25 years old. Many questions about the fundamental nature of transgenic plants are therefore still being raised. In part, this reflects the more recent emergence of the technologies but more importantly it reflects our limited understanding of the plant genome and the genetic mechanisms that govern how it works. Indeed, it was shown only as recently as 2005 how remarkably stable the transcriptional patterns and programming mechanisms are in plants and how impervious they are to the insertion of marker genes (El Ouakfaoui and Miki 2005). Such understanding is critical if transgenic plants are to be used as a vehicle to study the functions of unknown genes isolated from genomics studies. Furthermore, the biosafety of transgenic food or feed is evaluated by their equivalence to other plants that are already being used and known to be safe; therefore, it is important to determine whether nontargeted, unanticipated genetic changes could be induced during the transformation process. Marker genes allow scientists to study such fundamental processes and also provide a pivotal ingredient needed to generate transgenic plants.

Several kinds of marker genes have been developed and are needed for the diverse roles they play in biotechnology as well as in experimental plant science. These include the recovery of transgenic plants, the experimental manipulation of plant tissues, the assessment of plant gene regulatory mechanisms, the intracellular trafficking of proteins, and the assessment of biosafety of transgenic plants. The marker genes can be grouped into different categories depending on whether they (1) are selectable, (2) promote or suppress tissue growth and differentiation, and (3) are conditional on external substrates. As the complexity and needs of research increase, there might also arise a requirement to selectively remove marker genes from transgenic plants to create marker-free plants. This chapter also covers important promoters that are used in transgenic plants as they often represent the difference between a successful plant biotechnology project and one that remains in obscurity.

9.2. DEFINITION OF MARKER GENES

Various marker genes have played crucial roles in facilitating the production of transgenic plants, the subsequent identification of the transgenic plants, and the fine-tuning of procedures needed to increase the transformation frequencies. *Marker genes* fall into two categories: *selectable* marker genes and *nonselectable* (also referred to as *scorable marker* or *reporter*) genes (see Table 9.1 for examples).

9.2.1. Selectable Marker Genes: An Introduction

The first selectable marker genes enabled the production of transgenic plants by using chemicals in the plant growth media that allowed transgenic tissues to grow, but not non-transgenic tissues. Because only a few cells are transformed in a population of target cells in

TABLE 9.1. Categories of Marker Genes Used in Plants with Selected Examples

Classes of Marker Genes	Examples of Genes	Source of Genes	Selective Agent
Selectable marker genes			
Positive			
Conditional	*nptII, neo, aphII*	*Escherichia coli* Tn*5* (bacteria)	Kanamycin
	hpt, hph, aphIV	*E. coli* (bacteria)	Hygromycin
	bar	*Streptomyces hygroscopicus* (bacteria)	Phosphinothricin
	manA	*E. coli* (bacteria)	Mannose
Non-conditional	*ipt*	*Agrobacterium tumefaciens* (bacteria)	N/A[a]
Negative			
Conditional	*codA*	*E. coli* (bacteria)	5-Fluorocytosine
Nonconditional	*barnase*	*Bacillus amyloliquefaciens* (bacteria)	N/A
Non selectable (reporter) genes			
Conditional	*uidA, gusA*	*E. coli* (bacteria)	MUG, X-gluc
Nonconditional	*gfp*	*Aequorea victoria* (jellyfish)	N/A

[a]Nonapplicable.

explants, there would be little chance of recovering transgenic cells without selectable markers. Selectable marker genes fall into two separate families: those that provide positive and negative selection (Table 9.1). *Positive* selectable marker genes confer a selective growth advantage on plant cells so that transformed cells can outgrow the nontransformed cells. The mode of action may also be more severe through the use of toxic chemicals placed in the growth media. The chemicals are designed to selectively kill nontransformed cells, whereas transformed cells are allowed to live through the action of detoxification or resistance mechanisms encoded by the selectable marker genes. *Negative* selectable marker genes encode systems that are toxic to transgenic plant cells and selectively kill them. It is the positive selection systems that are used in plant biotechnology for the recovery of transgenic plants as the transformation process is seldom very efficient.

Conditional positive selectable marker genes are the most important family used in the recovery of transgenic plants and are key components of most transformation systems. Indeed, it is difficult to imagine how plant biotechnology could have progressed without these marker genes. They are usually incorporated into the basic design of transformation vectors used to insert genes into plant genomes and thus accompany other genes of interest or genetic elements fused to reporter genes in the transgenic plants (Fig. 9.1). They originate from a variety of sources, including plant and nonplant species but most are bacterial in origin and introduce a novel resistance trait (typically antibiotic resistance) into the plant (Table 9.1). To be effective, they should not interact with specific targets within the plant or alter signal transduction pathways in a way that changes the plant. If they create such changes it would be difficult to identify the phenotypes associated with the gene of interest or the factors affecting their expression. They also act as dominant genetic markers in the homozygous and hemizygous states.

For expression in plant cells selectable marker genes from bacteria must be extensively modified because the signals on the bacterial gene will not be correctly recognized by the

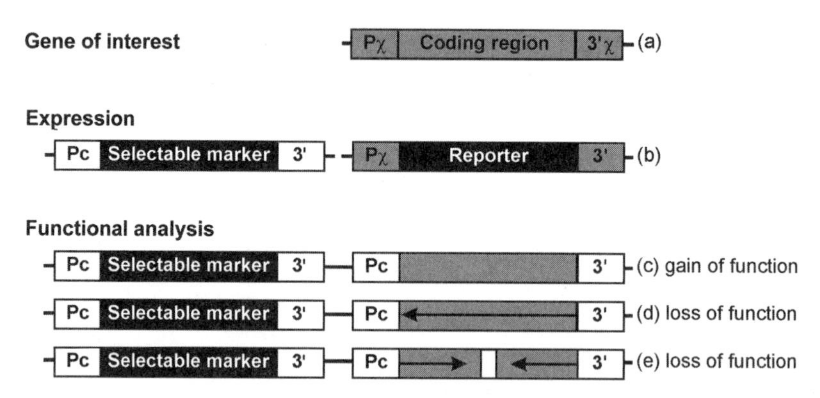

Figure 9.1. Functional organization of selectable marker genes and reporter genes on transformation vectors used to transfer DNA to plant cells. The selectable marker genes are a fundamental component of the transformation vectors as they are needed for the recovery of transgenic material. The vectors are used for many purposes, including the study of plant genes and their regulatory elements. Often the function of genes emerging from genomics studies are unknown and the transgenic plant provides an experimental model for gaining functional understanding. (a) The gene of interest can be examined in many ways. The regulatory elements are often found in the noncoding regions of the gene. For example, the promoter (Px) is found in the 5′ upstream region and includes a number elements needed for transcription, including the core promoter and often enhancer or repressor elements. Some of these elements may also exist in the 3′ end region. (b) By fusing the 5′ and 3′ noncoding regions to a reporter gene and inclusion of the chimeric gene in the transformation vector, the patterns of gene regulation can be assessed in transgenic plants. (c) Gain-of-function experiments can be performed by the overexpression of the coding region using a strong constitutive promoters (Pc), such as the 35*S* promoter, and 3′ ends needed for termination and polyadenylation (3′), such as those from the *nos* gene or 35*S* transcript. A phenotype in the transgenic plant may reveal function. (d) A mutant phenotype may also be mimicked by eliminating or reducing the expression of the gene of interest by creating an antisense transcript in the transgenic plant. (e) This may also be achieved by creating a vector with inverted repeats of the gene of interest, which may induce gene silencing. In each case, the selectable marker and the reporter genes serve different purposes.

plant gene expression machinery. The modifications could include changes to the codons favored by plants and elimination of cryptic sites that could result in aberrant processing of transcripts. They almost always include swapping of the upstream and downstream regulatory elements with plant sequences to create "chimeric genes" that will be recognized by the plant transcriptional and translational systems. Such chimeric genes are designed to express efficiently in plant cells (Fig. 9.1). The first chimeric genes created for plants were selectable marker genes, and these were directly responsible for the development of transformation technologies for plants (Fraley et al. 1983; Bevan et al. 1983; Herrera-Estrella et al. 1983). These include antibiotic- or herbicide-resistant markers.

Figure 9.2 compares the use of a typical conditional positive selectable marker gene system with newer nonconditional positive selectable marker gene systems being developed for the selection of transgenic plants. Cells that express the conditional positive selectable marker gene are supplied with a novel resistance trait, which is the ability to detoxify a toxic substrate in the tissue culture media used to culture plant cells and regenerate plants. Only transformed tissue can grow normally and regenerate into plants because the untransformed cells are prevented from growing or are killed by the substrate. The nonconditional systems allow transformed cells to be distinguished by alterations in growth and development

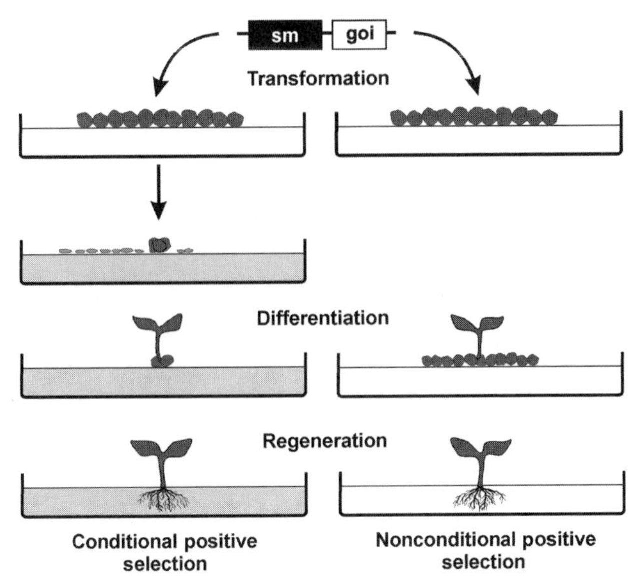

Figure 9.2. Comparison of a typical conditional positive selection system with a nonconditional positive selection system. The conditional systems are the most abundant and extensively used, whereas nonconditional systems are currently in early stages of research and development. The conditional systems introduce a novel resistance trait often taken from bacterial sources. An external substrate (indicated by shading of the medium) is added to the tissue culture media used to grow and regenerate transformed material to suppress the growth of the nontransformed tissues. This differs from the nonconditional systems, which introduce genes that alter the growth and differentiation of the transgenic cells and tissues in a manner that allows them to be separated from untransformed cells and tissues in the absence of external substrates. This latter approach fundamentally alters the plant material by intervening in basic cellular processes. In this illustration the selectable marker gene (sm) and gene of interest (goi) are linked in the transforming DNA. The process of plant regeneration depicted is organogenesis, which commonly involves the differentiation of the shoots followed by the roots separately. It may also occur in a single step through the process of embryogenesis, which may be induced in somatic cells or gametic cells.

without the use of external substrates. Both allow the recovery of the transgenic cells that may be present at a very low frequency in the cell population. The first selectable marker gene used in plant biotechnology and still the one most often used in research is the neomycin phosphotransferase II gene (*nptII*), which allows plant cells expressing it to survive and grow on culture media containing the antibiotic kanamycin (Fig. 9.3).

As most transformation events occur at very low frequencies the selectable marker genes are an essential component of most transformation protocols. This is particularly important for plant species where the tissue culture methodologies are poorly developed or the efficiency of transformation is low. The frequency of recovery of resistant tissues or plants provides a measure of transformation efficiency that can be used to optimize the components of the transformation protocols in a stepwise manner. Chapters 5 and 11 describe this process in greater detail. Once transgenic plants have been recovered, the selectable marker gene can act as a genetic marker for subsequent genetic studies as it is linked to the gene of interest (Fig. 9.1). For example, the selectable marker gene allows the researcher to predict the number of segregating insertion events that have occurred in a transgenic line and also to

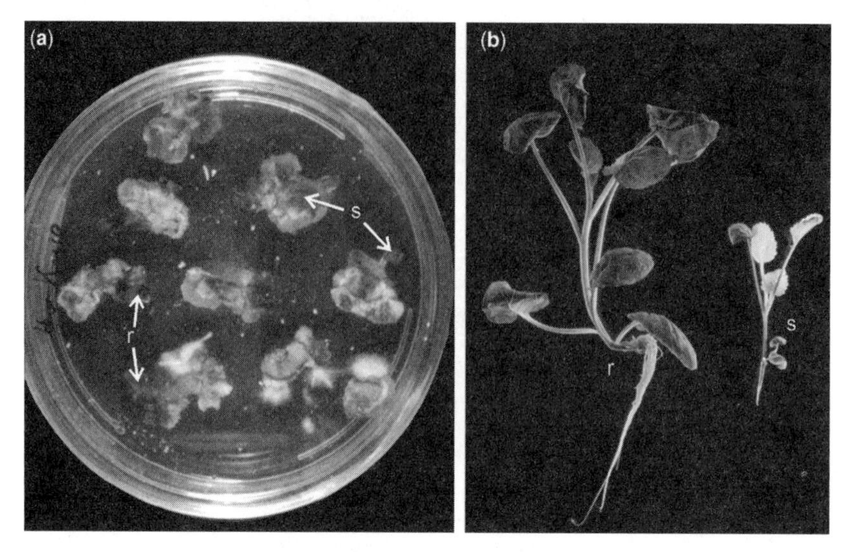

Figure 9.3. Selection of transgenic canola (*Brassica napus* cv Westar) on kanamycin-containing tissue culture media (reproduced with permission from Pierre Charest, PhD thesis, Biology Department, Carleton University, 1988). Stem explants were first infected with an *Agrobacterium tumefaciens* strain harboring a transformation vector with a chimeric *nptII* gene designed to confer kanamycin resistance on transformed plant tissue. (a) After cocultivation of plant tissue and *Agrobacterium* to allow transformation to occur, the plant tissues were transferred to tissue culture media containing kanamycin for growth of callus tissue and shoot differentiation. Much of the non-transformed tissues turned white and stopped growing because they were sensitive to the antibiotic(s). Transformed tissues remained green and continued to grow and differentiate because they were resistant to kanamycin (r). (b) Transgenic shoots that differentiated in the presence of kanamycin were excised from the callus and transferred to media for the regeneration of roots (r). Escapes that were not truly kanamycin-resistant were unable to regenerate roots in the presence of the antibiotic(s). See color insert.

monitor the transmission of the linked transgenes among the progeny of the plant. One hopes for a transgenic line or event with a simple 3 : 1 Mendelian segregation; however, it is common for multiple insertion events to occur during a transformation experiment. The genetic analysis of marker gene segregation is usually an important step in selecting the homozygous transgenic lines with single insertions used for detailed studies of the gene of interest. At this point the selectable marker gene is dispensable and can be eliminated from the plant. A number of methods have been developed to remove them, and these will be discussed later.

9.2.2. Reporter Genes: An Introduction

Whereas selectable marker genes help the researcher select transgenic tissue, reporter genes usually report which cells are transgenic. Figure 9.4 illustrates the use of a typical conditional, nonselectable (or reporter) gene. The *uidA* gene (also called *gusA*) codes for the enzyme β-glucuronidase (GUS), which can react with a chemical substrate, 5-bromo-4-chloro-3-indolyl glucuronide (X-gluc), to create a blue precipitate, thus providing a *histochemical* stain that reflects the location of transgene expression (Fig. 9.4a) (Jefferson et al. 1987). The staining patterns are used to understand the strength and

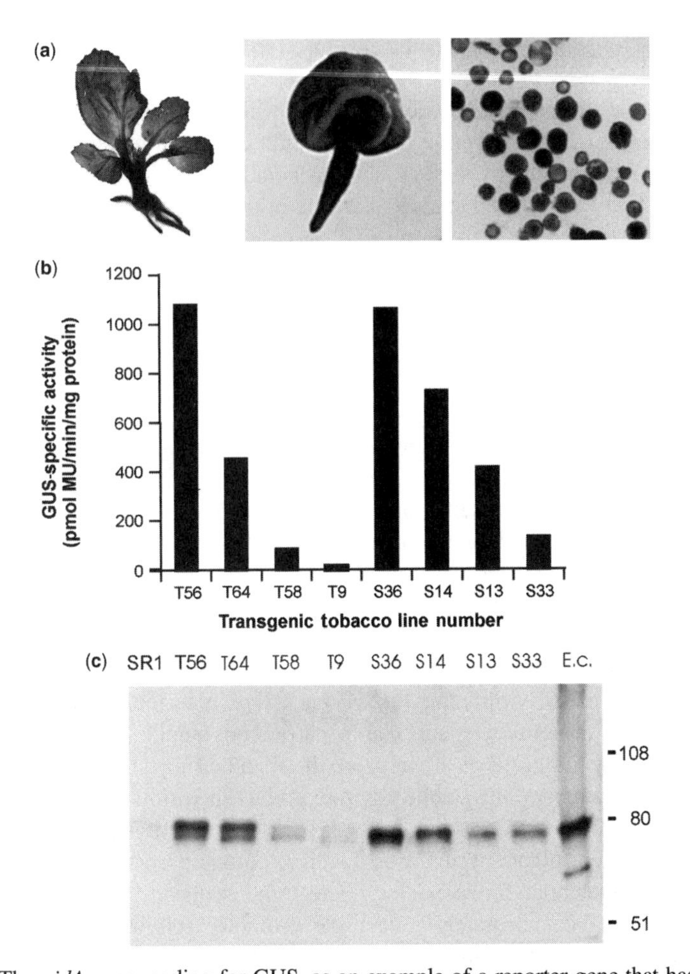

Figure 9.4. The *uidA* gene, coding for GUS, as an example of a reporter gene that has been extensively used in plants. (a) Histochemical staining for GUS activity using the substrate 4-methyl umbelliferyl glucuronide (MUG) allows detection of gene activity in specific tissues of transgenic plants. Shown are the staining of cauliflower plantlets in which constitutive expression of GUS is conferred by a strong constitutive promoter, tCUP (photo courtesy of Dan Brown, Agriculture and Agri-Food Canada, London, Ontario, Canada); excised embryos from transgenic canola seeds in which seed-specific expression is conferred by the napin promoter; and transgenic canola pollen in which cell-specific expression is conferred by the pollen-specific promoter (Bnm1 promoter). Note here that pollen cells are segregating as transformed and nontransformed cells indicated by the presence and absence of staining. (b) Measurement of GUS enzyme specific activity using the substrate 5-bromo-4-chloro-3-indolyl glucuronide (X-gluc). Each separate transgenic line of tobacco differs in the level of gene expression due to variation in the influences on the inserted genes from the genetic elements and chromatin environment at the different sites of insertion. These are often called *position effects* (also see Fig. 9.5). To compare differences among genes and elements introduced into transgenic plants, analyses must account for a large number of transgenic lines to reduce the influence of position effects. Reporter genes provide a valuable means for gathering large amounts of data. Here, a comparison of the promoter strengths of the 35*S* (plant lines with the *S* designation) and tCUP (plant lines with the *T* designation) constitutive promoters is inferred by comparing the activities of the reporter gene. (c) To ensure that the reporter gene reflects transcriptional activity, RNase protection assays are used to measure the relative amounts of GUS mRNA accumulating in the transgenic lines. This assay involves the formation of stable RNA duplexes with a radiolabeled antisense RNA probe followed by RNase digestion of the single-stranded RNA molecules so that the protected double-stranded RNA can be separated by gel electrophoresis and quantified. See color insert.

pattern conferred by gene regulatory elements fused to the chimeric reporter gene (Fig. 9.1) and how the gene is controlled in the various cells of the plant. The specific activity of the GUS enzyme can also be quantitatively measured, which reflects the relative level of gene expression, which might in turn indicate the strength of the promoter fused to the marker gene (Fig. 9.4b) (Jefferson et al. 1987). A generation of plant biotechnologists have looked at blue "GUS" spots and staining patterns in plants to answer questions such as

Are my plants transgenic?
Is the gene expressed?
How is my promoter working?
These are also discussed in Chapter 6.

Although any marker gene can be monitored using techniques for detecting and measuring specific DNA, RNA, or protein sequences (Chapter 11), the reason for using a reporter gene is the convenience of generating large volumes of data and working with large sample sizes. This is important as transgenic plants vary considerably in the expression of transgenes because of position effects or the varying influences of the insertion sites on the inserted genes (Fig. 9.4b). Large amounts of data are needed to overcome the variability to extract meaningful information. It is also useful when looking for rare events. For example, when plant transformation experiments are begun on a "new" plant species, researchers need help in optimizing the system. Rare transgenic events can be found using marker genes, and improvements can be carried out quickly. Because data generated through conditional reporter genes are indirect, they can be subject to factors that can affect interpretation. These include the stability of the product in various tissues, the presence of inhibitors in certain tissues, or varying levels of background activity. Such factors can vary greatly among different plant species. It is therefore common practice to confirm the findings using techniques that independently detect the sequence of interest (Fig. 9.4c). To ensure that the data accurately reflect the promoter activity, RNA blots or other methods are often combined with marker gene activity to confirm the patterns of mRNA accumulation (see Chapter 11). The advantages and disadvantages of the various reporter gene systems will be discussed later.

9.3. PROMOTERS

Chapter 6 illustrated the importance of promoter sequences in the spatio temporal regulation of plants genes. Similarly, all transgene constructs, including genes of interest (GOIs) and marker genes, require promoters to regulate their transcription reproducibly and predictably. Transgenes can be cloned with a variety of *heterologous* (from another source) promoters during construction of the chimeric genes within the transformation vectors to provide biotechnologists with a range of expression patterns to suit their experiments (see the previous chapter). For optimal performance in regenerating transgenic plants, it is important that selectable marker genes are expressed in all of the cells of the plant. Usually selection pressure is applied shortly after the transformation phase and maintained throughout growth and differentiation in tissue culture and during the stages of plantlet regeneration (Figs. 9.2 and 9.3). For genetic analysis of the inserted DNA, selection may also be applied during seed germination and seedling growth to measure the segregation of the marker gene among progeny. Only *constitutive* promoters can ensure expression in all tissues and at all stages of development (Figs. 9.1–9.4). The most frequently used constitutive promoters were those associated with genes found in the T-DNA of the Ti plasmids, in particular the nopaline synthase (*nos*) gene (Fraley et al. 1983; Bevan et al. 1983) or the promoter of the cauliflower mosaic virus (CaMV), in particular that associated with the 35*S* transcript. Although not of plant origin, both were among the best studied during the time when transformation technologies were first being developed. The 35*S* promoter is generally the stronger promoter of the two (Sanders et al. 1987) and provides an advantage in selection efficiency, particularly in species where the selection procedure is not optimal. These two promoters are generally effective over a very wide range of dicot species; however, they were shown to not be very good in moncot species. For cereals, the rice actin and maize ubiquitin 1 (Christensen and Quail 1996) promoters provide better alternatives. Today, many other constitutive promoters have been isolated, but the 35*S* and nos promoters are still extensively used. Why is that? There are two reasons: (1) a large body of knowledge on their performance and behaviors has accumulated since the mid-1980s, and (2) they are widely available in most of the plant biotechnology laboratories worldwide. If it works, and you have it, why use anything else? This question will be addressed later in the chapter.

As we saw in the previous chapter, when constructing transformation vectors, a number of factors must be considered in addition to the promoter selected to drive the expression of marker genes. For example, the orientation of the promoter within the transferred DNA is also extremely important. The 35*S* promoter may interact with neighboring promoters in the vector and plant sequences at the insertion site. It is known to radically alter the specificity of tissue-specific promoters. This should not be particularly surprising as it often used in activation-tagging experiments, in which it is randomly inserted into the genome to elevate the expression of genes within its range of influence (Fig. 9.5). Field studies

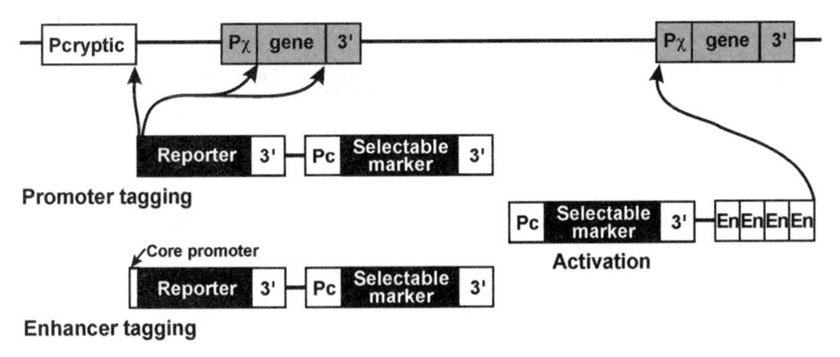

Figure 9.5. Interactions occurring between marker genes and elements in the plant genome. Various experimental strategies have been developed to probe and exploit the plant genome for functional elements and genes. This includes the use of vectors in which reporter genes are introduced into transgenic plants without key regulatory elements such as promoters and enhancers. Activation of the reporter is therefore dependent on the acquisition of the missing elements from the genome at the site of insertion. These are called *enhancer trap* or *promoter trap* experiments. The frequency of trapping such elements can be very high. The regulatory elements may be associated with expressed genes (Px) or may lie dormant in the genome as cryptic elements (Pcryptic). An alternate strategy used to activate genes of interest is by introducing strong constitutive enhancer elements alone. This is often referred to as *activation tagging* of genes. Interestingly, this strategy can be combined with selection and/or screening techniques to recover genes within specific functional groups.

have also shown that interactions could occur with infecting viral sequences resulting in transgene silencing. In transgene silencing, the transgene stops being expressed. For many years there have been observations of gene silencing mediated through sequence similarity between promoters introduced on transformation vectors and resident genes. This area of research is relatively new, but information is rapidly accumulating on the nature of the interactions that can modify expression and the orientation of gene elements that will mediate targeted gene silencing. For example, inverted repeats within the DNA sequence must be avoided (Fig. 9.1). Furthermore, vectors should be constructed so that the promoters associated with the selectable markers are located at maximal distances from promoters associated with the gene of interest or the borders with the plant DNA (Fig. 9.6). The genome has an abundance of promoters that could interact with incoming DNA on insertion. Indeed, this knowledge has been used for promoter discovery research through the use of promoter or enhancer trap strategies (Fig. 9.5). Interestingly, a new strong constitutive promoter, tCUP (Fig. 9.4), was discovered in this way and was found to be very useful for driving the expression of selectable marker genes because it did not interact with other promoters as extensively as the 35*S* promoter (Fig. 9.6).

Nonselectable marker (reporter) genes have been used extensively to study the specificity and level of plant promoter activity; therefore, the promoters combined with reporter genes have been much more diverse. It is common to have a selectable marker gene and a

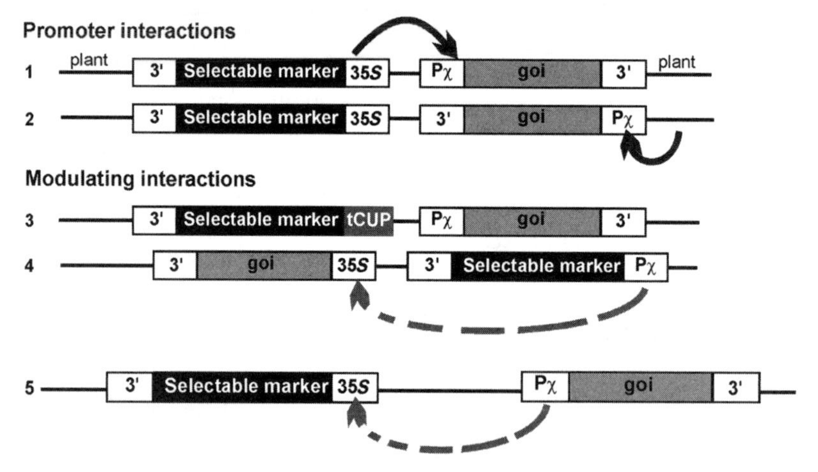

Figure 9.6. Construction of plant transformation vectors to avoid interactions among promoters used to drive selectable marker genes and genes of interest (GOI). Constitutive promoters, such as the 35*S* promoter, are frequently used to drive expression of the selectable marker genes. However, the 35*S* promoter will interact with other promoters (red arrows) within the transformation vector, particularly if they are situated near each other (configuration 1). This can lead to aberrant or unpredictable expression of the gene of interest (GOI). Similar interactions may occur with elements within the plant DNA that become positioned close to promoters within the transferred DNA (configuration 2). These interactions can be minimized by the design of the vector. The simplest approach is to use a constitutive promoter that does not tend to interact with other promoters, for example, the tCUP promoter (shown in configuration 3). The genes within the transferred DNA can also be positioned so that their promoters are spaced as far away from each other as possible through their orientation relative to each other (configuration 4) or by the insertion spacer DNA between them (configuration 5). These manipulations will reduce the extent of the interactions (indicated by the broken arrows).

reporter gene linked together in the same DNA fragment that is transferred to the plant cell (Fig. 9.1). As discussed above, the orientation and type of promoter fused to the reporter gene must be carefully balanced, with the promoter fused to the selectable marker genes (Fig. 9.6).

So far, we have been discussing insertions into the nuclear genome of plants and the transcription of marker genes by RNA polymerase II in the nucleus. Technologies for the insertion of genes into the chloroplast genome are also very advanced for some plant species, such as tobacco (Svab and Maliga 1993). Some of the selectable marker genes used in chloroplast transformation are functional in both nuclear and plastid transformation (Miki and McHugh 2004); however, different promoters are needed for their expression in the different genomes. For chloroplasts the ribosomal RNA operon promoter (Prrn) is particularly effective and allows transformation frequencies that are equivalent to those achieved with nuclear transformation (Svab and Maliga 1993).

9.4. SELECTABLE MARKER GENES

Over 50 selectable marker gene systems have been described in the literature, primarily for nuclear transformation (Miki and McHugh 2004), but only a small number have been adopted for routine use. Having many different systems is important as they vary in efficiency among plant species. Furthermore, experiments are often required in which different transgene insertions are combined in single lines through genetic crosses using separate parental transgenic lines or through consecutive transformation steps. Different selectable marker genes allow the researcher to follow the segregation of each insertion event independently. The underlying principles used to achieve selection differ widely among the selectable marker genes, and the terminology for describing them in the literature has been confusing. Table 9.1 provides a classification system, with selected examples, for the various marker genes used in plants. A more comprehensive list can be found in Miki and McHugh (2004).

9.4.1. Conditional Positive Selectable Marker Gene Systems

This category contains the largest number of and most widely used selectable marker gene systems developed for plants. The genes code for an enzyme or product which provides resistance to a substrate that selectively inhibits the growth and differentiation of the nontransformed tissues (Fig. 9.2). The toxic substrate may be an antibiotic, herbicide, drug, metabolic intermediate, or phytohormone precursor. The manner in which the substrate is applied is very important because the ease with which transformed cells are allowed to proliferate must be balanced with the stringency with which the nontransformed cells are suppressed or killed. The accumulation of toxins from dead tissues can adversely affect the ability of living tissues to survive, particularly if they are present in limited numbers within a larger population of dying or dead material. The optimal selection conditions tend to be specific for each plant species and tissue type. If not properly administered, the proportion of transgenic material recovered may be disproportionately low relative to the frequency of transformation events that actually occurred. Conversely, if the frequency of "escapes" (i.e., nontransgenic tissues that the researcher believes to be transgenic since they survived selection) is too high, then considerable effort and cost would be needed to sort them out from the nontransformed material later.

This category includes some recently developed systems that involve genes that provide access to a nutrient source that can be utilized only by the transformed tissue with the consequence that nontransformed tissues are eventually starved to death. An example is the *manA* gene, which confers on plant cells the ability to use mannose as a carbon source (Reed et al. 2001). The use of metabolic intermediates and drugs to achieve selection has also been demonstrated. In some cases, the systems can distinguish transformed from untransformed plants at the plantlet or whole-plant level but may not act efficiently during the tissue culture steps needed for the selection of transformed tissue.

The scientific literature shows that only three of these selection systems have been adopted routinely to generate transgenic plants for research or for commercialization. They include the *nptII* (Fraley et al. 1983; Bevan et al. 1983; Herrera-Estrella et al. 1983) and *hpt* (Waldron et al. 1985) genes, which confer resistance to the antibiotics kanamycin and hygromycin, respectively, and the *bar* or *pat* genes (De Block et al. 1989), which confer resistance to the herbicide phosphinothricin (Table 9.1). In field trials, the most frequently present selectable marker genes are the *nptII* and *bar*/*pat* genes (Miki and McHugh 2004).

9.4.1.1. Selection on Antibiotics. The aminoglycoside antibiotics include a number of different antibiotics, including *kanamycin*, neomycin, gentamycin, and paromomycin. Kanamycin is produced in the soil actinomycete, *Streptomyces kanamyceticus*. These molecules are very toxic to plant cells (Fig. 9.3) because they inhibit protein synthesis, but a number of enzymes are found among microbes that will detoxify them. One of these enzymes is a phosphotransferase that can confer resistance through the ATP-dependent-*O*-phosphorylation of the kanamycin molecule. *Neomycin phosphotransferase II* (NPTII) from *Escherichia coli* is the bacterial aminoglycoside 3′-phosphotransferase II [APH (3′) II, EC 2.7.1.95] most often used with plant cells to generate kanamycin resistance. It was originally selected for use in plants because prior work with mammalian and yeast cells demonstrated its effectiveness as a selectable marker in eukaryotic cells.

To function in plant cells the gene coding for NPTII (*nptII*, also designated *neo* or *aphII*) was fused to regulatory elements from the nopaline synthase gene (*nos*) from the T-DNA of the *Agrobacterium tumefaciens* Ti plasmid. The *nos* gene elements confer constitutive expression of the *nptII* gene and thus kanamycin resistance within all cells of the transgenic plant. A stronger upstream promoter sequence from the cauliflower mosaic virus (CaMV), which is responsible for transcription of the 35*S* RNA, generates a higher level of *nptII* gene expression, which results in a higher level of kanamycin resistance. The *nptII* gene can function as a selectable marker in both the nuclear and plastid genomes; however, a member of another class of selectable marker gene, namely, aminoglycoside-3″-adenyltransferase (*aadA*) is generally preferred for chloroplast transformation.

The popularity of kanamycin resistance, conferred by the *nptII* gene, is because it is very effective, functions in a wide range of plant species, and after extensive testing appears to be very safe for use in food and feed crops. As it also functions effectively in a wide range of microorganisms and eukaryotic cells, some initial concerns had been expressed about the potential transfer of antibiotic resistance to other organisms; however, it has been used since the mid-1980s in crops, and no adverse effects on humans, animals, or the environment have yet to appear (Flavell et al. 1992). It is also known that expression of the *nptII* gene in plants does not alter the patterns of transcription in plants, so that transgenic plants expressing it are essentially equivalent in composition to nontransgenic plants

(El Ouakfaoui and Miki 2005). This gene has been the most extensively studied among the selectable marker genes, and the largest body of information on its use in plants has been accumulated.

Hygromycin B is the second most commonly used antibiotic for selection of transformed plant cells (Waldron et al. 1985) after kanamycin; however, it is very toxic to plant cells relative to kanamycin and more difficult to apply without "overkill." It is an aminocyclitol antibiotic that also inhibits protein synthesis with a broad spectrum of activity. The bacterial gene, *aphIV* (also referred to as *hph*, *hpt*), codes for hygromycin phosphotransferase (HPT, EC 2.7.1.119), which acts by ATP-dependent phosphorylation of hygromycin B. It has been used extensively over a wide range of species.

Other antibiotics that have been used include streptomycin, bleomycin, streptothricin, and chloramphenicol. Generally, resistance is conferred by genes coding for enzymes that act by detoxification of the antibiotic through a modification to the molecular structures.

9.4.1.2. Selection on Herbicides. *Phosphinothricin* (PPT) or glufosinate ammonium is the active component of several commercial herbicides. As an analog of L-glutamic acid, it is a competitive inhibitor of glutamine synthase (GS) that is essential for the assimilation of ammonia into plants. By inhibition of glutamine synthase ammonia accumulates to toxic levels. The enzyme phosphinothricin *N*-acetyltransferase (PAT) will detoxify PPT by acetyl CoA-mediated acetylation and thus confer resistance. Two genes coding for the enzyme have been cloned: the *bar* gene (for bialophos resistance, where bialaphos consists of two L-alanine residues and PPT) from *Streptomyces hygroscopicus* and the *pat* gene (for phosphinothricin acetyltransferase) from *Streptomyces viridochromogenes.* Both have been used extensively as selectable marker genes, particularly among cereal species where kanamycin selection may be less efficient. Typically, kanamycin does not kill monocots very effectively, whereas bialophos or PPT does.

Plants containing the *bar* or *pat* genes have been among the first to receive regulatory approval for unconfined field production and have been assessed as safe by a number of international regulatory agencies. A number of other herbicides, including glyphosate, imidazolinones, and bromoxynil, can also be used in combination with their corresponding resistance genes for selection of transgenic plants.

9.4.1.3. Selection Using Nontoxic Metabolic Substrates. Most conditional positive selection systems use toxic substrates for selection of the transformed tissues; however, the use of nontoxic metabolic intermediates has emerged as an alternative. This type of system differs from the use of antibiotics, herbicides, or drugs in that the substrates are not inhibitors but rather carbon sources that are restricted from use by the plant cell unless provided with an enzyme that allows entry of the carbon source into primary metabolism. Examples of such selective agents include mannose and D-xylose. Bacterial genes coding for phosphomannose isomerase (PMI, EC 5.3.1.8) from *E. coli* (*manA*) or xylose isomerase from (EC 5.3.1.5) *Streptomyces rubiginosus* and *Thermoanaerobacterium thermosulfurogenes* (*xylA*) provide the enzymes that allow entry into glycolysis. The apparent advantage is that it works with a wide range of plant species and appears to yield higher transformation frequencies because the selection is not as harsh as with toxic substrates (Reed et al. 2001). This approach differs fundamentally from the others discussed so far in that the novel trait encoded by the selectable marker gene alters a basic aspect of plant metabolism.

9.4.2. Nonconditional Positive Selection Systems

A very new area of research is the development of positive selection systems that do not require any substrates for selection. The list was quite small as of 2002 [reviewed by Zuo et al. (2002)]. They are based on the use of genes that confer a growth advantage, distinguishable morphology or that selectively induce the differentiation of transformed tissues but do not necessarily kill nontransgenic tissues (Fig. 9.2). The use of shoot organogenesis to select for transformed tissues is the most advanced example at this time (2007). Shoot formation in culture depends on the presence of high cytokinin : auxin ratios. The T-DNA of the Ti plasmids from *A. tumefaciens* codes for the enzyme isopentenyl transferase (IPT), which catalyses the synthesis of isopentenyl adenosine-5′-monophosphate, which is the first step in cytokinin biosynthesis (Table 9.1). Expression of this gene alone in plant cells results in a higher frequency of shoot regeneration and recovery of transformed material. The difficulty is that the shoots have abnormal morphology due to the cytokinin imbalance and cannot produce roots (Ebinuma et al. 2001). To overcome this obstacle an inducible promoter, such as the β-estradiol-inducible promoter is needed to restrict the timing of expression of the *ipt* gene (Zuo et al. 2001). A number of alternatives have been demonstrated to have potential; however, these need time to be fully evaluated and developed. Again, this approach differs from most other systems in that it intervenes in the basic processes of plant cell growth and differentiation.

9.4.3. Conditional Negative Selection Systems

Conditional negative selection systems can play an important role in experiments by eliminating unwanted transformation events or when selecting against expression in specific tissues or under specific inducible conditions. Only a few systems have been described. One example is the bacterial *codA* gene (Stougaard 1993), which codes for cytosine deaminase (Table 9.1). It is interesting that it has been shown to be effective in nuclear and plastid transformation. This class of selectable marker genes codes for enzymes that convert a nontoxic substrate into a toxic substrate, thereby eliminating the transformed cells that express it.

9.4.4. Nonconditional Negative Selection Systems

Nonconditional negative selection systems have not been used widely as a selectable marker gene system, but they have been used effectively to *ablate* specific cell types in transgenic plants. These systems may code directly for toxins or enzymes that disrupt basic cellular processes causing cell death. Whereas most of the systems discussed so far are used for the production of transgenic plants, a nonconditional negative selection system kills cells in mature plants for biosafety or breeding purposes. An example is the ribonuclease, barnase, from *Bacillus amyloliquefaciens*. As shown in Figure 9.7, when it is expressed only in the tapetum, which gives rise to pollen cells, the tapetum is killed and therefore pollen cells cannot differentiate and mature. The consequence to the plant is the inability to produce pollen or male sterility (Mariani et al. 1990). The activity of *barnase* can be controlled by a specific protein inhibitor, *barstar*, which is also found in the same bacteria. Although this technology is recognized in plant sciences mainly as a molecular tool for generating male sterility and the commercial production of hybrid seed, it can also be used in the functional analysis of specific cell types and the gene

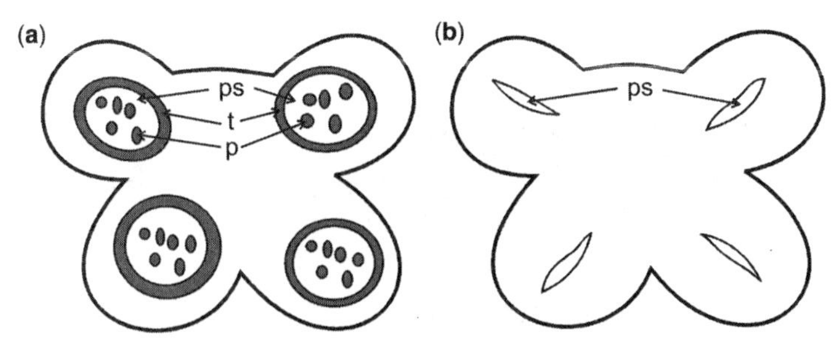

Figure 9.7. Diagrammatic illustration on the use of *barnase* as a negative selectable marker gene for the ablation of the canola tapetal cells: (a) wild-type; (b) transgenic. Barnase codes for a ribonuclease from *Bacillus amyloliquefaciens*. When it is expressed in plants under the control of the tapetum-specific promoter (*TA29*), expression was restricted to the cells of the tapetum (t) in transgenic plants. The ribonuclease activity in the tapetum resulted in failure of the tapetum to develop and collapse of the pollen sac (ps). Because the tapetal cells are the precursors of the pollen cells, pollen (p) cannot differentiate in the transgenic plants and the plants are therefore male sterile. Because the pattern of ribonuclease expression was tapetum specific the rest of the plant was unaltered. [Adapted from Mariani et al. (1990).]

regulatory elements that limit expression within them. As such, it may also be considered as a complement to the reporter genes described below.

9.5. NONSELECTABLE MARKER GENES OR REPORTER GENES

As important as the selectable marker genes have been for the development of transformation technologies, the nonselectable or reporter genes have played a different fundamental role in the growth of our understanding of gene regulation mechanisms in plants. These kinds of genes have formed an invaluable partnership with selectable marker genes in transgenic research. For reasons that will soon be apparent, reporter genes are also sometimes called "visible marker genes" since they change the appearance of plant tissues. Although several reporter genes have been described, three have been particularly influential and have dominated the scientific literature. These are the genes coding for *β-glucuronidase* (GUS), *luciferase* (LUC), and *green fluorescent protein* (GFP). GUS and LUC are conditional nonselectable marker genes as they require the use of an external substrate for detection of activity, whereas GFP is a nonconditional, nonselectable marker gene because the protein encoded by the gene is directly detectable without the use of a substrate.

9.5.1. β-Glucuronidase

The bacterial enzyme β-glucuronidase (GUS), which is coded by the *Escherichia coli* gene, *uidA* (also known as *gusA*) has been the most widely used reporter system in plants (Jefferson et al. 1987). It has a sensitive specific activity assay using 4-methylumbelliferyl glucuronide (**MUG**) as substrate and permits histochemical localization using 5-bromo-4-chloro-3-indolyl glucuronide (X-gluc) with specificity at the cellular level (Fig. 9.4). The enzyme is stable in plant cells and can accumulate to high levels without toxicity to the plant cell. It confers no apparent phenotype to plants in the absence of its substrates and therefore can be used to study plant processes without concern of artifacts resulting from

Figure 9.8. Fusion of a reporter and selectable marker gene to create a bifunctional gene: (a) GUS:NPTII fusion reporter system for plants that incorporates the *nptII* gene for kanamycin selection and the GUS reporter gene in a single module; (b) transformed tobacco shoots selected on kanamycin; (c) shoots with roots regenerated on kanamycin; (d) a transgenic seedling after two generations showing retention of GUS gene activity indicated by the histochemical staining with the GUS substrate X-Gluc (provided by Raju Datla, Plant Biotechnology Institute, National Research Council of Canada, Saskatoon, Canada). See color insert.

nontarget or pleiotropic effects (El Ouakfaoui et al. 2005). It is the most frequently used reporter gene in field trials so far. The greatest disadvantage is that both detection assays are destructive to the cells. The substrates are also quite expensive.

The GUS reporter gene system has been most important in the study of gene regulatory elements and mechanisms in plants. Generally, there is very little background activity in plants compared with other organisms; specifically, plants do not normally turn blue when exposed to X-gluc. It has been used in transcription fusions to study a wide range of regulatory elements cloned from the plant genome (Fig. 9.1) and also for promoter-trapping experiments (Fig. 9.5). It also forms stable translational fusions with proteins; for example, fusions with the *nptII* gene to generate bifunctional proteins that can be used as a selectable marker and as a reporter (Fig. 9.8).

9.5.2. Luciferase

The firefly (*Photinus pyralis*) enzyme luciferase (EC 1.13.12.7) was one of the first useful reporters for plants (Ow et al. 1986). Whereas GUS gives transgenic cells a blue color, luciferase produces light (Fig. 9.9), an example of bioluminescence; however, it has not been used as extensively as GUS. The enzyme catalyzes the ATP-dependent oxidative decarboxylation of luciferin as substrate; therefore, it is another conditional nonselectable marker system. A significant advantage is the sensitive, nondestructive monitoring system that allows real-time analysis. Furthermore, the half-life of the luciferase protein in plant cells is lower than that for GUS and may reflect transcriptional activity more accurately. It is often used as an internal control in experiments that require the use of more than one reporter system.

9.5.3. Green Fluorescent Protein

The Pacific jellyfish (*Aequorea victoria*) green fluorescent protein (GFP) is now becoming the most important reporter gene system for plants [reviewed by Stewart (2001)].

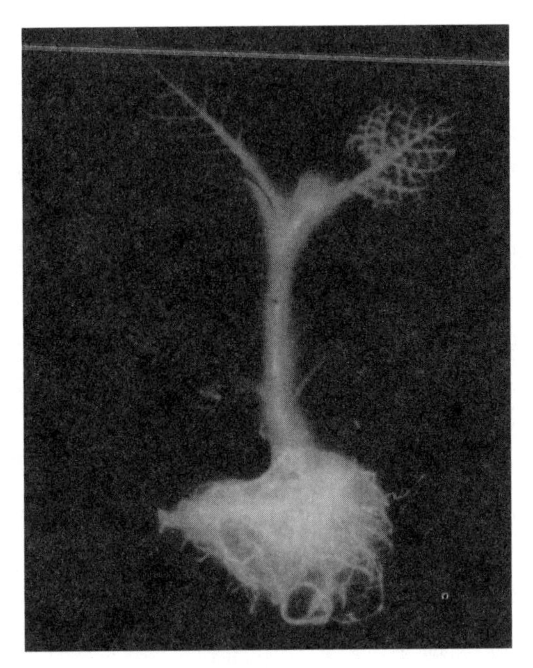

Figure 9.9. Luminescence detected in transgenic tobacco transformed with the firefly luciferase gene driven by the 35*S* promoter and watered with a solution of luciferin, the luciferase substrate. [Reprinted with permission from Ow et al. (1986), copyright 1986, AAAS.] See color insert.

GFP requires no external substrate for detection, and there have been no reports of detrimental effects on the fitness of plants that express it; therefore, it is a nonconditional nonselectable marker gene. The novelty of GFP and other fluorescent proteins is that they combine great sensitivity at the subcellular level using bioimaging technologies made available through confocal laser scanning microscopy with real-time detection in living cells (Fig. 9.10). A wide variety of fluorescent proteins are being developed to extend the range and complexity of processes that can be simultaneously monitored in living cells (Stewart 2006). For example, fluorescent proteins can be fused to various plant proteins and used as a tag to monitor their trafficking and interactions (Fig. 9.10). In field studies, GFP also permits the rapid and easy detection of transgenic plants or plant parts such as pollen (Fig. 9.11). In tissue culture, GFP has been used in combination with selectable marker genes to identify and enrich the content of transformed material to improve the recovery of transgenic plants from species where the current transformation and selection systems are inefficient. New fluorescent proteins are being discovered, and we can expect a rainbow of colors in the near future (Stewart 2006). Especially useful will be fluorescent proteins that emit in the orange and red spectra (Fig. 9.12). This is because natural autofluorescence is less in the orange and red spectra in most plants.

9.6. MARKER-FREE STRATEGIES

Because the roles of the selectable marker genes are often served during the generation of transgenic plants and not after they have been developed, there is generally no need to

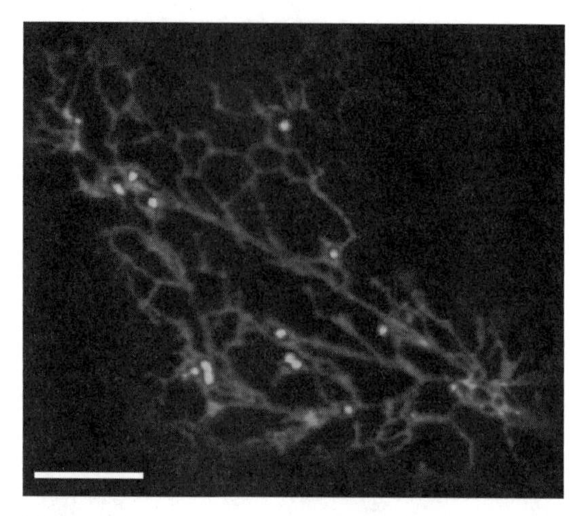

Figure 9.10. Confocal laser-scanning microscopy of leaf mesophyll cells transiently expressing peptides fused to green fluorescent protein (green image) and yellow fluorescent protein (red image). Green fluorescent protein is fused to the HDEL tetrapeptide (spGFP-HDEL) to achieve ER retention and thus reveals the cortical ER network in leaf cells. The proximity of the Golgi to the ER network is revealed by the yellow fluorescent protein fused to a Golgi glycosylation enzyme (ST-YFP). (Bar = 10:m.) [Reprinted from Brandizzi et al. (2004), with permission.] See color insert.

maintain them in the transgenic plant. Their removal may provide some advantages if the plant is to be used for another round of transformation because the same selectable marker gene, if effective, can be used repeatedly. If the safety of the selectable marker to health or environment is a concern, then it may be useful to have a method to remove it from the plant before commercialization. Furthermore, it would be essential to remove the selectable marker gene if it altered plant growth and differentiation. In rare cases, the transformation frequency may be high enough to recover transgenic plants through screening techniques

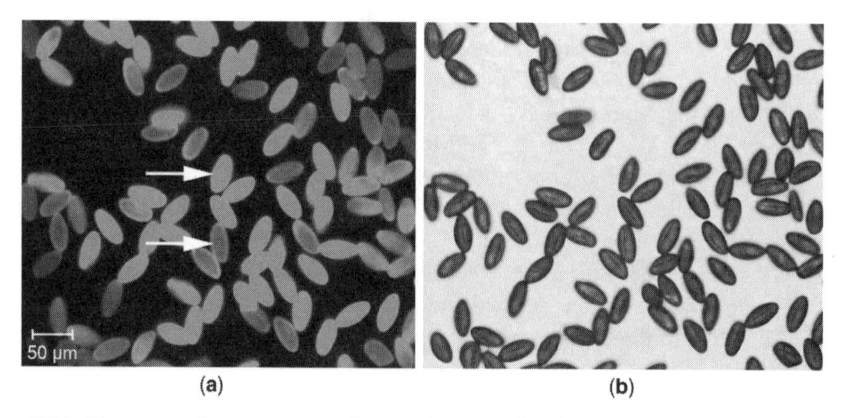

Figure 9.11. The green fluorescent protein has been useful for marking whole plants using a 35*S*-GFP construct and plant parts such as pollen using a GFP under the control of a pollen-specific promoter (LAT59) from tomato: (a) 867 ms, 200× under blue light; (b) 1.7 ms, 200× under white light. The arrows in (a) show GFP fluorescence of pollen cells. (Photos courtesy of H. S. Moon and Neal Stewart.) See color insert.

Figure 9.12. Novel fluorescent proteins whose genes were recently cloned from corals and expressed in tobacco (a) and *Arabidopsis* (b) plants. See color insert.

without selection. With certain traits, such as herbicide resistance, the gene of interest may be directly selectable without the need for a separate selectable marker gene.

The easiest method for generating marker-free plants is by cotransformation of the gene of interest with a marker gene followed by segregation of the unlinked genes into separate lines [Fig. 9.13; reviewed by Ebinuma et al. (2001)]. Although effective, this requires the production of many transgenic lines initially because so many are discarded. Furthermore, the technology is restricted to transgenic plants that are propagated through seeds. This would exclude vegetatively propagated species, such as trees. Cotransformation can be achieved in many ways. For instance, the two genes can be introduced on two separate plasmids. If *Agrobacterium*-mediated transformation is used, this can be achieved by infecting tissues with separate plasmids in separate *Agrobacterium* strains or by separate plasmids in one strain. Cotransformation could also be achieved using a single strain carrying a single plasmid with two separate T-DNA regions. The frequencies of cotransformation may be very high (>50%, depending on the situation) allowing the selection of transgenic material carrying both the selectable marker and the gene of interest. The segregation of transgenic lines carrying the gene of interest from lines carrying the selectable marker gene may occur at frequencies sufficiently high to be practical in species that are efficiently transformed.

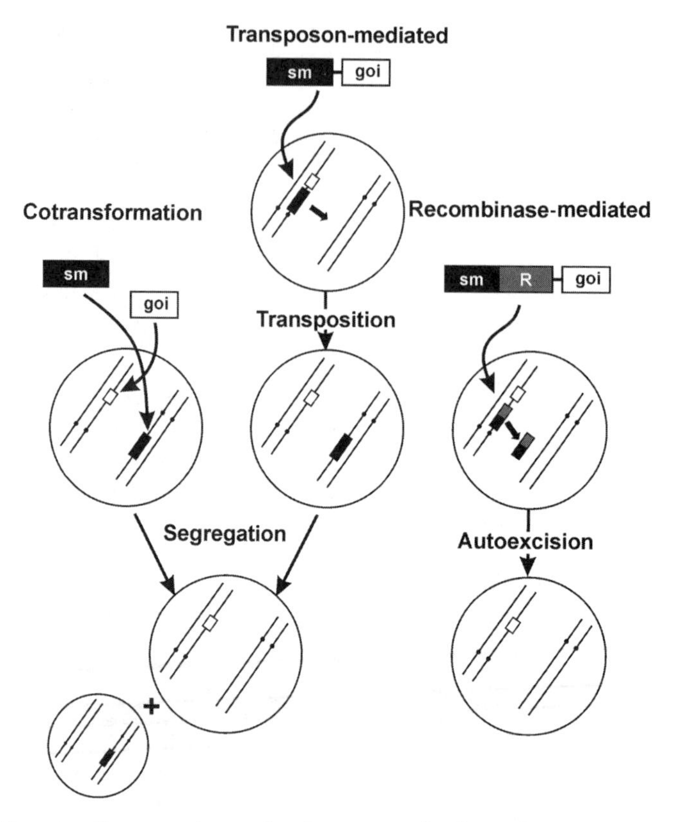

Figure 9.13. Processes for generating marker-free transgenic plants. Cotransformation is a practical process for generating marker-free transgenic plants. It depends on the integration of the selectable marker gene (sm) and gene of interest (goi) at separate chromosomal sites that can segregate away from each other in the next sexual generation. This can also be achieved by the introduction of both the selectable marker and gene of interest on the same vector and therefore insertion at the same site followed by the subsequent transposition of the marker gene to a separate locus that can segregate away from the gene of interest. A more recent advance is the use of excision recombinases (R) under inducible promoters to autoexcise itself along with the selectable marker gene. This process does not require a segregation step and has the potential for broader applications.

Only a few studies have emerged in which transposons have been used to translocate genes within the plant genome to break the linkage between selectable marker genes and genes of interest (Fig. 9.13). An interesting strategy has been developed using the nonconditional positive selectable marker gene (*ipt*) in combination with the *Ac* transposase element to remove the *ipt* gene. In this multiautotransformation (MAT) process, the *ipt* gene first acts positively to generate a proliferation of morphologically abnormal shoots with the "shooty" phenotype. They cannot regenerate because of the overproduction of cytokinin; however, normal shoots emerge at low frequency several weeks later following transposition of the *ipt* gene to a distant locus that can segregate away in somatic cells or it may be directly lost if not reinserted into the genome. [reviewed by Ebinuma et al. (2001)].

More recently, the use of site-specific recombinases has emerged as a versatile strategy for the selective removal of marker genes from an insertion site. The recombinases and their target sites include Cre/*lox* from bacteriophage P1, FLP/*FRT* from yeast

Saccharomyces cerevisiae, and R/*RS* from *Zygosaccharomyces rouxii* [reviewed by Ebinuma et al. (2001)]. The target sequences are placed around the genes targeted for excision followed by the introduction of the recombinase in a second round of transformation (Dale and Ow 1991). Again, this approach suffers from the criticism that it is restricted to seed-propagated plants to segregate the recombinase gene from the gene of interest. This has been partially overcome by the introduction of the recombinase by transient expression. Although excision occurs at a lower frequency, the recombinase gene is not integrated into the genome. Another promising approach incorporated the gene of interest along with the selectable marker genes and recombinase gene on one vector (Fig. 9.13). This strategy overcomes many of the earlier limitations by using an inducible promoter to express the recombinase, resulting in the autoexcision of the recombinase and the selectable marker genes simultaneously (Zuo et al. 2001; Ebinuma et al. 2001). This approach eliminates the need for successive rounds of transformation or crossing and minimizes the period of exposure of plants to the action of recombinases. Prolonged exposure to recombinases is a concern as unpredictable deletions in the genome may occur due to the action of the recombinases on cryptic target sites. Although the extent of such deletions is uncertain in the nuclear genome, examples have been reported in the plastid genome with Cre. With time the excision systems and gene regulatory systems will improve, and the technology is likely to be refined to practical levels. The strategies discussed above as well as other methods for marker gene removal in plants have been reviewed by Darbani et al. (2007).

9.7. CONCLUSIONS

It is difficult to imagine plant biotechnology without marker genes. As we have seen, effective marker genes have demonstrated little to no effect on the plant except their intended effect. Nonetheless, marker genes have been somewhat controversial, especially antibiotic resistance genes, because of the concern about horizontal gene transfer (HGT). HGT is the movement of DNA from one species to an unrelated species—in this case from transgenic plants to bacteria. In the event that an antibiotic resistance gene were to be horizontally transferred to bacteria, some people worry that new antibiotic resistance problems could be created that could harm human or ecosystem health. Even though HGT has not been demonstrated from transgenic plants to bacteria in a realistic experimental system, it has affected the politics of regulation and the perception of transgenic plants, which will be covered in later chapters.

LIFE BOX 9.1. DAVID W. OW

David W. Ow, Principal Investigator, Plant Gene Expression Center, USDA-ARS/UC Berkeley

David Ow while visiting the Cotton Research Institute, Anyang, China (June 2007).

Chance events shape a career. In Spring 1977, while an undergraduate in Rich Calendar's lab, I heard good reviews about Hatch Echols' graduate seminar course on Genetics and Society. An extensive list of topics for presentation was available, but as a lowly undergraduate, I got left with the topic of least interest to others: agriculture and society. I was so worried that I might not measure up to the graduate students that I spent countless hours at the library reading up the green revolution and promising technologies in plant tissue culture, nitrogen fixation, and photosynthesis. For a 1-unit pass/fail course, it turned out to be more work than any of my other classes. Unexpectedly, after boning up on this topic, I actually got excited about genetics for agriculture[1]. After my presentation, Hatch Echols in his usual tie-dye tee shirt had a chat with me about plants, agriculture and the third world. He saw that I might want some practical experience and suggested I see Renee Sung. So I ended up moonlighting in Renee's lab to learn plant tissue culture. When it came time for graduate work, however, I decided on bacterial genetics; plant cell culture work was too slow for my liking. Fred Ausubel's lab was cloning bacterial nitrogen fixation genes, and also doing petunia cell culture, so I ended up at Harvard.

The summer before grad school, I took a month off to visit the Orient. In 1978, China was off limits except for a 3-day tour to Guangzhou. On the train from Hong Kong, I sat next to a Mrs. Bogorad.[2] Apparently, Lawrence Bogorad just left for an official delegation to Beijing, and put her on the Guangzhou tour until they could join up again. I don't know whether meeting her had any relevance, but in Fall 1978, during my rotation in Fred's lab, Dr. Bogorad called me to a reception for a Chinese delegation. Being a first year student, I was a bit nervous but managed a good exchange with the Chinese visitors, who were all quite aged as they had received their PhDs from the West before communism. When we parted, some of them even invited me to visit, which I thought was just a polite gesture.

A member of that delegation was San-Chiun Shen, who did his Ph.D. with Norman Horowitz at Caltech. He would

[1]A graduate student of that class, Sally Leong, also ended up with a career in USDA.
[2]Also on the same tour was a to-be Harvard classmate Donny Strauss.

come to Boston every so often as his son was doing graduate work at the University of Massachusetts. As his lab also worked on nitrogen fixation, he would take the opportunity to drop in on Fred. Each time he saw me he extended his invitation, but I had the alibi that I was in the midst of my thesis work. By 1982, however, Fred told him I was near completion, and so he got quite serious about having me teach his lab molecular techniques. What initially was supposed to be a short visit somehow developed into a one-year plan. About that time, the folks moving on or scheduled to leave were Sharon Long for a Stanford faculty position, Jonathan Jones to Advance Genetic Sciences, Gary Ruvkun to Wally Gilbert's lab, Venkatesan Sundaresan to Mike Feeling's lab, and Fran DeBruijn to Jeff Schell's empire. Well, you can imagine the response when folks heard of my postdoc in China. Not only did everyone think I was nuts, some even suggested (trying to be helpful) that this might mean an end to my career in the big league. Only Boris Magasanik, a member of my thesis committee, offered supportive advice.

I had to wait for my wife to graduate from Columbia, so I became a postdoc for six months in Fred's new lab at the Massachusetts General Hospital. The higher postdoctoral pay was necessary considering my next position. I met Stephen Howell at the Keystone Conference and was impressed with his science and personality, so I sent off a bunch of postdoctoral fellowship proposals to join his lab upon my return from Shanghai. The best holiday greeting I received in the winter of 1983 was a telegram from the Damon Runyon Foundation. The Helen Hay Whitney Foundation wanted an interview, but would not pay for my international airfare, so I couldn't go. With a monthly salary of ¥200 (equivalent to ~US $100, but not convertible to foreign currency), it was out of consideration. By early 1984, I also heard from NSF and since it paid more, I declined the Damon Runyon fellowship.

I managed to teach molecular biology techniques through a research project with a graduate student, Yue Xiong, and a lab assistant, Qing Gu. We did functional analysis of nitrogen fixation gene promoters by site-directed mutagenesis and DNA sequencing. The story behind the story could fill pages, but in short, we completed the work and published in early 1985, surprisingly before a similar paper by a British group later that year. The Editor Rich Losick thought it was the first paper from China in the *Journal of Bacteriology*.

From August 1983 to July 1984, my wife and I lived out of a hotel. With room, board and roundtrip airfare covered, my monthly ¥200 RMB was just spending money. By Chinese standards, it was a high salary, but we often had to pay tourist prices at hotel stores. Outside of the foreigner-only hotels, many items were rationed, especially food, so money (our nonconvertible type) was worthless without the coupons that were rationed to the Chinese citizens. Of course, while we muse at this, at the heart of the rationing was poverty. What I saw working in the midst of the system was quite a contrast to what I experienced growing up in San Francisco, and I am not from privileged background. As a scientist, you just couldn't help but to question the purpose of science and as well as appreciate what food production can mean for others. China today, at least the coastal cities, is much better (largely due to economical reforms). The rural area with ~70% of China's 1.3 billion, however, still has a very long way to go.

After China, I did my NSF postdoc with Stephen Howell at UC San Diego and by the end of 1986, moved to the newly formed USDA Plant Gene Expression Center that is affiliated with UC Berkeley. The scientific career since then has been rewarding and the stories behind them equally interesting. However, the later years were just segments of the journey, on a path that was charted by earlier experiences. Despite living in a publish-or-perish, grant-or-starve environment, I have done my best not to deviate too far from that path, and it is gratifying to know that some of the work bearing my participation have made tangible contributions—the luciferase gene as a research tool, and a commercial transgenic corn product derived from site-specific excision of its antibiotic resistance gene.

Doing well in a career can be less about innate ability, education and opportunity than with motivation and commitment. Had I not got stuck with the presentation on genetics and agriculture in Hatch Echols' class, I doubt whether this city boy would have taken an interest in agricultural research. Had I not run into Mrs. Bogorad on a sightseeing tour, I might not have spent a year in Shanghai and come away with an experience that has solidified my commitment and priorities in science. Over the years, I had suggested to many graduate students that they ought to consider some postdoctoral time in a less developed country, but few gave it a second thought—aside from thinking that I was nuts.

LIFE BOX 9.2. C. NEAL STEWART, JR.

C. Neal Stewart, Jr. Professor and Racheff Chair of Excellence in Plant Molecular Genetics, University of Tennessee

Neal Stewart, pondside, where most of this book was edited.

My childhood years in the 1960s were spent on a small family farm not far from the proverbial Mayberry in North Carolina. My grandfather was the farmer and all of his daughters, including my mother, built houses on adjoining property, like satellites around the home planet. As suburbia encroached and grandpa grew ill and died, the 1970s rolled in and life went on. Farmlife and nature were chief interests in childhood and formative years, but also were hot rods. Biotechnology is kind of like that too—a combination of nature and technology that is somewhat of a paradox.

In college I majored in horticulture and agricultural education. I was either going to grow flowers or teach. I had told myself that I was not smart enough or a good enough student to really go into science. A spiritual awakening at the end of my college years coupled with a few-year stint of teaching middle school (in-school suspension of all things!) convinced me that science might be out of reach after all. My fairly recent (and pregnant) wife and I decided to pack our bags and head off to graduate school where I was fortunate enough to work with ecophysiologist Erik Nilsen at Virginia Tech for masters then PhD degrees. I still wonder why he took me under his wing—I was a babe in biological research, with no experience in science. His nuturing and the support of my wife got me through the MS in ecology. With a bit more confidence, I decided to add DNA into the fix of ecology and studied the population genetics and phenotypic plasticity of cranberry. I can still recall the laughter among my peers when I said in the early 1990s that I wanted to be a molecular ecologist. No one there had ever heard of such a thing, but my choice of phraseology was validated when the journal "Molecular Ecology" was begun.

Severely bitten by the DNA bug by this time, I was also fortunate to gain entrance into Wayne Parrott's lab at the University of Georgia. I think he was having a hard time finding a well-qualified postdoc, and I was foolish enough to naively launch into a soybean transformation project. Soybean transformation was notoriously difficult and I had absolutely no experience in transgenics. But again, I was fortunate to team up with Donna Tucker in Wayne's lab. She is one of those few gifted scientists who has the "golden hands" in the craft of tissue culture and an eye to select the right stuff and throw the wrong stuff away.

Biosafety research was then a natural area for me to combine transgenics and ecology—something I began in Wayne's lab and have continued on during my career as a faculty member. Initially, I was a GMO sceptic—I was convinced that there would be ecological downsides of releasing trillions of transgenic plants into the environment. But by 2004 when I had written *Genetically Modified Planet* I had become convinced by reams of data that there were far more current and potential environmental benefits from biotechnology than risks. I still loved nature and could clearly see how the technology could make farming more environmentally friendly. My lab is now full of exciting young scientists who are researching environmental biotechnology projects ranging from the ecology of transgenic plants to plants designed to detect contaminants, such as the explosive TNT, in the environment. I've also become very interested in the genomics of weedy and invasive plants. It seems to me that these plants, with genomes adapted to competing against crops and natural vegetation are a much larger threat to the environment than GMOs, yet it is a severely underfunded and understudied field. I've become convinced that we worry about the wrong things.

In my own life I was worried that I was "not smart enough" for science. I've been worried that research funding would dry up. Worried about lab personnel. Worry is a waste of time, since the things I've worried about have been moot. My advice to students is to follow their dreams, focus, and find the right people to help make their dreams come true.

REFERENCES

Bevan MW, Flavell RB, Chilton M-D (1983): A chimaeric antibiotic resistance gene as a selectable marker for plant cell transformation. *Nature* **304**:184–187.

Brandizzi F, Irons SL, Johansen J, Kotzer A, Neumann U (2004): GFP is the way to glow: bioimaging of the plant endomembrane system. *J Microsc* **214**:138–158.

Christensen AH, Quail PH (1996): Ubiquitin promoter-based vectors for high-level expression of selectable and/or screenable marker genes in monocotyledonous plants. *Transgenic Res* **5**:213–218.

Dale E, Ow D (1991): Gene transfer with subsequent removal of the selection gene from the host genome. *Proc Natl Acad Sci USA* **88**:10558–10562.

Darbani B, Eimanifar A, Stewart CN Jr, Camargo W (2007): Methods to produce marker-free transgenic plants. *Biotechnol J* **2**:83–90.

DeBlock M, De Brower D, Tenning P (1989): Transformation of *Brassica napus* and *Brassica oleracea* using *Agrobacterium tumefaciens* and the expression of the *bar* and *neo* genes in the transgenic plants. *Plant Physiol* **91**:694–701.

Ebinuma H, Sugita K, Matsunaga E, Endo S, Yamada K, Komamine A (2001): Systems for the removal of a selectable marker and their combination with a positive marker. *Plant Cell Rep* **20**:383–392.

El Ouakfaoui S, Miki B (2005): The stability of the Arabidopsis transcriptome in transgenic plants expressing the marker genes *nptII* and *uidA*. *Plant J* **41**:791–800.

Flavell RB, Dart E, Fuchs RL, Fraley RT (1992): Selectable marker genes: Safe for plants. *Bio/Technol* **10**:141–144.

Fraley RT, Rogers SG, Horsch RB, Sanders PR, Flick JS, Adams SP, Bittner ML, Brand LA, Fink CL, Fry JS, Gallupi GR, Goldberg SB, Hoffman NL, Woo SC (1983): Expression of bacterial genes in plant cells. *Proc Natl Acad Sci USA* **80**:4803–4807.

Herrera-Estrella L, De Block M, Messens E, Hernalsteen J-P, Van Montagu M, Schell J (1983): Chimeric genes as dominant selectable markers in plant cells. *EMBO J* **2**:987–995.

Jefferson RA, Kavanaugh TA, Bevan MW (1987): GUS fusions: β-glucuronidase as a sensitive and versatile gene fusion marker in higher plants. *EMBO J* **6**:3901–3907.

Mariani C, De Beuckeleer M, Truettner J, Leemans J, Goldberg RB (1990): Induction of male sterility in plants by a chimaeric ribonuclease gene. *Nature* **347**:737–741.

Miki B, McHugh S (2004): Selectable marker genes in transgenic plants: Applications, alternatives and biosafety. *J Biotechnol* **107**:193–232.

Ow D, Wood KV, DeLuca M, DeWey JR, Helinski DR, Howell SH (1986): Transient and stable expression of the firefly luciferase gene in plant cells and transgenic plants. *Science* **234**:856–859.

Reed J, Privalle L, Powell ML, Meghji M, Dawson J, Dunder E, Suttie J, Wenck A, Launis K, Kramer C, Chang Y-F, Hansen G, Wright M (2001): Phosphomannose isomerase: An efficient selectable marker for plant transformation. *In Vitro Cell Dev Biol-Plant* **37**:127–132.

Sanders PR, Winter JA, Zarnson AR, Rogers SG, Schaffner W (1987): Comparison of cauliflower mosaic virus 35S and nopaline synthase promoters in transgenic plants. *Nucleic Acids Res* **15**:1543–1558.

Stewart CN Jr (2001): The utility of green fluorescent protein in transgenic plants. *Plant Cell Rep* **20**:376–382.

Stewart CN Jr (2006): Go with the glow: Fluorescent proteins to light transgenic organisms. *Trends Biotechnol* **24**:155–162.

Stougaard J (1993): Substrate-dependent negative selection in plants using a bacterial cytosine deaminase gene. *Plant J* **3**:755–761.

Svab Z, Maliga P (1993): High-frequency plastid transformation in tobacco by selection for a chimeric *aadA* gene. *Proc Natl Acad Sci USA* **90**:913–917.

Waldron C, Murphy EB, Roberts JL, Gustafson GD, Armour SL, Malcolm SK (1985): Resistance to hygromycin B. *Plant Mol Biol* **5**:103–108.

Zuo J, Niu Q-W, Moller SG, Chua N-H (2001): Chemical-regulated, site-specific DNA excision in transgenic plants. *Nature Biotechnol* **19**:157–161.

Zuo J, Niu Q-W, Ikeda Y, Chua N-H (2002): Marker-free transformation: Increasing transformation frequency by the use of regeneration-promoting genes. *Curr Opin Biotechnol* **13**:173–180.

CHAPTER 10

Transgenic Plant Production

JOHN FINER and TANIYA DHILLON

Department of Horticulture and Crop Science, OARDC/The Ohio State University, Wooster, Ohio

10.0. CHAPTER SUMMARY AND OBJECTIVES

10.0.1. Summary

Foreign genes are transferred into plants primarily using one of two methods: *Agrobacterium tumefaciens*—or particle bombardment-mediated transformation, Agrobacterium-mediated transformation relies on a natural genetic engineer: the causal agent of crown gall disease in plants, which is one of the most intriguing stories in plant pathology particle bombardment-mediated transformation relies on accelerating DNA-coated microscopic particles into plant cells to deliver DNA to the genome. There are other less widely used methods and plenty of potential technological development opportunities to improve transformation efficiency. Finally, there are scores of methods that researchers have reported to be successful, but with little follow-up data, which might indicate that they are not very effective.

10.0.2. Discussion Questions

1. What is a transgene/transgenic plant?
2. What part or parts of the plant cell provide the most resistance to DNA introduction?
3. In the case of a successful DNA introduction, where in the target cell does the foreign DNA end up?
4. What are some differences between physical and biological methods for DNA introduction into plant cells?
5. What are some ways that the biological method for DNA introduction (*Agrobacterium*) has been improved over the years?
6. How is gene introduction performed with the model plant, *Arabidopsis*? Is this technique widely applied to other plants?
7. What are the size and composition of the particles that are used for the particle bombardment method?

Plant Biotechnology and Genetics: Principles, Techniques, and Applications, Edited by C. Neal Stewart, Jr.

8. How do the DNA integration patterns differ in plant cells, transformed via *Agrobacterium* and particle bombardment?
9. Can you think of additional methods for DNA introduction into plant cells?

10.1. OVERVIEW

Transgenic plants can be simply defined as plants that contain additional or modified genes that were introduced using specific physical or biological methods. The introduced DNAs or transgenes are typically very well defined and are precisely manipulated in the laboratory prior to delivery into the target plant cells. The methods for DNA introduction into plants cells are quite varied and are dependent largely on the plant selected for study and the background of the scientist performing the work. Over the years, tremendous efforts have been placed in development of gene introduction or "transformation" technology and, for many, if not most plants, the procedures have become almost routine. The efficiency of transgenic plant production is still being improved, and new methods for DNA delivery are still needed.

Although numerous methods have been developed for production of transgenic plants, *Agrobacterium* and *particle bombardment* are the two main methods used by most transformation laboratories. *Agrobacterium* has been called a "natural genetic engineer" and relies on this *biological* vector for transgene introduction. Particle bombardment is a *physical* method for DNA delivery and utilizes DNA-coated microscopic metal particles that are accelerated toward a suitable target tissue. Although emphasis in most laboratories has been placed on use of *Agrobacterium*, most of the acreage of transgenic crops represents the success of particle bombardment. Other procedures for DNA delivery do exist, and each has benefits and drawbacks. In order to better understand the challenges of producing transgenic plants and the overall process, one must first try to visualize DNA delivery to a single target plant cell and have a basic understanding of how to eventually recover a whole genetically engineered plant from that single targeted cell.

10.2. BASIC COMPONENTS FOR SUCCESSFUL GENE TRANSFER TO PLANT CELLS

10.2.1. Visualizing the General Transformation Process

Prior to the successful production of transgenic plants in the mid-1980s (Horsch et al. 1985), efforts to improve plants relied extensively on classical plant breeding through sexual hybridization and evaluation of spontaneous or induced mutations. Although plant breeding remains the foundation of plant improvement, a typical sexual cross results in the mixing of tens of thousands of genes and requires sorting through progeny to find the individuals with traits of interest. Through transgenic plant production, a single gene of interest can be introduced into a plant, improving a previously productive plant by a single, preselected gene or trait of interest. The basic concept is extremely simple; introduce a single gene into a single cell and generate a whole plant from that cell that will express the well-defined trait. The plant should be *exactly* the same as the starting material with the exception of the introduced transgene, which should impart a precise new and improved characteristic to the plant. So, how do you get a gene into the plant cell and target it to the nucleus? Also, where are the cells that need to be targeted; ones that can either give rise to whole plants directly or give rise to the pollen or egg (germ line) for successful transmission of the introduced DNA to progeny?

10.2.2. DNA Delivery

To consider DNA delivery into plant cells for the production of transgenic plants, the plant cell wall, cell membrane, and the nuclear membrane represent formidable barriers. The cell wall surface can be visualized as a stainless-steel scouring pad, with the steel fibers representing cellulose fibers. The cell wall (especially the young cell wall) has some level of flexibility to allow cell elongation and movement but is a fairly rigid structure, held together with cement of pectin and other crosslinking materials. Although there are "holes" in the cell wall, the plasmodesmata connect the protoplasm of adjacent cells and do not provide open access for DNA introduction. In order to deliver DNA across the cell wall, the cell wall must first be physically breached. This is also the case for passing DNA across the cytoplasmic membrane. Holes or breaks in the cell wall cannot be so severe that the target cell is irreparably damaged, but damage at some level must be done, to get a relatively large molecule of DNA into the cell. To complicate matters, plant cells are almost always hypertonic, which means that there is internal pressure pushing the cytoplasm against the cell wall, keeping plant tissues rigid. The pressure can be temporarily reduced or eliminated by lowering the osmotic pressure within the plant cells, causing the plant tissue to "wilt." By either drying the tissue or placing it on a medium containing sugars, the osmotic potential of the tissues can be temporarily lowered and DNA introduction efficiencies are then improved by reducing leakage of cytoplasm from holes or breaks in the cell wall (Vain et al. 1993).

Introduction of DNA into the cell is only part of the story as the nucleus is the desired destination in most cases. The chloroplast and mitochondria also contain genetic information and can take up and incorporate DNA, separately from the nucleus. However, we will focus on nuclear transformation here since it is the predominant mode to produce transgenic plants. So, how does the introduced DNA get to the nucleus, and what happens when it finally arrives? With the physical methods of DNA delivery, it appears that the DNA is actually delivered to an area either adjacent to the nucleus or into the nucleus itself. Naked DNA (introduced DNA is almost always uncoated and unprotected as opposed to native chromosomal DNA, which is specifically folded, organized, and coated with proteins) probably does not survive long outside the nucleus. For biological methods of DNA introduction, the DNA is naturally coated with proteins, which protect the DNA from degradation and escort it to the nucleus. Even if the introduced DNA reaches the nucleus, it is not precisely known what happens to this foreign DNA or how exactly it is incorporated into the plant genome. It appears that the natural machinery of the cell, which repairs, modifies, and replicates DNA, is involved with sewing the foreign DNA into the genomic fabric of the target cell. Regions of native DNA are constantly being stripped of protective proteins, unfolded, accessed, and reassembled. DNA can be tightly coiled and precisely ordered, but access to chromosomal DNA is needed for it to function. If foreign DNA is in the right place at the right time, it may slip into the reassembly process and become incorporated into the native DNA. Although presented here as a moderately haphazard process, foreign DNA must be precisely configured and introduced, show a necessary functionality, and behave as endogenous plant DNA in order to be retained.

10.2.3. Target Tissue Status

For successful production of transgenic plants, plant cells, which have the ability to grow (differentiate) into whole plants, should be targeted. The ability of a single cell to grow into a whole plant is called *totipotency*, and the cell that is naturally totipotent is the fertilized egg. Although it is probably true that all plant cells have the *potential* to grow into whole plants, that potential has not yet been reached for most cells. At this point in transgenic plant history, scientists can regenerate plants only from specific cell types in most plants. With a few plants, many different cell types are more easily manipulated to grow into whole plants through the tissue culture process (see Chapter 5). Successful production of genetically engineered plants is dependent on the coordination of DNA delivery with generation of a whole plant from the single cell, which is targeted for DNA introduction.

An ideal target would therefore be the fertilized egg or even the pollen that gives rise to the fertilized egg. Unfortunately, these ideal targets do not appear to be responsive for almost all plants with the exception of the model plant, *Arabidopsis thaliana*. The next most suitable target for DNA delivery might be the shoot meristem, which gives rise to the aboveground parts of the plant. Although the meristem has been successfully targeted for DNA introduction, it is a complex multicellular structure, and the most appropriate target cells are located in the center of the structure, buried under quite a few cell layers. Surface cells are obviously more accessible for DNA delivery. In the clear majority of cases, the target tissue used for production of transgenic plants consists of rapidly growing specialized plant cells, which have been induced to form whole plants. These cells should be physically accessible, actively dividing (DNA replication accelerates DNA integration into the genome), and able to give rise to whole plants. These cells should also be resilient enough to tolerate the breach of the cell wall and membrane by the DNA, which is truly an intrusive event in the life of a plant cell.

10.2.4. Selection and Regeneration

Because of the nature of DNA introduction, only a small percentage of plant cells can usually be successfully targeted. The clear majority of cells therefore just get in the way. How do scientists pick out the rare cell that contains the foreign DNA? For almost all transformation efforts, selection is the key. Along with the gene of interest, another gene, encoding resistance to an antibiotic or herbicide, is introduced (see Chapter 9 for a more thorough description). The mixture of transformed and nontransformed cells is then exposed to the antibiotic or herbicide, and only those cells containing the resistance gene will survive and grow. *Selection* refers to the ability of the transformed cells to proliferate in the presence of otherwise toxic selective agents. Resistance genes will encode for proteins that either detoxify a toxin or produce an alternate form of a target enzyme that is insensitive to the toxin. The most commonly used antibiotic resistance genes are neomycin phosphotransferase and hygromycin phosphotransferase, which provide resistance to the antibiotics kanamycin and hygromycin, respectively. The most commonly used herbicide resistance gene is the *bar* gene (sometimes referred to as the *pat* gene), which encodes for phosphinothricin acetyl transferase. This enzyme inactivates the herbicides glufosinate and bialaphos, which were originally discovered for their antibiotic properties. Selection for growth in the presence of toxic agents is the most common form of selection and is often called "negative selection" (as noted in the last chapter, the terminology used to describe selection of marker genes is variable among researchers).

Transformed cells can, however, be selected in other ways. "Positive selection" refers to the ability of a cell to survive by utilizing nutrient sources that are unavailable to nontransformed cells. As an example, a sugar such as mannose cannot be metabolized by most cells, unless mannose can be converted to the useful form of fructose, using a transgene that encodes phosphomannose isomerase. Cells containing this gene can grow on a medium containing mannitol as the sole carbon source, while the nontransformed cells will starve. A toxin is not required for selection. Another selection scheme utilizes genes that allow cells to be identified and physically isolated from nontransformed cells. Introduction of "reporter genes" allows scientists to identify transformed cells through a unique characteristic, such as a new color or emission of fluorescence or phosphorescence. Introduction of the gene encoding the green fluorescent protein imparts a fluorescent green color to plant cells, when viewed under high-energy blue or UV light. Cells or clusters of cells containing the green fluorescent protein can be visually detected and physically isolated (see previous chapter).

Once transformed cells have been recovered and purified from nontransformed cells, whole genetically engineered plants can be recovered through the tissue culture regeneration process (Chapter 5).

10.3. AGROBACTERIUM

Agrobacterium is a soilborne bacterium that has been rightfully called the "natural plant genetic engineer." Over its evolutionary journey, this bacterium has developed the unique ability to transfer part of its DNA into plant cells. The DNA that is transferred is called the *T-DNA* (for *t*ransferred *DNA*) and this DNA is carried on an extrachromosomal plasmid called the *Ti* (*t*umor-*i*nducing) plasmid. Through intervention of scientists, the Ti plasmid no longer causes tumor formation in infected plant cells, but the T-DNA region is still transferred. As opposed to DNA transfer methods that utilize direct uptake of DNA into plant cells, the use of *Agrobacterium* may appear to be more complex because two different biological systems (bacteria and the target plant cells) are involved. This might have been true in the early years of plant transformation, but today, *Agrobacterium* provides the method of choice for most plant transformation efforts. With methods utilizing introduction of DNA without a biological vector (direct DNA uptake), it appears to be necessary to deliver the DNA to the nucleus of the target cell, but with *Agrobacterium*, the T-DNA itself possesses the necessary signals for delivery there.

Most direct DNA introduction systems require expensive instrumentation, but *Agrobacterium* is simply prepared by growth on an appropriate medium and inoculated on the plant tissue. Additional claims of simpler foreign DNA insertions and more consistent transgene function in plants transformed with *Agrobacterium* may or may not be valid, and this appears to depend more on how the DNA is delivered with direct DNA introduction systems than on any inherent problem with the method. Considering primarily overall transformation efficiency, advances since the mid-1980s in our understanding of the *Agrobacterium*-mediated DNA transfer process have led to tremendous increases in efficiency and use of this transformation vector.

10.3.1. History of *Agrobacterium*

Agrobacterium tumefaciens is naturally occurring bacterium that causes a disease in plants called *crown gall*. Crown gall disease remains a problem with many horticultural plants, notably roses, grapes, and euonymus. The main symptom of the disease is a gall or tumor that forms on the cut stem (Fig. 10.1) or crown of a plant. The *crown* is the part of the plant that lies at the soil–air interface. This disease was a mystery to plant pathologists for many years as it does not always follow Koch's postulates, which specify that an extract from an infected organism should cause the disease when reinoculated on a healthy plant. Also, the tumors that were formed on plants would continue to grow in the absence of any microorganisms. For some time, the plant tumors were thought to be similar to some types of human cancer, but this was an incorrect assumption. Why do plant cells infected with wild-type *Agrobacterium* grow as a tumor?

Unraveling the mystery of the disease is a fascinating story in itself and has led to the use of the bacterium for genetic engineering research. For crown gall disease, wild-type *Agrobacterium* invade wounded tissues of dicotyledonous plants. The crown is a suitable entry point as the stem is often split or torn here. The bacteria may either colonize dead and dying cells or simply attach themselves to the outside of a wounded living cell (Fig. 10.2). Through a series of chemical signals that are sent from the plant cell to the bacterium, *virulence* genes are activated in the bacterium that cause the bacterium to enter its virulence mode. Some of the more important mechanisms are outlined in the next section. In the end, the T-DNA is excised from the bacterium and delivered to the genome of the target plant cell.

For wild-type bacteria, the T-DNA contains only a few genes, which encode for enzymes leading to the production of plant hormones and an opine, which is a nitrogen-rich organic compound that is a suitable food source for the bacterium. Tumors are formed as a result of hormone production in the plant cells, and the opines that are produced in the tumor are used by bacteria on the tumor or in the soil after being washed from the tumor by rain. The bacteria do not colonize living, dividing tumor cells, and these tumor cells can be grown in tissue culture without added hormones. Generation and analysis of some *Agrobacterium* mutants that contained disrupted hormone synthesis genes helped

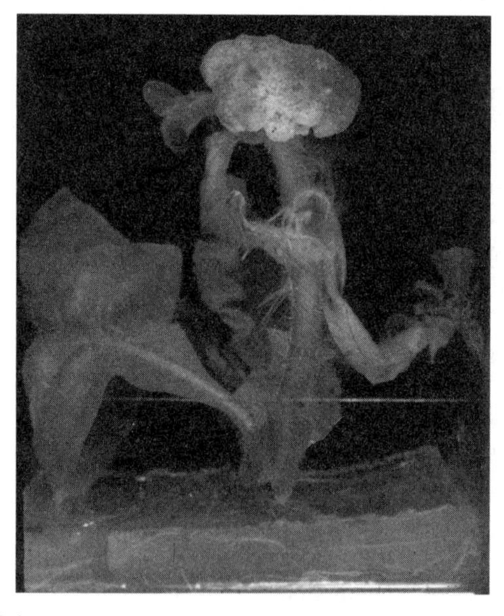

Figure 10.1. *Agrobacterium*-induced tumor formation on tobacco stem.

Figure 10.2. *Agrobacterium* growing on soybean tissue.

clarify parts of the story. If one of the hormone biosynthesis genes was disrupted and the bacterium was inoculated onto tobacco plants, the tumor would produce a mass of misshapen roots. If the other hormone biosynthesis gene were disrupted, a shooty tumor would result from tobacco inoculation. If both genes were disrupted, no tumor would form (Fig. 10.3). This hormone effect was suspiciously similar to results obtained with tobacco callus in tissue culture, and these different tumor phenotypes were correctly identified as resulting from an altered hormone balance in this tissue. Much of the research that showed the transfer of DNA from the bacterium to the plant cell and even speculation on the use of this process to improve plants was put forward by the "*Agrobacterium* Queen," Mary-Dell Chilton (see Life Box 1.2). Other contributions from Monsanto scientists and Marc van Montague and colleagues from Ghent University were also very significant.

10.3.2. Use of the T-DNA Transfer Process for Transformation

The transition from forming tumors on tobacco stems to routinely transforming wheat and corn with specific genes of interest resulted from multiple advances in the understanding of both the T-DNA process and the interaction of bacteria with plant cells. Since there are

Figure 10.3. Sunflower seedling hypocotyls inoculated with *Agrobacterium* without (a) and with (b) hormone biosynthesis genes.

many thorough reviews on the mechanism of *Agrobacterium* T-DNA transfer (Binns and Thomashow 1988; Zambryski 1992; Tzfira and Citovsky 2006), only the basic features as relating to transformation are presented here (Fig. 10.4). To start, the plasmid that is used as a vector for *Agrobacterium*-mediated transformation has been whittled down to contain only the essential components. *Agrobacterium* vectors are called "binary vectors" because they are the second of two plasmids that are involved in the overall process. Many of the transfer functions are retained on a modified Ti plasmid with the T-DNA removed. The second binary vector contains the T-DNA, but the hormone and opine biosynthesis genes have all been deleted. What is left on the binary vector, aside from the components that allow the plasmid to be retained in the bacterium, are the left and right "borders" of the T-DNA region. Genes of interest are cloned between the borders, which are recognition sequences for the T-DNA processing machinery. The genes for the T-DNA processing machinery are still located primarily on the modified Ti plasmid where they direct T-DNA processing on the binary plasmid. After *Agrobacterium* is inoculated on the appropriate plant tissue, the bacteria may recognize the target tissue as a suitable host; remember that this bacterium is a pathogen that infects plant tissue. Chemical signals are put out by both the plant tissue and the bacteria. Wounded plant tissues from appropriate plant tissues produce acetosyringone, which activates the bacterial *virulence* (*vir*) genes, which initiates the T-DNA transfer machinery. Not all wounded plant tissues produce acetosyringone, and the lack or poor production of acetosyringone by monocot cells originally made it difficult to impossible to produce transgenic monocots using *Agrobacterium*. Addition of synthetic acetosyringone to the inoculated plant tissues allows *Agrobacterium*-mediated transformation of monocots to proceed and tremendously enhances transformation of other moderately susceptible target plants. Once the bacteria infect plant tissue, most plants will respond by trying to fight off the invasion by either producing antipathogenic compounds or sacrificing cells adjacent to the

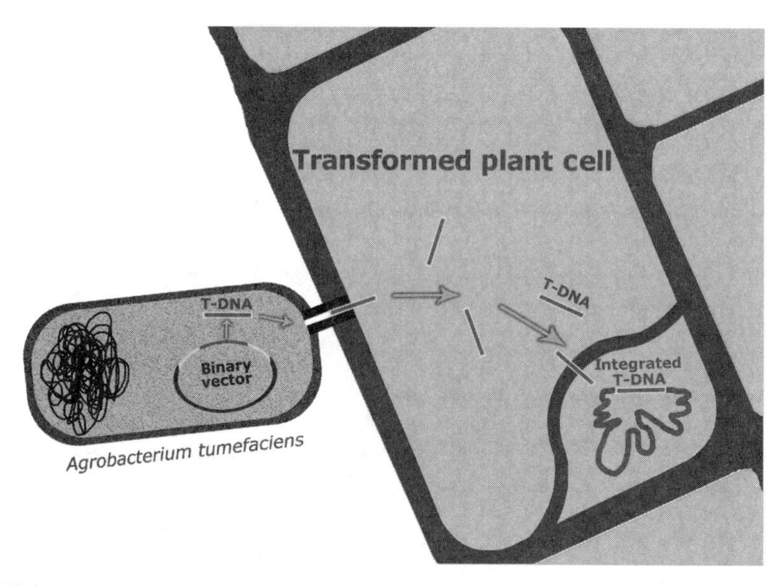

Figure 10.4. Schematic of *Agrobacterium*-mediated transformation of a plant cell, showing production of the T strand from the binary vector, transport through the bacterial pillus, and integration into plant chromosomal DNA. See color insert.

infected region to prevent spread of the invasion. Pathogens, in turn, have developed methods to introduce regulatory compounds into plant cells, in an attempt to shut down the defensive machinery of the target cell. Although some of these mechanisms are known, some are still being investigated, and a more thorough understanding of the infection process will allow further increases in the efficiency of *Agrobacterium*-mediated transformation.

Once the *vir* genes are activated, the T-DNA on the binary vector is processed for transport to the target plant cell. Some of the *vir* gene products excise the T-DNA from the binary plasmid as a single-stranded DNA molecule, while other *vir* gene products coat the T-DNA to prevent degradation. Yet additional *vir* gene products bind to the T-DNA to act as navigators or signals to direct the DNA out of the bacterium, through the plant cytoplasm, and to the nucleus. Through the action of other *vir* genes, the bacterium produces a *pillus*, which is the conduit for transfer of the T-strand (the single-stranded, coated, signal containing T-DNA is the "T-strand") from the bacterium to the target plant cell. The pillus is essentially a protein tube, which extends from the bacterium through the cell wall and into the cytoplasm of the target cell. After the T-strand is delivered to the nucleus, the last role of the signal protein on the T-strand is to find and nick the host DNA as an insertion point for the T-DNA. The T-DNA appears to insert primarily into gene-rich and transcriptionally active regions of DNA that are more exposed and accessible.

10.3.3. Optimizing Delivery and Broadening the Range of Targets

As more is learned about the mechanisms underlying *Agrobacterium*-mediated transformation of plant cells, the efficiency of the process will undoubtedly increase. The three main approaches for improving transformation are (1) increase delivery of the bacteria, (2) induce the *vir* genes, and (3) minimize defense responses of the target tissue.

Numerous methods have been developed to increase the delivery of the bacteria to the target plant tissue. Since the bacteria infect though wounded tissues, most of these methods strive to either increase overall wounding or call for precision wounding. The most common tool for wounding of the target tissue is the scalpel, which is simply used to excise plant tissues. When the tissue is cut, this presents a suitable binding/entry point for the bacterium. Wounding can be increased by scoring the target tissue multiple times, with a scalpel blade. Severe wounding of this sort eventually leads to a loss of the ability of the plant tissue to regenerate. Precision wounding using either sonication or particle bombardment (described later in this chapter) results in the generation of large numbers of extremely small wounds. Precision microwounding, if done properly, does not extensively damage the tissue structure and tremendously increases the number of entry points and attachment sites for the bacteria.

Induction of the *vir* genes through the addition of acetosyringone has led to routine transformation of plants that were initially not thought to be susceptible to *Agrobacterium*-mediated transformation. Although acetosyringone may not improve transformation of very susceptible plants (which already produce sufficient levels), it is routinely added during the coculture period for most other plants. *Coculture* is the time period during which bacteria are permitted to invade, infect, and transform plant cells. The coculture period ends when appropriate antibiotics are added, to eliminate the bacteria after their job is done. Results, similar to acetosyringone addition, can be obtained with the use of *vir* gene mutants, which were modified to be active in the absence of acetosyringone.

The method that may hold the most promise for future increases in efficiency of *Agrobacterium*-mediated transformation is to alter the response of infected plant cells to the bacterium. During the interaction between *Agrobacterium* and plant cells, elevated peroxidase activity and subsequent oxidation may cause tissue browning and cell death. Improvements have been made in transformation frequencies following the addition of reducing agents, which minimize the effects of oxidizing agents produced by infected plant tissues. The most commonly used agents are cysteine, silver nitrate, and ascorbic acid. In addition to reducing agents, enormous potential exists for using agents and genes that eliminate or reduce programmed cell death (PCD) in target tissues. Although PCD is a good natural defense mechanism for sequestering or localizing an infection and preventing spread, a reversal of this defense leads to higher transformation efficiency. Certainly, additional optimization strategies are possible, and the basic evaluation of compatible plant germplasm with different *Agrobacterium* strains is always the best place to start. Timing of coculture periods along with determination of media for optimum plant tissue growth is always important.

10.3.4. Agroinfiltration

Certain situations exist where rapid manipulation of gene expression is needed, but it is not necessary to transform a cell and take the time to recover a whole transformed plant. Why? In some cases, the effects of introducing a new gene or lowering the levels of expression of a native gene can be very quickly determined using *agroinfiltration* (or agroinfection). For agroinfiltration (Vaucheret 1994), *Agrobacterium* is injected or infiltrated into leaves of a suitable target plant, notably *Nicotiana benthamiana*, where large numbers of leaf cells are transformed. For this method, an *Agrobacterium* suspension is forced into the internal leaf airspace by tightly holding a syringe (without the needle) to the leaf and pushing the plunger. A variation of this method requires dipping the plant into an *Agrobacterium* suspension to wet the leaves and then applying vacuum to force the bacterium into the internal leaf airspace. To enhance the levels of gene delivery and spread, the T-DNA can be modified to contain viral gene components to launch the viral amplification and transfer machinery, making this method very efficient for production of transgene product (Fig. 10.5) in plants without transfer to the next generation.

10.3.5. Arabidopsis Floral Dip

Arabidopsis has become and remains the model for plant genomics. The genome and the plant itself are small, the generation time is rapid, and it is ridiculously easy to transform. The floral dip method was developed for *Agrobacterium*-mediated transformation of *Arabidopsis*, and no other plant currently responds similarly, even after extensive research efforts. Floral dip results in generation of independent transgenic seed, probably as a result of *Agrobacterium*-mediated transformation of the female gametophyte or the egg. For floral dip, *Arabidopsis* plants are grown to the flowering stage or just prior to flowering. The plants are simply immersed in a suspension of *Agrobacterium*, containing the wetting agent, Silwet®, a detergent that reduces surface tension and allows good access of the bacterial suspension to the cracks, crevices, and pores on the plant. After the dipping treatment, plants are maintained under high humidity for a few days and allowed to eventually flower and set seed. Since *Arabidopsis* produces so many seed and the plants are so small, seeds can be easily planted on selective media or seedlings/plants can be screened for a certain characteristic or phenotype to recover whole transgenic plants.

Figure 10.5. Agroinfiltrated *Nicotiana benthamiana* plants showing high levels of GFP expression. The aerial parts of the tobacco plant were submerged in an *Agrobacterium* suspension, and the plant was then placed under vacuum for infiltration. (Image provided by John Lindbo, Department of Plant Pathology, University of California-Davis.) See color insert.

It appears that bacteria proliferate or multiply at low levels and coexist within the tissues of the plant. They do not invade the cells of the plant as most other pathogens do, but they bind to suitable target cells for DNA delivery. If *Agrobacterium* transforms a somatic or vegetative cell within the plant, this is probably a terminal event. In the plant, a transformed leaf or stem cell will not give rise to anything other than another leaf or stem cell. For recovery of whole, transgenic plants, the goal is to target germline cells, cells that will contribute to the fertilized egg. For *Agrobacterium*-mediated transformation of *Arabidopsis*, the egg is targeted, leading to the production of transgenic seed. Usually each seed is from a different transformation event.

Why is *Arabidopsis* so easy to transform? Why don't corn and soybeans work the same way? (Soybean transformation has been very inefficient since the mid-1980s.) As we learn more about the transformation process, it may eventually be possible to recover transgenic corn and soybeans with the same ease as transgenic *Arabidopsis*. For now, inefficiencies in transformation remain both a mystery and a reason for transformation scientists to continue working on improvements.

10.4. PARTICLE BOMBARDMENT

10.4.1. History of Particle Bombardment

Particle bombardment refers to a method where heavy-metal particles (~1 μm gold or tungsten) are coated with DNA, accelerated toward the target tissue, and penetrate the cell wall to rest either adjacent to or directly in the nucleus. The DNA on the particles somehow finds its way to the native DNA of the target cell, where it becomes integrated into the chromosome to become a permanent addition to the genome.

The term *particle bombardment* can be used interchangeably with the similar terms *microprojectile bombardment*, *Biolistics®*, *particle acceleration*, and *gene gun technology.* The term that is currently most often used is *particle bombardment.*

As opposed to *Agrobacterium*, which is a biological vector for DNA introduction, particle bombardment is a purely physical method for DNA delivery. The DNA is physically precipitated onto metal particles, and those particles are then rapidly accelerated toward the target tissue. The particles penetrate through the cell wall by punching holes in that rigid structure and continue to do so until being stopped by the density of the target tissue. To visualize what is occurring, imagine bullets penetrating a thin piece of wood to enter water beneath. The wood is the cell wall and slows down the particles abruptly while the water gradually slows them down further until they stop. The analogy to bullets above is no coincidence.

Particle bombardment was invented by John Sanford and colleagues in the mid-1980s. The approach was further developed and optimized by Ted Klein (see Life Box 10.2), a postdoc at Cornell University in John Sanford's laboratory. Conceptually, a 22-caliber rifle, loaded with blanks, was first used to evaluate the damage to plant tissue from the shockwave resulting from an ignited powder load. The "gun" with "bullets" concept was further perpetuated with the introduction of the first commercial device, which used a 22-caliber powder load to generate a controlled explosion to accelerate small tungsten particles down the barrel of a modified gun. Between "shots," the particle bombardment device had to be cleaned with gun-cleaning swabs and brushes. Later versions of particle guns used other types of forces to generate the energy required to accelerate the small particles (Finer et al. 1999). The required violent forces, needed to accelerate the particles, could be created by generating high-voltage arcs across a gap or by using high-pressure air or CO_2. It was not unusual at the time to perform bombardments with muffling headsets, to dampen the sound from the early devices. Today, in most laboratories, high-pressure helium is used to generate the force needed to accelerate small gold particles (Fig. 10.6) toward the target tissue. DNA is first precipitated onto the particles, which are then placed as a monolayer on a Mylar carrier sheet, called a *macrocarrier* (this term refers to the structure that carries the

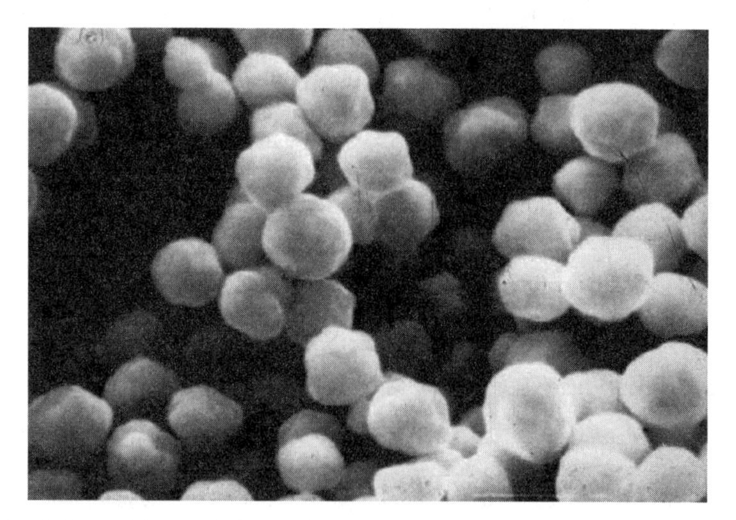

Figure 10.6. Gold particles used for particle bombardment, prior to DNA precipitation. Gold particles are more uniform and spherical than tungsten particles.

particles, while *microcarrier* was the term originally designated for the particles, as they are small and carry the DNA). The controlled explosion used to accelerate the macrocarrier is provided by high-pressure helium, which is released from a small chamber following the breakage of a ruptured disk, designed to break at specific pressures. The macrocarrier, with the particles on one side, travels a short distance and smashes into a screen, stopping the macrocarrier and allowing the particles to continue along their path. In most cases, the whole procedure is performed under partial vacuum, because the presence of air slows down the particles. A partial vacuum, applied for a short duration, does not appear to damage the biological targets.

Although there are numerous versions of particle bombardment devices (Fig. 10.7), they all utilize the same basic approach and are similarly patent-protected. The main manufacturer of particle bombardment devices is Bio-Rad, who offers two different versions of the device. While one version is a large, heavy, vacuum-utilizing, research lab unit, the other version is handheld and moderately portable. The handheld Helios device has received more attention for gene therapy work, while the large, benchtop unit is standard in most plant transformation laboratories. There is certainly a cost associated with all of these devices, which limits the use of particle bombardment for those academic research laboratories with insufficient resources.

10.4.2. The Fate of Introduced DNA

For DNA introduction using particle bombardment, DNA is first precipitated onto the particles using either calcium chloride or ethanol, which are commonly used for DNA precipitation. When the DNA precipitates, it sticks to whatever is at hand. It is unclear how "tightly" the DNA is bound to the particles, but it must be able to withstand the incredible force of acceleration and cell wall/cytoplasm penetration and also come off the particles after delivery.

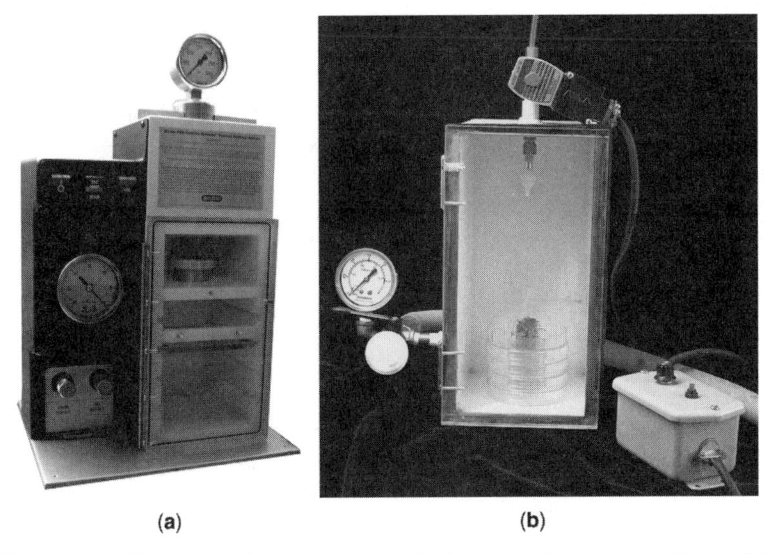

(a) (b)

Figure 10.7. Two different particle bombardment devices: (a) the commercially available PDS-1000/He (Bio-Rad) and (b) the noncommercial particle inflow gun (right).

During bombardment, the majority of the metal particles do not find their target. Most of the particles either embed in the cell wall, enter the vacuole, or end up somewhere else in the cytoplasm; only a few reach the nucleus. After all, thousands of particles are delivered using literally a "shotgun" approach. Evaluation of those cells that express the introduced DNA shows that the overwhelming majority of cells (>90%) have particles either adjacent to or in the nucleus (Fig. 10.8) (Yamashita et al. 1991; Hunold et al. 1994). Unlike *Agrobacterium*, where integration of the introduced DNA into the plant chromosomal DNA is orchestrated by bacterial proteins that are bound to the T-strand, particle bombardment results in the introduction of naked DNA. Clues to the fate of the introduced DNA can be taken from studying the final arrangement of the integrated DNA within plant chromosomal DNA.

In general, the patterns of DNA integration in the plant chromosome are very complex. To be more specific, it can be a real mess. Usually, the introduced DNA integrates into a single site (locus) on the chromosomal DNA. However, the introduced DNA can also integrate at multiple sites, which makes analysis more difficult. To complicate the situation further, it is common to obtain multiple copies of the transgene in each integration site. And (it gets worse) the copies can be partial copies, with varying orientations. In addition (last thing), the introduced DNA appears to be mixed or interspersed with plant genomic DNA (Pawlowski and Somers 1998). Imagine the replication and repair machinery of the nucleus as an army of overworked, frantic, multiarmed, DNA tailors. The DNA tailors are supposed to make exact copies of chromosomes and fix any small mistakes, while they are sewing huge amounts of new DNA strands. They are working fine until the whoosh of this huge boulder (1-μm particle) overhead that is carrying DNA. It looks like plant DNA, so they take what they can and use it in their sewing operation. It is not a perfect fit, but they are frantic and under the time constraints to get all of the chromosomal DNA replicated before the cell divides. For particle bombardment, it is unclear whether the particles actually physically break the chromosomal DNA or merely deposit DNA in the proximity of the replicating parts of chromosomes. It is clear that the introduced DNA can integrate into chromosomal DNA, with very complex patterns. However, complex

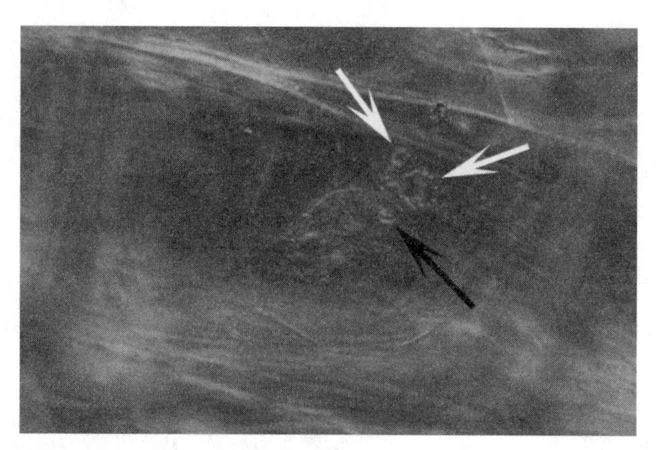

Figure 10.8. For successful transformation, particle bombardment results in the delivery of heavy-metal particles either next to (white arrows) or directly to the nucleus (black arrow). (Photo provided by Joseph Chiera, Department of Horticulture and Crop Science, OARDC/The Ohio State University.)

integration patterns can be largely controlled by manipulating the configuration of the introduced DNA (see next section).

10.4.3. The Power and Problems of Direct DNA Introduction

As particle bombardment is a physical method for DNA introduction, complications from biological interactions with the plant (as with *Agrobacterium*-mediated transformation) are avoided. A wide variety of plant tissues can be used as targets for particle bombardment. These range from embryos, seedlings, shoot apices, leaf disks, microspores, and immature pollen grains to potato tubers and nodes (Altpeter et al. 2005). Although the foreign DNA integration patterns (discussed above) can be very complex, this mechanism for DNA recombination and integration can be an advantage. Various DNAs can be mixed and cointroduced using a method called *cotransformation*. Reports of 12–15 different DNAs have been successfully cotransformed into soybean (Hadi et al. 1996) and rice (Chen et al. 1998). This is potentially useful for pathway engineering, where it is necessary to introduce multiple genes simultaneously.

Particle bombardment remains the only method that can be used for transformation of chloroplasts and mitochondria. Plastid transformation (Bock and Khan 2004) is useful in cases where large amounts of the transgene product are needed. The integration of foreign DNA into plastid DNA is also simple because integration events are less complex compared to nuclear transformation. For plastid transformation, the foreign DNA is precisely constructed so that it combines with similar sequences in the target plastid DNA, using *homologous recombination*. Another advantage of plastid transformation is that transgene escape via pollen is avoided since plastids are only maternally inherited in most plant species. However, like the floral dip method, this technique is currently limited to a small number of species.

In hand with the numerous merits of particle bombardment, there are certain drawbacks that limit the use of particle bombardment. The main perceived limitations are the randomness of DNA integration and the high copy number of introduced DNAs. As with most methods of DNA introduction, the position and orientation of the transgene in the plant chromosome will differ with every transformation event. The location of the transgene within the target chromosome will influence the expression of that gene. Transgenes in more active regions of genomic DNA will express at higher levels, while integration in less active areas will lead to lower expression: *position effects*. More importantly, the number of copies of introduced DNA can be incredibly high, leading to inactivity of the introduced DNAs (Taylor and Fauquet 2002). One might think that the presence of many copies of a particular transgene would result in very high expression. But expression of the transgene is often downregulated by the plant, a phenomenon known as *cosuppression*, *homology-dependent silencing*, *RNA interference* (RNAi), or *RNA silencing* (Zhong 2001; Butaye et al. 2005). Selection of plant cells or tissue showing uniform transgene expression is critical. Several techniques have been developed to minimize variation in transgene expression from particle bombardment. These methods are similarly applicable to other direct DNA introduction methods (discussed later in this chapter).

10.4.4. Improvements in Transgene Expression

Variation in transgene expression resulting from particle bombardment can be reduced to some extent by modifying the introduced DNA. Since high-copy-number integration

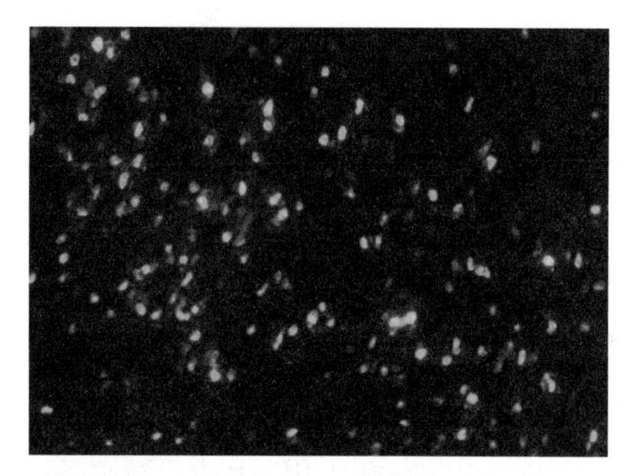

Figure 10.9. Particle bombardment-mediated transient GFP expression in lima bean cotyledonary tissues. This target tissue is flat, nonpigmented, and ideally suited for tracking GFP expression in individual transiently transformed cells. See color insert.

appears to lead to transgene silencing, any method of controlling copy number would lead to an improvement in consistency of transgene expression. To start, reducing the concentration of DNA appears to reduce the copy number of the transgene in the target cell. High concentrations of DNA are still used in many cases and are a remnant of early optimization strategies. Use of high concentrations of DNA results in high levels of *transient expression*, which is used to optimize DNA delivery conditions and involves the rapid expression of the some transgenes, often marker genes, which can be measured and quantified within 24 h of DNA introduction (Fig. 10.9). As a result of these levels of high transient expression, DNA concentrations that are much higher than necessary are often used for stable transformation studies. The beneficial effects of lower concentrations of DNA on stability of transgene expression should be evaluated for each different target tissue. Copy number of the introduced transgene can also be lowered by simplifying the form of the introduced DNA. Simple integration patterns result if a fragment of DNA containing only the gene of interest is used. When the backbone of the cloning plasmid is eliminated from the bombardment precipitation mix, this results in low copy transgene integration (Agrawal et al. 2005).

Transgene expression can also be stabilized using genes for certain viral proteins termed "suppressors of silencing." Suppressor proteins are produced by viruses to suppress the defense system of plants against viruses. After a virus invades a plant cell, the plants try to shut off or silence invading viral genes. The virus, in turn, evolved genes to turn off or suppress the silencing mechanism. Viral silencing suppressors can potentially be used to allow high levels of expression of transgenes through a similar mechanism. However, the effect of these viral proteins on normal plant development is still unclear.

10.5. OTHER METHODS

10.5.1. The Need for Additional Technologies

With the two main methods for DNA introduction, why are additional methods needed? Isn't this enough? In the scientific community (and for humanity in general), the theme is "bigger, better, stronger, faster." Certainly, plant transformation is achievable, and transgenic plants have been obtained using all of the plants of major economic importance. But efficiencies of existing methods can always be increased, and new methods may yield even higher transformation rates. Floral dip of *Arabidopsis* is very straightforward and efficient, but there is room for improvement in the recovery of more transgenics, and more importantly, application of this method to other plants would be quite useful. In addition, all of the methods that have been presented, including those that will be presented below, are protected by patents. The status of intellectual property drives much of plant biotechnology, and the methods for transgenic plant production are no exception. New transformation technologies will probably be protected by patents, but the availability of more choices is always beneficial (see Chapter 14). The additional technologies presented here do not represent a complete or thorough list. The methodologies are presented to provide a sampling of the types of ideas that have been generated since the dawn of transgenic plant production in the mid-1980s.

10.5.2. Protoplasts

For DNA introduction into plant cells, the cell wall represents the major barrier. When the plant cell wall is enzymatically or physically removed, protoplasts (Fig. 10.10) are the end result. Protoplasts are very fragile single cells that must be maintained in a osmotically and nutritionally balanced medium to prevent lysis. They are typically generated using enzyme mixtures of cellulases and pectinases to digest cell walls, and mannitol is often used to maintain the osmotic integrity of these naked cells. Protoplasts can be generated from many different types of tissue, but young leaf mesophyll tissues and embryogenic cultures are the most common. Although protoplasts can be manipulated in a number of

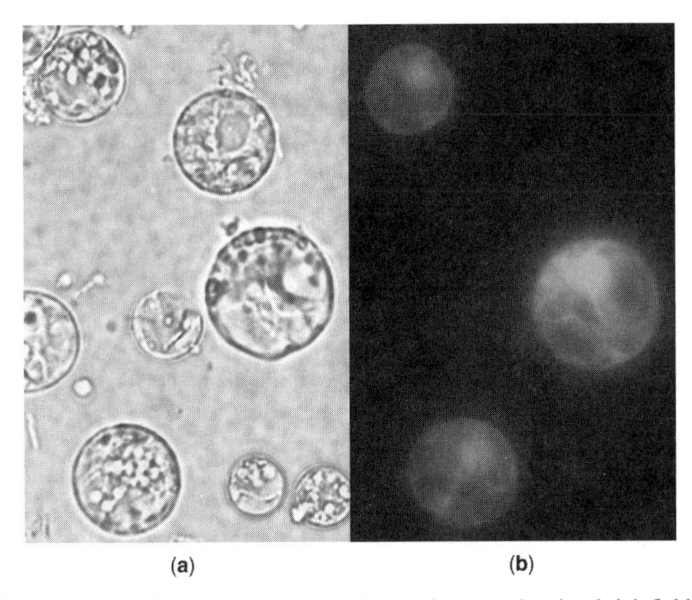

(a) (b)

Figure 10.10. Maize protoplasts, electroporated with a *gfp* gene, showing brightfield (a) and with GFP filters (b). (Illustrations provided by Pei-Chi Lin and JC Jang, Department of Horticulture and Crop Science, OARDC/The Ohio State University.) See color insert.

ways in the laboratory, they are most often used either for DNA introduction or to generate fusion hybrids.

The two main methods used to introduce DNA into protoplasts are electroporation and polyethylene glycol (PEG) treatment. For *electroporation*, protoplasts are placed in a DNA solution between two electrodes and exposed to brief pulses of high-voltage current. The pulses cause pores to form in the membrane and the DNA then enters the cells. *PEG treatments* are also performed in the presence of DNA and probably also result in the formation of pores, from membrane destabilization.

With their cell walls removed, protoplasts can be manipulated in additional ways that are not possible with intact plant cells. DNA can be introduced into protoplasts using microinjection, which is the most common method used for transformation of animal cells. Microinjection utilizes precisely drawn and cut glass needles, which will shatter if pushed into an intact plant cell. Surprisingly, protoplasts can also be very efficiently transformed using *Agrobacterium*. The bacteria are able to very effectively adhere to and transform protoplasts as the protoplasts are regenerating new cell walls.

Although protoplast transformation can be extremely efficient with more than 50% of the cells receiving DNA, tremendous problems are encountered when attempting to recover whole plants from these single cells. Whole transgenic plants have been recovered from a variety of plants using protoplast transformation, but this method is seldom used today for generation of transgenic plants. Because DNA introduction efficiency can be very high, transient expression in protoplasts is routinely used for analysis of factors that modulate gene expression (Sheen 2001).

10.5.3. Whole-Tissue Electroporation

Although electroporation can be used for very efficient transformation of protoplasts, application of electric pulses to whole tissues can also result in DNA introduction, although at reduced rates of efficiency (D'Halluin et al. 1992). With the cell wall intact, formation of pores in the cell membrane is of limited value for DNA introduction. Whole-tissue electroporation has been successfully used with rapidly growing tissues which contain thin, newly formed cell walls. Partial enzymatic digestion of whole tissues using cellulases and pectinases can remove enough of the cell wall to allow DNA introduction using electroporation of "intact" tissues.

10.5.4. Silicon Carbide Whiskers

Developed originally for DNA introduction into insect eggs, use of silicon carbide whiskers have been successfully applied for DNA introduction into plant cells (Kaeppler et al. 1990). Silicon carbide whiskers are long, rigid two-pointed microscopic "spears" that are added to plant cells and DNA and then vortexed. The spears or whiskers are approximately 1 μm thick and 15–50 μm long. Although the analogy of "being in a Jacuzzi with porcupines" has been used to describe this technology, the shaking motion is much more violent and is probably more closely akin to a paint mixer found in hardware stores. It is unclear whether the whiskers enter the cell with DNA as they are thrown about, or whether they penetrate the cells after being wedged between two cell clusters as they collide. The low efficiency of transformation using silicon carbide whiskers along with disposal under conditions similar to those for asbestos render this method unsuitable for most laboratories.

10.5.5. Viral Vectors

Since most plants can be infected by numerous viruses, viral vectors could potentially be used as another "natural" DNA introduction method for plants. Using their own transport mechanism, viruses can spread on their own throughout their host, so introduction of a virus into a single cell can eventually lead to the presence of virus genes in almost every cell of the inoculated plant. Although viral vectors can be used for extremely efficient introduction and transport of virus genes, these genes do not integrate into the genome of the host cell. Therefore, they will not be transmitted to the next generation through the pollen and egg.

However, inoculation of viruses into plant cells can be as simple as rubbing the leaf in the presence of the virus, and a single site of inoculation can lead to expression of viral genes in most cells of the plant (which is similar to production of a transgenic plant but is not quite the same). For successful introduction and expression, the gene of interest must be appropriately packaged in the viral genome, which tends to be less cooperative in accepting foreign DNA. Viral vectors are useful for very rapid production of proteins in plants without the need to generate a whole plant from a single, transformed cell.

10.5.6. Laser Micropuncture

For direct DNA introduction into plant cells, the use of microlasers continues with the theme of creating holes in the cell wall (Badr et al. 2005) for DNA delivery. This is perhaps one of the more elegant and least often utilized methods for DNA introduction into plant cells. Lasers are very precise in targeting certain cells, but the instrumentation required for this method is quite involved, and the number of cells that are targeted is very small. In contrast, for particle bombardment, the number of cells that transiently express an introduced transgene will be 5000 (higher on occasion) per shot. Many more cells are actually targeted—this is the number of cells that receive the DNA close to or in the nucleus and transiently express the introduced DNA. For laser micropuncture (and protoplast microinjection, discussed above), cells are targeted one at a time. It is doubtful that the use of microlasers for DNA introduction will increase tremendously, but it is a noteworthy method for DNA introduction into plant cells.

10.5.7. Nanofiber Arrays

Successful use of nanofiber arrays (Melechko et al. 2005) for DNA introduction into plant cells has not yet been consistently obtained, but convincing results have been demonstrated using animal cells (McKnight et al. 2003). Nanofiber arrays can best be described as a microscopic "bed of nails" (Fig. 10.11). Although not a new concept, the ability to precisely generate properly proportioned arrays is relatively new. Early attempts to generate nanofiber arrays resulted in the formation of nanoscale pyramid-shaped structures on a silicon chip (Hashmi et al. 1995). In this early work, the surface of the chips was precisely etched away, to leave the nanofiber pyramids. The newer arrays are composed of long, thin structures, and they hold much more promise for success with DNA introduction into plant cells. Nanofiber arrays are actually grown on chips, with very precise composition, height, and spacing possible. DNA can be chemically bound to the fiber or simply precipitated onto it. For successful DNA introduction into animal cells (McKnight et al. 2003), the arrays were stationary and the animal cells were propelled toward the chip. Cells were then allowed to grow, while still impregnated with fibers, on the chip. Although the cell wall

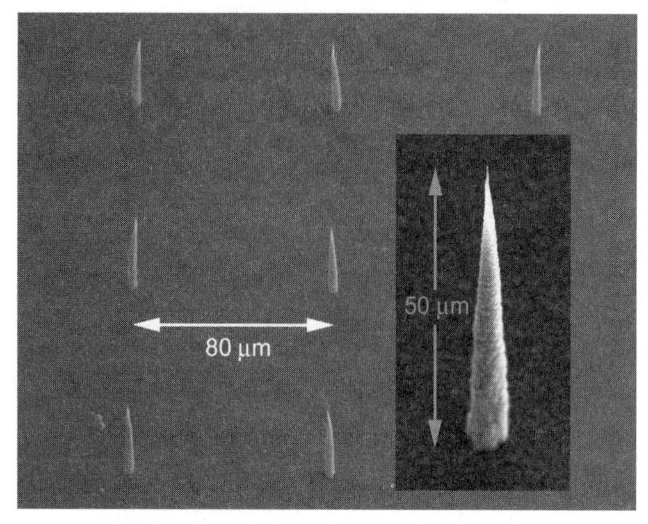

Figure 10.11. Nanofiber array with single fiber at higher magnification (inset). (Illustration provided by Tim McKnight, Oak Ridge National Laboratory, Oak Ridge, TN.)

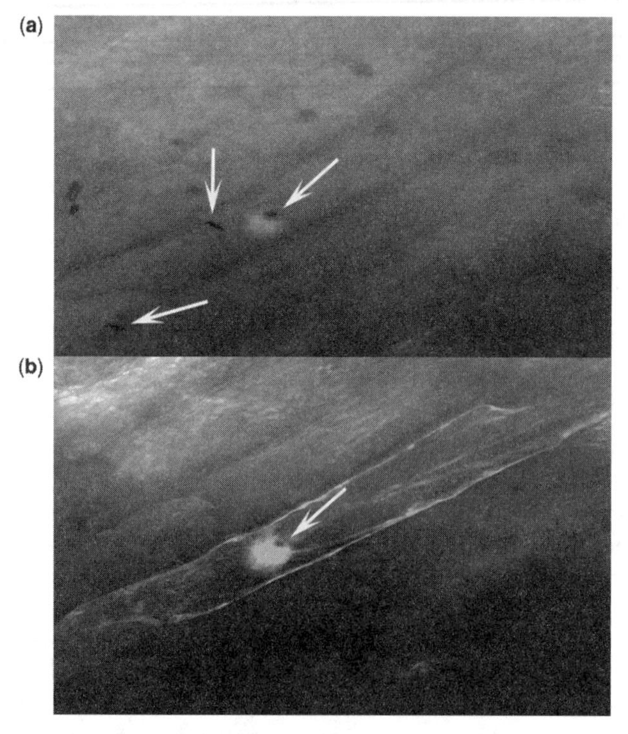

Figure 10.12. Nanofiber array introduction of DNA into onion cells. The white arrows in panel (a) point to dislodged nanofibers, while the arrow in panel (b) shows one fiber embedded in the nucleus of a cell, expressing an introduced *green fluorescent protein* gene. (Photos provided by Joseph Chiera, Department of Horticulture and Crop Science, OARDC/The Ohio State University and Tim McKnight, Oak Ridge National Laboratory, Oak Ridge, TN.)

is certainly much more of a barrier than the animal cell membrane, the fibers are sufficiently strong and rigid to penetrate the plant cell wall. Also, because the chip surface is covered with fibers, many cells can be targeted using a single chip. Early results with onion epidermal cells show the utility of this approach (Chiera and McKnight, unpublished; Fig. 10.12), but the high-efficiency delivery of DNA-coated nanofibers directly to the nucleus of multiple plant cells remains a challenge.

10.6. THE RUSH TO PUBLISH

10.6.1. Controversial Reports of Plant Transformation

As with most areas of the sciences, breakthrough technologies are highly prized and quite valuable. In addition to the notoriety that comes along with new discoveries, patent protection can provide a reasonable source of additional income, at least for university scientists. Truly new ideas in the sciences are actually not very common, and most of the advances that are reported in the scientific literature represent incremental improvements in preexisting technology or small steps in our understanding of processes. When something really new and novel does come along, it should be critically evaluated. Unfortunately, even in science, this does not always happen. As a result, there are numerous reports in the literature that initially cause quite a stir and then disappear because they did not work or worked with such a low efficiency that they were deemed impractical. The plant transformation literature is filled with reports like this. At the risk of alienating colleagues, some controversial reports of plant transformation methods are listed below.

10.6.1.1. DNA Uptake in Pollen. For one of the first reports of plant transformation (Hess 1980), pollen from a white-flowering petunia was soaked in DNA extracted from a red-flowering petunia. When this soaked pollen was used to pollinate the white-flowering petunia, some of the resulting seeds produced plants with either partially or fully red flowers. The author concluded that the DNA must have been taken up by pollen and passed onto the seedling from the fertilization process. The authors were cautious about the interpretation of their work and reached their conclusions of transformation as the most probable explanation of their results. They did the appropriate controls and noted that a small amount of red pigmentation could occur in white flowers at certain times of the year and in response to various stresses. Since that work was published over 25 years ago, no one has been able to repeat this work, even after extensive efforts. The tools to test for the presence of foreign DNA were not in existence at the time this work was done and the red flower color was the only evidence for transformation. The most plausible explanation for these results is pollen contamination, which the author discounted as they had never observed this with any of their controls.

10.6.1.2. Agrobacterium-Mediated Transformation of Maize Seedlings. Although *Agrobacterium*-mediated transformation of maize (Graves and Goldman 1986) is now fairly routine, this early report of *Agrobacterium*-mediated transformation of maize tissues remains quite controversial. At the time of this report, there were a few claims of *Agrobacterium*-mediated transformation of monocots and no reports for the economically important cereals. In addition to a scientific publication, this work led to the issuance of numerous patents. In this work, maize seedlings were wounded and inoculated with

Agrobacterium. Although transgenic plants were not recovered, the authors reported that the seedlings tested positive for the presence of opine synthase enzymes. These specific enzymes can be produced only after successful T-DNA transfer, and opine synthase analysis was one of the only tests for successful transformation at the time. This work was done before the optimization treatments, which were described earlier in this chapter, were even known. Moreover, the transformation efficiency was 60–80%, which is high by today's standards. If photocopies of this paper are inspected, inconsistencies cannot be detected. However, if the original paper is carefully examined, one can see that the differences between the control and experimental treatments disappear when the images showing opine production are digitally lightened or darkened to provide similar background levels. This paper is continuously referenced in the transformation literature, but it probably should not be.

10.6.1.3. Pollen Tube Pathway. The pollen tube pathway method (Luo and Wu 1988) for transformation is different from pollen transformation method (discussed above) as the pollen is not transformed, but the pollen tube is used as a vehicle for the delivery of DNA. The basis of this method is the inoculation of DNA into the hollow pollen tube, where it finds its way to the freshly fertilized egg for incorporation into the DNA of the young zygote. Timing was reported to be critical as the pollen is first placed on the stigma for germination. After the pollen tube grows down the style to the ovary, the stigma is severed, leaving a narrow hollow pollen tube as an open pathway to the fertilized egg. DNA is then inoculated onto the open pollen tube, where it was believed that capillary action drew the DNA in solution to the zygote. On the surface, this method appears to have some merit, but the pollen nuclei are the only things to enter the egg, and the fertilized egg or zygote has the same barriers as any other young plant cell, notably the cell wall. From the 1980s through the 1990s, there were many additional reports in the literature of the successful use of the pollen tube pathway for many different crops; almost all of these reports originate from China. Although it is very difficult to publish negative results, Shou et al. (2002) performed a very extensive study of the pollen tube pathway method in soybean and concluded that it was not reproducible. It appears that the pollen tube pathway method for DNA introduction is invalid. In the first published report (Luo and Wu 1988), transformation was confirmed using reliable molecular techniques but the patterns of DNA hybridization (see Section 10.6.2, later in this chapter) were a little unusual and may have been misinterpreted.

10.6.1.4. Rye Floral Tiller Injection. In this early report of plant transformation, young floral tillers of rye were injected with DNA carrying a kanamycin resistance gene (de la Peña et al. 1987). The authors speculated that the DNA was transported through the plant's vascular system to the germ cells, where it was taken up and incorporated. They suggested that the cells that ended up forming pollen were probably transformed with this injected DNA. The end result from floral tiller injection was the production of seeds carrying a kanamycin resistance gene. Molecular analysis seemed to show the presence of an intact gene in the rye DNA, but the most important results were only briefly described in the paper and presented as "data not shown." The "data not shown" term is used in situations where it may not be necessary to present data or images, but these data should have been presented for this work. In this paper, the authors also claim that the experiment was repeated (again, repeatability is expected for scientific reports) with similar results of recovery of transgenic rye plants. The authors write, "We are confident that this simple transformation procedure can be extended to other cereals," but this work was never even repeated with rye. It is unclear what exactly led the authors to their conclusions, but the idea of transporting DNA through the vascular system to target the male germ cells causes one to question the stability of the rye genome itself.

10.6.1.5. Electrotransformation of Germinating Pollen Grain. If the ideal transformation system were available, it would be pollen transformation (Smith et al. 1994). What could possibly be more convenient than simply introducing DNA into pollen and then pollinating a plant to generate transgenic seed? Here is yet another report of pollen transformation that has not been pursued or repeated in over 10 years. In this report, pollen from tobacco was germinated, washed, and subjected to electroporation. Although electroporation clearly works well for protoplasts and some actively growing plant tissues, it may have its limitations for stable DNA introduction into pollen. DNA in the growing pollen tube is not actively dividing and may not be receptive for foreign DNA. The authors report the optimization of DNA delivery through transient expression of gene activity, which is quite feasible as introduced DNA does not have to be incorporated into the host DNA to be functional. Transient expression in germinating pollen is described in this paper, along with molecular analysis of some of the recovered plants. The authors report that 40–70% of the surviving pollen (electroporation kills 35% of the pollen) displayed transient expression and that one-third of the 743 plants, which were eventually recovered, showed some activity from the transgene. This recovery rate is very high. Although proper molecular analysis of *one* plant appears valid, comparative analysis of more plants seems feasible and should have been presented, considering the large number of plants recovered. See the next chapter for transgenic plant analysis methods—we can see why these are so important in this section.

10.6.1.6. Medicago Transformation via Seedling Infiltration. Although a relatively unknown plant outside of the plant sciences community, *Medicago truncatula* has been presented as a "model" for legumes: the plant family, which includes alfalfa, peas, and all of the "beans" (soybeans, lima beans, green beans, etc.). As a legume model and potential counterpart to *Arabidopsis* (which is the unquestionable model for all plants), large amounts of resources were placed toward the development of comparable transformation technologies for *Medicago truncatula*. These efforts resulted in a publication describing the development of the *Arabidopsis* floral dip method for this plant (Trieu et al. 2000). Although most of the plant scientists on the planet have successfully used the *Arabidopsis* floral dip method, replication of the work described in this paper for this legume model have been nonexistent. Transformation efficiencies of 3–76% were reported, but it remains unclear to this day whether any transgenic plants were actually recovered. The appropriate molecular analyses were set up and are accurately presented in this paper but they were grossly misinterpreted. As opposed to the one plant analyzed from the pollen grain electrotransformation (discussed above), many different plants were analyzed in this report. The difficulty lies in the patterns of DNA hybridization that were presented in the paper. In most cases, hybridization patterns in transgenic plants should be unique; in this paper, most of the plants displayed the same single band (see Chapter 11 for details). The criteria to be considered in evaluating the success of transgenic plant production are not that complex. It is surprising that so many scientists are not fully aware of them.

10.6.2. Criteria to Consider: Whether My Plant Is Transgenic

Successful transgene introduction in plants can be confirmed in a number of different ways (see the following chapter). Validation of transformation is based on either the presence of foreign DNA in the plant genome or the expression of the transgene in the form of a new enzyme or protein. Few of these validation methods are reliable on their own; often analysis at a number of different levels is required. Some considerations for the main methods that are used to confirm the transgenic nature of transgenic plants are described below.

10.6.2.1. Resistance Genes. One of the most common methods for false confirmation of transgene expression is to evaluate plant tissues and seedlings for resistance to herbicides (any compound that is toxic to plant tissues). Although herbicide resistance genes are almost always used as a selective agent, the levels of herbicide used for selection are often at the lower end of toxicity. This means that there is the possibility of allowing escapes, which may not contain the transgene but could still survive in the presence of the herbicide. It is rare that transformation experiments give rise to plant tissue and plants that either grow unaffected or die in the presence of the herbicide. In most cases, the recovered tissues show some yellowing or browning, indicating slight toxicity effects. The ability for plant tissue to survive in the presence of toxic agents depends on the density and vigor of the plant tissue, the medium used for growth of the target cells, and the stability of the selective agent. Some selection systems that have been thoroughly worked out and optimized may be very trustworthy. However, growth of tissues or seedlings on selective media is not enough to confirm the presence and expression of a herbicide resistance transgene.

10.6.2.2. Marker Genes. Expression of marker genes results in the direct or indirect formation of a product that can be either chemically analyzed or visually confirmed. The most common marker genes are those that can be visualized. The presence of the β-glucuronidase (GUS) enzyme encoded by the *uidA* gene is analyzed by placing the plant tissue in the presence of an artificial substrate that is broken down by the enzyme to yield a blue product. When the GUS enzyme is present, the tissues expressing the transgene will turn blue. Often, the blue product is difficult to see in green plant tissues. The chlorophyll can be removed from the tissue after treatment, for clarification. If the solution containing the artificial substrate is incorrectly modified or the plant tissue is incubated for too long, everything can turn blue, leading to false-positive results.

Another commonly used marker gene encodes the green fluorescent protein (GFP), which emits a fluorescent green light, if the tissue expressing the gene is illuminated with UV or high intensity blue light (Figs 10.5, 10.9, 10.10, 10.12). Special instrumentation is needed to detect GFP, and filter sets are required. If black lights or UV lamps are used without filter sets, detection of this fluorescent protein is impossible unless the amounts of GFP protein are quite high (Fig. 10.5). The main problem in detecting GFP in plants is the presence of other plant compounds that either interfere with detection or fluoresce themselves. For example, chlorophyll fluoresces bright red under UV or strong blue light. Waxes, materials in leaf hairs or trichomes, and even dirt on the leaves can fluoresce a similarly to GFP, and some filter sets can make everything resemble GFP expression. The presence of the appropriate color for these marker genes must be carefully evaluated and then compared with an expected pattern for gene expression for the most accurate results.

10.6.2.3. Transgene DNA. Ultimately, the transgenic nature of a plant relies on the detection of the new transgene through DNA analysis. In some cases, DNA analysis has become so sensitive that small amounts of contaminants in the laboratory can yield false-positive results. Use of the polymerase chain reaction (PCR) must be cautiously weighed as false positives are so common with this method. In addition, PCR does not test for the integration of the transgenic DNA, only its presence in the sample. So, if there is some DNA on the leaves from an adjacent plant or the *Agrobacterium* remains in/on the plant, there will be a positive signal. PCR is a great screening tool in the laboratory, but PCR results should never be presented as proof of transformation.

The best method for molecular analysis of integrated transgenic DNA is Southern hybridization analysis (see Chapter 11 for details). Many publications present Southern blots showing the same-sized band for all clones. If enzymes are used that cut a fragment out of the transgene, a single band will be generated. A single band will also be generated if the starting DNA is from a bacteria or DNA that is contaminating the sample. If a restriction enzyme is used that cuts the foreign DNA at only one location, it will also cut somewhere in the plant DNA, producing different sized fragments from each different transformation event. More bands are typically generated from plants obtained using direct DNA introduction, while *Agrobacterium*-mediated transformation yields fewer and less complex banding patterns. Regardless of the method for DNA introduction, the presence of unique band sizes and band numbers should be used to confirm transgene integration resulting from each different transformation event. It is also important to analyze the progeny of putatively transgenic plants. A transgenic plant should pass the transgene on to progeny with Mendelian-expected frequencies. Non-Mendelian inheritance of transgenes suggests problems at some level.

10.7. A LOOK TO THE FUTURE

In the early days of transgenic plant production, the major difficulty was the actual production of transgenic plants. As transformation science progressed, the procedures for gene delivery, gene selection, and transgenic plant production became more standardized for most plants. Transformation systems for even the most difficult to transform plants can now be termed, "consistent but inefficient." This means that, if you know what you are doing, you can count on the production of a few transgenic plants for each experiment. Many plants that used to be difficult to transform are no longer even considered "difficult." So, for many plants, transformation is no longer limiting and the analysis of transgenics is the new bottleneck. Can we even analyze fewer plants if we eliminate the variation in transgene expression by developing more reliable methods to introduce the transgene into exactly the same spot (in the genome) each time? Can we follow the lead of the automotive industry by automating more of the process? Transformation science, as with science in general, moves forward through the systematic optimization of known systems and the discovery of new approaches. Hopefully, one of the young scientists reading this chapter will take the lead to optimize or develop a new transformation technology that will eliminate one or more of the remaining bottlenecks in transgenic plant production.

LIFE BOX 10.1. MAWD HINCHEE

Maud Hinchee, Chief Technology Officer, ArborGen, LLC

Maud Hinchee

I certainly did not plan on being a plant biologist. However, my mother always thought I would be a botanist, because I eliminated her eggplant yield from her backyard garden by sterilizing the flowers without her knowing it (I hated eggplant). It wasn't until I took a college course that captured my imagination that I decided to be come a botanist. The class was Plant Anatomy, which in some universities can be quite dry. However, this course was taught by Dr. Tom Rost at the University of California, Davis—a young professor who taught using an experimental approach to understanding the form, structure and function of plant cells, tissues and organs while allowing us to appreciate the esthetic beauty of plant cells. As a somewhat artistic type, I liked this blend of scientific discovery and microscopic art. I went on to receive my B.S. degree in Botany from the University of California, Davis, (UCD) in 1975 and then my M.S. in Botany from the University of Washington (UW). In my undergraduate research, I studied the development of roots. At UW, I compared and contrasted the anatomical and growth characteristics between aerial and soil roots of *Monstera deliciosa*. I returned to UCD where I received my Ph.D. degree in 1981 in plant morphogenesis. My project was to determine what effect the cotyledons of pea had on the development and distribution of lateral roots in young seedlings.

How these various research projects enabled me to become a plant biotechnologist is probably a matter of being in the right place at the right time. Researchers were just starting to make some headway in developing methods for inserting genes into plant cells. Since so little was known at that time as to what controlled which cells successfully incorporated DNA, and which of these cells subsequently could develop into a whole plant, I was able to provide valuable insights to the process as a plant morphogeneticist. I did my first training in plant transformation techniques during a postdoctoral research associate position at the Hawaiian Sugar Planters' Experiment Station in Hawaii, working on the incorporation of DNA into sugarcane protoplasts. I then was hired at Monsanto Co, in St. Louis, MO in 1982. My first role was determine why regeneration and transformation experiments in soybean weren't leading to the expected results. This activity provided me much insight into the cellular basis for the regeneration process and allowed me to design methods to specifically target our genetic engineering tool, *Agrobacterium*, to the right cells at the right time. The result was a successful and reproducible soybean transformation protocol that yielded the first transgenic soybean containing the Roundup Ready gene. Today, 90% of the soybeans grown in the United States have this trait. It gives me great pride still to drive by a soybean field that is clean of weeds and think of the provided to farmers that is due in some small way to my research efforts. Working at Monsanto was the greatest learning experience of my life. Besides the opportunity to develop transformation methods for a variety of crops that included sugarbeet, flax, potato, strawberry, cotton and sweetpotato, I also learned how a biotech product was "built" from the ground up—from conception of the gene construct all the way through to regulatory approval of a transgenic plant. Another rewarding experience I had in this time was leading a team of Monsanto and African scientists to develop virus-resistant sweetpotato for subsistence farmers in Kenya and other parts of Africa. All this experience served me well, in my next role as a the technical lead for a business team dedicated to developing biotech collaborations in specialty crops world wide in crops such as forestry, sugarcane and fruits and vegetables.

I left Monsanto in 2000 to become the Chief Technology Officer of ArborGen LLC, a forestry biotechnology company which currently develops genetically improved planting stock for the pulp, timber and bioenergy industries. As much as I enjoyed my time at Monsanto, I truly enjoy the ability I have now to help guide a young company towards successful product development. I foresee in the future that transgenic technologies will play a very important role in sustaining our environment by providing solutions to the worsening energy crisis. ArborGen will be marketing trees that will require a relatively small land "footprint" due to its high productivity and which can supply a renewable and sustainable source of biomass for the production of cellulosic ethanol. A woody biomass feedstock for biofuel production will help enable the United States and other countries to lessen their dependence on the world's dwindling petroleum supply. I am proud of what I have contributed to in crop biotechnology in the past, but I believe that transgenic industrial crops may provide some of the greatest benefits in the future.

LIFE BOX 10.2. TED KLEIN

Ted Klein, Senior Scientist, Pioneer Crop Genetics Research, DuPont Agriculture & Nutrition

Ted Klein

When asked how I decided on a career in plant molecular biology, I often answer by saying that even as a student in high school I knew that I wanted to become a soybean genetic engineer. Given that I graduated DeWitt Clinton High in the Bronx in 1972, this is a highly unlikely scenario. Of course, I am trying to make the point that it is very difficult to predict the course of one's career. I would never have predicted that I would be involved with breakthrough science that changed the course of agriculture.

Biology was my real focus in high school and I truly enjoyed learning about the intricacies of organisms. I went on to attend McGill University in Montreal and was fortunate to major in Plant Science at the agriculture campus (Macdonald College). My thought was that the most important and practical aspects of biology were related to agriculture and plant development and that I would pursue a career in this area. I was drawn to learning about the interactions between organisms, especially those between plants and microbes. I found the courses in plant pathology, microbiology and microbial ecology particularly interesting. Soil seemed to be where the real action was. I went on to do graduate work at Cornell University with Martin Alexander, the noted soil microbiologist. My research focused on aspects of the nitrogen cycle and the organisms responsible for converting ammonium to nitrate in acid environments. As I was finishing my degree, my goal was to continue on in microbial ecology and hopefully obtain a faculty position after a postdoc. However, I had the good fortune of meeting John Sanford and learned about his concepts for genetic engineering of crops. John worked at the New York Agricultural Experimental Station in Geneva, about 50 miles from Ithaca. Driving home to Ithaca after our meeting, I was convinced that he was on to something totally new and extremely exciting.

For the next three years, I worked with John on the development and implementation of the gene gun for DNA delivery to cells and tissues. Our process evolved from using a real gun (air pistols and rifles) to a specially designed apparatus fabricated at Cornell's Submicron Facility with Nelson Allen and Ed Wolf. We tried to deliver small tungsten particles into anything that wouldn't move (onions, paramecia, *Drosophila* eggs). This was before simple reporter genes (such as GUS) with strong plant promoters were available. Eventually with the help of Ray Wu, we were able to bombard onion cells and show that genes could be delivered and expressed. At that time, the goal of a number of labs was to introduce genes into important crop species such as corn, rice and soybean. We went on to collaborate with scientists at Pioneer to show that maize cells could be transformed. After working with John, I decided to do additional postdoctoral work at the Plant Gene Expression Center in Albany, California with Mike Fromm. These were exciting times with the gene gun being applied to a number of important biological questions. We were able to directly deliver DNA into intact tissues to study transcription factors, phytochrome regulation of gene expression, and tissue specific expression. We were also able to stably transform maize, an important breakthrough for agriculture.

The gene gun is now an accepted tool in biological research with many applications in animal cell biology. Virtually all of the transgenic corn and soybean grown by farmers was engineered with the gene gun. So as should be apparent, it is very difficult to predict the course of one's career.

REFERENCES

Agrawal PK, Kohli A, Twyman RM, Christou P (2005): Transformation of plants with multiple cassettes generates simple transgene integration patterns and high expression levels. *Mol Breed* **16**:247–260.

Altpeter F, Baisakh N, Beachy R, Bock R, Capell T, Christou P, Daniell H, Datta K, Datta S, Dix PJ, Fauquet C, Huang N, Kohli A, Mooibroek H, Nicholson L, Nguyen TT, Nugent G, Raemakers K,

Romano A, Somers DA, Stoger E, Taylor N, Visser R (2005): Particle bombardment and the genetic enhancement of crops: myths and realities. *Mol Biol* **5**:305–327.

Badr YA, Kereim MA, Yehia MA, Fouad OO, Bahieldin A (2005): Production of fertile transgenic wheat plants by laser micropuncture. *Photochem Photobiol Sci* **4**:803–807.

Binns AN, Thomashow MF (1988): Cell biology of *Agrobacterium* infection and transformation of plants. *Annu Rev Microbiol* **42**:575–606.

Bock R, Khan MS (2004): Taming plasmids for a green future. *Trends Biotech* **22**:311–318.

Butaye KMJ, Cammune BPA, Delauré SL, De Bolle MFC (2005): Approaches to minimize variation of transgene expression in plants. *Mol Breed* **16**:79–91.

Chen L, Marmey P, Taylor NJ, Brizard JP, Espinoza C, D'Cruz, Huet H, Zhang S, de Kochko A, Beachy RN, Fauquet CM (1998): Expression and inheritance of multiple transgenes in rice plants. *Nat Biotechnol* **16**:1060–1064.

de la Peña A, Lörz H, Schell J (1987): Transgenic rye plants obtained by injecting DNA into young floral tillers. *Nature* **325**:274–276.

D'Halluin K, Bonne E, Bossut M, De Beuckeleer M, Leemans J (1992): Transgenic maize plants by tissue electroporation. *Plant Cell* **12**:1495–1505.

Finer JJ, Finer KR, Ponappa T (1999): Particle bombardment mediated transformation. In *Plant Biotechnology: New Products and Applications*, Hammond J, Mc Garvey PB, Yusibov V (eds). Springer-Verlag, Heidelberg, pp 59–80.

Graves ACF, Goldman SL (1986): The transformation of *Zea mays* seedlings with *Agrobacterium tumefaciens*. *Plant Mol Biol* **7**:43–50.

Hadi MZ, MD McMullen, JJ Finer (1996): Transformation of 12 different plasmids into soybean via particle bombardment. *Plant Cell Rep* **15**:500–505.

Hashmi S, Ling P, Hashmi G, Reed M, Gaugler R, Trimmer W (1995): Genetic transformation of nematodes using arrays of micromechanical piercing structures. *BioTechniques* **19**:766–770.

Hess D (1980): Investigations on the intra- and interspecific transfer of anthocyanin genes using pollen as vectors. *Z Pflanzenphysiol* **98**:321–337.

Horsch RB, Fry JE, Hoffmann NL, Eichholtz D, Rogers SG, Fraley RT (1985): A simple and general method for transferring genes into plants. *Science* **227**:1229–1231.

Hunold R, Bronner R, Hahne G (1994): Early events in microprojectile bombardment: Cell viability and particle location. *Plant J* **5**:593–604.

Kaeppler HF, Gu W, Somers DA, Rines HW, Cockburn AF (1990): Silicon carbide fiber-mediated DNA delivery into plant cells. *Plant Cell Rep* **9**:415–418.

Luo Z, Wu R (1988): A simple method for the transformation of rice via the pollen-tube pathway. *Plant Mol Biol Rep* **6**:165–174.

McKnight TE, Melechko AV, Griffin GD, Guillorn MA, Merkulov VI, Serna F, Hensley DK, Doktycz MJ, Lowndes DH, Simpson ML (2003): Intracellular integration of synthetic nanostructures with viable cells for controlled biochemical manipulation. *Nanotechnology* **14**:551–556.

Melechko AV, Merkulov VI, McKnight TE, Guillorn MA, Klein KL, Lowndes DH, Simpson ML (2005): Vertically aligned carbon nanofibers and related structures: Controlled synthesis and directed assembly. *J Appl Phys* **97**:41301.

Pawlowski WP, Somers DA (1998): Transgenic DNA integrated into the oat genome is frequently interspersed by host DNA. *Proc Natl Acad Sci USA* **95**:12106–12110.

Sheen J (2001): Signal transduction in maize and Arabidopsis mesophyll protoplasts. *Plant Physiol* **127**:1466–1475.

Shou H, Palmer RG, Want K (2002): Irreproducibility of the soybean pollen-tube pathway transformation procedure. *Plant Mol Biol Rep* **20**:325–334.

Smith CR, Saunders JA, VanWert S, Cheng J, Matthews BF (1994): Expression of GUS and CAT activities using electrotransformed pollen. *Plant Sci* **104**:49–58.

Taylor NJ, Fauquet CM (2002): Microprojectile bombardment as a tool in plant science and agricultural biotechnology. *DNA Cell Biol* **21**:963–977.

Trieu AT, Burleigh SH, Kardailsky IV, Maldonaldo-Mendoza IE, Versaw WK, Blaylock LA, Shin H, Chiou T-J, Katagi H, Dewbre GR, Weigel D, Harrison MJ (2000): The transformation of M*edicago truncatula* via infiltration of seedlings or flowering plants with *Agrobacterium*. *Plant J* **22**:531–541.

Tzfira T, Citovsky V (2006): *Agrobacterium*-mediated genetic transformation of plants: Biology and biotechnology. *Curr Opin Biotechnol* **17**:147–154.

Vain P, McMullen MD, Finer JJ (1993): Osmotic treatment enhances particle bombardment-mediated transient and stable transformation of maize. *Plant Cell Rep* **12**:84–88.

Vaucheret H (1994): Promoter-dependent trans-inactivation in transgenic tobacco plants; kinetic aspects of gene silencing and gene reactivation. *CR Acad Sci* **317**:310–323.

Yamashita T, Lida A, Morikawa H (1991): Evidence that more than 90% of β-glucuronidase-expressing cells after particle bombardment directly receive the foreign gene in their nucleus. *Plant Physiol* **97**:829–831.

Zambryski PC (1992): Chronicles from the *Agrobacterium*-plant cell DNA transfer story. *Annu Rev Plant Physiol Mol Biol* **43**:4645–4690.

Zhong G-Y (2001): Genetic issues and pitfalls in transgenic plant breeding. *Euphytica* **118**:137–144.

CHAPTER 11

Transgenic Plant Analysis

JANICE ZALE

Department of Plant Sciences, University of Tennessee, Knoxville, Tennessee

11.0. CHAPTER SUMMARY AND OBJECTIVES

11.0.1. Summary

Once transgenic plants are produced, they must be analyzed using methods common to molecular biology. The transgene might initially be probed by using the polymerase chain reaction, but eventually DNA hybridization (Southern blot) analysis must be performed to assess transgene integration. To gauge transgene expression, RNA hybridization (northern blot) analysis is done, but real-time reverse transcriptase PCR is commonly used. So, Southern blot tells whether and how many copies of the DNA transgene are integrated into the genome, and northern blot is useful for determining how much of the transgene is transcribed. However, analysis eventually centers around how much transgenically expressed protein is produced: using western blot analysis or enzyme-linked immunosorbent assay (ELISA).

11.0.2. Discussion Questions

1. Through what type of chemical bond does the complementary probe bind to the nucleic acid?
2. Nucleic acids and proteins are separated according to size in agarose and sodium dodecyl sulfate–polyacrylamide gel electrophoresis (SDS-PAGE) gels, respectively. Why do both types of macromolecules migrate toward the anode in an electrical current?
3. What is gene expression, and how can you measure it?
4. Explain why phenotypic data provide evidence of transformation but not conclusive proof of transformation.
5. What factors are most important when designing a Southern blot experiment to test for transgenic status?

Plant Biotechnology and Genetics: Principles, Techniques, and Applications, Edited by C. Neal Stewart, Jr.

11.1. INTRODUCTION

After a particle bombardment or *Agrobacterium*-mediated transformation experiment to generate transgenic plants is conducted (see Chapter 10), several tests must be conducted to determine whether the experiment yielded transgenic plants and to determine a host of other features about the plants. In the perfect world, an experiment will yield hundreds of *transgenic events* from which to choose, and at least one event with a single insertion and high level of gene expression. The unique placement of the transgenic construct within the plant genome constitutes a transgenic event, along with the progeny of that plant. Therefore, each transformation event is said to be *independent* of the other events. Of equal importance is the fact that the phenotype of the transgenic plant will be changed only in the way expected. Independent transformants (or transgenic events) must be identified among the nontransformed plants and the validity of each transformation event established (e.g., Are my plants really transgenic?). This is accomplished through a series of physical, phenotypic, and genetic analyses (Birch 1997; Potrykus 1991). New tools have been developed to validate transformation events. Throughout this chapter, we will explore the various tests biotechnologists use to analyze transgenic events.

11.2. DIRECTIONALLY NAMED ANALYSES: AS THE COMPASS TURNS

A number of requirements must be met before declaring that a plant is transformed, and the best way to do this is through molecular analyses. The analyses that molecular biologists have devised to characterize segments of DNA, RNA, and protein in organisms have unique names that are often confusing to a newcomer in the field because they are in no way descriptive of the molecules described: a secret nomenclature of molecular biologists. In 1975 Edwin Southern at the University of Oxford described a way to transfer DNA to a membrane and then probe it with homologous DNA of interest (a *probe*); hence *DNA gel blot* analysis is also known as *Southern blot* analysis (Southern 1975). Southern blot analysis is performed to demonstrate that the transgene is physically integrated into the genome, usually nuclear, which also determines the number of *insertions* or *copy number* of transgenes. It also shows whether the entire gene was integrated into the plant genome. A truncated or rearranged copy will not be functional. After determining that a transgene is stably integrated, the next step is often to see if its specific mRNA is produced. There are several ways to do that. *RNA gel blot analysis*, also known as *northern blot*, is the conventional method. This test is not named after a person, but is a cute derivative of its seemingly opposite direction on a compass. Not that RNA is opposite of DNA, but in a moment of scientific frivolity, northern blots must have seemed like a nifty name to refer to RNA blots. Another assay for mRNA transcript detection and quantification is *real-time reverse transcriptase* (RT)-PCR. *Real-time RT-PCR* does not need as large a sample and is more rapid. *Immunoblot* analysis, also known as *protein blot* or *western blot* analysis (or did you think it would be "eastern"?) identifies transgenic proteins with antibodies that indicates bands of discernible sizes that can be semiquantified by their intensities. An alternative way to see if the specific transgenically produced protein is produced is by enzyme-linked immunosorbent assay (ELISA), which uses a specific antibody to probe a protein mixture in a microtiter plate.

Without providing Southern blot analysis showing integration of the transgene, phenotypic data are only partially qualifying evidence and not proof that a transformation event actually occurred. Gene expression may be due not to stable integration of a transgene, but due to transient expression (Bellucci et al. 2003; Wenck et al. 2003). For annual, sexually reproducing species, genetic segregation of the transgene in T_2 progeny will provide additional evidence for Mendelian inheritance (Birch 1997; Potrykus 1991). The original transgenic plant is called the T_0 generation. Suitable controls such as nontransformed, wild-type plants and plasmid should be included where appropriate.

11.3. INITIAL SCREENS: PUTATIVE TRANSGENIC PLANTS

Initial screens to identify *putative transformants* should be simple, economical, and high-throughput. Biotechnologists use "putative" in an optimistic way—the transgenic events remain putative until significant evidence mounts from definitive molecular analyses. Screens are designed to minimize detection of false positives later, thereby decreasing the number of samples to be analyzed by the labor-intensive gel blots. Putative transformants are usually identified as those that withstand selection agents and are positive with polymerase chain reaction (PCR) and/or ELISAs. Selectable and scorable markers (see Chapter 10) are also helpful in initial screens of transgenic plants. None of these screens are typically publishable in the scientific literature without accompanying in-depth molecular and phenotypic data.

11.3.1. Screens on Selection Media

Transformed plants regenerated from callus or after *in planta* transformation of germinating seedlings are termed T_0 plants. T_0 plants are always *hemizygous*—meaning that there is a copy of the transgene at a novel locus in the plant genome—the DNA gets integrated into one chromosome but not the homolog at the same locus. T_0 plants produce T_1 seeds, which, in turn, develop into T_1 plants that carry the transgene in either a hemizygous, homozygous positive, or homozygous negative state in an expected 2 : 1 : 1 Mendelian ratio if there is a single-copy insertion of the transgene. Similarly, transformed seed produced after transforming T_0 ovaries using the floral dip method (see the previous chapter) are T_1 seeds, which will be hemizygous for the transgene. In this case, the T_2 generation is the generation in which there will be segregation and recessive phenotypes will be unmasked.

Screening transgenic seeds can be accomplished by aseptically plating seed on solidified media with negative selection agents such as kanamycin, hygromycin, or phosphothricin in Petri plates to identify plants that withstand the selection agent (Weigel and Glazebrook 2002). Surviving plants are transplanted to pots with potting mix to grow to maturity for sample and seed collection. Initially, it may be necessary to determine the minimum lethal dose of selection agent that kills 100% (LD_{100}) of wild-type plants but allows transformed plants to survive. Nontransformed plants may become chlorotic, or bleached, or fail to develop roots (Fig. 11.1). For comparative purposes, wild-type samples should be plated on media without selection agent. In some species, optimal concentrations of selection agent may be cultivar-specific.

Let's imagine that an individual T_0 transgenic plant is self-pollinated to produce T_1 seed. The T_0 plant will have the transgene present in the hemizygous condition. The T_1 seeds and plants will segregate in a 3 : 1 phenotypic ratio for the transgene if there is a *single insertion* of the transgene. Single-insert transgenic events are desirable because the Mendelian inheritance is simple, and expression patterns are often more predictable than with multiinsert

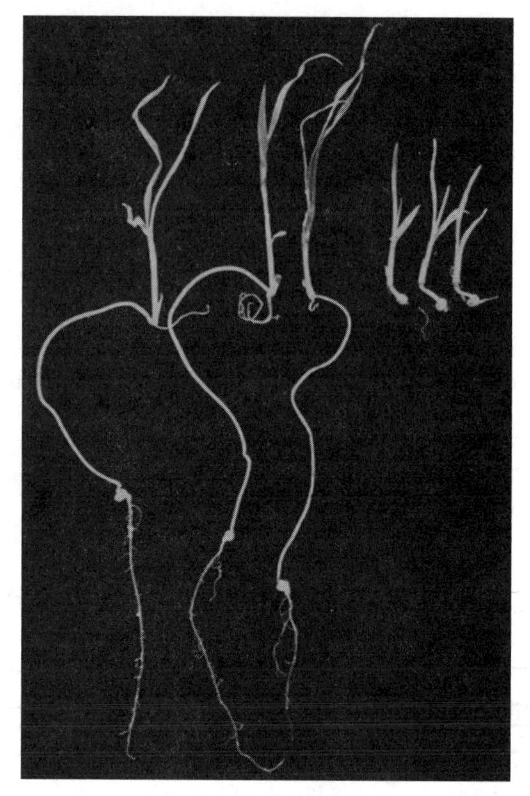

Figure 11.1. Comparison of wild-type millet seedlings grown on media with and without selection agent. Wild-type millet seedlings grown on MS media with 0 mg/L geneticin (left) versus wild-type millet seedlings grown on MS media with 60 mg/L geneticin (right). Seedlings grown on geneticin failed to develop roots and are more bleached in appearance.

events. Therefore, if 100 T_1 millet seeds segregating for a single insertion are plated onto tissue culture media containing a selection agent, 3 : 1 segregation might appear as approximately 75 of the seedlings appearing as the ones in Figure 11.1 left, and the approximately 25 nontransgenic segregants appearing as those in Figure 11.1 right. Note that the word "approximate" is used here since by chance the actual numbers can vary. Because of this variation, multiple replicate plates would be used and a statistical test, such as the chi-squared (χ^2) test would be employed to see if the observed results do not differ from the expected result of 3 : 1 segregation.

11.3.2. Polymerase Chain Reaction

Polymerase chain reaction (PCR) (Mullis 1990; Saiki et al. 1986) is a highly sensitive method that can be used to screen for genes of interest (GOIs) or selectable markers (SMs) using relatively crude DNA extracts. It is the amplification of a DNA sequence in a microcentrifuge tube with two primers that are complementary to the 5′ and 3′ ends of the DNA sequence and Taq DNA polymerase from the thermophile bacterium *Thermus aquaticus*. Through repetitive heating and cooling cycles, the DNA sequence between the two primers is amplified (Reece 2004). PCR results are invalid if positive and negative controls are omitted or do not work.

Extra precautions should be taken when assaying putative transformants produced using *Agrobacterium*-mediated gene transfer as the bacteria harboring the plasmid may be a source of nonintegrated DNA in the plant cell and produce false positives in PCR screens. *Agrobacterium* can systemically infect grapevine, rose, and fruit trees, but there are methods available to distinguish systemic infection (Cubero and Lopez 2005) from transgene integration. PCR should not be attempted in young plantlets or seedlings regenerated directly after an *Agrobacterium* treatment without an intervening antibiotic treatment, and even then, there might be occasional *Agrobacterium* cells living among plant tissue. Regenerating plantlets and whole seedlings can be treated with an antibiotic such as carbenicillin (Cheng et al. 1997) or cefotaxime (Broothaerts et al. 2005), respectively, to minimize bacterial contamination. These antibiotics kill *Agrobacterium* but do not harm plant tissue. Seed produced after *in planta* (floral dip) transformation of developing inflorescences should be washed in a dilute sodium hypochlorite solution before plating and/or planting (Weigel and Glazebrook 2002). T_1 plants have a much lower chance of *Agrobacterium* contamination compared with the original putatively transformed parent, and therefore PCR data are more reliable on DNA extracted from progeny plants.

PCR reactions are easily contaminated at setup by aerosolized DNA (Scherczinger et al. 1999); therefore, cleanroom assembly, separate from the site where DNA extractions are performed, is ideal. Aerosol tips and dedicated pipettors are additional safeguards against contamination. In addition, in a lab that works extensively with the same GOI, PCR can sometimes amplify contaminating DNA.

11.3.3. Enzyme-Linked Immunosorbent Assays (ELISAs)

For some selectable markers, qualitative or semiquantitative ELISAs are simple, sensitive, high-throughput screens (e.g., Agdia™ *NPTII* assay). Relatively crude plant protein samples are extracted, and samples are incubated in the wells of microtiter plates that are precoated with an antibody that binds the transgenic protein (antigen), primarily through

Figure 11.2. Positive *NPTII* ELISA (Agdia™). Duplicate samples of NPTII standards are shown in rows on the left, and duplicate samples are shown in rows in the rest of the microtiter plate. Plant samples that are positive for NPTII protein are shown.

hydrogen bonds (Memelink et al. 1994; Sambrook and Russell 2001). The wells are washed and incubated with an enzyme-linked conjugate that will produce a color change when flooded with substrate. The degree of color change can be judged by the naked eye or quantified on a spectrophotometer (Fig. 11.2). Semiquantitative ELISAs may also be used to estimate transgenic protein production.

11.4. DEFINITIVE MOLECULAR CHARACTERIZATION

11.4.1. Intact Transgene Integration

Copy number, or the number of times a transgene is inserted into the plant genome, is demonstrated by Southern blot analysis. Southern blot analysis is a multistep process that takes several days. It entails isolating sufficient (tens of micrograms) quantities of genomic DNA per plant, digesting the DNA to completion using a carefully chosen restriction enzyme or enzymes, and separating the fragments on a gel according to size. The DNA is transferred to a membrane (blot), and then the immobilized genomic DNA is hybridized with a radiolabeled probe (Feinberg and Vogelstein 1983, 1984) or a nonisotopically labeled probe (Langer et al. 1981) (Fig. 11.3).

Southern blots require high-quality high molecular weight genomic DNA (e.g., $\geq$20 kb). Extraction of high molecular weight DNA is critical; if the DNA is degraded or sheared, the bands on a Southern blot will not be distinct, since there will be smearing of DNA that is hybridized with probe. If the DNA is uncut, it will produce an unacceptable high molecular weight artifact after hybridization (Birch 1997; Potrykus 1991). After the DNA is digested to completion, the DNA fragments are separated on an agarose gel (Fig. 11.4a). DNA has a negative charge because of the phosphate groups and will travel toward the anode, which has a positive charge. Once the fragments have separated, the gel is placed on a nylon membrane (blot), under added weight and absorbent towels, and the fragments migrate into the blot by capillary action (Sambrook and Russell 2001). This process is termed "blotting," and the nylon filter is easier to handle in subsequent applications than the agarose gel, and is also more conducive to probing. In the traditional Southern blot, the membrane is

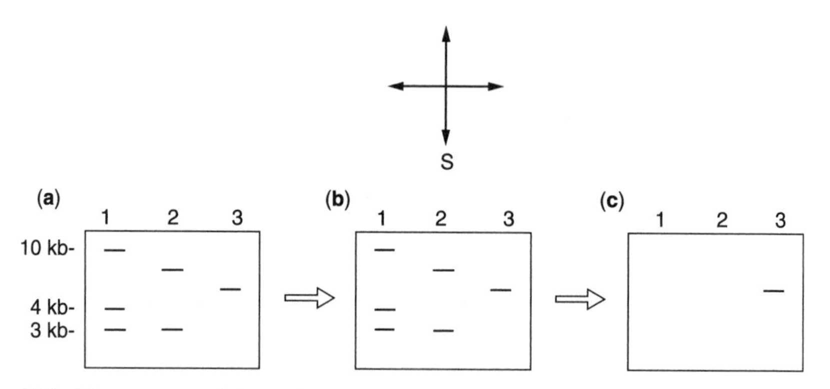

Figure 11.3. The process of DNA gel blot analysis (Southern blotting and hybridization): (a) digested DNA fragments are separated according to size on an agarose gel; (b) DNA fragments are transferred to a blot; (c) autoradiograph of the blot after hybridization with radiolabeled, single-stranded complementary DNA probe.

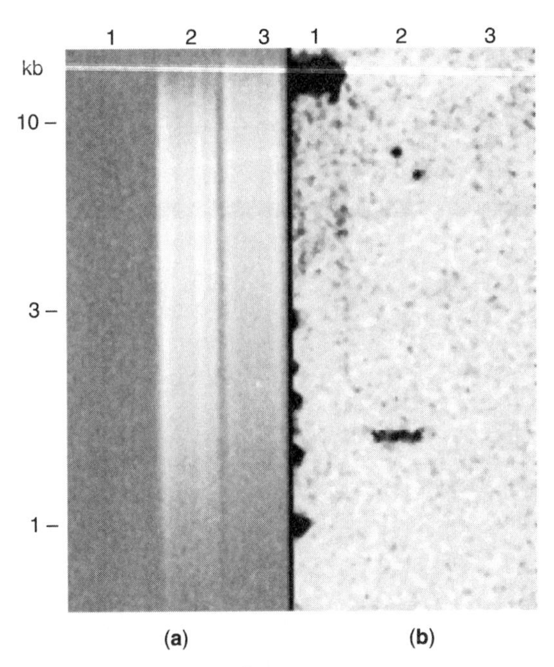

Figure 11.4. Digestion of plant genomic DNA that is electrophoretically separated for Southern blot analysis (a) and a phosphorimage of the blot after hybridization with radiolabeled probe (b). The samples are as follows: Lane 1, plasmid control; lane 2, putative transgenic plant; lane 3, non-transgenic plant control. Genomic DNA was extracted and digested to completion and run on a 0.8% agarose gel (a). The DNA was blotted to a membrane, hybridized with a radiolabeled probe, and exposed to a phosphor screen (b). There is a plasmid band (>10 kb) in lane 1 and one band in lane 2 at 1.8 kb that have sequences complementary to the probe.

placed on the gel and the DNA molecules migrate upward into the membrane via capillary action. However, a more recent protocol describes a downward Southern blot in which the gel is placed on the membrane and the molecules move quickly into the membrane as a result of capillary action and gravity (Fig. 11.5) (Chomczynski 1993). The blot is hybridized with a probe complementary to the GOI or SM, and the probe binds to these homologous sequences through hydrogen bonds. Probes are usually derived by restriction digestion of plasmid DNA or PCR amplification of the GOI (Reece 2004; Sambrook and Russell 2001) and gel purification of the fragment. After hybridization, nonspecifically bound probe is washed away at high stringency (i.e., high temperature and/or low salt) and the blot is exposed to X-ray film in autoradiography (Reece 2004; Sambrook and Russell 2001) or a phosphor screen (Johnston et al. 1990) and scanned to produce a digital image.

Two different kinds of genomic digests are commonly used to show transgene integration in the genome:

1. Copy number can be determined relative to a border of the introduced DNA. For example, this type of analysis has been used in *Agrobacterium*-mediated transformation experiments of *Arabidopsis* in which few T-DNA insertions are transferred (Katavic et al. 1994). It is important to know how many insertion sites are present. Single copies

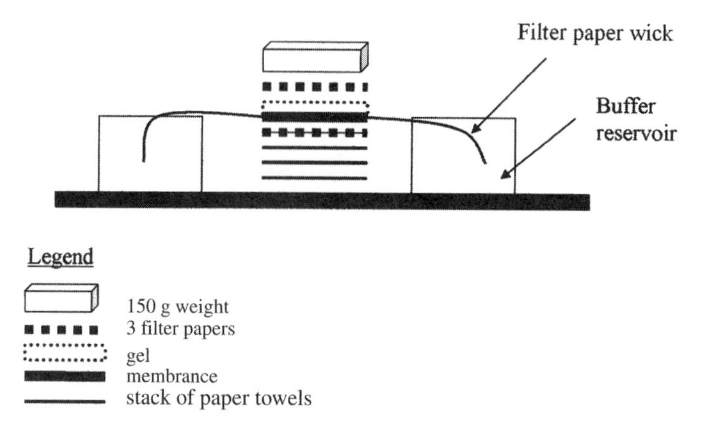

Figure 11.5. The downward Southern blot apparatus. Stacks of paper towel are topped with filter paper, membrane, the gel, a filter paper wick, more filter paper, and a weight. DNA travels into the membrane via wicking, capillary action and gravity.

are desirable because of simple segregation patterns, among other reasons. As an example, Figure 11.6 shows a plasmid used in *Agrobacterium*-mediated transformation. Figure 11.7 shows the T-DNA from right border (RB) to left border (LB) that would be transferred to the plant genome using *Agrobacterium.* Genomic DNA of the transgenic plant would be digested with *Eco*R1, an enzyme that cuts relative to the left T-DNA border and would cut into the plant genome, and the blot would be hybridized with the selectable

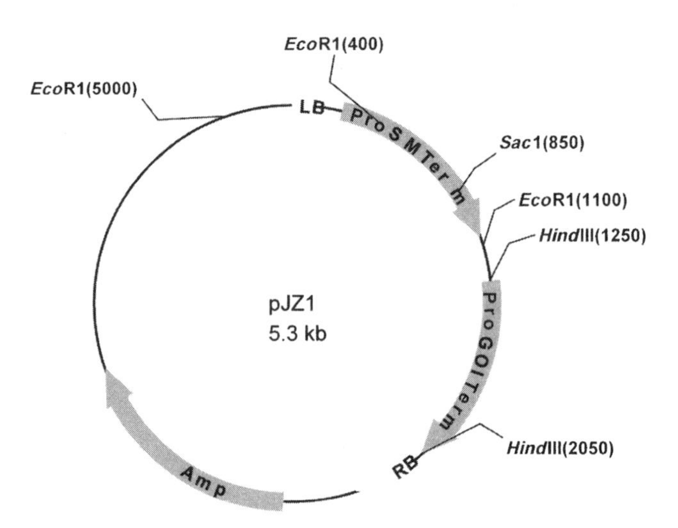

Figure 11.6. Map of plasmid pJZ1. The map shows the location of Pro:SM:Term (Promoter:Selectable Marker:Terminator), the Pro:GOI:Term (Promoter:Gene of Interest: Terminator), and the ampicillin resistance gene (*amp*). Location of the *Eco*R1 and *Sac*1 sites are shown for the purpose of restriction enzyme digestion for Southern analysis. (LB = left border of T-DNA; RB = right border of T-DNA.) The T-DNA between the left and right borders would be transferred in an *Agrobacterium*-mediated transformation event. A particle bombardment vector would not carry the LB and RB sequences.

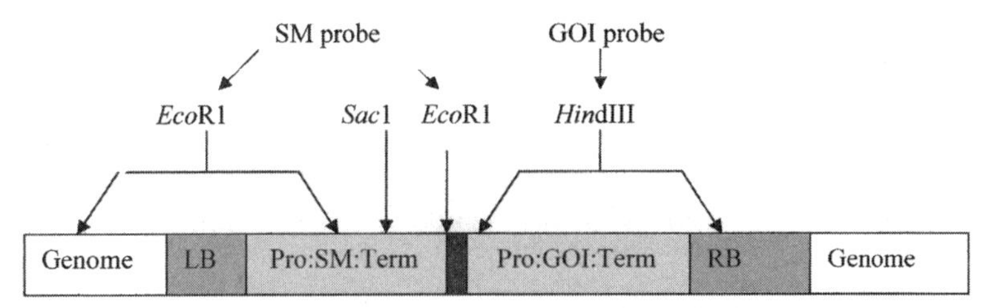

Figure 11.7. Integration of T-DNA in plant genome. The T-DNA from LB to RB is integrated in the plant genome. Left border analysis would involve digestion with *Eco*R1 that cuts in the selectable marker (SM) and into the plant genome and probing with SM cDNA or PCR fragment. A *Hin*dIII digest could be used as a diagnostic Southern blot to test whether the entire promoter: coding region: terminator of the gene of interest (GOI) was transferred in its entirety and is the correct size. (LB = left border; RB = right border; Pro:SM:Term = Promoter:Selectable Marker:Terminator; Pro:GOI:Term:Promoter:Gene of Interest:Terminator.)

marker (SM) probe (Fig. 11.7). Copy number can be determined relative to either T-DNA border, but right border analysis is the best determinant of whether the T-DNA has transferred to the plant genome because it tends to transfer more faithfully than does the left border (Caplan et al. 1985). However, the nature of the markers in the genetic construct and their usefulness as probes will ultimately determine which border is appropriate to analyze.

2. Copy number can be determined by using an enzyme that cuts once in the introduced DNA, and then the genomic blot is probed with sequences on either side of the restriction site. As an example, *Sac*1 cuts once in the introduced DNA, and the blot could be probed with either the selectable marker (SM) or the gene of interest (GOI) (Fig. 11.7). This method has been used in Biolistics, where multiple copies often integrate into one chromosomal location (Taylor and Fauquet 2002). See Jordan (2000) for an additional example.

Both means of determining copy number yield the same information. Band sizes will be different from the plasmid control if the introduced DNA has integrated into the genome. The researcher needs only to count the number of bands to determine the number of times the DNA was integrated into the plant genome. If possible, independent transformants produced using the same plasmid should be shown on a single blot, in addition to wild-type control and plasmid DNA samples, to lessen the likelihood of misinterpreting data from a single lane (Birch 1997). Using the same kind of analysis with T_1 plants or more advanced generations, we can also determine which copies of the transgene are inherited in individual progeny plants.

Real-time PCR (Higuchi et al. 1992, 1993) is a relatively new, high-throughput procedure for determining transgene copy number in plants (Bubner and Baldwin 2004; Ingham et al. 2001; Mason et al. 2002), but it should be used as a supplement to Southern analysis (Bubner and Baldwin 2004). It is an automated PCR procedure that monitors amplification of a DNA sequence in a microcentrifuge tube during repetitive PCR cycles by use of a fluorescent dye that is bound to primer or the PCR product itself whose signal increases in direct proportion to the PCR product (Reece 2004). Bubner and Baldwin (2004) present an in-depth discussion of this method, including its applications and limitations.

11.4.2. Determining the Presence of Intact Transgenes or Constructs

DNA gel blots are also used to determine whether a particular transgene (e.g., promoter: coding region:terminator) has been transferred in its entirety, using an enzyme digest that cuts on either side of the transgene (see Fig. 11.6, *Hind*III digest). This type of analysis will augment copy number but is not proof of transformation on its own (Birch 1997; Potrykus 1991).

11.4.3. Transgene Expression: Transcription

11.4.3.1. Northern Blot Analysis. RNA gel blots or northern blots (Thomas 1980) are used in hybridizations with complementary probes to detect, quantify, and size transcripts and monitor tissue-specific transgene expression at the mRNA level (Figs. 11.8 and 11.9). Like Southern blots, northern blots require large amounts of intact nucleic acids, but unlike Southern blots, no restriction digestion is necessary—individual transcripts are already naturally size-fractionated. Unlike DNA, RNA is rather unstable; for instance, it can be degraded by enzymes that naturally chew up RNA—RNases—the curse of all sorts of RNA analysis for researchers. RNase contamination of samples can occur easily; the ubiquitous RNase enzymes are present on hands, plasticware, glassware, and in distilled water. All solutions, glassware, tubes, and tips should be RNase-free to prevent degradation (Sambrook and Russell 2001). In RNA gel blots, total RNA or mRNA is separated according to size in agarose gel electrophoresis. RNA also has a negative charge and behaves like DNA in gel electrophoresis. There is a tendency; however, for RNA to form secondary structures (it can fold back on itself from hydrogen bonding between complementary bases); therefore, formaldehyde or glyoxal are denaturants added to the gel and sample buffer to eliminate intermolecular interactions (Memelink et al. 1994; Sambrook and Russell 2001). If formaldehyde is used, it must be washed from the gel before the RNA will transfer to the membrane. The gel is blotted to a nylon membrane to which the RNA migrates through capillary action. The transcript is detected by hybridization of the membrane with a complementary radiolabeled probe (Feinberg and Vogelstein 1983, 1984) or nonisotopically labeled probe (Langer et al. 1981; Memelink et al. 1994; Sambrook and Russell 2001); the probe may be single-stranded DNA or single-stranded RNA (Memelink et al. 1994; Sambrook and Russell 2001).

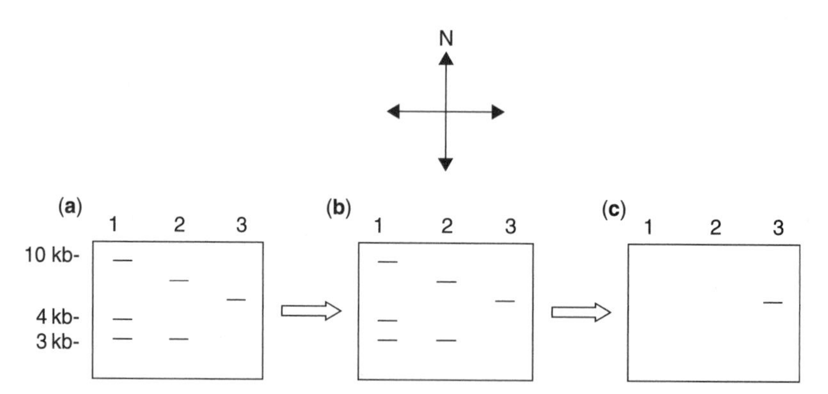

Figure 11.8. The process of RNA gel blot analysis (northern blotting and hybridization): (a) single-stranded RNA fragments are separated according to size on an agarose gel; (b) RNA fragments are transferred to a blot; (c) autoradiograph of the blot after hybridization with radiolabeled complementary single-stranded DNA or RNA probe.

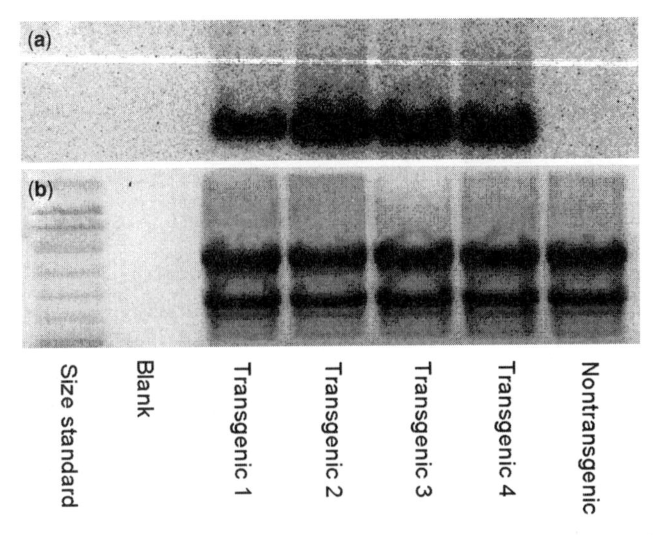

Figure 11.9. Northern blot: (a) specific cDNA from a arsenic repressor element is hybridized to transgenic plant mRNA, indicating that certain transgenic events highly express the transgene; (b) contains stained total RNA bands indicating approximately equal loading. (Figure courtesy of Jason Abercrombie.)

Nonspecifically bound probe is washed from the membrane (Memelink et al. 1994; Sambrook and Russell 2001), and the membrane is exposed to X-ray film in autoradiography (Reece 2004; Sambrook and Russell 2001) or a phosphor screen (Johnston et al. 1990) to develop the image (Fig. 11.9).

There are numerous ways to increase signal on an RNA gel blot. If the RNA is degraded during extraction, the signal will be weak. Temperatures should be kept low ($<4°C$) at most steps of RNA extractions. RNA gel blot sensitivity can be increased by enriching for mRNA (Memelink et al. 1994; Sambrook and Russell 2001), and this also decreases the problematic nonspecific binding of probe to the ribosomal bands. There are in vitro transcription kits widely available for the production of single-stranded RNA probes that can be radiolabeled to high specific activity for increased sensitivity. Loading more RNA will also increase the sensitivity of the procedure. Typically, a "housekeeping gene" such as actin or ubiquitin is also used as a probe along with the GOI to control for differential loading of RNA among lanes. In addition, or alternatively, the total mRNA can be quantified on the gel prior to blotting as a loading control procedure for subsequent quantitation of specific transcript (Fig. 11.9).

11.4.3.2. Quantitative Real-Time Reverse Transcriptase (RT)-PCR. The power of quantitative real-time PCR has been coupled with the reverse transcriptase (RT) reaction as a tool to monitor and quantify gene expression. In this procedure, total RNA or mRNA is extracted and used as a template with a complementary primer, 2′-deoxynucleoside 5′-triphosphates (dNTPs), and reverse transcriptase to generate single-stranded complementary DNA (cDNA) (Reece 2004). The cDNA is replicated in normal PCR using Taq DNA polymerase. Repetitive cycles of quantitative real-time PCR amplify the sequence of interest, while a fluorescent dye monitors the accumulation of PCR product (Reece 2004). This method is sensitive and high-throughput and does not require gel fractionation. It does not require a lot of starting material, but the sequence of the GOI is required for primer design.

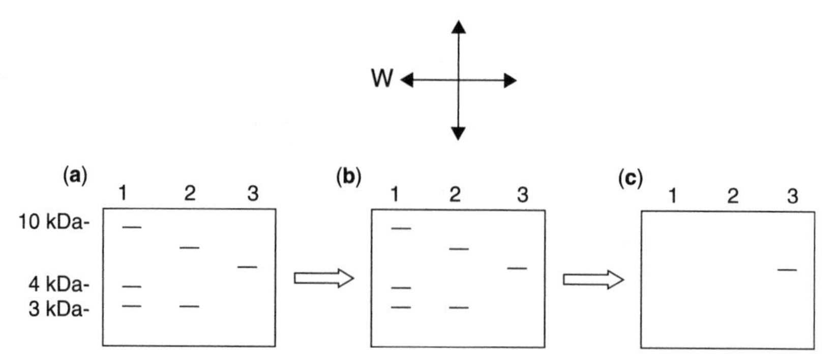

Figure 11.10. The process of western blot analysis. (a) polypeptides are separated according to size on a polyacrylamide-SDS gel; (b) the polypeptides are transferred to a PVDF membrane; (c) the membrane is incubated with an antibody raised to the transgenic protein, and the antibody is detected with an enzyme conjugate secondary antibody that causes a color or chemiluminescent change detectable on film.

One of the most important aspects of specific transcript quantitation is use of the appropriate statistical analyses (Yuan et al. 2006).

11.4.4. Transgene Expression: Translation: Western Blot Analyses

Protein gel blot analyses or western blots show in a semiquantitative manner that the introduction and transcription of a GOI is causing a production of a specific recombinant protein in the transgenic plant. The researcher can see if the intact protein is produced or whether it is not processed correctly. In western blotting (Burnette 1981), total protein is isolated and fractionated on a polyacrylamide gel, transferred to a membrane, and then detected with primary and secondary antibodies (Reece 2004) (Figs. 11.10 and 11.11). Sodium

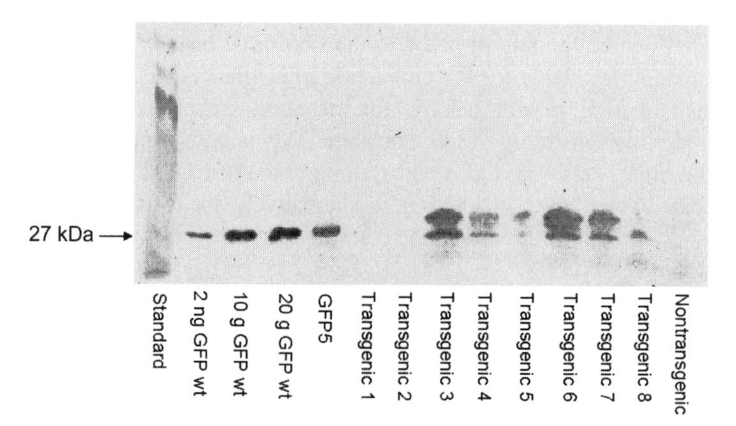

Figure 11.11. Western blot. This experiment used a GFP-specific polyclonal antibody to detect GFP in transgenic plants. A semiquantitative assay can be performed by running purified GFP at various concentrations, in this case, from 2 to 20 ng (see the first three lanes to the right of the protein size standard). The next lane is from a mGFP5-ER transgenic plant. The next lanes are from individual transgenic plant events with an earlier GFP variant—mGFP4. Note that some transgenic events are extremely low-expressing and that there are two bands where only one is expected. (Data courtesy of Staci Leffel and Neal Stewart.)

dodecyl sulfate polyacrylamides (SDS-PAGE) gels are most commonly used for separating proteins on the basis of size (Laemmli 1970). In these gels, disulfide reducing agents such as β-mercaptoethanol or DTT are added to denature the proteins into *polypeptides*, and SDS swamps the polypeptides with a net negative charge so that they migrate toward the anode (Laemml 1970; Memelink et al. 1994). The proteins are transferred to polyvinylidene fluoride (PVDF) membrane by electroblotting, the membrane is incubated with an antibody raised to the transgenic protein, and the antibody binds primarily through hydrogen bonds (Memelink et al. 1994; Sambrook and Russell 2001). The antibody is detected with an enzyme conjugate secondary antibody that causes a color or chemiluminescent change detectable on film (Memelink et al. 1994) (Fig. 11.11). Protein bands can be semiquantified by band intensity or relative to a standard protein.

11.5. DIGITAL IMAGING

In the olden days of molecular biology (pre-1990s), simple nondigital photographs were taken to document raw data from gels and blots. No one does that anymore with the advent of digital imaging. Most journals have guidelines stating which software imaging adjustments are acceptable for enhancing digital images of nucleic acid and protein blots to avoid appearances of research misconduct. For example, Rossner and Yamada 2004 offer a set of guidelines for the *Journal of Cell Biology* (http://www.jcb.org/cgi/content/full/166/1/11). Linear adjustments to brightness and contrast must be made to the entire gel image—never to only a portion of the gel, and never to obscure or alter the original data. For example, any nonlinear gamma adjustments must be disclosed in the figure legend. Selective removal of background within lanes must be performed judiciously, if at all. It is prudent to consult the specific journal for its instructions in handling digital images.

11.6. PHENOTYPIC ANALYSIS

After screening and molecular characterization, transgenic plants should be grown and their *phenotypes* assessed to determine whether they differ from wild-type. The phenotype is the genetic make-up (*genotype*) as influenced by the environment. Traits altered with respect to the SM used in transformation experiments such as resistance to herbicides and resistance to antibiotics should be tabulated. A transgenic plant expressing a new GOI may also have an altered phenotype with respect to seedling emergence, growth habit, days to flower, days to maturity, seed color, disease resistance, and other parameters, in comparison to wild-type.

11.7. CONCLUSIONS

Initial screens on selective media, with ELISAs or PCR will determine whether one has putative transformants. Copy number Southern gel blots will show transgene integration. Real-time PCR can augment but not substitute for copy number Southern gel blots. Analysis of intact genes will show whether the GOI or SM has been transferred in its entirety to the plant or is truncated. Northern gel blots, RT-PCR, and western gel blots will demonstrate transgene expression. Phenotypic traits other than the GOI may be altered in a transformed plant.

REFERENCES

Bellucci M, De Marchis F, Mannucci R, Arcioni S. (2003): Jellyfish green fluorescent protein as a useful reporter for transient expression and stable transformation in *Medicago sativa* L. *Plant Cell Rep* **22**:328–337.

Birch RG (1997): Plant transformation: Problems and strategies for practical application. *Annu Rev Plant Physiol Plant Mol Biol* **48**:297–326.

Broothaerts W, Mitchell HJ, Weir B, Kaines S, Smith LM, Yang W, Mayer JE, Roa-Rodriguez C, Jefferson RA (2005): Gene transfer to plants by diverse species of bacteria. *Nature* **433**:629–633.

Bubner B, Baldwin IT (2004): Use of real-time PCR for determining copy number and zygosity in transgenic plants. *Plant Cell Rep* **23**:263–271.

Burnette WN (1981): "Western blotting": Electrophoretic transfer of proteins from sodium dodecyl sulfate-polyacrylamide gels to unmodified nitrocellulose and radiographic detection with antibody and radioiodinated protein A. *Anal Biochem* **112**:195–203.

Caplan AB, Van Montagu M, Schell J (1985): Genetic analysis of integration mediated by single T-DNA borders. *J Bacteriol* **161**:655–664.

Cheng M, Fry JE, Pang S, Zhou H, Hironaka CM, Duncan DR, Conner TW, Wan Y (1997): Genetic transformation of wheat mediated by *Agrobacterium tumefaciens*. *Plant Physiol* **115**:971–980.

Chomczynski P (1993): One-hour downward alkaline capillary transfer for blotting of DNA and RNA. *Anal Biochem* **201**:134–139.

Cubero J, Lopez M (eds) (2005): *Agrobacterium Persistence in Plant Tissue after Transformation.* Humana Press, Totowa, NJ, pp 351–364.

Feinberg AP, Vogelstein B (1983): A technique for radiolabeling DNA restriction endonuclease fragments to high specific activity. *Anal Biochem* **132**:6–13.

Feinberg AP, Vogelstein B (1984): A technique for radiolabeling DNA restriction endonuclease fragments to high specific activity. Addendum. *Anal Biochem* **137**:266–267.

Higuchi R, Dollinger G, Walsh PS, Griffith R (1992): Simultaneous amplification and detection of specific DNA sequences. *Bio/Technology* **10**:413–417.

Higuchi R, Fockler C, Dollinger G, Watson R (1993): Kinetic PCR analysis: Real-time monitoring of DNA amplification reactions. *Bio/Technology* **11**:1026–1030.

Ingham DJ, Beer S, Money S, Hansen G (2001): Quantitative real-time PCR assay for determining transgene copy number in transformed plants. *Biotechniques* **31**:132–141.

Johnston RF, Pickett SC, Barker DL (1990): Autoradiography using storage phosphor technology. *Electrophoresis* **11**:355–360.

Jordan MC (2000): Green fluorescent protein as a visual marker for wheat transformation. *Plant Cell Rep* **19**:1069–1075.

Katavic V, Haughn GW, Reed D, Martin M, Kunst L (1994): *In planta* transformation of *Arabidopsis thaliana*. *Mol Gen Genet* **245**:363–370.

Laemmli UK (1970): Cleavage of structural proteins during the assembly of the head of bacteriophage T4. *Nature* **227**:680–685.

Langer PR, Waldrop AA, Ward DC (1981): Enzymatic synthesis of biotin-labeled polynucleotides: Novel nucleic acid affinity probes. *Proc Natl Acad Sci USA* **78**:6633–6637.

Mason G, Provero P, Vaira AM, Accotto GP (2002): Estimating the number of integrations in transformed plants by quantitative real-time PCR. *BMC Biotechnol* **2**:20.

Memelink J, Swords KMM, Staehelin LA, Hoge JHC (eds) (1994): *Southern, Northern and Western Blot Analysis*. Kluwer Academic Publishers, Dordrecht, pp 1-F1:1–23.

Mullis KB (1990): Target amplification for DNA analysis by the polymerase chain reaction. *Ann Biol Clin (Paris)* **48**:579–582.

Potrykus I (1991): Gene transfer to plants: Assessment of published approaches and results. *Annu Rev Plant Physiol Plant Mol Biol* **42**:205–225.

Reece RJ (2004): *Analysis of Genes and Genomes*. Wiley, Chichester, UK.

Rossner M, Yamada KM (2004): What's in a picture? The temptation of image manipulation. *J Cell Biol* **166**:11–15.

Saiki RK, Bugawan TL, Horn GT, Mullis KB, Erlich HA (1986): Analysis of enzymatically amplified beta-globin and HLA-DQ alpha DNA with allele-specific oligonucleotide probes. *Nature* **324**:163–166.

Sambrook J, Russell DW (2001): *Molecular Cloning: A Laboratory Manual*, 3rd ed. Cold Spring Harbor Laboratory Press, Cold Spring Harbor, NY.

Scherczinger CA, Ladd C, Bourke MT, Adamowicz MS, Johannes PM, Scherczinger R, Beesley T, Lee HC (1999): A systematic analysis of PCR contamination. *J Forens Sci* **44**:1042–1045.

Southern EM (1975): Detection of specific sequences among DNA fragments separated by gel electrophoresis. *J Mol Biol* **98**:503–517.

Taylor NJ, Fauquet CM (2002). Microparticle bombardment as a tool in plant science and agricultural biotechnology. *DNA Cell Biol* **21**:963–977.

Thomas PS (1980): Hybridization of denatured RNA and small DNA fragments transferred to nitrocellulose. *Proc Natl Acad Sci USA* **77**:5201–5205.

Weigel D, Glazebrook J (eds) (2002): *Arabidopsis: A Laboratory Manual*. Cold Spring Harbor Laboratory Press, Cold Spring Harbor, NY, pp 1–354.

Wenck A, Pugieux C, Turner M, Dunn M, Stacy C, Tiozzo A, Dunder E, van Grinsven E, Khan R, Sigareva M, Wang WC, Reed J, Drayton P, Oliver D, Trafford H, Legris G, Rushton H, Tayab S, Launis K, Chang YF, Chen DF, Melchers L (2003): Reef-coral proteins as visual, non-destructive reporters for plant transformation. *Plant Cell Rep* **22**:244–251.

Yuan JS, Reed A, Chen F, Stewart CN Jr (2006): Statistical analysis of real-time PCR data. *BMC Bioinformatics* **7**:85.

CHAPTER 12

Regulations and Biosafety

ALAN McHUGHEN

University of California, Riverside, California

12.0. CHAPTER SUMMARY AND OBJECTIVES

12.0.1. Summary

Transgenic crops are the most regulated and tested plants ever produced, and much of the regulation is a response to concerns about biosafety issues. There are two areas of biosafety concerns: food safety and environmental safety, each with corresponding regulatory issues.

12.0.2. Discussion Questions

1. What are regulations supposed to achieve?
2. With GM crops spreading so quickly, how are we assured of their health and environmental safety?
3. How is genetic engineering (biotechnology) regulated?
4. How do the risks posed by products of biotechnology compare to those posed by conventional technologies?
5. How does biotechnology threaten biosafety?
6. How do different countries regulate products of biotechnology?

12.1. INTRODUCTION

This chapter explores how governments regulate food and agriculture emanating from one group of technologies, *genetic engineering* (also called *genetic modification*, *rDNA*, or simply "biotechnology"), and investigates the scientific validity of such regulations.

Our human ancestors began the serious art of agriculture about 10,000 years ago. In those days and until the near-present time, the major concern was simply getting enough food. Today's agriculture issues still include, for approximately 800 million people,

Plant Biotechnology and Genetics: Principles, Techniques, and Applications, Edited by C. Neal Stewart, Jr.

getting enough to eat, but also a range of other concerns, such as food safety and nutrition. In addition, other economic and political issues can occupy the minds of those who typically show few signs of hunger or malnutrition. The planet supports a burgeoning human population well over 6 billion, but agriculture can sustainably provide for only about 3–4 billion of us. For good or for bad, the success of humans at procreation now demands that we turn increasingly against nature in order to provide enough food to maintain the increasingly unnatural human population. As agriculture becomes increasingly technological, and less and less traditional, many people become increasingly vocal in expressing concern for safety in food production systems.

Our prior history shows little mass interest in the safety of food production, especially if there was sufficient safe food to go around. But societies have always suffered from local, regional, or widespread food famines and adulterations, and these scourges continue today. With public interest in food and agriculture increasing within affluent societies, newer technologies are coming under scrutiny as potentially hazardous.

The transition from traditional farming practices and food production systems to the application of modern technologies in all aspects of agriculture and food in the early twentieth century was accompanied by a mass exodus of farm folk to urban centers. As a result, unlike a century ago, few urban people in affluent societies have a direct personal or family connection to farming and consequently have little comprehension of how food is produced. This unfortunate ignorance leads to gross misconceptions and a rather romantic aura of "traditional" farming. The anxiety fostered by beliefs that the agricultural technology is suspect also leads to demands that government assume a greater role in ensuring the safety and security of the food supply, even when there is little or no scientific justification (on the basis of actual harm) for doing so.

A large number of technologies—all of which pose some degree of risk to health or environment—have been introduced to farming and food production in the last 100 years. Many of these, such as mechanization, farm management (agronomy), and genetic modification through plant and animal breeding, have had a dramatic and positive impact on both the quantity and quality of food produced. In addition, technological advances and applications in food storage, processing, and transport allowed human society to eat, flourish, and expand well beyond natural limits to the sustainable population and allowed individuals to enjoy an expected average lifespan nearly double that of our grandparents.

Nevertheless, all technologies do carry risks, and in modern risk-averse society, those risks must be identified, assessed, and managed. Because of the long history of relatively "safe" introductions of technology to agriculture and food, most city dwellers paid little heed to risks associated with adoption of, for example, tractors on the farm, although many farmers (and family members) suffered death or dismemberment from mechanical accidents involving the powerful machines, and such accidents continue today. Through the twentieth century, governmental regulations evolved to ensure the safe application of almost all technologies and innovations in farming. However, in the 1970s and 1980s, many people began to question the safety of food production systems and the efficacy of regulations governing them. Spurring this anxiety, in the absence of any true problems with the food supply, was the increasing awareness of chemical fertilizers, pesticides, and the general feeling that farming was becoming "high tech," and not the way it was in the old days. One manifestation was a common wariness and subsequent demand to increase regulation on plant, animal, and microbial breeding, where genes were modified using recombinant DNA (rDNA) technologies, often called *genetic engineering* (GE) or *genetic modification* (GM) to produce genetically engineered/modified organisms

(GEOs or GMOs). In response to this, governments around the world rushed to assure the public that "something was being done to protect the public and the environment from the hazards of genetic engineering" and establish regulatory mechanisms to oversee GE as applied to agriculture and food production.

12.2. HISTORY OF GENETIC ENGINEERING AND ITS REGULATION

Genetic engineering, recombinant DNA, is much older than most people realize. The first successful DNA "recombination" or human-mediated hybridization between two specific but diverse DNA strands was reported by Boyer and Cohen in 1973 (Cohen et al. 1973). At first, the scientific community itself recognized that the great power of the new technology also implied risk (Berg et al. 1974), and in 1975 a group of leading scientists convened at Asilomar, California to discuss the issues. They called for a largely self-regulated set of guidelines to cautiously assess the risks with the emerging technologies. In the United States the National Institutes of Health (NIH) in 1976 took the next step when it formalized and established strict rules to regulate rDNA research activities. Although the NIH guidelines applied only to federally funded rDNA research programs, many agencies [including the Environmental Protection Agency (EPA), Food and Drug Administration (FDA), and the US Department of Agriculture (USDA)] adopted the rules as sensible precautionary policy. The voluntary NIH guidelines thus became, in effect, mandatory for virtually all rDNA research conducted in the United States and internationally.

With the scientific community enthusiastic about the applications of rDNA and other forms of biotechnology, bureaucracies recognized the impending certitude that biotechnology would not remain an academic and laboratory novelty, and that manufacturers of products developed using the new technologies would eventually be seeking market and environmental release. Consequently, they began gearing up to deal with potential hazards. One of the first papers was from the Organization for Economic Cooperation and Development (OECD), which provided a standardized and workable definition of "biotechnology . . . the application of scientific and engineering principles to the processing of materials by biological agents to provide goods and services" (OECD 1982). Although the definition is unwieldy and captures virtually everything involving biological systems, including products of conventional breeding and food production systems, it remains widely used today and provides the basis for regulations in many countries. The OECD report also noted the necessity of regulating products of biotechnology, assuming that they, like everything else, were not inherently risk-free. By the mid-1980s, the living organisms generated as a result of rDNA research [also known as *transgenic organisms*, *genetically modified organisms* (GMOs), or *genetically engineered organisms* (GEOs)] were being generated and attracted attention because of their own potential for risk, particularly as potential threats to the environment and as food/feed safety hazards. In 1986, the US Office of Science and Technology Policy (OSTP) investigated the regulatory milieu and compiled a *Coordinated Framework for Regulation of Biotechnology*. This document coordinated the existing regulatory bureaucracy with relevant studies coming from the scientific community. They recommended adapting existing legislation and regulatory authority to encompass products of biotechnology, tapping existing regulatory expertise in relevant agencies, particularly the USDA, FDA, EPA. Thus, GMO plants would be regulated for food and feed safety concerns by regulators with appropriate expertise in FDA, those GMO plants with pesticidal properties by EPA, and those with plant pest potential (environmental risks) by USDA. This coordinated effort and allocation of responsibility to different agencies continues today in the United States.

At about the same time, the OECD released a major study (based on its own recommendation in the earlier 1982 report) on biosafety related to biotechnology, often called simply the "Blue Book" (OECD 1986), which also remains widely quoted and cited today for its fundamental commonsense approach to risk assessment. It was the first scientific analysis to consider environmental hazards that might be posed by transgenic organisms, and served as a standard from which many governments and regulatory agencies have based their procedures for assessing risks with products of biotechnology. It remains, even after 20 years, "fresh" in the sense that it was prescient, identifying legitimate risk concerns with rDNA technologies even before transgenic organisms were let loose on the environment.

In contrast, some other jurisdictions, notably those in the European Union (EU), believing rDNA to be so novel and potentially hazardous that existing legislation and regulatory expertise was not capable of handling it, created entirely new bureaucracies to regulate GMOs.

By the end of the 1980s, the US National Academy of Sciences issued a "white paper" declaring, among other things, that rDNA produced no new categories of risk, and that risk assessment should be based on the physical features of the product, not on the process by which it was developed (NRC 1987). Subsequent studies from the National Academies of Science [via the National Research Council (NRC)] on increasingly specific points dealing with risks posed by rDNA all came to the same general conclusion, that all methods of genetic manipulation can generate potentially hazardous products, that rDNA is not inherently hazardous or invariably generates products with higher risk than do other methods, and that risk assessment should focus on the final product, regardless of the method of breeding (NRC 2000, 2002, 2004).

Back at the lab, the techniques of gene splicing, as it has become known, have been applied to a wide range of products, including medical, industrial, and, yes, agriculture and food production. In the late 1970s, the early experimental successes saw genetically engineered microbes produce proteins from rDNA transferred genes, and the technical advances were quickly adapted to commercial applications, including generating human therapeutics. Human insulin produced by rDNA from the human gene transferred to bacteria was reported in 1978. This development led to the first approval for the first commercial application of rDNA technology, the diabetes drug insulin (trade name: Humulin, from Genentech), in 1982. Many other pharmaceutical products developed using rDNA quickly followed.

Transgenic plants made their lab and greenhouse appearance in 1983, as three independent groups reported their developments at the Miami winter symposium, and other groups followed quickly.

In Belgium, Jeff Schell and Marc Van Montagu produced tobacco plants resistant to kanamycin and methotrexate (Schell et al. 1983, Herrera-Estrella et al. 1983). At Monsanto in St. Louis, USA, Robert Fraley, Stephen Rogers, and Robert Horsch generated transgenic petunia plants resistant to kanamycin (Fraley et al. 1983a, 1983b). And in Wisconsin, John Kemp and Timothy Hall inserted a gene from beans into sunflower (Murai et al. 1983).

The first open-air field trials of transgenic plants were planted as early as 1985, but the numbers of trials, species, traits, and countries climbed dramatically in the late 1980s and early 1990s.

However, it took 10 years (to 1993) before the first whole plant was commercialized and grown unregulated in the field, a virus-resistant tobacco in China (Jia and Peng 2002; Macilwain 2003), followed by the first transgenic food crop, Flavr Savr tomato, in 1994. Neither GM product remains on the market today. The Flavr Savr failed because of inconsistent production capacity and delivery to market (Calgene, the company developing Flavr Savr, claims that they sold every tomato delivered to the stores and that they were unable to keep up with demand); the Chinese tobacco was withdrawn because of pressure from smokers worldwide who feared that smoking GM tobacco (but not regular tobacco?) might pose a health risk.

The first GE food product, the milk coagulating agent Chymosin, was developed in 1981 and, after various improvements, testing, and safety assessments, was approved and reached the market in 1988 (in the UK) and 1990 (in the USA). Most of the hard cheese now made uses this genetically engineered protein in place of rennet from calf stomach. Such cheeses are popular yet remain unlabeled, even in places where labeling based on the process of rDNA is mandated. Although only trace amounts of the enzyme remain in the final food product, it is disingenuous and misleading to consumers to claim that the cheese is "non-GMO," at least not without an explanation. More on this later.

The subsequent deployment and adoption of GE crop varieties has been impressive. According to ISAAA (James 2005), the one-billionth acre of commercial GE crop was grown in 2005, with the total acreage spread across 21 countries. In 2005 alone, according to James, GE crops covered 222 million acres [~90 million hectares (ha)] worldwide. This represents an impressive growth within an industry, namely, agriculture, not known for quick adoption, particularly of controversial technologies. The major players remain fairly constant, with the United States, Argentina, Canada, and China leading the way, but also significant acreages in some smaller countries, including such diverse lands as South Africa, Philippines, Iran, and Romania. Some 14 countries are now growing over 100,000 acres of GE crops (James 2005).

In the United States, the major GE crops include soybeans, corn, and cotton; biotech cultivars of these crops captured 89%, 61%, and 83% of their respective market acreages (USDA/NASS 2006). Minor commercialized GE crops include potato, tomato, and flax (all no longer grown), plus virus-resistant papaya and some squash. GE alfalfa has recently been approved, and GE crops currently under development for US farmers, include sugarbeets, plum pox-resistant plums, disease-resistant citrus, and a broad array of others. For a complete listing of US approved crops, see http://www.aphis.usda.gov/brs/not_reg.html and, for a combined (USDA, EPA, and FDA) searchable database, http://usbiotechreg.nbii.gov/database_pub.asp.

Internationally, GE crops under development include improved versions of locally important crops, such as GE brinjal (eggplant or aubergine) and high-protein potato in India; corn in South Africa; broccoli; tomato; sweet potato; papaya; banana; winter melon; watermelon; rice; several tree events; and even transgenic animals (pigs) in Taiwan; rice, turfgrass, potato, and various local species of vegetables and produce in Korea; oil palm in Malaysia; and cassava in Kenya and other countries of east Africa. An exhaustive listing of GE species and traits in development around the world would be both extensive and quickly outdated. Those interested in the technical and regulatory progress of agricultural biotechnology in developing countries should consult www.isaaa.org frequently.

Most of the GE crops commercialized to date carry input traits such as disease or herbicide resistance, or pest control, but newer products are focused on output traits, such as enhanced nutritional profiles and removal of allergenic or other antinutritional proteins and substances. One reason why these "consumer-oriented" GE products are not available today is the long and expensive regulatory process. One point worth remembering is that all GE crop cultivars receive far more regulatory oversight and safety assessments than do similar crops with similar traits and therefore posing similar risks.

12.3. REGULATION OF GE

Effective regulations protect the public and the environment from threats of harm. Also, because all regulatory bureaucracies have limited financial, human, and other resources, they must, in order to be effective, apply the regulatory maxim: things posing the greatest threat receive the greatest scrutiny.

Of course, regulatory bureaucracies, like all bureaucracies, do not always work efficiently or effectively. Political expediency often interferes with the strict adherence to the scientifically sound maxim. Below we explore how some regulatory bureaucracies apply their allocated resources to agricultural biotechnology.

12.3.1. United States

Discussion of regulatory policy for products of biotechnology in the United States started relatively early. As mentioned above, the Office of Science and Technology Policy (OSTP), recognizing that potential risks and regulatory expertise were distributed across several bureaucracies, developed a coordinated framework to assign responsibilities to those relevant agencies (OSTP 1986). Within this, *regulated articles* (as they are called) were assigned to the different agencies according to their intended use, but also recognizing that some articles—and, in practice, most—were captured for regulation by more than one agency. As a result, FDA was given primary responsibility for regulating risks to food and feed, EPA to regulating products with pesticidal properties, and USDA to biotechnologically derived plants with potential to become agricultural pests. In many cases, all three agencies evaluate a product; for example, a food crop with rDNA-mediated novel herbicide resistance would trigger review by USDA for plant pest potential, EPA for the new herbicide aspects, and FDA for any changes to the quality of the derived food and feed.

Other products, for example, an ornamental (nonfood/feed) plant with an altered flower color, might avoid regulatory review by EPA and FDA, but will be captured by USDA. In fact, to date all rDNA plants seeking deregulation were captured and regulated by USDA. All commercialized rDNA-derived food crops were also reviewed by FDA, even though the food itself was unchanged and thus the FDA assessment was considered "voluntary" (much to the dismay of some, who believe that FDA should regulate all biotech products as a mandatory exercise).

The United States conducts regular evaluations of its own regulatory procedures, to ensure that the regulators remain aware of the most recent developments in the technology, and may adapt regulatory procedures to account for those developments. The scientific foundations are often reviewed by committees ("panels") of the National Research Council under the administration of the National Academies of Science. Administrative procedures are also frequently reviewed, usually involving solicitation of public input and suggestions for improvement. In addition, public input is sought at several stages of the regulatory review, usually after an announcement in the *Federal Register* detailing a particular product under review.

12.3.1.1. USDA. The office within USDA responsible for regulatory oversight of agricultural products of rDNA is the Animal and Plant Health Inspection Service (APHIS), office of Biotechnology Regulatory Service (BRS). Regulators in BRS claim legislative authority to capture and regulate rDNA-derived plants under the Plant Protection Act of 2000. The main concern in BRS is that the "regulated article" (i.e., rDNA-derived plant) might become a "plant pest" (defined broadly) and negatively impact the environment, so they focus their assessments on pest characteristics. USDA assesses whether the regulated article (product of rDNA breeding) might directly or indirectly cause disease or other damage to a plant. Regulated articles can include rDNA-produced plants, microbes, and animals. Of course, the primary regulated articles to date are herbicide-tolerant crops, insect-protected crops, and a handful of other transgenic plants (late-ripening tomatoes, disease-resistant papaya, potatoes, etc.), as well as some transgenic microbes. BRS controls not only prospective releases to the open environment but also the international importation and interstate transport of transgenic organisms.

BRS allows environmental releases of transgenic plants through two routes: notification and permit. Notifications are used for specified low-risk crops and traits, while a permit is required for those transgenic organisms posing greater apparent risk, such as species less familiar to BRS or those producing pharmaceutical compounds. Eventually, after the evaluations are complete, and if the data support it, the developer may petition for "unregulated" status. BRS conducts an environmental assessment to ensure that the product is indeed not a potential plant pest, and also seeks public comment before issuing the decision. Once a "regulated article" acquires "nonregulated" status, it can be grown, sold, and distributed much like any other nontransgenic variety.

12.3.1.2. FDA. The Food and Drug Administration (FDA), operating within the department of Health and Human Services (HHS), concerns itself with the safety of foods and feeds. Interestingly, unlike the case in USDA, in FDA the trigger for mandatory capture for regulatory assessment is not the process of rDNA but the physical composition of the food or feed in question. This is the basis of considerable debate, as some people demand that FDA conduct safety assessments of all foods derived from biotech plants, animals, and microbes, even those with chemical compositions identical to those of current foods of the same type.

The FDA review focuses on three questions:

1. Does the novel food or feed contain any new allergens?
2. Does the novel food or feed contain any new toxic substances?
3. Has the novel food or feed changed the nutritional composition in any way, either increasing or decreasing nutrients, antinutritional substances, or other components?

Problems from ingesting food result from the presence of damaging substances such as allergens in sensitive people, or toxicants. In the long term, problems can also arise from the absence or diminution of nutrients ordinarily present in a given food. For example, many people enjoy orange juice and benefit from the rich source of vitamin C. If for some transgenic reason oranges ceased to produce ascorbic acid (vitamin C), some consumers might inadvertently develop symptoms of vitamin C deficiency, namely, scurvy. As this is an undesirable effect, FDA would check a new orange for ascorbic acid content, just to ensure that it was still present in appropriate concentrations. To date, biotech-derived (biotechnologically derived) foods have not been found to unexpectedly lose normal nutrients, and all commercialized biotech-derived foods have the same nutritional content as do similar conventional foods. Newer transgenic foods might be specifically modified to enhance nutritional composition. In those cases, the FDA review becomes mandatory, and the new food will have to be labeled as such, because it would no longer fit the definition of the traditional, unmodified food.

A bigger concern is the possibility that the novel food carries an unexpected allergen. Such an event has occurred, although the product was never commercialized and no one was harmed. In this situation, a gene to enhance the nutritional status of soybean (which is naturally deficient on the amino acids methionine and cysteine) was cloned from Brazil nut and transferred to the legume. Tests showed that the transgenic soybean did indeed express the Brazil nut gene and generate the expected protein rich in these amino acids, thus successfully increasing the nutritional balance of the bean. Subsequent premarket tests showed that the new soybean was, unfortunately, also allergenic to consumers allergic to Brazil nut, indicating that the storage protein in Brazil nut responsible for the good desired amino acids was also a major allergen, even when expressed in soybean (Nordlee et al. 1996). Since the Brazil nut transgenic soybean was found to be a likely source of allergens during the course of evaluation, it is heralded as a case showing that regulations are effective.

Even without a mandatory premarket food safety assessment, every commercialized rDNA crop was reviewed by FDA regulators under a voluntary consultation. In other words, the developers of the new crops and foods wanted the FDA to review the safety even though it was not legally required. The reasons are clear enough; developers want help from FDA to ensure that their new products are safe before putting them on the market. Without that safety check, a new food released onto the market and later found to have, for example, new toxic substances would face (1) regulatory action from FDA for releasing an adulterated food and (2) litigation from unsuspecting consumers harmed from ingesting the adulterated food. With the dire consequences, especially of the latter, and with the simple and sensible procedures in the "voluntary" FDA consultation, any biotech food developer who bypassed the FDA review would be nothing short of foolhardy.

12.3.1.3. EPA. The US Environmental Protection Agency (EPA) is concerned with risks posed by pesticides (including herbicides). According to EPA, a pesticide can be any substance or combination of substances intended to prevent damage by any pest, or intended for use as a plant growth regulator. For transgenic plants, this usually means herbicide-tolerant or insect-protected cultivars, but can include others also. Importantly, EPA claims that it does not regulate the transgenic plant per se, but rather any pesticidal properties associated with the transgenic plant. Because of this pesticidal properties trigger, not all transgenic plants trigger EPA regulatory purview. For those transgenic plants with pesticidal properties, EPA issues permits for large-scale (>10-acre) field trials and seed increase plots, and also regulates commercial registration for any such plant varieties sold with pesticidal claims, such as *Bt* corn or herbicide-tolerant soybeans.

12.3.2. EU

The European Union (EU) seems most confused on the issue of biotechnology. Many leading scientific technical developments in biotechnology have occurred within the borders of EU nations, but the application and deployment of the technologies is chaotically skewed, with seemingly rapid commercialization of medical, food, and industrial biotech applications, while lagging in GM crop approvals and releases. It seems contradictory that hard cheese in the EU emanating from a GMO, albeit lacking in detectable GE DNA or protein, is exempt from regulations or special labeling, but corn, canola, or soybean oil, similarly lacking detectable GE DNA or protein, is so captured for both extensive regulatory oversight and product labeling. Internally, the European Union is not so united with regard to their views of regulation. Several member states appear at least somewhat supportive of agricultural applications of biotechnology, others are more hesitant, and some remain rigidly hostile.

Regulations are split among several pieces of legislation. In the early days of agricultural biotechnology, the EU split their regulations between two regulatory Directives, 90/119/EC covered contained use of genetically modified microorganisms, and 90/220/EC, which covered deliberate environmental release of genetically modified organisms. Both of these were later substantially amended; 90/119/EC was superseded in 1998 by 98/81/EC, and 90/220/EC was superseded by 2001/18/EC in 2001. In addition, Regulation (EC) No. 1830/2003 amended Directive 2001/18/EC, outlining traceability and labeling provisions for GMOs and their derived foodstuffs. Regulation EC 1829/2003 provides specific details for labeling requirements.

A listing of the 18 GMOs authorized under Directive 90/220/EC is available at http://ec.europa.eu/environment/biotechnology/authorised_prod_1.htm, and those proceeding under 2001/18/EC can be found at http://ec.europa.eu/environment/biotechnology/authorised_prod_2.htm. The EU also provides a listing of biotech products pending under 2001/18/EC at http://ec.europa.eu/environment/biotechnology/pending_products.htm.

In addition to these primary regulatory documents, Regulation EC 258/97, superseded by EC 1829/2003, covers approvals for "food and feed consisting of, containing or produced from genetically modified organisms," and Regulation EC 1946/2003 (http://ec.europa.eu/environment/biotechnology/pdf/regu1946_2003.pdf) provides the EU procedures governing the transboundary movements (i.e., international trade) of GMOs, effectively implementing the Cartagena Protocol, as well as unintentional transboundary movements.

Complicating this already complicated bureaucracy is the "safeguard clause," which allows member states to essentially opt out of accepting GMOs deemed safe under the various regulatory directives. This escape clause has been used liberally by member states hostile to GMOs. Member states invoking the safeguard clause are required to submit scientifically sound justification for rejecting the determination of safety, but in every case the scientific committee failed to find justification. In spite of this conflict within the vast European bureaucracy, GMOs remain relatively scarce in the farmers' fields (to date, although five member states—Czech Republic, France, Germany, Portugal, and Spain—are cultivating GMOs, and those are on tightly limited acreages). Foods derived from GMOs are even rarer, except for such examples as the hard cheeses produced with enzymes from GMOs, which escape regulatory scrutiny and labeling due to a convenient semantic distinction between foods produced *from* GMOs (which are captured for regulatory scrutiny) and those foods produced *with* GMOs (which, like the cheeses, are curiously exempt). Considering the public anxiety in Europe surrounding GMOs, it seems odd that the general public would appreciate the distinction between *from* and *with* to the extent that a food made *from* GMOs faces a heavy regulatory burden while a similar one, posing similar (insignificant) risk, made *with* GMOs gets a free pass without so much as a label.

Unfortunately, although the EU documents all claim to stem for a concern to protect health and the environment from risks associated with GMOs, nowhere are such risks documented and ascribed to GMOs specifically. Nevertheless, the EU strictly regulates almost all aspects of agricultural biotechnology and resulting products, making EU-approved products the most scrutinized and likely safest products ever to reach the commercial marketplace.

Interestingly, the United States, Canada, and Argentina brought suit against the EU in the World Trade Organization (WTO), claiming that these regulatory measures were illegal because they appeared to discriminate against "foreign" products of biotechnology and served as an illegal trade barrier. The WTO agreed, but the final resolution, if there is one, will probably take several years. A major issue is the focus on the assumed risks posed by biotechnology and its products. With scientific studies worldwide unable to document any health or environmental risks unique to GMOs, the EU was hard-pressed to justify their position in establishing regulations to protect against health and environment against "the risks inherent in GMOs." Indeed, European scientists have been actively busy searching for such risks for several years. According to Kessler and Economidis (2001), the European Commission itself spent 70 million euros to fund 81 research projects employing 400 teams of scientists between 1984 and 2000 to characterize risks associated with GMOs. Not one risk unique to GMOs was documented.

12.3.3. Canada

Canada remains unique worldwide for recognizing that risk is posed by potentially hazardous products, not by the process by which the products are made, and captures for regulatory oversight "novel" products, even some not developed using rDNA or other forms of biotechnology. Currently, all other jurisdictions use a process-based trigger for regulatory capture, and that process is rDNA (although the legal definitions of "rDNA" and "biotechnology" do vary considerably). To date, Canada remains the only country where the conclusions of the scientific community (viz., that breeding process is unrelated to risk) have been adopted into the regulatory practice. Once regulatory action is triggered, however, differing jurisdictions are remarkably consistent in their scientific risk assessments.

Canada assigns regulatory responsibility to three main federal agencies: the Canadian Food Inspection Agency (CFIA), Health Canada, and the Canadian Environmental Protection Agency (CEPA). Health Canada is responsible for food safety exclusively, while offices within CFIA are concerned with environmental issues and threats to animal feed. CEPA provides an insurance "catchall," capturing anything that appears to "fall through the cracks" or find "loopholes" to avoid regulatory scrutiny altogether. The "novel plant" approvals by CFIA in Canada can be found at http://www.inspection.gc.ca/english/plaveg/bio/dde.shtml.

12.3.4. International Perspectives

Regulatory agencies worldwide recognize that products of biotechnology can pose risks, the same as can similar products from other means of genetic manipulation, including traditional breeding. Simply because they are generated using rDNA does not make them benign; they may have food or feed safety issues, and they may have features enabling them to threaten ecosystems.

Food safety is a common fear, and food safety regulatory agencies worldwide consider the possibility that the regular food has intentionally or unintentionally become adulterated, toxic, or allergenic during the breeding process, or has significantly reduced (or enhanced) nutrients. All such agencies question the source of an introduced gene, to determine whether, for example, that source is allergenic. Because of the earlier work showing the allergenic Brazil nut storage protein to be allergenic even after gene transfer and expression in soybean (Nordlee et al. 1996), we know that allergenic proteins do not need their "home" genetic or physiological background to elicit an allergenic response. Similarly, food safety agencies are concerned with the potential for the transfer of potential toxic and other antinutritional substances from donor species, and the possibility that the transfer of even benign genes and proteins might exacerbate production of endogenous toxins, allergens, and antinutritional substances in the recipient species and foodstuffs.

Fortunately, both traditional and biotech crop and food developers also appreciate these real risks and conduct premarket testing to convince themselves (if not everyone else) that their new variety carries no additional toxic, allergenic, or antinutritional substances. Any breeding lines exhibiting such problematic substances are eliminated from consideration for commercialization long before any regulatory agency sees them. No company wishes to face the liability of releasing a true threat to health.

Also fortunately, human physiology being what it is, a toxin, allergen, or other antinutritional substance will pose the same risk to virtually everyone, worldwide. While there may be some differences in exposure, due to cultural or cuisine preferences or preparation methods, a toxin to western Europeans will also be toxic to Indians. This means that the basic safety testing will be common to all, so the questions asked and answers demanded by the US FDA or EFSA in Europe will be of interest to consumers worldwide, and food safety regulators need not duplicate the entire (and expensive) food safety bureaucracy, but may instead concentrate on local variations in cuisine, including consideration for method of preparation (e.g., cooked vs. raw) or overall exposure (e.g., a food may be a major dietary component in one culture and minor elsewhere).

The other main scientific concern for regulatory action, environmental risks, is more variable. Consensus in the scientific (if not always in the regulatory or political) community recognizes the factors in environmental risk is not the method of breeding but the species in question, the trait, and the region of release. While human physiology is much the same worldwide, ecosystems vary widely, such that a plant deemed benign by USDA APHIS for release in the United States might be wreak ecological havoc when released in the Amazon basin. Because of the environmental variation, regulatory agencies worldwide concerned with ecological effects cannot rely entirely on determinations made by regulators in a different environmental region. As not all countries or regions enjoy the regulatory resources of the United States, Canada, or the European Union, international efforts and regional coalitions attempt to economize biosafety review of potential threats to the environment.

One predominantly scientific society devoted to assessing environmental risk from products of biotechnology is the International Society for Biosafety Research (ISBR), which holds biennial symposia to discuss various scientific and regulatory developments concerning biosafety and how biotechnology may affect the biosphere, and is particularly concerned with the issues as they relate to developing countries. The proceedings of the last several years of these symposia are available online at http://www.ISBR.info. ISBR also publishes the scientific journal *Environmental Biosafety Research*, consisting of articles, commentaries, and editorials on research relevant to the ecological impacts of products of biotechnology. It is available online at http://www.isbr.info/journal/.

Another attempt to consolidate information on the risks of products of biotechnology and their potential effect on biodiversity, particularly in poorer countries, is the Cartagena Protocol, which emanates from the Convention on Biological Diversity (CBD); see http://www.biodiv.org/biosafety/. The objective is "to protect biological diversity from the potential risks posed by living modified organisms resulting from modern biotechnology" (http://www.biodiv.org/biosafety/background2.aspx).

Over 130 countries have signed the protocol, which obligates signatories to establish bureaucracies to identify, monitor, document and track living modified organisms (LMOs). The agreement covers international trade of the designated LMOs, which means viable, nonprocessed products of biotechnology. Essentially, this means grains and oilseeds such as soybeans, maize, canola, and cottonseeds, but not vegetable oils or food products derived from the commodities.

One useful provision of the Protocol is the Biotechnology Clearing House, a repository of information on living modified organisms (which, unfortunately, is defined by the process of biotechnology, not to actual threats to the environment). The Protocol, now ratified by 134 countries (although, to date, no major agricultural exporters) and the clearinghouse database allow countries access to information on particular GM crops and assist in making regulatory decisions on the degree of risk to local ecosystems. The portal to the clearinghouse is available online at http://bch.biodiv.org/. Unfortunately, the Cartagena Protocol is founded on the assumption that products of biotechnology presents a threat to biodiversity (see above paragraph), but no evidence to support this assumption is provided. Particularly unfortunate is the corollary assumption that biodiversity is threatened *only* by biotechnology, as all non-LMO grains oilseeds and other viable commodities in international trade are exempt. The many scientific studies of environmental risks posed by products of biotechnology invariably conclude that products of biotechnology do not pose any greater threat to environment than do conventional products, thus invalidating the underlying assumption of Cartagena Protocol. There remains not a single documented case where a GMO (or LMO) has caused harm to biodiversity (McHughen 2006). This means that the true threats to biodiversity, the things that have wreaked havoc in our planetary ecosystems over the years, will continue unabated, because Cartagena directs all regulatory resources to protecting against hypothetical risks (in LMOs) and no effort to stop the things that actually cause real harm.

12.4. CONCLUSIONS

Most current regulatory systems are scientifically flawed, in spite of assertions from politicians and regulators claiming that their system is indeed "scientifically sound". The scientific flaws are several, and each of them invalidates the entire regulatory structure:

1. Most regulatory bureaucracies assume that "traditional" means of genetic modification are risk-free, but that the processes of biotechnology inherently pose risk (Fig. 12.1). This assumption is rarely challenged, in spite of scientific studies over several years and from many countries establishing that the processes of biotechnology are not inherently more hazardous than other breeding methods [see, e.g. OECD (1986), NRC (1987, 2004), and Kessler and Economidis (2001)]. Of course, challenging the assumption opens the door to potential regulation for all products of plant breeding, not just those derived from rDNA. And since there is little or no public demand to launch risk assessments for conventional agriculture, the only scientifically valid position is to relax the strict regulation of at least some benign GMOs to the level of that imposed on conventional agricultural products of similar risk. In many parts of the world, relaxing regulatory oversight of GMOs is politically unpalatable, even if scientifically justified.

2. A major motivation in some jurisdictions to regulate GMOs exclusively is the assumption that transferring genes across the species barrier is unnatural and potentially hazardous. However, the concept of a rigid species barrier is itself inherently flawed, as there are countless examples, both in nature and under human manipulation, of moving

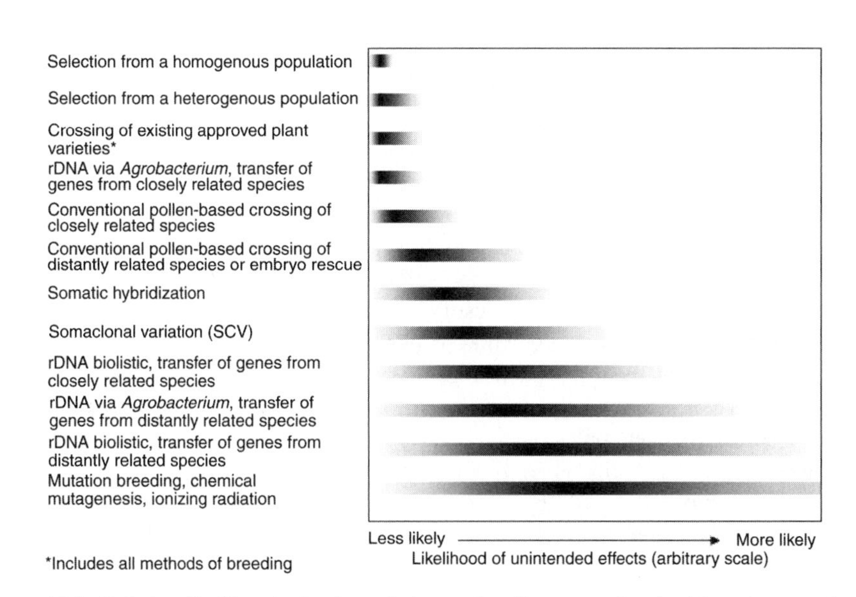

Figure 12.1. Relative likelihood of unintended genetic effects associated with various methods of plant genetic modification. The gray tails indicate the conclusions about the relative degree of the range of potential unintended changes; the dark bars indicate the relative degree of genetic disruption for each method. It is unlikely that all methods of genetic engineering, genetic modification, or conventional breeding will have equal probability of resulting in unintended changes. Therefore, it is the final product of a given modification, rather than the modification method or process, that is more likely to result in an unintended adverse effect. For example, of the methods shown, a selection from a homogenous population is least likely to express unintended effects, and the range of those that do appear is quite limited. In contrast, induced mutagenesis is the most genetically disruptive and, consequently, most likely to display unintended effects from the widest potential range of phenotypic effects. [From National Research Council (NRC) (2004); used with permission.]

genes from one species to another without added hazard. *Agrobacterium tumefaciens* is a natural genetic engineer, moving short pieces of prokaryotic bacterial DNA into eukaryotic nuclei and having the transferred genes integrate into the host genome. Many popular bread wheat (*Triticum*) cultivars carry fragments of rye (*Secale*) chromosomes.

3. No jurisdiction has sufficient resources to "test everything for everything," so a sensible system to prioritize regulatory resources evolved the maxim that products posing the greatest risk should face the greatest regulatory scrutiny. But this sensible approach has been abandoned in the case of biotechnology. Consider two canola cultivars, one made resistant to acetolactate synthase (ALS) inhibitor herbicides using rDNA, and the other with identical herbicide resistance, except that it was developed using induced mutagenesis. The two similar cultivars pose similar risks, yet the biotech cultivar faces far greater regulatory scrutiny. Similar cultivars posing similar risks should face similar degrees of regulatory scrutiny.

4. There is an unsubstantiated assumption that the risks posed by biotechnology are unique and should be evaluated as absolutes. However, risk is relative or comparative. Instead of asking "What are the risks associated with this GMO crop cultivar?" a scientifically valid question is "What are the risks associated with this GMO crop cultivar relative to the risks associated with the conventional cultivar that it will displace?" By focusing exclusively on the "new" thing and ignoring the status quo or current counterpart cultivar, any identified risk with the GMO cultivar (and everything poses some degree of risk) can be and has been used as an argument to justify banning the GMO, even though a proper, relative risk assessment might show it to be substantially superior to the riskier but currently grown cultivar.

5. The assertions that regulations are scientifically sound are invariably buttressed by scientific documentation showing the technical validity of the various assays, tests, measurements, and other criteria required by the risk assessors. But this line of argument merely supports the technical, not the overall, scientific validity. Technical skill in conducting a technically sound assay is not sufficient to satisfy scientific validity; it is necessary in addition to scientifically justify the rationale for conducting the test in the first place. For example, testing an extracted purified protein from a GM plant for possible allergenicity might seem a prudent regulatory requirement. But conducting the allergenicity trials is not scientifically valid, even if the trials themselves are conducted in a technically sound manner, unless there is a hypothesis or evidence suggesting that the protein may actually be allergenic. If the gene were cloned from a known allergenic source, or if the protein shared amino acid sequence homology with a known allergen, then yes, the technical allergenicity assays might be scientifically valid and prudently required. But to demand and conduct such trials merely to show the public that scientific tests for potential allergenicity were being conducted, or to exercise control over the developer, fails to increase real confidence in the safety of the product and jeopardizes public trust when the test was later found to be unnecessary, done only to appease public concerns.

6. Most risk assessments of GM plants are overly onerous and unnecessary in terms of informing risk management policies. Once sufficient data are collected to reach a determination of relative safety (or otherwise) of the GM cultivar, the law of diminishing returns kicks in; input resources escalate dramatically, but the additional data gleaned from the expenditure are usually superfluous and unconstructive. These additional data requirements

undermine public confidence without adding any compensating features or increased assurance of product safety. The curious public wonders why so many additional and apparently unnecessary tests are demanded, and speculates that perhaps this product really is more hazardous than the developer and regulators are letting on. So the barrage of demanded tests and assays, instead of increasing public confidence, has the opposite effect—the public becomes even more suspicious and distrustful of both the product and the regulatory system.

The lesson is simple; to increase public trust as well as to increase confidence in product safety, before imposing and requiring any test or assay, the regulator should be able to answer "How will the information from this test/assay help inform or increase confidence in the safety (or otherwise) of this product?" If the answer is simply "more of the same," the regulatory demand is not scientifically valid and disrespects the public right to effective regulation.

Worldwide, regulations governing products of biotechnology purport to protect health and environment from the risks posed by the "new" genetic technologies. However, there is a crucial disconnect between the regulatory approach and actual protection of health and environment, because the regulations, almost invariably, assume that products of biotechnology pose greater risk than do similar products generated using other methods of gene manipulation. Reports from scientific bodies in the United States, European Union, and elsewhere, going back to the mid-1980s, establish that GMOs (i.e., products of biotechnology) are not inherently more hazardous. Therefore, regulations that specifically capture GMOs for regulatory scrutiny and exempt similar products produced using non-GM methods of breeding are scientifically unjustified and thus scientifically invalid.

The ramifications of this approach are clear. Health is sometimes threatened by food-borne hazards (BSE, dioxins, *Salmonella* and foot-and mouth disease outbreaks in EU; diarrhea-causing strains of *Escherichia coli* in organic produce in USA; etc.). Ecosystems and biodiversity have clearly suffered from human agricultural activity and breeding, such as the introductions of invasive species, particularly in Australia and North America. To date, there are no verified cases of damage to human or animal health, or to the environment, from GMOs. All recorded threats come from non-GMOs. Yet almost all regulations capture for scrutiny only those products resulting from the process of biotechnology, and explicitly exempt non-GMOs, the sources of all known damage. The disconnect between the regulatory practice and the scientific recommendations ensures continued threats and damage to health and environment, and will do so until regulations capture and scrutinize those products posing the greatest risk, regardless of breeding method.

Recognizing that regulatory resources are limited even in affluent societies means rigid adherence to the regulatory maxim; regulatory resources should focus on the highest-risk products. In other words, to be effective, regulatory scrutiny should correlate with degree of risk. In that situation, regulators will concentrate their expertise to scrutinize true threats to health or environment, and place less emphasis on those products posing little or no threat.

Adopting this policy will mean capturing some products currently exempt from regulatory scrutiny, including so-called traditional means of gene modification such as induced mutagenesis using ionizing radiation. It will also mean exempting some currently captured low-risk products of rDNA technology. In essence, it means abandoning the popular but scientifically flawed approach of capturing only those products derived from the processes of biotechnology. Such an approach does not necessarily mean an increase in overall regulatory burden, as the newly captured, more hazardous products are offset by the exemption of less hazardous, currently captured products. Also, the elimination of excessive or redundant assays beyond those necessary to reach a scientifically valid safety determination will reduce the burden all around. If politicians and regulators adopt their own maxims, perhaps we can enjoy some true protection from the dangerous things currently eroding our health and planetary ecosystems.

LIFE BOX 12.1. ANTHONY SHELTON

Anthony Shelton, Professor of Entomology, Cornell University/NYSAES

Tony Shelton trying to catch up on email, a never-ending job in academics.

I took a circuitous route in my formal studies as a scientist, but don't regret it. Although I was accepted into a premed program, during my first week on campus I transferred into the Great Books of the Western World Program. In this unique program we had no formal lectures but rather a Socratic dialogue in all our classes. We only read original works, no textbooks. People interpreted the texts, argued their views and came up with a better understanding based on the discussion. Freshman year we studied Euclid and the Greek classics and worked our way up to Einstein and Joyce in our senior year. After receiving my B.A. in classics and philosophy, I went into the biological sciences. Entomology was particularly appealing since it combined my love of ecology, biology, food systems and the environment. Like many young people at the time, I was tremendously influenced by Rachel Carson's seminal book, *Silent Spring*. There had to be a way of producing our food and fiber in a more environmentally responsible manner, and the idea of integrated pest management (IPM) was becoming a buzz word. IPM focused on understanding insect–plant interactions within the environment and using host plant resistance and biological control as the foundations for managing pests. Over the years, this concept of IPM has become the standard practice. However, we never really had any food plants that were strongly resistant to caterpillars (Lepidoptera) or beetles (Coleoptera) and in most agricultural systems biological control couldn't cut it alone, so insecticides continued to play a key role in IPM. One interesting insecticide was the bacterium, *Bacillus thuringiensis* (Bt), which could be sprayed on the plant and was strongly promoted by Rachel Carson as an alternative to broad-spectrum insecticides. When caterpillars took a bite of foliage treated with Bt they were killed by a protein produced by Bt, but this protein did not affect mammals and most other organisms. The problem was that it was impossible to treat all the surfaces where an insect would feed and growers had to treat often since sunlight quickly broke down Bt. With the advent of genetic engineering beginning in the 1970s, scientists began to see many different uses of this new technology in agriculture. One of the first was to insert Bt genes into plants so the plants would produce essentially the same Bt proteins that were in the foliar spray. In 1996, the first genetically engineered Bt plants were commercialized and, by 2005, they were grown on 26.3 million hectares. Finally, we had some plants that were resistant to some caterpillars and beetles! Perhaps we were on the road forward that Rachel Carson had advocated.

However, the road forward with genetically-engineered insect resistant plants has had a few bumps in it. On the one hand, the adoption of Bt plants has risen incredibly quickly in several countries and has led to dramatic reductions in the use of "harder" insecticides, fewer pesticide poisonings, and improved farmer income. Additionally, the fear that insects would rapidly become resistant to Bt plants has not materialized even after more than 10 years (this is in stark contrast to nearly all other insecticides). Lack of resistance to date may be due to the wisdom of creating Bt plants with a high enough dose that heterozygosity for resistance would be controlled (it is the heterozygous individuals that drive resistance in a population) and the requirement of having refuges of non-Bt plants so that susceptible alleles would be maintained in the population.

Additional fears that Bt genes would spread to wild and weedy relatives and cause environmental havoc and that non-target organisms, especially biological control agents, would be negatively impacted have proven to be unfounded. In fact, Bt plants have advanced the use of biological control because they have reduced the use of broad-spectrum insecticides that are harmful to many biological control agents. However, regulatory issues and acceptance of Bt plants in some countries has been problematic. Bt plants and other products of biotechnology have been called everything from "unnatural and playing God" to "Frankenfoods." If you asked 100 people in the general public who were opposed to genetically engineered plants their reasons for their position, you'd likely get many different answers including questions about long term food safety issues, corporate control of agriculture, and globalization. Few would be knowledgeable enough to ask or interpret the technical issues and to analyze the risks and benefits of using this new technology compared to continuing with older technologies for insect management, many of which are far more hazardous. From a scientific standpoint, the environmental and health benefits of Bt plants have been well documented. However, these benefits often get lost in the bigger discussion on biotechnology, and this presents a serious dilemma in a democratic society.

I strongly believe that scientists have an obligation to make their voices heard on important issues such as genetically engineered plants for pest management, but we must do so in a responsible manner. Isn't it our obligation to help inform the public dialogue on these issues? Who else is more qualified to do so?

LIFE BOX 12.2. RAYMOND D. SHILLITO

Raymond D. Shillito, Regional Manager External Technical Services—Americas Molecular and Biochemical Analytical Services of Bayer CropScience LP

Ray Shillito

How did I end up doing what I am doing now? I followed my instincts, stayed open to possibilities, made mistakes, collaborated with good people, and never stopped learning. My advice is to find something you enjoy doing, as you will usually be good at it. One major thing I learned along the way was, in research, to only try to do one difficult thing at a time. Another is that traditional biochemistry was a great basis for work in this field.

My entry into this field was through tissue culture, a discipline which is still way underestimated by most. I studied Quantitative Biochemistry and became interested in obtaining auxotrophic mutants in plants. I was fortunate enough to get a position to do a Ph.D. at the University of Leicester with Professor H. E. Street, a major force in plant tissue culture. Towards the end of my studies, we discussed using insertion mutagenesis with *Agrobacterium* to make mutants and H. E. suggested the laboratory of Prof. Schilperoort. Sadly, H. E. Street died at Christmas 1977. Lyndsey Withers and Bill Cockburn helped me complete my thesis, and I obtained a grant to study in Schilperoort's laboratory at Leiden in the Netherlands.

When I arrived, Loci Marton had just left, having shown that regenerating *Nicotiana* protoplasts could be transformed by *Agrobacterium*. Thus I was introduced to *Agrobacterium* and protoplasts, and to the worlds of gene transfer and molecular biology. Eventually we moved to Basel Switzerland where Ingo Potrykus assembled a team at the Friedrich Miescher Institut to work on transformation and protoplasts; Jurek Paszkowski and Mike Saul completed the core team. Work at the FMI was very collaborative, and I enjoyed collaborating with the groups led by Pat King, and Barbara and Tom Hohn. In 1983, we were able to transform protoplast directly using DNA without any *Agrobacterium* sequences. This was a major contribution to the field, and we had an excellent experimental system to test other ideas. It led to transforming protoplasts at high efficiency and studies of co-transformation and expression of selectable markers, and of inheritance of introduced genes. My interest in *Agrobacterium* led to collaboration with Szdena Nicola-Koukolikova and others to investigate the structure of DNA transferred during transformation. This was a very stimulating time, and I was lucky to experience working with a fun and successful group of people.

After 5 years Mary-Dell Chilton gave me the opportunity to move to the USA, to Ciba-Geigy's Biotechnology effort. Due mainly to the efforts of Catherine Cramer and Gleta Carswell, we were able to show regeneration of elite maize protoplasts by 1988. Next my traditional biochemistry training helped in developing a novel selection method for PPT resistance using a pH indicator. We were able to use it to select transformed colonies arising from maize protoplasts and thus regenerate a transformed maize plant. The pH indicator method was used by Martha Wright and her team to obtain what eventually became event 176.

When I moved to AgrEvo to build and manage a group to do regulatory studies I moved from research to development. This opportunity gave me the chance to get to know crop plants in the real world. My role increasingly became one of technical expert in the testing field and dealing with outside laboratories and agencies. My present work involves establishing and maintaining contacts with a multitude of stimulating people on a daily basis. I get to use the skills and knowledge I have learnt along the way, including the quantitative approach to analysis, which means I have in a way come full circle.

Overall I can say I have enjoyed most of the journey. I still retain my links with the tissue culture community, as this is the basis of modern agricultural biotechnology. I have been fortunate to have the company, advice and mentoring of many people as mentioned above, including those who were authorities in their field. I have been lucky enough to follow and grow with a technology from its inception through to its implementation in agriculture, and have got a great deal of enjoyment out of it. I will leave it to others to judge my contribution.

As to where this discipline and plant biotechnology is headed; we are gradually overcoming misinformation and misconceptions. I hope that we can be allowed to use the technology to benefit those who need it most, without over-restrictive regulation; where delivery of critical traits to small subsistence farmers in a safe form of seed which is easily used is a daily occurrence rather than a dream. However, this will take time, maybe too much time for some.

REFERENCES

Berg P, Baltimore D, Boyer HW, Cohen SN, Davis RW, Hogness DS, Nathans D, Roblin R, Watson JD, Weissman S, Zinder ND (1974): Letter: Potential biohazards of recombinant DNA molecules. *Science* **185**:303.

Cohen SN, Chang AC, Boyer HW, Helling RB (1973): Construction of biologically functional bacterial plasmids in vitro. *Proc Natl Acad Sci USA* **70**:3240–3234.

Fraley RT, Rogers SB, Horsch RB (1983a): Use of a chimeric gene to confer antibiotic resistance to plant cells. *Advances in Gene Technology: Molecular Genetics of Plants and Animals. Miami Winter Symposia*, Vol 20, pp 211–221.

Fraley RT, Rogers SG, Horsch RB, Sanders PR, Flick JS, Adams SP, Bittner ML, Brand LA, Fink CL, Fry JS, Galluppi GR, Goldberg SB, Hoffmann NL, Woo SC (1983b): Expression of bacterial genes in plant cells. *Proc Natl Acad Sci USA* **80**:4803–4807.

Herrera-Estrella L, Depicker A, van Montagu M, Schell J (1983): Expression of chimaeric genes transferred into plant cells using a Ti-plasmid-derived vector. *Nature* **303**:209–213.

James C (2005): *Global Status of Biotech/GM Crops in 2005*. ISAAA Briefs No. 34-2005: Executive Summary http://www.isaaa.orgkc/bin/isaaa_briefs/index.htm.

Jia S, Peng Y (2002): GMO biosafety research in China. Environ *Biosafety Res* **1**:5–8.

Kessler C, Economidis I (2001): *EC Sponsored Research on Safety of Genetically Modified Organisms. A Review of Results*. European Communities, Luxembourg. Also online at http://europa.eu.int/comm/research/quality-of-life/gmo.

Macilwain C (2003): Chinese agribiotech: Against the grain. *Nature* **422**:111–112.

McHughen A (2006): Problems with the Cartagena Protocol. *Asia Pacific Biotech* **10**:684–687.

Murai N, Sutton DW, Murray MG, Slightom JL, Merlo DJ, Reichert NA, Sengupta-Gopalan C, Stock CA, Barker RF, Kemp JD, Hall TC (1983): Phaseolin gene from bean is expressed after transfer to sunflower via tumor-inducing plasmid vectors. *Science* **222**:476–482.

National Research Council (NRC) (1987): *Introduction of Recombinant DNA-engineered Organisms into the Environment, Key Issues*. White paper, National Academies Press, Washington, DC.

National Research Council (NRC) (2000): *Genetically Modified Pest Protected Plants: Science and Regulation*. National Academies Press, Washington, DC.

National Research Council (NRC) (2002): *Environmental Effects of Transgenic Plants*. National Academies Press, Washington, DC.

National Research Council (NRC) (2004): *Safety of Genetically Engineered Foods*. National Academies Press, Washington, DC.

Nordlee JA, Taylor SL, Townsend JA, Thomas LA, Bush RK (1996): Identification of a Brazil-nut allergen in transgenic soybeans. *New Engl J Med* **334**:688–692.

Organization for Economic Cooperation and Development (OECD) (1982): *Biotechnology, International Trends and Perspectives*.

Organization for Economic Cooperation and Development (OECD) (1986): *Recombinant DNA Safety Considerations: Safety Considerations for Industrial, Agricultural and Environmental Applications of Organisms Derived by Recombinant DNA Techniques* (http://www.oecd.org/dataoecd/45/54/1943773.pdf).

Schell J, van Montagu M, Holsters M, Zambryski P, Joos H, Inze D, Herrera-Estrella L, Depicker A, de Block M, Caplan A, Dhaese P, Van Haute E, Hernalsteens J-P, de Greve H, Leemans J, Deblaere R, Willmitzer L, Schroder J, Otten L (1983): Ti plasmids as experimental gene vectors for plants. *Advances in Gene Technology*: Molecular Genetics of Plants and Animals. Miami Winter Symposia, Vol 20, pp 191–209.

US Office of Science and Technology Policy (OSTP) (1986).

USDA/NASS (2006): http://usda.mannlib.cornell.edu/usda/nass/Acre//2000s/2006/Acre-06-30-2006.pdf.

CHAPTER 13

Field Testing of Transgenic Plants

DETLEF BARTSCH and ACHIM GATHMANN

Federal Office of Consumer Protection and Food Safety, Berlin, Germany

CHRISTIANE SAEGLITZ

Institute of Environmental Research, Aachen University, Germany

ARTI SINHA

Department of Biology, Carleton University, Ottawa, Ontario, Canada

13.0. CHAPTER SUMMARY AND OBJECTIVES

13.0.1. Summary

When companies or academic labs develop transgenic plants with proven traits, they must assess growth and yield under field conditions. In addition, environmental risk analysis experiments are also commonly performed in the field. A tiered assessment is recognized as being the most appropriate and rigorous approach to assess environmental and economic effects from both scientific and regulatory standpoints. Field design and statistical considerations are described here to assess the performance of transgenic plants using transgenic corn as an exemplary case study.

13.0.2. Discussion Questions

1. What are the two overarching objectives for field testing of transgenic plants?
2. What two factors are determined for risk assessment?
3. What are some important and appropriate controls for field testing—say, for Bt crops?
4. Give some examples of lower-tier experiments versus upper-tier tests. Why bother with lower-tier tests?
5. Discuss what factors would be needed for the risk assessment of a nonagronomic trait, such as a pharmaceutical. Where would the risk assessor begin, and how would we know when the risk assessment is over—that is, a decision between safe and not safe?
6. Which is more important: that a field test be performed for grain yield or environmental biosafety?

Plant Biotechnology and Genetics: Principles, Techniques, and Applications, Edited by C. Neal Stewart, Jr.

13.1. INTRODUCTION

Field testing is an important last step in the creation of transgenic plants. Two important and interrelated aspects are discussed here: agronomical performance and biosafety. If a company wants to commercialize a transgenic crop variety (and typically they do), it is important to show that it performs as well as its *parent* or *isogenic* variety under a number of locations and types of fields. To be viable, it cannot have any genetic or phenotypic malformations. So, experiments must be performed to compare growth and yield, as well as test for the durability and robustness of the transgenic trait in the field. For example, an insect-resistant plant would be required to adequately express the transgene to kill target insects. Robustness of expression under field conditions is needed to guarantee farmers economic benefits. The second part of field testing, *biosafety*, is more complicated, and it is important to show that a transgenic product is environmentally benign. This has proved to be crucial for placement on the market since worldwide regulation requires a number of tests to convince regulators to approve transgenic plants as safe (see Chapter 12). This chapter focuses on field testing to evaluate the environmental safety (exemplifying any risks for nontarget butterfly caterpillars) and the economic benefit (on yield) posed by genetically modified Bt maize. Thus, we will use this particular case as an example, since it covers much of the ground that is needed for field testing. Some examples, such as those for herbicide resistance, might be simpler, and there are, no doubt, more complicated cases. We use Bt maize since there is a large body of knowledge that has been accumulated over the past 10 or so years and it is a success story of sorts. Since the mid-1990s, genes of *Bacillus thuringiensis* (Bt) that encode butterfly-specific toxins (Cry1Ab, Cry1Ac, and Cry9) were engineered into maize for protection against the European corn borer (*Ostrinia nubilalis*). Bt maize has been hailed as a success, having essentially passed all the regulatory tests in the United States and the European Union (EU), and is grown widely across North America, part of the EU, and other areas.

13.2. ENVIRONMENTAL RISK ASSESSMENT (ERA) PROCESS

Environmental risk assessment (ERA) is particularly significant in the context of genetically modified organisms (GMOs). There are some good reasons to be careful when introducing new technologies, in particular when new biopesticides are introduced into the environment. However, it is believed by some concerned people that any (as yet unperceived) effects they have on the environment could be adverse, if not downright "catastrophic." Whatever the starting point is, a scientifically sound ERA factors in the following aspects:

1. Biological properties of the parental unmodified organism (maize in our example)
2. Source of the introduced gene(s), expression, and nature of the gene product ("Bt" protein kills pest caterpillars, but may also affect "lovely" nontarget butterflies)
3. Characteristics of the genetically modified organism, including its performance and impact on the environment, taking into account the information of points 1 and 2 (above)

Environmental risk assessment has a conceptual framework consisting of four steps described briefly below: evaluation of need for ERA, problem formulation, controlled experiments and gathering of pertinent information, and finally the risk evaluation.

13.2.1. Initial Evaluation (ERA Step 1)

The initial evaluation of need determines whether a risk assessment is required for a specific case. Clearly defining the need as it meets the expectations of the final audience will help in designing the overall risk assessment and determining how the information will be used and communicated. Common reasons for conducting an ERA include regulatory requirements, scientific inquiry, and scientific responses to public concerns.

13.2.2. Problem Formulation (ERA Step 2)

Once the need for the ERA has been clearly defined, the risk assessment moves forward to the problem formulation phase. In this stage, appropriate risk hypotheses are defined in order to address the scope of the assessment (e.g., whether Bt maize harms lovely nontarget butterflies more than does conventional pest control). Biological aspects of the system, such as the specificity of the mode-of-action and expression (of the particular genetic trait), the spectrum of Bt activity, and Bt susceptibility of caterpillar, as well as relevant exposure profiles are considered while formulating the hypotheses. Other points to consider while identifying potential risks include the intended scale of cultivation (total USA or only a few states) as well as other ecological considerations that might affect the environmental impacts (e.g., protected areas with lovely butterflies near cultivation sites).

13.2.3. Controlled Experiments and Gathering of Information (ERA Step 3)

The next step in the ERA involves conducting tests and experiments to gather data pertaining to the study. For example, only a selected group of lovely butterflies can be feasibly studied at one time under laboratory and later field conditions. Hence, species selection must be done very carefully—ensuring that the butterfly species represent both ecologically and economically important taxa.

13.2.4. Risk Evaluation (ERA Step 4)

The overall assessment of the risks is a complicated process. Evaluation of risk would involve the consideration of several perspectives, and can easily go haywire. What is known as the *tiered risk assessment* model was introduced to enable a standardized scientific evaluation of risks internationally. This method consists of several tiers, each consisting of a description of the "problem" at a specific level and the approach to be followed in dealing with it.

13.2.5. Progression through a Tiered Risk Assessment

A *tiered risk assessment* is recognized as being the most appropriate and rigorous approach to assess nontarget effects from both scientific and regulatory standpoints. Both hazards and exposure can be evaluated within different levels or "tiers" that progress from worst-case scenarios framed in highly controlled laboratory environments to more realistic conditions in the field. Lower-tier tests serve to first identify and test potential hazards, and they are conducted in the laboratory to provide high levels of replication and control, which increase the statistical power to test hypotheses. Where no hazards are identified and the transgenic crops are not different from conventional crops, the new product is regarded as "proven safe." Where potential hazards are detected in these early tier tests, additional information is required. In these cases, higher-tier tests can serve to confirm whether an effect might still be detected at more realistic rates and routes of exposure. Higher-tier studies, including semifield or field-based tests, offer greater environmental realism, but they may have lower statistical power. Lower statistical power means that the there is a greater likelihood that real effects will not be observed (false negative). One reason for lower power is the high variability of environmental conditions (e.g., climate) that might counteract GM trait-specific effects. Nevertheless, these higher-tier tests are triggered only when early tier studies in the laboratory indicate potential hazards at environmentally relevant levels of exposure. In exceptional cases, higher-tier studies may be conducted at the initial stage when early tier tests are not possible or meaningful. For example, plant tissue might be used because purified protein is not available for lower-tier work. Higher levels of replication or repetition may be needed to enhance statistical power in certain circumstances. In cases where a potential hazard is detected in a lower-tier test, the tiered approach provides the flexibility to undertake further lower-tier tests in the laboratory to increase the taxonomic breadth (e.g., testing more insect species) or local relevance of test species, thus avoiding the costs and uncertainties of higher-tier testing. Depending on the nature of the effect, one may also progress to higher-tier testing anyway, particularly in cases where there is no previous experience with the crop or protein under investigation. The various tiered approaches that have been described for *nontarget risk assessment* differ in their specific definitions of individual tiers, but they all follow the same underlying principles. Higher-tier tests usually involve semifield or field tests and sometimes are conducted when lifecycle (especially reproduction parameters) or *tritrophic* evaluations are warranted. In general, these tests are problematic because of their complexity and high intrinsic uncertainty. Higher-tier tests require expertise and care in experimental design, execution, and data analysis. As a consequence they are subject to problems of low statistical power, particularly if they are used for "proof of hazard." These tests should therefore be conducted only when they can further reduce uncertainty in the risk assessment, and only when justified by detection of unacceptable risk at the lower tiers of testing. For further reading, see the paper by Romeis et al. 2008. Statistical power has been mentioned several times, and this concept requires clarification. Multiple samples and replicates of experiments are needed for high statistical power, which we can define here as the ability to detect real differences that might exist. Biological systems are highly variable, and statistical tests help researchers test hypotheses, for example, it the differences observed are due to chance variation or result from expression of a transgene. Lower-tiered experiments that can be tightly controlled offer higher capacity to detect real differences than when we layer field effects on higher-tiered experiments. The ground rule is that the more lifelike the experiment, the bigger and more expensive it will be to truly understand natural variability and variability caused by the transgene addition.

13.3. AN EXAMPLE RISK ASSESSMENT: THE CASE OF Bt MAIZE

Let us examine the scenario that has garnered the most attention in the risk assessment world: Bt maize pollen exposure. During flowering, maize pollen might land on leaves of host plants (hosts or food for insects) growing in and around maize fields, and these plants might be consumed by caterpillar larvae. Fields and field margins are important habitats for some butterfly species. As a consequence of the intensification of agricultural practices and the loss of (semi)natural habitat types, field margins have become increasingly important habitats for conserving biodiversity (ERA step 1). *Risk* is defined as a function of the adverse effect (*hazard* or *consequence*) and the likelihood of this effect occurring (*exposure*). For butterfly species the potential hazard is the toxicity of pollen containing Bt protein, and the likelihood of the event is the environmental exposure of caterpillars to the pollen (ERA step 2). Laboratory studies show that monarch butterfly caterpillars that consume Bt maize pollen from the transgenic event Bt 176 had higher mortality, slower development, and lower pupae weights than did those fed non-Bt control pollen (ERA step 3). This result caused a great deal of angst, which was accompanied by media and regulatory attention. This case shows that extrapolation of laboratory data to field scenarios can be quite controversial; this case has been among the most (if not *the* most) controversial of all from GM plants. Laboratory tests provide information on toxicity and fitness parameters, but they often represent "worst-case scenarios," which do not reflect field conditions or population processes that operate over farming landscapes. For example, maybe under tier 1 tests caterpillars were force-fed too much pollen compared with realisitic field exposures. Therefore, adverse effects identified in laboratory studies must be verified under field conditions because spatial, temporal, and environmental factors can alter possible adverse effects from, for example, exposure to the Bt protein or temporal overlap between pollen shed and phenology of butterfly caterpillar. One experimental exposure to Bt protein study under field conditions was performed by the authors of this chapter in Germany. In a database survey it was shown that approximately 7% of the German butterflies (macrolepidoptera species) occur mainly in farmland areas where maize is grown [for further reading on how this was done, see Schmitz et al. 2003]. The case study summarized below addresses some of the issues discussed above. In particular, this study attempted to compare the effect(s) (if any) of Bt maize on nontarget lepidopteran larvae, with that of conventional insecticides. The suitability and efficacy of the experimental designs and methods used for ERA were also evaluated. It is important that proper comparisons and control treatments be used in ERA experiments to ensure that results are relevant to real agriculture. Since most farmers would spray insecticide instead of simply letting insects eat their entire crop, it is important that ERA for insect resistant transgenic plants such as Bt maize include comparisons using chemical insecticides, since this is what most farmers use to control damaging insects. There are a few researchers who would like to use, as the main baseline, idyllic conditions that do not exist in much of real agriculture, but that would not be a fair, realistic, or useful comparison.

13.3.1. Effect of Bt Maize Pollen on Nontarget Caterpillars

An experimental maize field in Germany was studied over a 3-year period from 2001 to 2003. The field was divided into plots about 0.25 ha in size surrounded by a strip of conventionally grown corn with a minimum of 4.5 m in width (Fig. 13.1). There were 24 plots in total, on which corn was cultivated in three different ways (or in more precise terminology, *treatments*). The maize treatments were used in a randomized pattern to avoid side effects from the surrounding environment. A conventional variety, "Nobilis" with a similar genetic background but no transgene was used. Recall from Chapter 3 that this is called *near-isogenic*. ISO (O for control) was the control treatment using the near isogenic plant with no insecticide spray. This treatment provided a baseline for any assessment of effects. In the second treatment the near-isogenic variety was sprayed with the chemical insecticide Baytroid (this treatment is abbreviated INS), which simulated classical pest

... some field site data

- Cultivation of Bt corn event Mon 810
- Plot size 54 × 46 m
- 72 rows per plot
- Spacing: 75 cm between rows, 17 cm within rows
- Sowing date: between end of April and beginning of May
- Herbicide treatment (Callisto), three-leafe stage
- Insecticide treatment (Baytroid) of
  INS plots beginning of July depending on phenology of ECB
- Harvest date in October

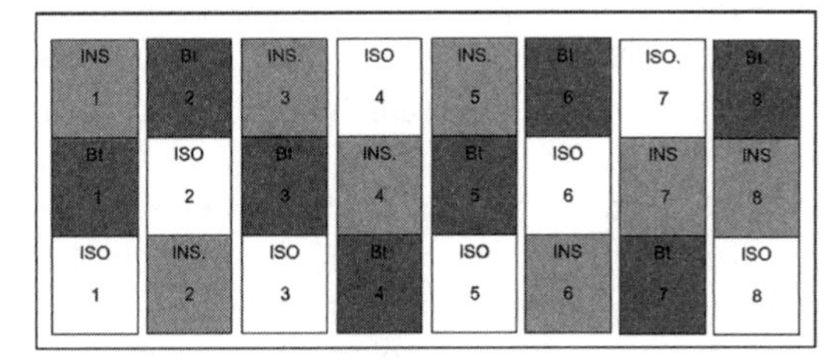

Figure 13.1. Field trial design for testing the environmental impact of bioinsecticidal (Bt) corn pollen or chemical insecticide use on nontarget butterflies. The cornfield trial was performed in an area with European corn borer (ECB) infestation. The trial consisted of eight replications and three treatments (INS = chemical insecticide on a conventional ECB-susceptible variety, ISO = conventional ECB-susceptible variety without any pest ECB control, Bt = GM corn with internal biopesticide protection against ECB).

control. Bt maize, variety Novelis transgenic event MON810 (abbreviated Bt), which synthesized the Cry1Ab protein insect control, was used in the third treatment. As host plants (weeds) for standardized attraction of butterflies, we used the artificial plantation of goosefoot (*Chenopodium album*) and mustard (*Sinapis alba*) within corn rows (Fig. 13.3).

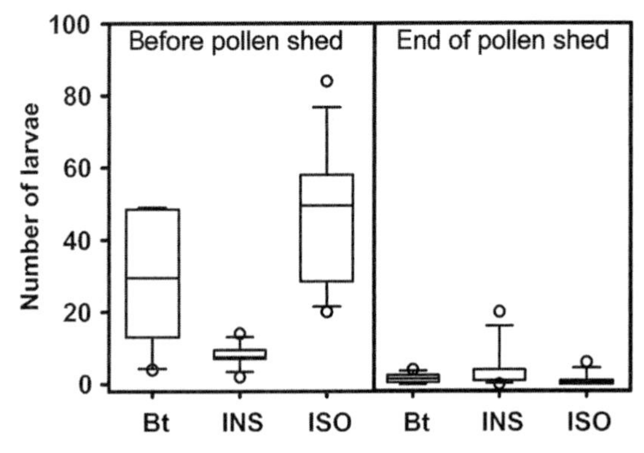

Figure 13.2. Collection of caterpillar larvae on weeds in cornfields. Presented are the median number of *Plutella xylostella* (diamondback moth caterpillars) on their host plant *Sinapis alba* (white mustard) in the three different treatment Bt corn (Bt) and the near-isogenic variety with (INS) and without (ISO) chemical insecticide application. Larvae were collected before and at the end of pollen shed. Results are pictured as box plots. Horizontal line indicates median of individuals per plot (eight replications for each treatment); box represents 75% of all values; upper and lower dashes represent 90% of all values. Dots indicate extreme values.

Pollen densities on the host leaves were estimated using a double-sided adhesive tape glued onto microscope slides. Caterpillars were sampled from the plants at the beginning and end of pollen shed. They were carefully replaced back on the plant after identification. Conventional bioinsecticides based on Bt protein have been used for several years in the control of pests, even before the development of Bt transgenic crops. Studies on their nontarget effects have generally shown no negative impact on predator (parasitoids to the insects) populations. However, these results cannot be fully extrapolated to plants expressing the Bt protein. The microbial Bt products contain Bt protoxins, which are activated in the insect's midgut by proteases (see Chapter 8). Some of the transgenic plants, on the other hand, express partially activated Bt proteins, which could have a potentially different impact on the insect populations. Hence, it can be argued that there is a need to investigate whether the unique delivery system, and the constant exposure of the protein to the insects, has an effect on natural enemies.

13.3.2. Statistical Analysis and Relevance for Predicting Potential Adverse Effects on Butterflies

Field testing requires careful analysis. For the German field trial on caterpillars, we used a statistical evaluation called the "proof of safety" between Bt maize and the nearisogenic variety (ISO). Maize pollen density was estimated to be 52–972 pollen grains/cm^2 on *Chenopodium album* and 100–894 pollen grains/cm^2 in *Sinapis alba*. No significant differences were observed in pollen densities between plant species. Note the wide range of potential exposures. Of the nine butterfly species recovered from the field, only two—*Plutella xylostella* L. and *Pieris rapae* L.—were abundant enough to be considered for statistical analysis. Caterpillars in both of these specieas are considered to be pests on mustard crops, such as canola, cabbage, and broccoli, but not on maize. Throughout the study period, the numbers of caterpillars (of both *P. xylostella* and *P. rapae*) were lower in plots with insecticide treatment (Fig. 13.2). Pollen density on the plant leaves can be affected by several factors, including relative humidity, growth stage, and distance from maize fields, as well as shape and structure of host leaves (e.g., waxy or hairy surfaces). It was observed that more pollen was shed (as inferred from pollens/cm^2) from Bt maize in comparison to the conventional maize; however, this could be attributed to the better health of the plants themselves. The Bt plants were observed to be more robust than their isogenic counterparts because they were not damaged by European corn borer as were the ISO plants, which would be expected to lead to the production of more pollen. Hence, no reliable conclusions about the (possible) more adverse effects that they could have on butterfly species could be deduced. No statistically significant detrimental effects of the Bt pollen on the larvae were found (Fig. 13.3). The most important reason for the differences in laboratory results and those from field testing (the latter indicating low overall risk) is the very low level of Bt protein exposure to caterpillar in the field as Bt corn pollen is a much rarer food source under realistic environmental conditions. A less important reason is the temporal overlap between caterpillar development and pollen shed. By the beginning of pollen shed, caterpillars often develop to the final instar stages (Fig. 13.4). Susceptibility to Bt protein is known to decline with older caterpillars, thereby reducing the effect that Bt pollen could have on them. Similar studies were done on the monarch butterfly to estimate the potential risk under field conditions in the United States. After considering distribution data of the monarch butterfly and their host plants, overlap between pollen shed and development of larvae, and exposure of larvae

Figure 13.3. Weed strip of white mustard (*Sinapis alba*) in a cornfield for collection of butterfly larvae on top of weed leaves.

to Bt pollen, risk of Bt pollen to monarch butterflies was determined to be negligible (Sears et al. 2001) (ERA step 4). Again, the amount of Bt maize pollen force-fed to monarch butterfly larvae was much higher than in field exposures, therefore indicating overestimation of risks. It is interesting to note here that the degree of hazard would not change

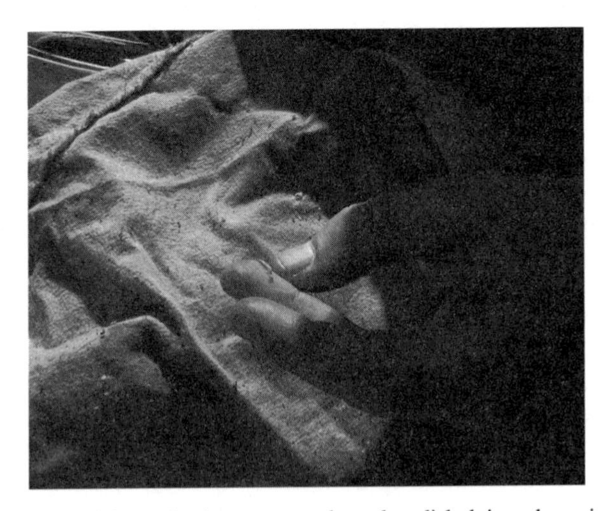

Figure 13.4. Collection of butterfly larvae was done by dislodging them into a cotton tray. Taxonomic knowledge and experience is needed for identifying the right caterpillar species because of the small size of the larvae.

because of exposure. Bt Cry1Ac endotoxin will kill monarch larvae and certain other larvae if there is enough of it. The job of risk assessment is to determine *relevant* exposures and hazards.

13.4. PROOF OF SAFETY VERSUS PROOF OF HAZARD

A proof of safety (=equivalence) between Bt maize and the near-isogenic variety was performed using a two-sided $(1-2\alpha)$ confidence test (learn more about this approach by reading Chow and Shao 2002). The percentage change of abundance is easier to interpret than the species-specific absolute difference of arthropods. Therefore, confidence intervals for Bt/ISO ratios were estimated. A ratio >1 for a taxon is equivalent to an x-fold increase in abundance in the treatment; a ratio <1 is equivalent to a decrease in abundance in the treatment down to $x\%$. According to the risk assessment objective, the demonstration of no meaningful change for selected nontarget species in Bt maize relative to the near-isogenic variety should be proven. The nontarget species can be considered as not meaningfully changed, if the lower and upper limits of the confidence interval for the abundance ratio are close to and encompass the value 1. Otherwise, the compared treatments cannot be seen as being "equivalent." Abundances can vary in all three treatments; therefore, the confidence intervals for the ratios INS/ISO and Bt/INS were also estimated.

13.5. PROOF OF BENEFITS: AGRONOMIC PERFORMANCE

When companies or academic labs develop transgenic plants with traits to be proved environmentally safe, they still need to pass a performance test where growth and yield are assessed under practical field conditions. Again we use transgenic maize as an exemplary case study together with a variety of conventional comparators. Variety registration is a substantial requirement for any new crop brand and varieties in many countries. The evaluation process is governed by independent bodies like the Federal Variety Registration Office in Hanover, Germany. Companies need to send seeds of the new varieties that can be grown on several contracted farms in representative locations in Germany. The comparative approach of yield and performance ensures that farmers get the best available varieties and information. Any new GM corn varieties must be tested in the same way as are non-GM conventional varieties. They will be registered only if their agricultural performance is improved in comparison to standard varieties. Here we present a representative dataset of field performance of several candidates for variety registration in Germany (Fig. 13.5). In these studies many plant attributes were tested. Resistance against European corn borer and kernel yield serve as examples, but many traits are considered. The data show that superior yield of GM corn is not always evident in comparison to conventional varieties. Transgenic events need to be integrated into elite variety lines by classical breeding (see Chapter 3). As we have seen, the genetic background is very important and may also lead to very different performance levels depending on the environmental and climatic conditions at a given site. However, corn borer infestation was dramatically reduced in Bt varieties, leading to various slight yield increases compared to those of three conventional maize varieties. An additional benefit might also stem from the fact that high-cornborer infestations can increase the chance of infections with plant pathogens, such as fungi that produce mycotoxins. Thus, decreasing incidence of pathogens and other indirect

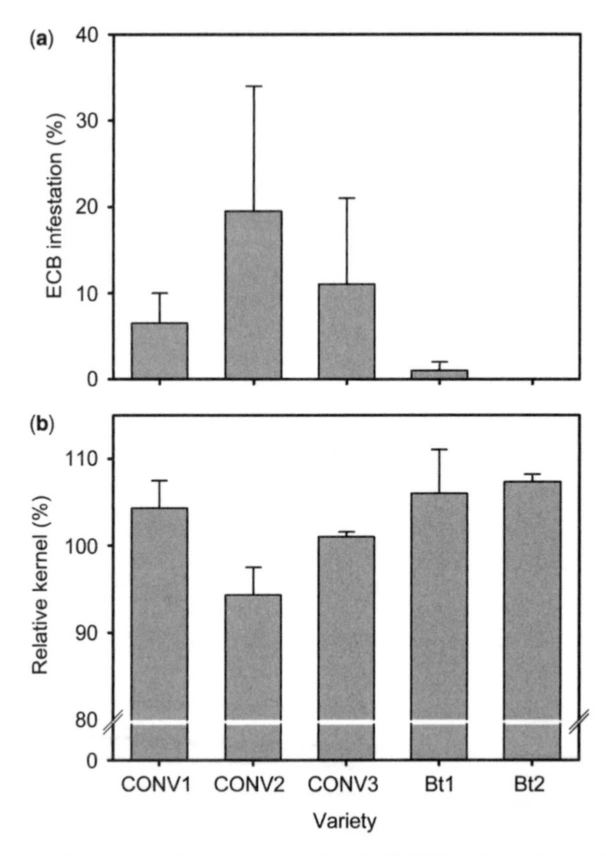

Figure 13.5. Yield performance of three conventional (CONV1–3) and two Bt (Bt1–2) maize variety candidates in a field testing of the Federal German Variety Testing Office [Bundessortenamt-(BSA)]. The data on (a) ECB infestation and (b) kernel yield were pooled from fields at three locations in ECB infestation areas spread over Germany. The ECB infestation is specified as percent of infested plants. Relative kernel yield is specified as percentage of yield compared to three standard maize varieties. The bar graph presents mean value and standard error.

factors must be considered as important criteria when evaluating agricultural performance and benefits.

In summary, farmers will buy and cultivate only those elite varieties that fit best their local needs. From the data presented here it is clear that not only does the level of pest infestation and control determine yield; optimal adaptation to local cultivation conditions, such as soil characteristics, climate, and planting and harvest time, as well as other cultivation practices such as fertilizer and herbicide management, might influence the agronomic benefits for farmers.

13.6. CONCLUSIONS

Both biosafety and benefits are important to regulators and consumers. Proof of safety is more important for the regulatory side. Proof of benefits is more important for the economic viability for the company selling the transgenic crop and the farmer who grows it.

The example of testing nontarget butterfly species showed the value of field experiments in environmental risk assessment. Laboratory and semi–field studies far overestimated any adverse effect of Bt maize on caterpillars. Field experiments were needed for a comprehensive evaluation of the real environmental effects of Bt maize. As every plant protection practice has an impact on agroecosystems, the overall risk–benefit evaluation needs to compare the impact of both chemical and GM pesticide treatment on nontarget organisms (in this case butterflies) and the yield performance. Field testing for variety registration demonstrates on one hand the sensitivity of the testing system to agricultural management practices and on the other hand, the environmental impact of conventional and biotechnological pest management strategies. Both types of environmental and economic studies did not indicate any adverse effect of Bt maize per se. However, not every Bt corn variety had the power to provide necessary (yield) benefits to the farmer.

LIFE BOX 13.1. DETLEF BARTSCH

Detlef Bartsch, Professor, RWTH University (Aachen) and Regulator at Federal Office of Consumer Protection and Food Safety (Berlin)

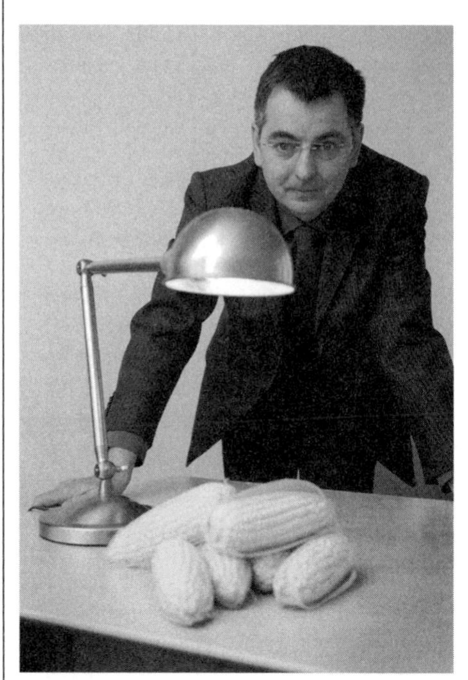

Detlef Bartsch with GM maize.

It was 1977 when I started my involvement in "science" as a 16 year old political rebel. I got a flysheet on the potential environmental impact of nuclear power stations and immediately felt that I need[ed] to take action based on the dramatic type of selected information, which was provided by—I must say looking backward—a group of concerned citizens. During the following two personal years of storm and stress I joined several environmental and political initiatives trying to protect nature and the environment. Taking part in public discussions I soon became wary of "official experts"—sometimes professors—who explained in scientific terms that . . . "there is no reason to worry . . . everything is safe . . . trust me I am the expert." As my innocent intuition told me the opposite, I decided that I myself should become an environmental expert for protecting the public against unscrupulous industry-paid footmen.

I thought the best discipline for that purpose was biology, and I started my first semester in college in 1980. The "no nukes" time was soon exchanged by the age of "forest decline due to acid rain," which triggered a specialization in the second half of my study towards ecology. I joined ecosystem research in the Geobotany Institute of the University of Göttingen and took courses in forest ecology, soil sciences and phytopathology. It was 1985 when I discovered genetic engineering to be a potential threat to the environment. A small group of students started to critically overview the foundation of the research in my Univerity's Genetics Institute, and I joined their discussion in my free time. At that time there was a strict distinction between molecular biology and ecological sciences. I basically expanded my masters and PhD time (1986–1990) in the field of ecophysiology on the scientific question: Is soil iron availability the driving force for vegetation differentiation into calcifuge and calcicole ecotypes? My answer was: probably yes, but there are more multiple cause–effect relationships. This draws my attention to a personal experience: I am sure that to a large extent in public perception "*the example seems to be everything*" in ecology. That's probably the reason why one can find for every real or fictional environmental concern support by some ecological data. It is still the great challenge in ecological sciences to find generalizations and rules, which is difficult as there are so many influencing factors.

Anyhow, I continued to be (politically) interested in the environmental consequences of genetically modified organisms (GMOs). It was a lucky random event that paved my way towards environmental biosafety research: I went in the middle of my PhD work in 1988 for a six week internship to the German Parliament where I accompanied a parliamentarian engaged in environmental politics. One small task I got was to evaluate a new draft legislative act on GMOs. I made a phone call to the leading German environmental expert in the National GMO Biosafety Committee: Prof. Herbert Sukopp, who was an expert in exotic plant species ecology. A 20 minute chat with him resulted in my being offered a postdoc position in his lab at the Berlin Technical University two years later, in 1990. My task was to organize a conference and ecology expert database for the interdisciplinary assessment of GMOs. During the next 15 month I gained experience and made the right contacts to become involved in the first biosafety research projects with sugar beet in cooperation with plant breeding industry in field experiments starting in 1993. This was the first time that a GM crop was released into the environment in Germany, and I was the first ecologist to study competitiveness and GMO out-pollination with wild-type plant relatives. My next 10 years were characterized by intensive experimental studies, teaching and field trips with students, and highly polarized public discussions with concerned citizens.

In a world of simply black-and-white views, I suddenly was pushed to the "pro-GMO" side as some people were not able to see why it was important to collect scientific data that enables science-based decisions. What a change! I became the opposed official expert myself who is mistrusted by gut-feeling driven opponents of a new technology. This was one of the most pervasive experiences in my life. It was now my problem to tell people that the truth in the GMO world is colorful and not a black-and-white story. Anyhow, I tried my best both in communicating my research as well as improving my scientific knowledge. A great time in this respect was my sabbatical study on the origin of Californian wild sea beet (*Beta vulgaris* ssp. *maritima*) in the lab of Prof. Norm Ellstrand at UC Riverside in1998.

Back in Germany, the environmental impact of Bt corn became a new object of interest. I spent three more fruitful years with my scientific mentor Prof. Ingolf Schuphan at the Aachen University of Technology. The German university system has a narrow window of opportunity to obtain a professorship. Even though I was near the final cut, I had no luck in the end to get a full professorship position.

But as luck would have it, I left in 2002 for a full time regulator job in a Federal German Agency. My job is highly inspiring since I combine scientific background information, political implications, and cost-benefit considerations into regulatory decisions. I've learned during my scientific career that good decisions are mostly those that are taken based on knowledge and not on uncertainty. Now I am an expert working for the German government and as independent expert for the European Food Safety Authority and sit in front of sceptic young rebels who want to save the world against evil techniques, but I think plant biotechnology could potentially offer more advantages than disadvantages for better and more environmentally-friendly agriculture. Plant biotechnology is based on my on scientific experience really not black-and-white, but is as colourful as life. I hope to be an honest mediator and decision maker.

REFERENCES

Chow SC, Shao J (2002): A note on statistical methods for assessing therapeutic equivalence. *Contr Clin Trials* **23**:515–520.

Gathmann A, Wirooks L, Hothorn LA, Bartsch D, Schuphan I (2006): Impact of Bt maize pollen (MON810) on lepidopteran larvae living on accompanying weeds. *Mol Ecol* **15**:2677–2685.

Romeis J, Bartsch D, Bigler F, Candolfi MP, Gielkens MMC, Hartley SE, Hellmich RL, Huesing JE, Jepson PC, Layton R, Quemada H, Raybould A, Rose RI, Schiemann J, Sears MK, Shelton AM, Sweet J, Vaituzis Z, Wolt JD (2008): Non-target arthropod risk assessment of transgenic insecticidal crops. *Nat Biotechnol* **26**:203–208.

Schmitz G, Pretscher P, Bartsch D (2003): Selection of relevant non-target herbivores for monitoring the environmental effects of Bt maize pollen. *Environ Biosafety Res* **2**:117–132.

Sears MK, Hellmich RL, Stanley-Horn DE, Oberhauser KS, Pleasants JM, Mattila HR, Siegfried BD, Dively GP (2001): Impact of Bt corn pollen on monarch butterfly populations: A risk assessment. *Proc Natl Acad Sci USA* **98**:11937–11942.

CHAPTER 14

Intellectual Property in Agricultural Biotechnology: Strategies for Open Access

ALAN B. BENNETT, CECILIA CHI-HAM, GREGORY GRAFF, and SARA BOETTIGER

Public Intellectual Property Resource for Agriculture, Department of Plant Sciences, University of California, Davis, California

14.0. CHAPTER SUMMARY AND OBJECTIVES

14.0.1. Summary

Patenting of intellectual property is of critical importance in biotechnology, and plant biotechnology is no exception. DNA sequences, proteins, transformation techniques, and in fact, the transgenic plants themselves, all have utility and can be considered inventions, and therefore patentable. Nearly all transgenic plants have numerous patented or patentable inventions associated with them, and some prominent examples are examined in this chapter. In addition, there is a fairly recent movement to work around patents to deliver biotechnology and transgenic crops to the poorer farmers of the world. Taking cues from the open-source software movement, a relatively small group of biotechnologists are developing and dispersing inventions that are not constrained by patents and can be further improved on, and then used for humanitarian purposes.

14.0.2. Discussion Questions

1. What is intellectual property, and how does it differ from tangible property? Discuss ways in which intellectual and tangible property rights can be transferred to third parties.
2. What is a patent, and what are the limitations on patent rights?
3. Contrast the "tragedy of the commons" and "tragedy of the anticommons" metaphors. How do the metaphors relate to intellectual property, particularly in agricultural biotechnology?

Plant Biotechnology and Genetics: Principles, Techniques, and Applications, Edited by C. Neal Stewart, Jr.

4. What is "freedom to operate" (FTO) in the intellectual property context? What are the main issues in considering FTO when developing an improved crop variety using agricultural biotechnology?
5. In the E8 case study, how does prior art preclude patenting? Discuss ways research scientist could use publications as a means to place inventions in the public domain.
6. While patent law has presented opportunities to protect intellectual property in the field of biotechnology, it has also generated a struggle to reconcile public and private interests. How are the emerging models represented by PIPRA and CAMBIA trying to stimulate innovation and promote open access while avoiding the tragedies of the anticommons?

14.1. INTRODUCTION

Scientific advances in many fields have been treated historically as public goods, and this was particularly true in agriculture. Universities and other public-sector institutions were the leaders in developing improved crop varieties that were transferred to farms through cooperative extension services in the United States or equivalent organizations internationally (Conway and Toenniessen 1998). This model, however, has changed rapidly in the last few decades, primarily because of greater utilization of formal *intellectual property* (IP) protection of agricultural technologies and plant varieties by the public sector, as well as the development of a research-intensive private sector that now makes major contributions in enhancing the productivity of US agriculture (Kowalski et al. 2002). In particular, the expanded use of formal IP rights for agricultural biotechnology-based products can be understood by considering the significant amount of time and financial resources needed to develop a new transgenic crop and the high costs of obtaining regulatory approval to market such a crop. In the face of these costs, the time-limited exclusivity provided by patents allows the investor an opportunity to recoup the costs of research and development. Indeed, it is very likely that the agricultural biotechnology industry would not have developed in the absence of a strong framework for IP protection.

The growth in patents related to agricultural biotechnology can be seen in Figure 14.1. These data indicate a strong growth in the issuance of patents by the US Patent and Trademark Office (USPTO), and similar trends are also apparent in patent applications internationally, suggesting that this is a global trend. The scope of inventions represented by the data in Figure 14.1 is quite broad but can be conceptually divided into two main categories: those that cover research tools or *enabling technologies* that are required to produce transgenic plants or to discover new gene functions and those that cover *trait technologies* that confer specific attributes to genetically modified plants. This distinction is important because all researchers and research institutions need access to the fundamental tools of agricultural biotechnology if the greatest benefits of the technologies are to be realized, whereas exclusive access to specific trait technologies is an effective means of ensuring that the new crop varieties expressing these attributes are developed. As a consequence, there is a delicate balance in the overall innovation framework between exclusive access to certain technologies while at the same time ensuring broad access to other technologies. A very similar situation was addressed in the early 1980s when Stanford University and the University of California, San Francisco patented the basic methods of recombinant DNA manipulations (Cohen and Boyer 1980). This patent, covering the fundamental tool of

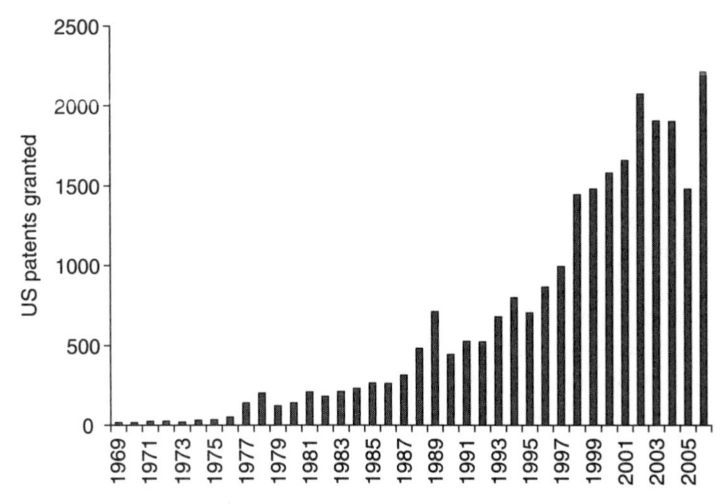

Figure 14.1. Annual trends in plant biotechnology patents issued by the US Patent and Trademark Office between 1969 and 2005.

modern biotechnology, was nonexclusively licensed under reasonable financial terms, a strategy that resulted in licenses to 468 companies (Feldman et al. 2005). Broad innovation was encouraged through access to the key enabling technologies; and many companies became successful by using these tools to develop proprietary products based on other patented technologies that they exclusively held. Ultimately, the licensing strategy enabled \$35 billion in worldwide product sales and brought in \$255 million in licensing revenues (Feldman 2005).

In agricultural biotechnology today, the innovation system needs a balance of both exclusive and nonexclusive access to patented technologies to effectively support new crop development and to provide both commercial growers and subsistence farmers with the best genetic technology possible for their crops. The ownership of critical intellectual property and the rights to practice or use certain technologies is becoming a major issue confronting researchers in this area. Even purely fundamental academic research is not protected by an "experimental use" exception from patent infringement and may become increasingly entangled in issues involving access to IP rights (Eisenberg 2003). While the importance of intellectual property in agriculture is becoming better recognized in both the public and private sectors, many researchers, businesspeople, R&D decision-makers, and policymakers are still relatively uninformed about how to find, understand, and utilize IP information, including published patents and patent applications. In this chapter we will provide an overview of the major issues and what a research scientist needs to be aware of when navigating the IP landscape of agricultural biotechnology.

14.2. INTELLECTUAL PROPERTY DEFINED

Intellectual property is a legally created form of property that applies to your ideas or the "products of your mind" and gives the owner a set of rights that are comparable to tangible property rights. The concept of intellectual property was insightfully addressed by Thomas Jefferson when he said

> If nature has made any one thing less susceptible than all others of exclusive property, it is the action of the thinking power called an idea, which an individual may exclusively possess as long as he keeps it to himself.... Inventions then cannot, in nature, be a subject of property. (However) society may give an exclusive right to the profits arising from them, as an encouragement to men to pursue ideas which may produce utility (Jefferson, 1987).

Jefferson's concept of society providing legal mechanisms for inventors to have exclusive rights to profits from their ideas was subsequently integrated into the United States Constitution, Article I, Section 8, which states that "The Congress shall have power... to promote the progress of science and useful arts by securing for limited times to authors and inventors the exclusive right to their respective writing and discoveries." This forms the basis for IP rights and has become the cornerstone of the innovation process in the United States and more recently in many other countries throughout the world.

There are several forms of intellectual property, including plant and utility patents, copyright, trademarks, and trade secrets. In agricultural biotechnology, the dominant forms of intellectual property are *patents*, and these are the primary focus of this chapter. Patents provide just what the constitution promised—the right to exclude others from using your invention. Importantly, this right is conferred by a national government for a specified time period, usually 20 years. The patent is enforceable only in those countries in which it was specifically awarded and after the 20-year term of the patent expires, the invention can then be used by anyone without restrictions. So, in general, a patent provides an intellectual property right that is geographically limited to the specific countries in which patent protection is obtained and it is time-limited by the term of the patent. This is a significant way in which patent-protected intellectual property differs from tangible or real property, where ownership is rarely limited by either geography or time. These differences between intellectual property and tangible property often have an impact in biological research since research materials (vectors, genes, cell lines, etc.) are usually obtained under the terms of a *material transfer agreement* (MTA), which likely contains provisions on how the material is used. Because the MTA governs the transfer of tangible or real property, the terms of the agreement typically do not contain geographic or temporal limitations and, as a result, the restrictions imposed by MTAs can become particularly problematic.

The monopoly that a patent provides to an *inventor* is a very powerful economic right, and, as a consequence, the invention must meet a standard of *novelty*, *non-obviousness*, and *utility*—that is, the invention must be original and not previously known, it must not be an obvious extension of previously known information, and it must have some useful purpose. The standard of novelty has an important implication for researchers since the primary means of scientific communication is through broad publication, which, if done carelessly, can destroy the patentability of an invention. The section of US patent law relevant to novelty states that a patent application can be rejected on lack of novelty grounds if "the invention was... patented or described in a printed publication in this or a foreign country, before the invention thereof by the applicant for patent, or.... More than one year prior to the date of the application for patent in the United States." In most other countries, the one-year grace period provided in the United States does not exist and public disclosure of an invention immediately bars patentability in those countries. In addition to the timing of public disclosures, a researcher also needs to consider the meaning of the words *printed* and *publication*. For example, is a document or a slide presentation posted on the Internet considered *printed*, such that the document bars future patentability? Any disclosure of a potentially patentable idea should be made thoughtfully and/or in consultation with an attorney or technology transfer office. In some cases, a clear public disclosure can be purposely designed to bar patentability in order to ensure that an invention remains in the public domain and available for everyone to use without restriction (Boettiger and Chi-Ham, 2007).

The patenting of plant and animal genes has been particularly controversial, and critics have argued that genes are not patentable because they exist in nature. However, the US Patent and Trademark Office (USPTO) concluded that an isolated and purified DNA molecule that has the same sequence as a naturally occurring gene is eligible for patent protection because it does not occur in its isolated form in nature. The USPTO did, however, modify and adopt a higher standard of "utility" in its guidelines for evaluating gene patents, requiring that the applicant demonstrate that the "utility is specific, substantial, and credible" (http://www.uspto.gov/web/offices/com/sol/notices/utilexmguide.pdf). In spite of this specific utility requirement, a number of patent applications claim the sequences of hundreds of genes for which the utility is only broadly defined. For example, US Patent Application 20,070,022,495 defines the utility of several hundred claimed genes as conferring an

> "improved trait relative to a control plant" and that
>
> "The improved trait is selected from the group consisting of larger size, larger seeds, greater yield, darker green color, increased rate of photosynthesis, more tolerance to osmotic stress, more drought tolerance, more heat tolerance, more salt tolerance, more cold tolerance, more tolerance to low nitrogen, early flowering, delayed flowering, more resistance to disease, more seed protein, and more seed oil relative to the control plant."

Time will tell how these broad patents are treated by patent offices and patent examiners and ultimately whether such broad gene patents are enforceable.

14.3. INTELLECTUAL PROPERTY IN RELATION TO AGRICULTURAL RESEARCH

The impact of public-sector research in agriculture has been very significant. In the United States, this dates back to the establishment of the Land Grant College system of universities that have led the development of improved crop varieties that were transferred to farms and to the agricultural industry through cooperative extension services in this country. Internationally, the system of crop research centers sponsored by the Consultative Group on International Agricultural Research (CGIAR) has a similarly large impact in developing new crop cultivars and agronomic practices that were delivered as a public good to support global food production. This model has been slowly changing, and the rate of change is now accelerating. At the core of this change is the increasing role of IP protection over agricultural inventions, as well as the development of a research-intensive private sector in agricultural biotechnology. Thus, both US and global agricultural systems are experiencing a change from research results being developed primarily in the public sector and the resulting technologies delivered for free as a public good to a system that is increasingly dominated by private companies who protect and treat results as a private asset. This has been accomplished through a much more intensive use of the patent system to protect agricultural innovations than was previously the case. The trend in patents awarded related to plant biotechnologies between 1980 and 2000 clearly illustrates the overall increase in patenting activity in this sector (Fig. 14.1).

Since the early 1980s, other fundamental changes in the nature and ownership of innovations in basic and applied agricultural research have complicated the mission of public research institutions. The primary change was the passage of the *Bayh–Dole Act*, which encouraged US universities to patent their innovations and license them to private-sector companies in order to encourage their commercial use. Since that time, patenting by public research institutions and universities as well as the development of formal technology transfer mechanisms have accelerated. Figure 14.2 illustrates the public sector's contributions to patented inventions in the area of plant biotechnology as compared with patents across all technology sectors. These data show that while public-sector institutions contribute to only about 2.7% of patents overall, their contribution to agricultural biotechnology patents is nearly an order of magnitude greater—contributing approximately 24% of all patents (Graff et al. 2003). While this trend has contributed to many positive economic outcomes, these new policies have also created challenges for public research institutions and universities in supporting broad innovation, particularly for agricultural applications that address small markets such as specialty crops or that support humanitarian, rather than commercial, purposes.

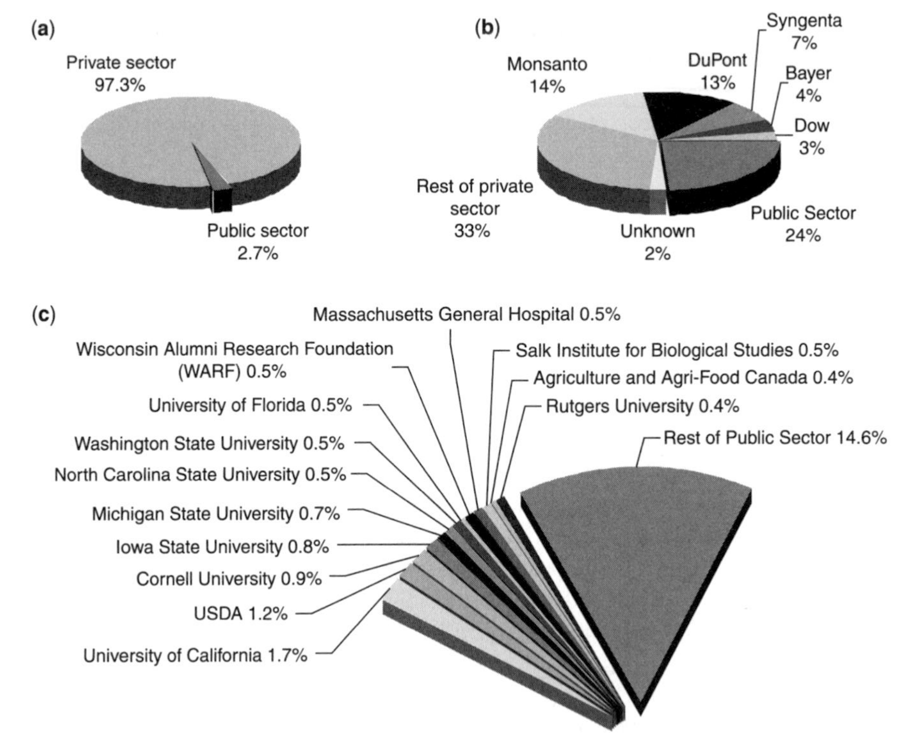

Figure 14.2. Distribution of assignment of US patents from 1982 to 2001 to private and public institutions: (a) all patents; (b) agricultural biotechnology patents; (c) public-sector agricultural biotechnology patents [adapted from Graff et al. (2003); used with permission]. See color insert.

14.4. DEVELOPMENT OF AN "ANTICOMMONS" IN AGRICULTURAL BIOTECHNOLOGY

The proliferation of patents in biotechnology led to the development of a metaphor to explain why people overuse or under use resources. The "tragedy of the commons" was a term coined by Garrett Hardin to explain why people overused shared resources, such as common pastures, because they had no incentive to conserve or extend the life of the resource (Hardin 1968). By analogy, Heller and Eisenberg (1998) described the "tragedy of the anticommons," which, as the result of a proliferation and fragmentation of IP ownership across multiple owners, prevents any single institution or company from assembling all of the necessary rights to produce a product, resulting in the underuse (or nonuse) of resources. Interestingly, whereas patents and IP generally are intended to encourage investment in research and development, the development of an anti-commons has the opposite effect of blocking innovation. Although this concept of the anticommons was initially described in relation to biomedical research, it also has direct relevance to agricultural biotechnology.

A prominent example of the complexity resulting from fragmented technology ownership and the potential for anticommons to arise was exemplified in the development of β-carotene-enriched rice by public-sector researchers who used at least 40 patented or proprietary methods and materials belonging to a dozen or more different owners in the gene transfer process (Kryder et al. 2000). Some examples of the types of patented technologies that are required for developing a genetically engineered crop are examined below.

14.4.1. Transformation Methods

The development of transgenic varieties typically relies on either *Agrobacterium*-mediated or particle-bombardment-mediated gene transfer methods (Herrera-Estrella et al. 1983; Klein et al. 1987) (see also Chapter 10). Fundamental methods related to both processes of gene transfer into plant cells were invented in the public sector (Barton et al. 2000; Sanford et al. 1991), but key patents for *Agrobacterium*-mediated transformation were licensed exclusively to Ciba–Geigy (now Syngenta), and the particle bombardment technology was licensed exclusively to DuPont for most fields of use. In addition, both private- and public-sector R&D organizations have patented a number of fundamental transformation methods, as well as improvements including vectors, species-specific protocols, and novel strategies to remove selectable markers and other "foreign" DNA from the plant to be commercialized (Hoekema et al. 1992; Hamilton 1998; Yoder and Lassner 1998; Pray and Naseem 2005; Fraley et al. 1991; Rogers and Fraley 2001). As a result of a variety of transactions, fundamental methods of gene transfer to plant cells were invented by either private-sector companies or by the public sector but then licensed exclusively to private companies and represent a key technology area where patents have the potential to block new innovations. Even though these fundamental technologies were invented in the 1980s, a key patent covering *Agrobacterium*-mediated gene transfer has not yet issued in the United States and will have a long life (17 years) when it is ultimately awarded. Some of the key patents covering particle bombardment have already expired and others will expire soon, so this technology may be more widely accessible in the near future.

14.4.2. Selectable Markers

The most commonly utilized plant selectable marker genes include the *nptII* and *hpt* genes that confer antibiotic resistance as a basis to select for cell transformation (Miki and McHugh 2004) (see also Chapter 9). Several other selectable markers conferring herbicide resistance or positive selection on the basis of novel carbon utilization pathways provide important alternatives to the antibiotic-based selection strategies (Roa-Rodriguez and Nottenburg 2003). Broad patents cover all of these selectable markers (Rogers and Fraley 2001; Santerre and Rao 1988; Bojsen et al. 1998). Selection strategies do not appear to have been the topic of public-sector research programs, and there are just a few examples of either public-domain or public-sector-patented selectable markers for use in plant transformation (Dirk et al. 2001, 2002; Hou et al. 2006; Mentewab and Stewart 2005). While there is potential to invent new selectable markers for plant transformation, at this point this represents another key enabling technology where patents have the potential to block new innovations.

14.4.3. Constitutive Promoters

Genetic regulatory elements are required to drive the expression of selectable marker genes and of specific transgenes. Selectable marker genes are typically driven by high-level constitutive promoters, with the most common constructs utilizing the CaMV 35*S* promoter derived from a viral genome and owned by Monsanto (Odell et al. 1985; Fraley et al. 1994). Many alternative promoters that confer constitutive gene expression were developed in public-sector organizations and are either in the public domain or can be licensed for nominal fees. These alternatives include monocot and dicot actin promoters (McElroy et al. 1990; McElroy and Wu 2002; An et al. 1996; Huang et al. 1997), a FMV 34*S* promoter (Comai et al. 2000), mannopine/octopine synthase (Gelvin et al. 1999), or FMV and PCLVS FLt promoters (Maiti and Shephard 1998, 1999). The FMV 34*S* has been used to drive constitutive gene expression and reported to be essentially equivalent to the CaMV 35*S* promoter (van der Fits and Memelink 1997; Romano et al. 1993), but has not been widely distributed to the public-sector research community. Each of these promoters provides a strategy for driving constitutive transgene expression using public-sector-derived or public-domain components.

14.4.4. Tissue- or Development-Specific Promoters

Although many genes can be expressed under the control of constitutive promoters, targeting of expression to plant organs or tissues is typically desirable to minimize nonspecific effects of the introduced gene. For example, seed-specific promoters (Blechl et al. 1999; Harada et al. 2001) have been patented with claims directed toward their use to drive expression of heterologous genes in developing seeds. Public-sector institutions have also patented a relatively large number of tissue- and/or developmental-specific promoters. Examples include the root-specific CaMV 35*S* fragment A promoter (Benfy and Chua 1992), a root cortex–specific promoter (Conkling et al. 1998), the Pyk10 root-specific promoter (Grundler et al. 2001), an epidermal cell–specific Blec promoter (Dobres and Mandaci 1998), and a vascular tissue–specific promoter RTBV (Beachy and Bhattacharyya 1998). In addition, there exists a large number of tissue and developmental specific promoters that have been characterized and placed in the public domain through publication. A wide range of constitutive and regulated promoters have been tabulated in a promoter database that includes information on expression characteristics as well as their IP status (the database will be hosted at PIPRA: www.pipra.org).

14.4.5. Subcellular Localization

In addition to specificity in tissue-level transgene expression, it is also often important to direct the targeting of the new protein to a specific subcellular location. For example, because β-carotene is produced in the plastids, the development of β-carotene-enriched rice utilized a transit peptide derived from the small subunit of Rubisco to target proteins to this subcellular compartment (Ye et al. 2000). This and other transit peptides have been the topic of intense study, and several companies have patented their use to direct proteins into plastids (Herrera-Estrella et al. 2000; Dehesh 2002). However, several early publications from public-sector research organizations described alternative transit peptides that were not subsequently patented and thus should be accessible in the public domain (Smeekens et al. 1986). Because transit peptides do not have a high degree of sequence similarity, it is likely that additional transit peptides will not be dominated by existing patent claims and alternative sources of functional transit peptides could be developed from public-domain information or from public-sector laboratories. Targeting to other subcellular locations has been the topic of intense research in both the public and private sectors, and there are many examples of public-sector research describing unpatented sequences targeting proteins to a variety of subcellular sites, including the cell wall, vacuole, plastids, and peroxisomes (Komarnytsky et al. 2000; Bednarek et al. 1989; Tague et al. 1989; Kato et al. 1996; Volokita, 1991; Hayashi et al. 1996).

Developing a new genetically engineered crop requires the assemblage of a number of patented technologies through *in-licensing* or, potentially, by a series of strategic mergers and acquisitions. Several companies have effectively done this and have used platforms of proprietary technologies to develop new varieties of major crops. However, work on crops of less commercial interest has progressed slowly, with few of the benefits of biotechnology having been realized in specialty crops (Clark et al. 2004). With the requirement for assembling a large number of patented technologies to produce genetically engineered crop and the fragmentation of IP ownership, it appears that the preconditions for the development of an anticommons exist in this technology sector. In addition, the observed slowdown in the development of new agricultural biotechnology products may be, at least in part, an effect of such an IP anticommons (Graff et al. in press).

14.5. FREEDOM TO OPERATE (FTO)

Navigating the complex IP landscape of a research project in agricultural biotechnology requires some analytical tools and specialized analytical capabilities (Fenton et al. 2007). The analysis requires both legal and scientific knowledge as well as access to both patent and literature databases and typically takes the form of what is known as a "freedom to operate" (FTO) opinion. The FTO opinion is a legal assessment of whether a research project or the development of a new product can proceed with a low, or tolerable, likelihood that it will not infringe on existing patents or other types of IP rights. It is important to note that the FTO determination is not absolute but reflects an evaluation of risk since there is typically some uncertainty around the interpretation of patent claims as well as uncertainty as to whether new IP may issue or be discovered at a later date. The FTO opinion may lead to a range of options: identifying in-licensing targets, considering the substitution of technologies, deciding to ignore the potential infringement, investing in workaround technologies, or perhaps deciding to abandon the project altogether. Although private firms are more likely to engage in FTO analysis because any infringement risk may directly affect their ability to develop new products and their ultimate profitability, public and not-for-profit private institutions are becoming increasingly aware of the need for better freedom to operate information. This is particularly true for research projects undertaken by universities or not-for-profit research centers for the specific purpose of developing new crops for developing countries. In these cases, it is critical that IP considerations be taken into account early in the research process.

While patents are the most common type of IP right encountered, a thorough FTO analysis will assess all types of existing property rights for the likelihood of infringement by the research project or the product being commercialized. Of particular concern are tangible property rights, such as cell lines, transgenic plants, germplasm, and plasmids, because, as described above, the transfer of tangible property often occurs under the terms of a material transfer agreement that has no geographic or temporal limitation. These terms can be particularly problematic and directly impact FTO.

Enabling technologies for plant transformation or transformation vectors combine several components such as promoters, selectable markers, marker removal systems, and more. Because of the fundamental role of these technologies, they have been extensively patented. In addition, the FTO surrounding plant enabling technologies is further complicated because these technologies are not used individually but are combined with a suite of related enabling technologies and specific trait technologies and are deployed in many different plant species. We can look at a relatively simple example of a single component of a transformation vector to illustrate the elements of an FTO analysis.

The target technology for this case study was a fruit-specific promoter from the tomato E8 gene. The E8 promoter has been used to improve fruit quality, extend fruit shelf life, and express edible human vaccines specifically in ripening tomato fruit. The first step in an FTO investigation is to clearly define the target technology. In this hypothetical case, the fruit-specific promoter will be used exactly as described in the initial publications by Deikman and Fischer (1988) and Giovannoni et al. (1989). The promoters in these publications are virtually identical and consist of about 2100 nucleotides upstream of the E8 structural gene. Further promoter characterization identifying the location and sequence of functional elements within the promoter and upstream nucleotide sequence was reported by Deikman et al. (1992). These publications draw the technical boundaries surrounding the target promoter technology and provide important prior art to subsequently filed patents.

To establish the relationship of publications and patents that describe or claim the E8 promoter, a patent landscape must first be established. The patent landscape should include patents and patent applications closely related to the technology. Keywords and authors of key publications are used to search for patents or patent applications. A separate search should then be conducted to identify patents or patent applications that referenced the scientific publications describing the technology. Additionally, in the E8 case, patented DNA and protein sequence databanks were searched using the promoter's DNA sequence as a query. The patent landscape will reveal "family" relationships among different patents and published patent applications. Patent families include later patent applications that claim the benefit of an earlier, related, application or later patent applications that arise from foreign filings of the parent application. Figure 14.3 illustrates a patent family

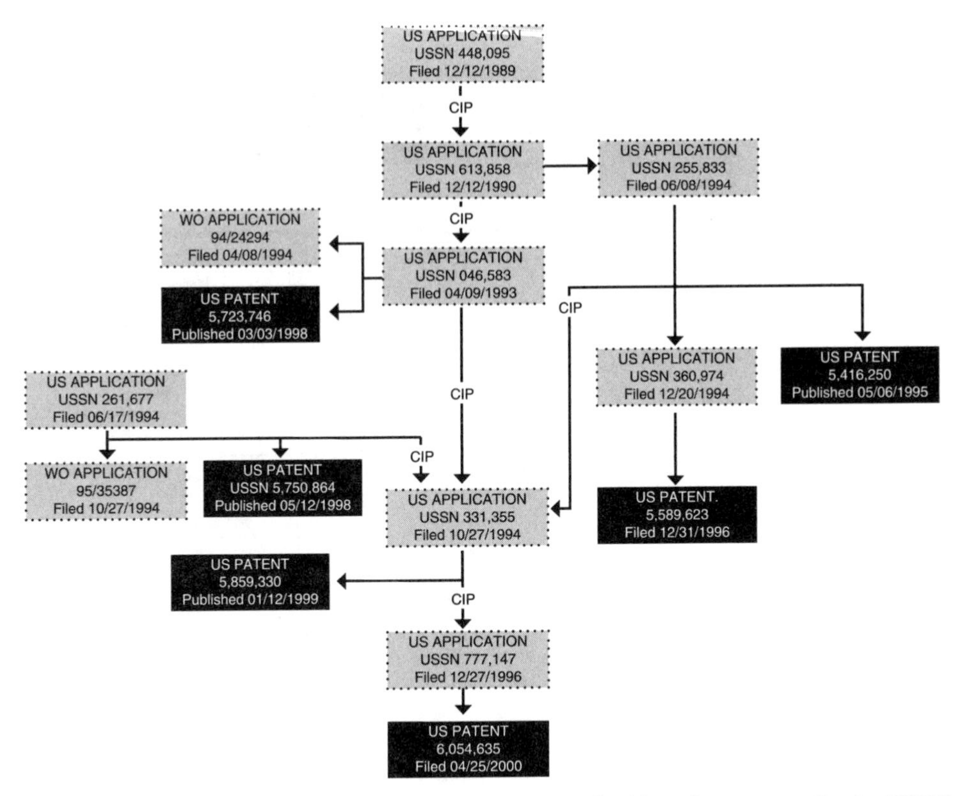

Figure 14.3. A family of related tomato E8-related patents derived from the parent application USSN 448,095 [from Fenton et al. (2007)].

arising from a 1989 patent application related to the E8 promoter filed by Agritope, an agricultural biotechnology company.

An informative way of analyzing the FTO search results is to construct a timeline of scientific literature, patent applications, and issued patents on the specific technology and on potentially overlapping subject matter. Ordering the patents and published applications according to their priority dates (also known as *effective filing dates*) reveals important relationships. For example, it reveals what publications or patents are prior art against newer patents. Since patents may only be granted if the claims are both novel and non-obvious over the *prior art*, this analysis reveals the relative dominance of earlier, broader patents over later, narrower patents. Figure 14.4 illustrates the IP priority timeline for the E8 promoter. A thorough FTO analysis may require direct contact with the researchers and, in this analysis, it was learned from the authors of the Deikman and Fischer (1988) publication that they did not apply for patent protection prior to their publication. This information was also confirmed by searching patent databases. On the basis of this information, it was presumed that the basic E8 promoter technology was in the public domain, but this conclusion required thorough review and documentation of the published literature or prior art relative to the subject matter of subsequent patents.

As shown in the priority timeline, the Deikman and Fischer (1988) and Giovannoni et al. (1989) publications initially describe the E8 promoter technology. This precluded the

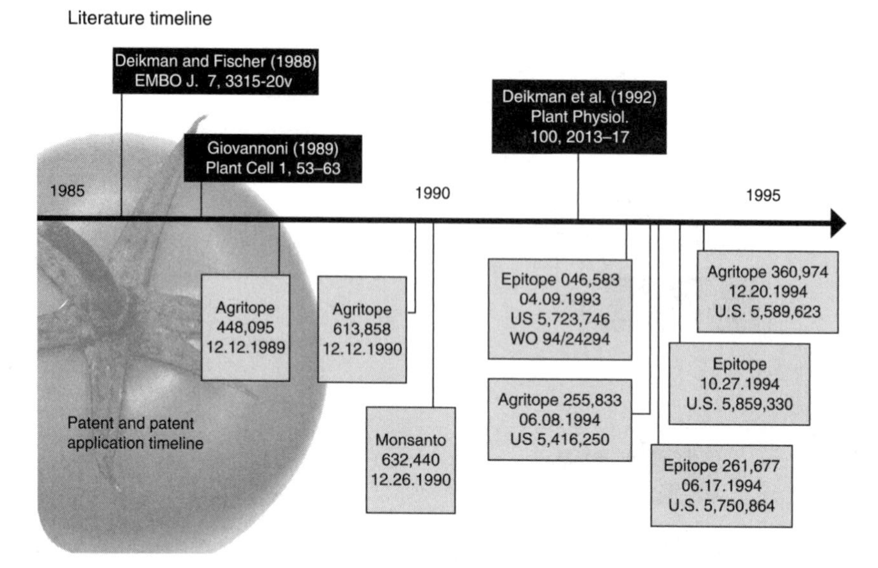

Figure 14.4. Timeline of tomato E8-related scientific publications and patents. The timing of the publication of scientific and patent literature illustrates the existing prior art at the time that related patent applications were filed [from Fenton et al. (2007), used with permission].

novelty of any subsequent patent claims on the E8 promoter per se (e.g., applications filed by Agritope and Epitope). Counsel concluded that the tomato E8 promoter constructs per se can be reasonably considered to be in the public domain. However, some of the subsequent patents claim chimeric constructs comprising the E8 promoter and heterologous genes and use of the E8 promoter in conjunction with other promoter elements. Thus, certain specific uses of the E8 promoter may infringe on these subsequent patents.

This example provides an overview of the data and information that should be considered in an FTO analysis. It is not difficult to imagine how the complexity of an FTO analysis would grow dramatically with the inclusion of multiple enabling technologies, one or more trait technologies, and proprietary germplasm. This is one of the challenges in understanding IP constraints and developing FTO strategies in plant biotechnology where multiple complementary technologies are necessarily integrated to develop new crop varieties.

14.6. STRATEGIES FOR OPEN ACCESS

The complex IP environment surrounding agricultural biotechnology research and development, exemplified by even a relatively simple FTO opinion, has spawned some new strategies and new organizations committed to lower the IP barriers to new crop developments and provide more open access to patented technologies. These issues are critical for small private companies attempting to enter this sector but can also be important for public or not-for-profit research institutions. For example, a Federal Circuit Court of Appeals ruling in the *Madey v. Duke* case emphasized that academic research is not protected by an "experimental use" exception from patent infringement, even when the research is purely fundamental (Eisenberg 2003). Most plant biotechnology laboratories routinely use patented technologies in their research without permissions. Although patent owners have rarely been concerned about academic research infringement in agriculture, in many instances fundamental biomedical research has been challenged because of IP issues (Marshall 2002). In addition to IP considerations in basic research, projects carried out in public or not-for-profit institutions that are targeted toward the development of crops for developing country farmers must consider the IP inputs to the project.

Most researchers are still relatively unfamiliar with how to find, understand, and utilize IP information, including published patents and patent applications. In addition, the ability to obtain the rights to use patented technologies has remained uncertain even for projects that have little commercial importance but, for example, may have large impacts in developing country agriculture. Several organizations have now emerged to address the relative inaccessibility of IP information and to provide a framework to ensure that IP does not block applications of agricultural biotechnology and, in particular, to facilitate projects that can have broad humanitarian benefits.

Several public-sector and not-for-profit agricultural research institutions, including the University of California, the Donald Danforth Plant Science Center, North Carolina State University, Ohio State University, Boyce Thompson Institute for Plant Research, Michigan State University, Cornell University, University of Wisconsin–Madison, University of Florida, the USDA, Rutgers University, Texas A&M University, and Purdue University, developed the Public Sector Intellectual Property Resource for Agriculture (PIPRA; www.pipra.org). These institutions made a public commitment to participate and promote strategies to collectively manage public-sector intellectual property in support of both US and developing-country agriculture (Atkinson et al. 2003). This initial founding group of PIPRA members has grown to over 45 institutional members in 13 countries, illustrating the widespread concern and interest in working collectively to remove and avoid IP barriers that might slow development of new crops.

A number of strategies have been implemented to enhance FTO using public-sector IP for agricultural biotechnology projects. For example, informed decisions regarding dissemination of new knowledge via open publication or protecting it with a patent are clearly important, and FTO can be improved if public-sector institutions systematically consider how, when, and if to use the patent system to support broad innovation (Boettiger and Chi-Ham 2007). Even when using the patent system, PIPRA encourages its members to reserve rights to use their newest and best technologies for humanitarian purposes, particularly when they issue exclusive commercial licenses (Bennett 2007). For US agriculture, it is also important to retain rights to use patented technologies in development of small specialty crops that are not currently within the commercial interests of large private-sector companies. The anticipated benefits of a collective IP management regime are to enable an effective assessment of FTO issues, overcome the fragmentation of public-sector IPR, reestablish the necessary FTO in agricultural biotechnology for the public good, and enhance private-sector interactions by more efficiently identifying collective commercial licensing opportunities.

Among PIPRA's core activities in developing a clearinghouse of public IP information and analytical resources, it is also developing consolidated technology packages, or patent pools, particularly in the area of enabling technologies for plant transformation. Patent pools have been used effectively by companies to expedite the development and diffusion of innovations that draw on many complementary technology components protected by multiple patents that are owned by multiple technology developers. In PIPRA's case, the development of a patent pool that provides FTO for plant transformation that incorporates patented technologies from multiple owners was facilitated by its broad membership and their shared commitment to make these technologies widely available. This strategy is likely to create commercial licensing opportunities as well to support humanitarian projects.

Another initiative has been spawned in an Australian organization, CAMBIA, to create a new approach to technology access in agricultural biotechnology modeled after the "open source" approach that is well developed in the IT software sector. The proprietary name for this initiative is Biological Innovation for Open Society or BiOS and this approach is built on a broad philosophical foundation to "to democratize problem solving to enable diverse solutions to problems through decentralized innovation" (http://www.bios.net/daisy/bios/home.html). At the heart of BiOS is licensing language designed to preserve a pool of patented technologies from private appropriation. The idea is to create a "protected commons" of enabling agricultural biotechnologies that are made freely available and whose use cannot be restricted by third-party patent rights. Even when a patent is granted on a new technology and broadly licensed, accessibility and utility can be reduced as subsequent patents stake proprietary claims for specific uses of the technology, or combinations of the technology with other technologies (these are often called "improvement patents"). By signing the BiOS license, a researcher or an institution agrees to contribute back to the pool, for free distribution, data on the use of the technology and the patent rights to any improvements made to the technology. Over time, with the improvements invented and contributed back to the pool by a set of researchers, the pool grows. Access to the original technologies that were donated by CAMBIA (Jefferson et al. 2002; Jefferson and Mayer 2003; Jefferson 2004, 2005) is preserved for a wide variety of applications, and the technologies become more valuable as improvements are shared and made accessible.

As one of several approaches designed to encourage broad-based participation in research in biotechnology in the face of constraints imposed by intellectual property rights (particularly those restricting the use of important enabling technologies), open source (OS) in biology is a new and controversial legal innovation. The model itself is often confused by evoking the principles of "openness" and "transparency," neither of which are simple goals to attain in the world of intellectual property rights. To agricultural researchers, OS appears to promise a return to the scientific environment of the past, where materials and ideas were exchanged with greater ease and collaboration was not circumscribed by a preoccupation with intellectual property.

In certain software applications, OS has provided a mechanism for achieving much easier collaboration with copyrighted software production. But plant biotechnology constitutes a field very different from that of software, primarily because IP protection relies on patents rather than copyright. The shift from copyright to patent law, the long timelines of research and development, and the expensive regulatory regime, provide great challenges for open, distributed innovation and the establishment of a protected commons of easily accessible technology in plant biotechnology. The development of a workable OS model in biotechnology is difficult to design, but BiOS represents one of the early pioneers that is seeking a practical implementation of a model in the life sciences.

14.7. CONCLUSIONS

Intellectual property as a tool to foster innovation has been important for over two centuries but has become a much more prominent feature of research in the life sciences and in agricultural biotechnology, in particular, only since the early 1980s. This trend is unlikely to be reversed, and, indeed, the importance of intellectual property as an intangible asset contributing to the value of life science companies continues to increase. However, robust and sustained innovation in agricultural biotechnology, as in many technology sectors, requires a balance of both exclusive and nonexclusive access to proprietary technologies. This balance should ensure that the fundamental research tools are broadly available to support research and development in many application areas and at the same time provide the exclusivity to specific trait or trait–crop combinations that will allow the developer of new varieties to recoup their substantial investment. The public sector has a role to play alongside agricultural biotechnology companies, particularly in providing research tools and broad enabling technologies and in addressing biotechnology applications in specialty crops whose market size may not justify commercial investment. The most powerful approaches, however, will come from public–private partnerships that mobilize proprietary technologies to address agricultural biotechnology product developments that have a high social but low commercial value, including strategies to feed some of the world's poorest populations.

REFERENCES

An YQ, McDowell JM, Huang S, McKinney EC, Chambliss S, Meagher RB (1996): Strong, constitutive expression of the Arabidopsis ACT2/ACT8 actin subclass in vegetative tissues. *Plant J* **10**:107–121.

Atkinson RC, Beachy RN, Conway G, Cordova FA, Fox MA, Holbrook KA, Klessig DF, McCormick RL, McPherson PM, Rawlings HR, Rapson R, Vanderhoef LN, Wiley JD, Young CE (2003): Public sector collaboration for agricultural IP management. *Science* **301**:174–175.

Barton KA, Binns AN, Chilton M-DM, Matzke AJM (2000): *Regeneration of Plants Containing Genetically Engineered T-DNA*. US Patent 6,051,757.

Beachy R, Bhattacharyya M (1998): *Plant Promoter*. US Patent 5,824,857.

Bednarek SY, Wilkins TA, Dombrowski JE, Raikhel NV (1989): A carboxyl-terminal propeptide is necessary for proper sorting of barley lectin to vacuoles of tobacco. *Plant Cell* **2**:1145–1155.

Benfy PN, Chua N-H (1992): *Plant Promoters*. US Patent 5,110,732.

Bennett AB (2007): Reservation of rights for humanitarian uses. In *Intellectual Property Management in Health and Agricultural Innovation: A Handbook of Best Practices*, Krattiger A (ed). pp. 41–46.

Blechl A, Anderson O, Somers DA, Torbert KA, Rines HW (1999): *Glutenin Genes and Their Uses*. US Patent 5,914,450.

Boettiger S, Chi-Ham C (2007): Defensive publishing and the public domain. In *Intellectual Property Management in Health and Agricultural Innovation: A Handbook of Best Practices*, Krattiger A (ed). pp. 879–896.

Bojsen K, Donaldson I, Haldrup A, Joersboe M, Kreiberg JD, Nielsen J, Okkels FT, Petersen SG (1998): *Mannose or Xylose Based Positive Selection*. US Patent 5,767,378.

Clark D, Klee H, Dandekar A (2004): Despite benefits, commercialization of transgenic horticultural crops lags. *Cal Agric* **58**:89–93.

Cohen SN, Boyer HW (1980): *Process for Producing Biologically Functional Molecular Chimeras*. US Patent 4,237,224.

Comai L, Sanger MP, Daubert SD (2000): *Figwort Mosaic Promoter and Uses*. US Patent 6,051,753.

Conkling MA, Mendu N, Song W (1998): *Root Cortex Specific Gene Promoter*. US Patent 5,837,876.

Conway G, Toenniessen G (1999): Feeding the world in the twenty-first century. *Nature* **402**:C55–C58.

Dehesh K (2002): *Plastid Transit Peptide Sequences for Efficient Plastid Targeting*. US Patent Application 200,20,178,467.

Deikman J, Fischer RL (1988): Interaction of a DNA binding factor with the 5′ flanking region of an ethylene-responsive fruit ripening gene from tomato. *EMBO J* **7**:3315–3320.

Deikman J, Kline R, Fischer RL (1992): Organization of ripening and ethylene regulatory regions in a fruit-specific promoter from tomato (Lycopersicon esculentum). *Plant Physiol* **100**:2013–2017.

Dirk LM, Williams MA, Houtz RL (2001): Eukaryotic peptide deformylases. Nuclear-encoded and chloroplast-targeted enzymes in Arabidopsis. *Plant Physiol* **127**:97–107.

Dirk LM, Williams MA, Houtz RL (2002): Specificity of chloroplast-localized peptide deformylases as determined with peptide analogs of chloroplast-translated proteins. *Arch Biochem Biophys* **406**:135–141.

Dobres MS, Mandaci (1998): Plant Promoter Useful for Directing the Expression of Foreign Proteins to the Plant Epidermis. US Patent 5,744,334.

Eisenberg R (2003): Patent swords and shields. *Science* **299**:1018–1019.

Feldman M (2005): Commercializing Cohen-Boyer 1980–1997. Rotman School of Management, Univ Toronto (http://www.kauffman.org/pdf/tt/Feldman_Maryann.pdf.)

Feldman M, Colaianni A, Liu K (2005): *Commercializing Cohen-Boyer 1980–1997*. DRUID Working Papers 05–21, Copenhagen Business School, Department of Industrial Economics and Strategy/Aalborg Univ, Department of Business Studies.

Fenton GM, Chi-Ham C, Boettiger S (2007): Intellectual property freedom to operate: The law firm's approach and role. In *Intellectual Property Management in Health and Agricultural Innovation: A Handbook of Best Practices*, Krattiger A (ed). pp. 1363–1384.

Fraley RT, Horsch RB, Rogers SG (1991): *Genetically Transformed Plants* EP0131620.

Fraley RT, Horsch RB, Rogers SG (1994): *Chimeric Genes for Transforming Plant Cells Using Viral Promoters*. US Patent 5,352,605.

Gelvin SB, Hauptmann R, Ni M, Cui D (1999): *Chimeric Regulatory Regions and Gene Cassettes for Expression of Genes in Plants*. US Patent 5,955,646.

Giovannoni J, DellaPenna D, Bennett AB, Fischer RL (1989): Expression of a chimeric polygalacturonase gene in transgenic rin (ripening inhibitor) tomato fruit results in polyuronide degradation but not fruit softening. *Plant Cell* **1**:53–63.

Graff G, Zilberman D, Bennett AB (in press): The contraction of product quality innovation in agricultural biotechnology. *Nat Biotechnol*.

Graff GD, Cullen SE, Bradford KJ, Zilberman D, Bennett AB (2003): The public-private structure of intellectual property ownership in agricultural biotechnology. *Nat Biotechnol* **21**:989–995.

Grundler F, Nitz I, Puzio P (2001): *Root-Specific Promoter*. PCT Application WO 01/44454.

Hamilton CM (1998): *Binary BAC Vector*. US Patent 5,733,744.

Harada JJ, Lotan T, Ohto M-A, Goldberg RB, Fischer RL (2001): *Leafy Cotyledon Genes and Their Uses*. US Patent 6,320,102.

Hardin G (1968): Tragedy of the commons. *Science* **162**:1243–1248.

Hayashi M, Aoki M, Kato A, Kondo M, Nishimura M (1996): Transport of chimeric proteins that contain a carboxy-terminal targeting signal into plant microbodies. *Plant J* **10**:225–234.

Heller MA, Eisenberg RS (1998): Can patents deter innovation? The anticommons in biomedical research. *Science* **280**:698–701.

Herrera-Estrella L, Depicker A, Van Montagu M, Schell J (1983): Expression of chimeric genes transferred into plant cells using a Ti plasmid-derived vector. *Nature* **303**:209–213.

Herrera-Estrella L, VanDen Broeck G, Van Montagu M, Scheier P, Schell J, Bohnert HJ, Cashmore AR, Timko MP, Kausch AP (2000): *Chimaeric Gene Coding for a Transit Peptide and a Heterologous Peptide.* US Patent 6,130,366.

Hoekema A, Hooykaas PJJ, Schilperoort RA (1992): *Process for Introducing Foreign DNA into the Genome of Plants.* US Patent 5,149,645.

Hou C, Dirk LM, Goodman JP, Williams MA (2006): Metabolism of the peptide deformylase inhibitor actinonin in tobacco. *Weed Sci* **54**:246–254.

Huang S, An YQ, McDowell JM, McKinney EC, Meagher RB (1997): The Arabidopsis ACT11 actin gene is strongly expressed in tissues of the emerging inflorescence, pollen, and developing ovules. *Plant Mol Biol* **33**:125–139.

Jefferson R (2005): *Biological Gene Transfer System for Eukaryotic Cells.* US Patent Application 10/953,392.

Jefferson RA (2004): *Biological Gene Transfer System for Eukaryotic Cells.* US Patent Application 10/954,147.

Jefferson RA, Harcourt RL, Kilian A, Wilson KJ, Keese PK (2002): *Microbial β-Glucuronidase Genes, Gene Products and Uses Thereof.* US Patent 6,391,547.

Jefferson RA, Mayer JE (2003): *Microbial β-Glucuronidase Genes, Gene Products and Uses Thereof.* US Patent 6,641,996.

Jefferson T (1987): Thomas Jefferson to Isaac McPherson. In *The Founders' Constitution*, Kurland PB, Lerner R (eds) Univ. Chicago Press (http://press-pubs.uchicago.edu/founders/documents/a1_8_8s12.html).

Kato A, Hayashi M, Kondo M, Nishimura M (1996): Targeting and processing of a chimeric protein with the N-terminal presequence of the precursor to the glyoxysomal citrate synthase. *Plant Cell* **8**:1601–1611.

Klein TM, Wolf ED, Wu R, Sanford JC (1987): High velocity microprojectiles for delivering nucleic acids into living cells. *Nature* **327**:70–73

Komarnytsky S, Borisjuk NV, Borisjuk LG, Alam MZ, Raskin I (2000): Production of recombinant proteins in tobacco guttation fluid. *Plant Physiol* **124**:927–934.

Kowalski SP, Ebora RV, Kryder D, Potter RH (2002): Transgenic crops, biotechnology and ownership rights: What scientists need to know. *Plant J* **31**:407–421.

Kryder RD, Kowalski SP, Krattiger AF (2000): *The Intellectual and Technical Property Components of Pro-vitamin A Rice (GoldenRiceTm): A Preliminary Freedom-to-Operate Review*, Brief No. 20, ISSAA; Ithaca, NY.

Maiti I, Shephard RJ (1998): *Promoter (FLt) for the Full Length Transcript of Peanut Chlorotic Strek Caulimovirus (PCLSV) and Expression of Chimeric Genes in Plants.* US Patent 5,850,019.

Maiti I, Shephard RJ (1999): *Full Length Transcript (FLt) Promoter from Figwort Mosaic Caulimovirus (FMV) and Use to Express Chimeric Genes in Plant Cells.* US Patent 5,994,521.

Marshall E (2002): DuPont ups the ante on use of Harvard's OncoMouse. *Science* **296**:1212–1213.

McElroy D, Wu R (2002): *Rice Actin 2 Promoter and Intron and Methods for Use Thereof.* US Patent 6,429,357.

McElroy D, Zhang W, Cao J, Wu R (1990): Isolation of an efficient actin promoter for use in rice transformation. *Plant Cell* **2**:163–171.

Mentewab A, Stewart CN (2005): Overexpression of an *Arabidopsis thaliana* ABC transporter confers kanamycin resistance to transgenic plants. *Nat Biotechnol* **23**:1177–1180.

Miki B, McHugh S (2004): Selectable marker genes in transgenic plants: Applications, alternatives and biosafety. *J Biotechnol* **107**:193–232.

Odell JT, Nagy F, Chua N-H (1985): Identification of DNA sequences required for activity of a plant promoter: The CaMV 35S promoter. *Nature* **313**:810–812.

Pray CE, Naseem A (2005): Intellectual property rights on research tools: incentives or barriers to innovation? Case studies of rice genomics and plant transformation technologies. *AgBioForum* **8**(2&3), article 7.

Roa-Rodriguez C, Nottenburg C (2003): *Antibiotic Resistance Genes and Their Uses in Genetic Transformation*. CAMBIA IP Resource, www.cambiaip.org, Canberra, Australia.

Rogers SG, Fraley RT (2001): *Chimeric Genes Suitable for Expression in Plant Cells*. US Patent 6,174,724.

Romano CP, Cooper ML, Klee HJ (1993): Uncoupling auxin and ethylene effects in transgenic tobacco and Arabidopsis plants. *Plant Cell* **5**:181–189.

Sanford JC, Wolf ED, Allen NK (1991): *Method for Transporting Substances into Living Cells and Tissues and Apparatus Thereof.* US Patent 5,036,006.

Santerre RF, Rao RN (1988): *Recombinant DNA Cloning Vectors and the Eukaryotic and Prokaryotic Transformants Thereof.* US Patent 4,727,028.

Smeekens S, Bauerle C, Hageman J, Keegstra K, Weisbeck P (1986): The role of the transit peptide in the routing of precursors toward different chloroplast compartments. *Cell* **46**:365–375.

Tague BW, Dickinson CD, Chrispeels MJ (1989): A short domain of the plant vacuolar protein phytohemagglutin targets invertase to the yeast vacuole. *Plant Cell* **2**:533–546.

van der Fits L, Memelink J (1997): Comparison of the activities of CaMV 35S and FMV 34S promoter derivatives in Catharanthus roseus cells transiently and stably transformed by particle bombardment. *Plant Mol Biol* **33**:943–946.

Volokita M (1991): The carboxy-terminal end of glycolate oxidase directs a foreign protein into tobacco leaf peroxisomes. *Plant J* **1**:361–366.

Ye XD, Al-Babili S, Kloti A, Zhang J, Lucca P, Beyer P, Potrykus I (2000): Engineering the provitamin A (beta-carotene) biosynthetic pathway into (carotenoid-free) rice endosperm. *Science* **287**:303–305.

Yoder JI, Lassner M (1998): *Biologically Safe Plant Transformation System*. US Patent 5,792,924.

CHAPTER 15

Why *Transgenic* Plants Are So Controversial

DOUGLAS POWELL

International Food Safety Network, Department of Diagnostic Medicine/Pathobiology, Kansas State University, Manhattan, Kansas

15.0. CHAPTER SUMMARY AND OBJECTIVES

15.0.1. Summary

Plant biotechnology has incited much protest in its relatively short commercial lifetime. Other than the scientific reasons (risk assessment) given in Chapter 14, why all the controversy? European groups have banned the planting of transgenic plants and refused to adopt them. Why? Some of the political and social reasons are explored in this chapter.

15.0.2. Discussion Questions

1. Why is Frankenstein's monster often used to illustrate the risks of biotechnology?
2. What are some of the factors that play into peoples' perception of risk?
3. What are the stigmas associated with plant biotechnology, and how can they be overcome?
4. What two major scientific stories prompted media attention in the 1990s?
5. What issues are still being debated? Should they be?

15.1. INTRODUCTION

In 1990 author Michael Crichton wrote in the novel *Jurassic Park* (Chrichton 1990) (which begat the film, which begat the sequel, and then the other sequel), that

> The late twentieth century has witnessed a scientific gold rush of astonishing proportions: the headlong and furious haste to commercialize genetic engineering. This enterprise has proceeded so rapidly—with so little outside commentary—that its dimensions and implications are hardly understood at all.

Plant Biotechnology and Genetics: Principles, Techniques, and Applications, Edited by C. Neal Stewart, Jr.

While not accurate, Crichton captured both the angst and ambitions that characterize public discussions of genetic engineering. Today, as farmers throughout North America, and increasingly the world, embrace the tools of agricultural biotechnology, environmental and activist groups continue to dub the products "Frankenfoods," consistent with the powerful Frankenstein science-out-of-control narrative that resonates deep within humans. The claims of "untested" and "Frankenfood bad" is rhetoric designed to alert rather than inform. Indeed, it can be claimed, and has been repeatedly, with a multitude of substantiation, that specific genetically engineered foods are better for the environment, contain lower levels of natural toxins, and are rigorously tested.

But that is not what this chapter is about. Health and environmental risks with any new technology will always be open to continual debate and refinement—that is the process of science and assessment of risk. Instead, this chapter attempts to highlight some of the broader questions about the interactions between science and society, using genetically engineered foods as a case study.

15.1.1. The Frankenstein Backdrop

First published in 1817, Mary Shelley's *Frankenstein* contained many warnings about science out of control. At a time when fundamental advances in organic chemistry were leading some scientific charlatans to say that they had discovered the secret of life, Shelley, a member of England's radical intellectual elite, had Professor Walden, Frankenstein's teacher, say that

> The ancient teachers of this science, promised impossibilities, and performed nothing. The modern masters promise very little; they know metals cannot be transmuted, and that the elixir of life is a chimera. But these philosophers, whose hands seem only made to dabble in dirt, and their eyes to pore over the microscope or crucible, have indeed performed miracles. They penetrate into the recesses of nature, and show she works in her hiding places. They ascend into the heavens; they have discovered how the blood circulates, and the nature of the air we breathe. They have acquired new almost unlimited powers; they can command the thunders of the heaven, mimic the earthquake, and even mock the invisible world with its own shadows.

Through the new-found wonders of chemistry, Professor Frankenstein creates a life, which pursues him, and finally, with his own life, he pays the price for scientific hubris. The centuries are filled with such tales of scientific hubris and calls for humility.

15.1.2. Agricultural Innovations and Questions

The use of chemical inputs into agricultural food production has a lengthy history. As early as 1000 B.C. the Chinese used sulfur as a fumigant; in the sixteenth century arsenic-containing compounds were utilized as insecticides, and by the 1930s the production of modern synthetic chemicals commenced. With the onset of World War II there was a rapid increase in the production and use of chemical substances such as DDT, used for control of insect-transmitting malaria. The postwar era marked the start of the modern agrochemical industry, and as a direct result of technical advancements in chemical production during this period, various insecticides, fungicides, and fumigants found their place in agriculture and food production (Powell and Leiss 1997).

Today, rather than spraying chemicals in fields to bolster crop production, genes from naturally occurring organisms with pesticidal or herbicide-tolerant properties are being engineered into plants. And just as in the earlier period of chemical pesticides use, the public discussion of agricultural biotechnology has been framed narrowly in terms of risk versus benefit, rather than in a more complete outlook where the objectives are to maximize benefits while minimizing risks.

The public discussion of genetically engineered foods has, since at least 1998, been characterized by seemingly simple questions that many have failed to adequately answer the following questions: Why are you messing with nature? Why don't you label everything? Can you guarantee there won't be any long-term risks? Why are you playing God?

A May 2007 review of a documentary, *The Future of Food* (available at http://www.newstarget.com/021827.html), while exaggerated, summarizes much of the concern regarding genetically engineered food:

> There is a cabal of power-hungry corporations that are systematically destroying humanity's future. These companies have taken over the food supply, injected pesticides, viruses and invading genes into staple crops, engineered "terminator" genes that make crop seeds unviable, destroyed the livelihood of farmers and used every tactic they could think of—legal threats, intimidation, bribery, monopolistic market practices and many more—to gain monopolistic control over the global food supply. One documentary brings you this astonishing story. Through the testimony of family farmers, ecological scientists, agricultural experts and numerous public documents, The Future of Food tells a horrifying, heart-stopping story of how Big Agriculture has sold out the future of human civilization for the almighty dollar.

Beginning in 1994 with the US introduction of the Flavr Savr tomato, the products of agricultural biotechnology—using the tools of molecular biology to move and alter specific genes to bolster crop productivity, extend the shelf life of fresh fruits and vegetables, and reduce the environmental stresses of food production—have been commercially available.

Since 1995, North American—and international—farmers have increasingly chosen to pay extra for genetically enhanced corn, soy, canola, and potato seed because, quite simply, it works: increased yields on the same amount of land, reductions in chemical use, more efficient farming systems. So, why has this technology engendered such deep hostility?

15.2. PERCEPTIONS OF RISK

How an individual perceives a risk—in this case the risk posed by genetically engineered food—has been the subject of extensive research. Sandman (1987) noted that the public generally pays too little attention to the hazardous nature of risks, and experts usually completely ignore those factors that fuel consumer unrest or outrage. Scientists, in general, define risks in the language and procedures of science itself; they consider the nature of the harm that may occur, the probability that it will occur, and the number of people who may be affected (Groth 1991). Most citizens seem less aware of the quantitative or probabilistic nature of a risk, and much more concerned with broader, qualitative attributes, such as whether the risk is voluntarily assumed, whether the risks and benefits are fairly distributed, whether the risk can be controlled by the individual, whether a risk is necessary and unavoidable or whether there are safer alternatives, whether the risk is familiar or exotic, whether the risk is natural or technological in origin, and so forth (Sandman 1987). But such generalizations are of limited value.

According to Covello (1983, 1992), research in the psychological sciences has identified 47 known factors that influence the perception of risk: issues like control, benefit, whether a risk is voluntarily assumed and, the most important factor, trust. While these factors can help explain why consumers are concerned about a potential risk such as genetically engineered food, differences in perception of risk only superficially explain the visceral outrage that has greeted the prospects of genetically engineered crops in some areas. By examining the various social actors and their tactics of public persuasion, a general picture emerges that helps explain the social controversy surrounding genetically engineered crops.

The current state of risk management and communication research suggests that those responsible for food safety risk management must be seen to be reducing, mitigating, or minimizing a particular risk. Those responsible must be able to effectively communicate their efforts and to prove that they are actually reducing levels of risk.

Stigma is a powerful shortcut that consumers may use to evaluate foodborne risks. Gregory et al. (1995) have characterized stigma as follows:

- The source is a hazard.
- A standard of what is right and natural is violated or overturned.
- Impacts are perceived to be inequitably distributed across groups.
- Possible outcomes are unbounded (scientific uncertainty).
- Management of the hazard is brought into question.

These factors of stigmatization certainly apply to the products of agricultural biotechnology. Stigmatization is becoming the norm for food and water linked to human illness or even death. The challenge then is, how to reduce stigma? The components for managing the stigma associated with any food safety issue seem to involve all of the following factors:

- Effective and rapid surveillance systems
- Effective communication about the nature of risk
- A credible, open, and responsive regulatory system
- Demonstrable efforts to reduce levels of uncertainty and risk
- Evidence that actions match words

Appropriate levels of risk management coupled with sound science and excellent communication about the nature of risk are required to further garner the benefits of any technology, including agricultural biotechnology.

The products of agricultural biotechnology began reaching mainstream status at the same time that the North American public was being exposed to massive amounts of microbial food safety information, beginning with the Jack-in-the-Box restaurant *Escherichia coli* O157:H7 outbreak of 1993 (Powell and Leiss 1997) leading to an unprecedented interest in the way food is produced. Consumer concerns about food safety—such as mad cow disease, *E. coli* O157:H7, and salmonella—have been pushed from the supermarket all the way back to the farm, such that any and all agricultural practices are coming under public scrutiny. This trend continues today and is reflected in increased sales of organic foods, books like the *100-Mile Diet* (Smith and Mackinnon 2007) and the growth of community-shared agriculture (CSA), as individuals seek to exert more control over the food that nourishes their bodies and souls.

Several of the practices of industry and governments during the mid-1990s promoted suspicion. The results of field trials of GE crops were often difficult to obtain—and still are—creating an atmosphere of distrust. These same groups often argued that genetic engineering of crops was an extension of bread- and winemaking in an attempt to make the unfamiliar familiar. John Durant has noted that attempts to characterize biotechnology as merely trivial extensions of the familiar techniques of baking, viniculture, and breeding are "pedantic" at best: "The technologies employed are completely different and it is the power and precision of the new molecular biology that drives both industrial growth and public concern" (Durant 1992). Comparisons to traditional breeding tend to magnify rather than soothe consumer concerns.

Many individuals supportive of genetically engineered crops argued that the term "genetic engineering" was alarming to the public, and instead terms like "crop improvement" and "biotechnology" should be used. Activists responded with "Frankenfood" and "GMO" (genetically modified organism), which now dominate public discourse. The topic of genetically engineered food was endlessly surveyed around the world, with public notions of agricultural biotechnology consistently articulated as concerns about uncertainty, "playing God,"and the involvement of powerful interests.

15.3. RESPONSES TO FEAR

In response to public risk controversies—like agricultural biotechnology—many politicians, company executives, and academics urge citizens to become better educated in matters scientific, to therefore overcome public fear as a barrier to "progress." This rhetorical strategy has been advocated by technology promoters in discussions of technological risk for the past 200 years. More recently, promoters of agricultural chemicals in the 1960s and nuclear energy in the 1970s have embraced the public education model. It has failed. Today, the notion of public education is the basis of dozens of communications strategies forwarded by government, industry, and scientific societies, in the absence of any data suggesting that such educational efforts are successful. As noted by Kelley (1995), voters in democracies routinely make decisions about policies about which they have no detailed academic understanding. Consumers have and will continue to make decisions about genetically engineered foods, whether they are "better educated" or not.

Genetic engineering is a powerful technology—and that is the source of potential benefit and unrestrained angst. It is also why the technology is regulated. As Norman Ball of the University of Waterloo noted (Ball 1992), all revolutionary technologies create three public responses, in succession: unrealistic expectations (all new technologies are oversold; there is an old saying that "bullshit is the grease on the skids of innovation"), confusion, and eventually finding a way to cope. Biotechnology has been, and continues to be, oversold, but as with other new technologies, a public discussion over time shifts from one of risks versus benefits to a more realistic approach of extracting whatever benefits a technology can bring while actively and prudently minimizing risks. But in many areas of the world, particularly Europe, the initial formulations of the public discussion of genetically engineered foods remains, and the products are thoroughly stigmatized, as in France, where risks such as nuclear energy are embraced (risk perception research would suggest a rejection of nuclear energy).

This suggests that the varying degrees of public controversy in various countries and within social groups within countries is directed by repetitious conversations about risks and benefits, beginning with media coverage, and what one is exposed too, and what one chooses to acknowledge. In 2007, the use of blogs, Wikipedia, YouTube, Facebook, and other Internet-based social networking activities allows not only for the democratization of information but also for the proliferation and regurgitation of unsubstantiated pap. Several stories have been repeatedly cited and have shaped the current view of genetically engineered foods. A few are dissected below:

Toward the end of 1996 the Natural Law Party mounted a cross-Canada book tour featuring Dr. John Fagan, which received extensive coverage; among the exaggerated or erroneous claims promulgated by Fagan and others, and one that deserves specific attention, is the statement that "40 people were killed and thousands were crippled by exposure to gene-tinkered food" (Graham 1996).

In 1989 there was an outbreak in the United States of a newly recognized fatal blood disease called *eosinophilia–myalgia syndrome* (EMS). The outbreak killed at least 27 people and sickened another 1500, and the cause was finally traced to certain batches of the amino acid L-tryptophan, manufactured in Japan by Showa Denko and widely available in the United States as a nutritional supplement. It has been estimated that, prior to this outbreak, $\leq 2\%$ of the US population took L-tryptophan in the belief that it helped manage insomnia, premenstrual syndrome, stress, and depression—this in the absence of any medical data supporting the effectiveness of the supplement.

L-Tryptophan is manufactured in a fermentation process using a bacterium, *Bacillus amyloliquefaciens*, in the same way that yeast ferments the sugars in barley into ethanol in beer. Subsequent investigations by US health authorities revealed that Showa Denko made two changes to its L-tryptophan manufacturing process in 1989, changes that allowed the contamination of L-tryptophan: (1) the company began using a strain of *B. amyloliquefaciens* that had been genetically engineered to produce larger amounts of L-tryptophan, and (2) they reduced the amount of carbon used to filter out impurities from the final product. Studies have shown that the disease-causing molecule appears only during purification and that cases of EMS have been linked to L-tryptophan produced by Showa Denko as early as 1983, long before the company used a genetically engineered bacterium. The risk information vacuum for GE food allowed such alarmist and erroneous versions of events to take root and flourish.

Powell and Leiss (1997) describe how a risk information vacuum arises when, over a long period of time, those who are conducting the evolving scientific research and assessments for high-profile risks such as genetically engineered foods, make no special effort to communicate the results obtained from these studies regularly and effectively to the public. Instead, partial scientific information dribbles out here and there and is interpreted in apparently conflicting ways, all of which is mixed with people's fears.

15.4. FEEDING FEAR: CASE STUDIES

Society as well as nature abhors a vacuum, and so it is filled from other sources. For example, events reported in the media (some of which are alarming) become the substantial basis of the public framing of these risks; or an interest group takes up the challenge and fills the vacuum with its own information and perspectives; or the intuitively based fears and concerns of individuals simply grow and spread until they become a substantial consensus in the arena of public opinion.

15.4.1. Pusztai's Potatoes

On August 10, 1998 Dr. Arpad Pusztai of the Rowett Research Institute in Aberdeen, Scotland, reported that after feeding five rats potatoes that were genetically engineered to contain one or two lectins, proteins that are known to be toxic to insects, they observed, over a 110-day period, that some of the rats manifested stunted growth and impaired immune systems. Dr. Pusztai reported the findings, not in a peer-reviewed scientific journal, but on the UK *World in Action* television program. After an internal review of the data, it emerged that not only had Dr. Pusztai ignored the conventional route of scientific peer review, but the experimental design lacked appropriate controls. Potatoes themselves are full of poisonous chemicals in quantities that vary depending on how they are grown, a phenomenon known as *somaclonal variation*, and must therefore be uniformly grown for any feeding trail to be informative. Moreover, rats do not like to subsist on raw potatoes, and their diets must be supplemented. By August 12, 1998, Dr. Pusztai was suspended and subsequently forced to retire.

The Pusztai affair, as it soon became known, spawned significant media coverage with numerous allegations. On February 12, 1999 a group of 20 international scientists released a letter supporting the work of Dr. Pusztai, and specifically charged that the process of genetic engineering itself, in particular the use of the 35*S* cauliflower mosaic virus promoter, was to blame. The 35*S* promoter is widely used in the genetic engineering of plants, to turn specific genes on and off. Because of this widespread use, regulators in Western countries already demand evidence that any 35*S* insertion is stable and well characterized. Other feeding experiments involving the 35*S* promoter have simply not found the problems described by Pusztai and supporters. Most importantly, though, the potatoes grown by Dr. Pusztai would never have been approved in Canada, the United States, or the United Kingdom. Subsequently, the UK Royal Society concluded that "Dr. Arpad Pusztai's widely publicised research into the effects of feeding rats genetically modified (GM) potatoes appears to be flawed, and it would be unjustifiable to draw from it general conclusions about whether genetically modified foods are harmful to human beings or not." The Pusztai affair is repeatedly cited as proof of harm from GE foods.

15.4.2. Monarch Butterfly Flap

On May 20, 1999 John Losey and colleagues from Cornell University published a brief letter in the scientific journal *Nature* (Losey et al. 1999) that drew intense national and international media coverage (PEW 2002). The report concerned a laboratory study in which the leaves of milkweed plants in a greenhouse were artificially dusted with pollen from conventional and genetically engineered Bt corn plants at levels approximating what the researchers thought occurred in nature. Bt corn has been genetically engineered to contain the protein from a common soil bacterium *Bacillus thuringiensis*. In this study, 3-day-old Monarch caterpillars were placed on the leaves and allowed to feed for 4 days. The researchers reported that 44% of the Monarch larvae fed leaves coated with Bt-pollen died. No caterpillar died that ate leaves dusted with regular corn pollen or the control leaves. Larvae feeding on the Bt-dusted leaves also ate much less and were less than half the size of larvae that fed on leaves with no pollen. No attempt was made, however, to compare the pollen coverage of the leaves in the lab to the coverage that might commonly exist in or near a cornfield.

The authors correctly recognized that the study was limited in applicability, and that field tests would be required to determine the significance of this finding in an artificial environment. On publication, Dr. Losey was quoted as saying "We can't forget that Bt-corn and other transgenic crops have a huge potential for reducing pesticide use and increasing yields. This study is just the first step, we need to do more research and then objectively weigh the risks versus the benefits of this new technology" (http://www.news.cornell.edu/releases/May99/Butterflies.bpf.html).

Losey soon found that his results had transformed into mutant tales of killer corn and sacred butterflies. *The New York Times* led on the front-page with a story entitled, "Altered Corn May Imperil Butterfly, Researchers Say" in which one researcher described monarchs as the "Bambi of the insect world" (Kaesuk Yoon 1999). To this date, Greenpeace demonstrators continue to dress in Monarch butterfly costumes and simultaneously drop dead at a prearranged time, usually for the convenience of television cameras. Great street theater, lousy public policy.

15.5. HOW MANY BENEFITS ARE ENOUGH

In October 2000, the US Environmental Protection Agency stated in a comprehensive report that corn, cotton, and potato crops genetically engineered to repel pests offered "significant benefits" to farmers and few risks, even for Monarch butterflies, giving an overwhelming stamp of approval to the technology as a way to boost yields, reduce farm chemicals, and lessen groundwater contamination. The report found that in 1999 alone, US farmers reduced pesticide costs by more than $100 million (www.epa.gov/scipoly/sap/).

The question "Do you want fish genes if tomatoes?" has been used repeatedly by Greenpeace and other activists in campaign literature and media accounts (Greenpeace 2001). Yet the actual experiment to transfer an antifreeze protein from cold-water flounder to enhance the cold tolerance of field tomatoes was attempted only once in 1991 and was unsuccessful (Powell http://www.foodsafety.ksu.edu/article-details.php?a=1&C=1&Sc=2&d=40). Nevertheless, when asked which foods in the supermarket are GE, consumers consistently cite vegetables, such as tomatoes, and fruit (IFIC 2002). While this is due partly to the short availability of the Flavr Savr tomato, it also demonstrates how memorable such evocative messages are to the public.

These are only a smattering of the dozens of examples of information intended to alarm rather than inform. By fall 1999, this combination of scientific naivety, media hype, and allegations of corporate conspiracy had come to characterize any and all public discussions of the role of genetically engineered foods. So Greenpeace and the Council of Canadians, two activist groups, hoping to build on the success in stigmatizing GE food in Europe, particularly the UK, held a public demonstration in front of a Loblaws supermarket in an affluent area of downtown Toronto, a Canadian beachhead into the United States (Fig. 15.1). Typical of the statements was that of Jennifer Story, health protection campaigner for the Council of Canadians, who asserted that "Genetically engineered foods have not been proven safe for human health and the environment. As the largest grocery chain in Canada, Loblaws has the obligation to take the lead, and take genetically engineered food off the shelf."

When public concern mounted in the UK and Europe in response to activist tactics, the scientific community, political leaders and opinion leaders were largely silent. Even if they had spoken out, the effects would have been marginalized by the fallout from the mad cow

Figure 15.1. Greenpeace demonstration in front of a Toronto grocery store (photo by Doug Powell).

crisis in Britain. On March 20, 1996 the British government announced what many already knew—that consumption of products from cattle with bovine spongiform encephalopathy, or mad cow disease, was leading to a new variant form of Creutzfeld–Jacob disease (vCJD) that struck the young and was particularly gruesome, leading to its inevitable death in the victim. Millions of animals were killed at a cost of billions of dollars in lost trade, and to date some 150 people have died from vCJD. Mad cow disease clearly represented modern agricultural practices as science out of control.

Unlike European farmers, North American farmers were eager to sample and adopt the newly available GE seeds, and were prepared to enter the public debate to retain and ensure access to those tools. As Thomas Jefferson wrote in a letter to William Charles Jarvis, dated Sept. 28, 1820, "I know of no safe depository of the ultimate powers of society but the people themselves; and if we think them not enlightened enough to exercise their control with a wholesome discretion, the remedy is not to take it from them, but to inform their discretion" (NRC 1989, p. 14).

This led to an active information campaign, not to educate, but to inform and let consumers decide for themselves. Farmers have no interest in providing a product that consumers don't want. That would be an expensive failure. A decade later, the debates seem so much ado about ... not much. Several of the debates are discussed below.

15.6. CONTINUING DEBATES

15.6.1. Process versus Product

Genetic variability is required to enhance traits deemed desirable by humans. Geneticists can travel the world searching for plants, animals, or microorganisms that posses a trait of interest such as increased productivity or disease resistance. Desirable variability can be selected over generations of breeding. Genetic engineering, using the tools of molecular biology, allows further sources of genetic variability to be introduced into a particular organism.

But there are other techniques for creating genetic variability between the black-and-white of traditional breeding and genetic engineering. Since the 1940s, mutagenesis breeding has been used to induce genetic variability, especially in cereals, by exposing seeds to doses of mutagens—compounds that induce mutations in DNA—such as ionizing radiation or mustard gas. The practice is still used today, as are other techniques. Should such products also be regulated? Or is it the process itself of genetic engineering that is inherently risky?

Proponents and critics have sparred on this point since the advent of genetic engineering, but the scientific community and North American regulators have consistently maintained that it is the endproduct, not the process, that should be regulated. Varieties of potatoes and celery, for example, have been produced through traditional breeding that were later discovered to contain unacceptably high levels of natural compounds. This view that the endproduct should undergo a safety assessment regardless of how it was produced has been enshrined in the Canadian Novel Food Act (1999) and was reaffirmed by an expert panel of the US National Academy of Sciences (2000).

15.6.2. Health Concerns

In 1994 the Flavr Savr tomato became the first whole, genetically engineered food to be approved by the US Food and Drug Administration (FDA) and, subsequently, Health Canada. Results of rodent feeding trials, submitted as part of the dataset that regulators reviewed, showed no difference between conventional and genetically engineered tomatoes. It also showed that rats do not like tomatoes.

The experiment highlighted one of the difficulties in assessing the safety of genetically engineered foods. For example, the genetically engineered field maize grown in North America (and now elsewhere) contains a gene from the common soil bacterium *Bacillus thuringiensis*, and is known as Bt corn. Regulators and several international scientific panels reasoned that because humans have been ingesting Bt without effect for decades (it is also widely used as an organic spray) and because the Bt toxins (in this case specific to the European cornborer) are proteins—and because any toxin protein remaining after processing would be quickly digested in the human gut—Bt corn is safe; or, in the words of the language-challenged, the Bt corn was found to be substantially equivalent to traditional corn. Any commercial concern wishing to sell a genetically engineered food, or indeed any new or novel food, must demonstrate substantial equivalence, based on molecular, nutritional, and toxicological data, to the appropriate regulatory body. If substantial equivalence is more difficult to establish, then the identified differences, or new characteristics, would be the focus of further safety considerations. The more a novel food differs from its traditional counterpart, the more detailed the safety assessment that must be undertaken. Future products of agricultural biotechnology, where complex pathways within a plant are altered to produce more nutritious foods, may require a more elaborate safety assessment. The genetically engineered foods available today are the result of relatively simple gene transfers, harnessing systems that are based in nature.

However, the attempt to improve any food can possibly lead to unexpected consequences. For example, in the laboratory, in one instance, a human allergen was transferred from one crop to another. During the preliminary assessment process, the company immediately discontinued the experiment. For the critics of biotechnology, the experiment proved that allergens could be transferred, and therefore, untold risks lay in the manipulation of food structure. For supporters, the incident showed that the regulatory system worked.

Indeed, molecular work in agricultural biotechnology has contributed significant knowledge to the database of food allergens.

15.6.3. Environmental Concerns

Biological systems are fluid and dynamic. Farmers have known for decades that when they overuse a particular agricultural tool, they create an evolutionary selection pressure, which, in many cases will lead to resistance, rendering the tool ineffective. The tools of genetic engineering are no different. Weeds in an agricultural setting can significantly reduce yields. Farmers have a number of options for controlling weeds in a cost-effective manner, including the use of approved and registered herbicides, crop rotations, and most recently, genetic engineering. In particular, several soybean and canola varieties are now available that contain genes, found naturally in bacteria, that confer herbicide tolerance, and that may allow producers to grow a bountiful crop with fewer chemicals.

One concern with herbicide-tolerant crops is that the gene responsible for such tolerance could move or transfer to neighboring weeds, thereby allowing such a weed to flourish as it becomes resistant to a particular herbicide (in which case the weed could still be controlled using other management practices such as tillage or alternative herbicides). The development of resistance is a common phenomenon in agriculture, and the transfer of genes from one plant to another is also known to occur, through either pollen or viruses that can naturally infect one plant and then move on to another. The same concern about resistance applies to insect-resistant crops, such as Bt corn. That is why corn producers who grow genetically engineered Bt corn are, for example, required to devote 20% of their acreage to non-Bt varieties.

15.6.4. Consumer Choice

Consumer choice is a fundamental value for shoppers, irrespective of science. Foods in Canada and the United States are labeled on the basis of health and nutritional data, but there are a variety of other voluntary labeling systems based on religious preference (e.g., kosher and halal meats), growing preferences (e.g., organic), or nutritional preferences (e.g., low-fat and low-salt). A market for biotechnology-free foods, labeled as such, might also emerge to meet consumer demand. However, many consumers will continue to base their food selections on taste, price, and nutritional content before other considerations. Labeling guidelines must accommodate all of these values.

15.7. BUSINESS AND CONTROL

Perhaps of greater public and even scientific concern is that the scientific and technological competence related to agricultural biotechnology has become concentrated within the private sector, particularly within multinational corporations such as Monsanto, Syngenta, DuPont, and Bayer. Such a concentration of expertise will advance the research priorities of industrialized countries while sacrificing the public good.

This is a debate that predates transgenic plants, since food production in general has a long history of corporate involvement. On June 29, 1912, following extensive newspaper advertisements, a prospectus for a new company, The Synthetic Products Company Ltd., was launched in Britain. A global rubber shortage from 1907 to 1910 had prompted

European researchers to search for a synthetic source, and that process, company backers believed, was on the verge of being discovered. A group at the Pasteur Institute in France had discovered a bacterium that converted starch into a fuel oil rich in both amyl alcohol and butanol. When the process was scaled up to industrial quantities by British scientists, the fermentation was altered, producing butanol, which had just been recognized as a key component of synthetic rubber manufacture; and acetone, a valuable component of explosives that had previously been imported.

As recounted by Robert Bud in *The Uses of Life: A History of Biotechnology*, the work had enormous commercial potential, and the scientists, far from being unworldly, "exploited the breakthrough to the hilt." The prospectus, which greatly exaggerated the scientific achievements, netted £75,000 despite stiff opposition from plantation rubber interests. Predictably, the process for converting starch from potatoes proved cumbersome, and the plant never realized the hopes expressed in the 1912 prospectus. But a pattern had been established, coupling scientific enthusiasm with the public's willingness to believe—at least the financial public—that would characterize efforts to profit from biology over the next century.

In a capitalist society, such involvement is to be expected. The challenge is to find a balance between private profit and public good, and to come to such conclusions in an open and democratic manner. Farmers, processors, distributors, and others in the farm-to-fork continuum are constantly striving to improve the safety, quality, and efficiency of the food supply. Genetic engineering is one additional tool that, with vigilance and oversight, can help achieve those goals.

15.8. CONCLUSIONS

After a decade of sometimes fierce public debate, what has been accomplished? Better oversight, changes in practices, shifting of entrenched attitudes? A little of all, but nothing of significance has been gained. People are for or against, maybe moving toward a public discussion of risks and benefits, but slowly.

Meanwhile, the World Health Organization estimates that up to 30% of all citizens of so-called developed Western countries will get sick from the food and water they consume each and every year and thousands will die. If the same energy and effort spent on GE foods could be harnessed to create a culture that values microbiologically safe food, there would far fewer sick people. In addition, there is also a technology trickle-down effect in general. Technology is typically created in the developed world and eventually used in the developing world. Needless debates and fearmongering can slow down innovation, which, in turn, negatively affects the people in the developing world. Is it possible that protests in well-fed Europe have led to starvation in Africa? This is certainly food for thought as we think about the future of plant biotechnology.

REFERENCES

Ball N (1992): Essential connections: Past and future, technology and society in Beyond the Printed Page: Online documentation. *Proc 2nd Conf Quality in Documentation*, Univ Waterloo, Waterloo, Ontario, pp 11–28.

Covello VT (1992): Trust and credibility in risk communication. *Health Environ Digest* **6**:1–5.

Covello VT (1983): The perception of technological risks: A literature review. *Tech Forecasting Social Change* **23**:285–297.

Crichton M (1990): *Jurassic Park*. Alfred A Knopf, New York, 416.

Durant J (1992): *Biotechnology in Public: A Review of Recent Research*. Science Museum London.

Graham D (Nov 20, 1996): Altering food called "dangerous experiment." *Toronto Star*, E1.

Greenpeace (Oct 4, 2001): Fishtomato.com. Retrieved Oct 4, 2001, from http://www.fishtomato.com/.

Gregory R, Slovic P, Flynn J (1995): Risk perceptions, stigma, and health policy. *Health Place* **2**: 213–220.

Groth E (1991): Communicating with consumers about food safety and risk issues. *Food Technol* **45**:248–253.

International Food Information Council (IFIC) (2002): *U.S. Consumer Attitudes toward Food Biotechnology: More U.S. Consumers Expect Biotech Benefits*. Wirthlin Group Quorum Survey.

Kaesuk Yoon C (May 20, 1999): Altered corn may imperil butterfly, researchers say. *New York Times*, A1.

Kelley J (1995): Public Perceptions of Genetic Engineering: Australia 1994. Australian Department of Industry, Science and Technology, Canberra. Retrieved Sept 10, 2007 from http://www.das.gov.au/~dist/home.html.

Losey JE, Raynor LS, Carter ME (1999): Transgenic pollen harms monarch larvae. *Nature* **399**:214.

PEW Initiative on Food Biotechnology (2002): *Three Years Later: Genetically Engineered Corn and the Monarch Butterfly*. Retrieved Sept 10, 2007, from www.foodsafetynetwork.ca/gmo/monarch.pdf.

Powell DA, Leiss W (1997): *Mad Cows and Mother's Milk*. McGill-Queen's Univ Press, pp 123–152.

Sandman PM (1987): Risk communication: Facing public outrage. *EPA J* **13**:21.

Smith A, MacKinnon JB (2007): *The 100-Mile Diet: A Year of Local Eating*. Random House Canada.

US National Research Council (NRC) (1989): *Improving Risk Communication. Committee on Risk Perception and Communication*. National Academy Press, Washington, DC, p 332.

CHAPTER 16

The Future of Plant Biotechnology

C. NEAL STEWART, Jr.

Department of Plant Sciences, University of Tennessee, Knoxville, Tennessee

DAVID W. OW

Plant Gene Expression Center, USDA-ARS/UC Berkeley, Albany, California

16.0. CHAPTER SUMMARY AND OBJECTIVES

16.0.1. Summary

Plant biotechnology has been wildly successful and has literally transformed plant agriculture. There are still undulating concerns about safety and sustainability that critics demand be addressed. In that light, there are some biotechnologies that are being developed that might improve not only the efficiency and precision of plant transformation but also public perceptions and biosafety as well. Site-specific recombination and zinc-finger nuclease systems are discussed in this chapter.

16.0.2. Discussion Questions

1. What is the main dichotomy between innovation and caution (or risk, or the perception of risk)?
2. How might site-specific recombination enhance biosafety?
3. Describe site-specific recombination and how it could lead to greater precision in plant transformation.
4. What are zinc-finger nucleases, and how might they alter the future of plant biotechnology?
5. How serious do you think that the problem of adventitious presence or admixture is now and will be in the future?

16.1. INTRODUCTION

The world's population of 6.5 billion is projected to reach 9.4 billion by 2050, requiring a doubling of 1980s agriculture. The 1980s are an important benchmark because this was

Plant Biotechnology and Genetics: Principles, Techniques, and Applications, Edited by C. Neal Stewart, Jr.

the decade when plant biotechnology emerged with a vision on how agriculture might be revolutionized. Although farming dates back some 10,000 years, the greatest production advances occurred during the past century. The question remains whether in the next 40 years, agriculture can continue to advance sufficiently to meet the world's projected needs. Most prognosticators have assumed that biotechnology must play a crucial role, wherein advances will be a combination of gene discovery that provides opportunities for genetic improvements, and new technologies for making engineering tasks more efficient and precise. That's the way the story is supposed to end, anyway.

In many ways, the appropriate ending to this book and answer to the question of the future of plant biotechnology is to heave a sigh, flip a coin, and declare emphatically, "Who knows?" Who knows, because the voices of skeptics and naysayers continue to grow louder with more respectability as the public grows increasingly careful about the adoption of technologies of all sorts. The fact remains that there are merits to biotechnology opponent's arguments; for instance, about transgene integration being imprecise, that we do not know everything about how genes are expressed and repressed, that there is absence of data about long-term risks, and that nature is a big place where ecological interactions are not completely characterized. Fair enough. Maybe plant biotechnology is doomed as too expensive to develop and deploy, rendering it unsustainable, especially if any of the predicted hazards are realized and the public finally makes up its collective mind that the technology is simply too risky to be sustained. Placed in the context of burgeoning organic agriculture markets that necessarily shun biotechnology, perhaps the transgenic plant movement in the late twentieth and early twenty-first centuries was a "one-trick pony." Maybe it is like putting a man on the moon in the late 1960s and early 1970s—cool for then but passé now. But on the other hand, given that transgenic plants have been adopted faster and on a more widespread scale than any prior agricultural innovation, and that the benefits of growing them clearly outweigh the seemingly minimum risks, it appears to have a future, and one that will likely be "transformed" from current plant biotechnology to something more precise and efficient. The future of biotechnology will certainly be driven by innovations and new applications. Although this chapter will not delve deeply into new uses, future applications of transgenes in crops and plants to realize explicit environmental benefits will likely be more compelling than will transgenically encoded herbicide resistance in row crops and other input traits. In that light, the future of transgenic plants looks rosy. However, the plant biotechnologist must resist the urge to put on the rose-colored glasses as has been done time and time again; transgenic plants will never be grown as a cure-all for everything. Conventional breeding, agricultural chemicals, process technology, and mechanization all have their places in feeding the world.

So, if we assume that plant biotechnology will follow the path of other successful technologies, what might we expect to happen next in technology development and deployment? When this book is published, the twenty-fifth anniversary of the production of the first stably transgenic plant will be history. *Agrobacterium*-mediated transformation and particle bombardment, the technologies used to produce plants that have now covered over 1 billion acres, are older than the Macintosh computer, the compact disk, and Kevlar. If we look at another transformative technology as an example—air travel—what might we expect to be the next step? Even though the Wright brothers first flew in a heavier-than-air machine in 1903, the first aerodynamically stable airplanes were not produced until 1910 or so. This timeline roughly parallels first technology developments leading to stable transgenic events; it took a few years of trial and error to render decent transformatation efficiencies. Air travel did not become a commercially reality until the

1920s. Even then, it was used to transport a fraction of long-distance passengers. Trains, ships, and automobiles were far more popular. There were many safety concerns in the early days of air travel, and safety considerations are still a big part of air travel. It is a highly regulated business, but one with clear markets and benefits. Aside from the invention and application of the jet engine, most of the improvements in air travel were incremental. Materials, fuel, avionics, and a whole host of other components gradually got better. Finally, there was a stasis of innovation. In the 1950s and the 1960s, commercial aircraft reached a level of form and function that differs little from that of currently used airplanes. Of course, the electronics of today's Boeing 737 are different from those of the one that first flew in 1965, but it is still a 737 in design and product. Innovation has not ceased in aircraft, but it no longer seems very transformative. Indeed, some revolutionary developments were not sustainable. For example, in the 1960s it was apparent that supersonic commercial air travel was feasible, and indeed, seemingly the future of airlines. Several companies raced to develop the first commercial supersonic jet, and the Concorde won. But few people could afford the high cost of airfare, and the last Concorde was parked earlier this decade as safety concerns combined with prohibitive expenses overtook innovation. Advanced technology in and of itself is not sufficient for commercialization. For subsonic jet travel, safety concerns are still with us, but the current air travel technology is generally accepted; indeed, it is deemed as a necessary component of modern life. That is because the 737 works quite effectively and is economically sustainable. One might argue that transgenic plants, in their current form, are much like the 737; at least the version of the 1960s. Both of these function fine, but key improvements do make a difference.

In that spirit, there are still some crucial incremental biotechnological innovations that are on the horizon worth mentioning here. Perhaps not all will be implemented into final products—they might be too "supersonic," but we can see patterns of technologies emerging that we think will make a difference. This chapter does not cover new and novel products. It is clear that pharmaceuticals will be produced in plants, transgenic plants will be used in phytoremediation and phytosensing, and plant biotechnology will be important for bioenergy production. These are all exciting, but what is happening on the technology side that bears watching? A few innovations will be introduced here as a finale to this book.

16.2. SITE-SPECIFIC RECOMBINATION SYSTEMS TO PROVIDE INCREASED PRECISION

As new technologies are developed, offering greater control over the placement and content of the DNA introduced into the plant genome, many of these innovations will eventually find their way into future generations of GM (genetically modified) plants. Greater engineering precision could help alleviate some biosafety concerns, but more importantly, advances in precision engineering will be adopted for its intrinsic value, since they enable complex engineering tasks to be achieved with greater speed and less effort.

Advances in the way DNA is introduced into a plant will likely benefit from breakthroughs in genetic *recombination-based* methods that were introduced in Chapter 7. Recall that when DNA is introduced into a cell nucleus, it may recombine into the genome by either homologous or nonhomologous processes, mediated by host proteins that repair DNA damage. In homologous recombination, the damaged DNA is repaired faithfully by incorporating or copying the template from a homologous source, such as that in a homologous chromosome or sister chromatid. In nonhomologous recombination, the DNA ends of two molecules are ligated together in a process also known as DNA end-joining or illegitimate recombination. In some lower eukaryotes, such as the budding yeast, the homologous recombination between introduced DNA and host DNA occurs at a high frequency. In most other higher eukaryotes, however, nonhomologous recombination is the predominant route by which new DNA is incorporated into the genome.

Homologous recombination has great potential for genome engineering. The homologous replacement of a host gene by an altered version of that gene can help elucidate gene function, as well as create new genetic varieties through the precise alteration of a host sequence. The use of homologous gene replacement is becoming more common in animal research, particularly in mice, although these are still rare events occurring in a small fraction of transgenic cells. Among plants, the engineering of host DNA by homologous recombination has been practical with a moss (*Physcomitrella patens*) that is becoming a model system for basic genetic studies (Schaefer 2001). With other higher plants, however, there have been only a few reports of successful gene replacement via homologous recombination that were detected at a percent or less of transformation events.

Some of the potentially most useful biotechnologies for increased precision are *site-specific recombination systems*, most of which are described from the viruses and plasmids of bacteria or yeast. Bacterial viruses, known as *phages*, often integrate their DNA into the genome of the bacterial host at a designated site. This site-specific integration process is highly efficient and depends on recombinase protein(s) encoded by the phage genome. Integration into the host genome enables a phage to hide within the bacterial genome for many generations until it elects to leave by a reversal of the integration process, or site-specific excision, which is also mediated by phage-encoded excisionase protein(s). Some plasmids also encode site-specific recombination systems. When plasmids replicate, mother and daughter plasmid chromosomes sometimes recombine as a result of host-mediated homologous recombination. The plasmid-encoded site-specific recombination system resolves these dimers back into monomers, thus ensuring the partitioning of plasmid molecules to dividing cells. Although these recombination systems originate from prokaryote or lower eukaryotes, many of them also function in higher eukaryotic cells. Since the mid-1980s, scientists have been using these site-specific recombination systems to cause a variety of site-specific deletion, inversion, or integration events in animal or plant cells.

The types of site-specific recombination systems that function in higher eukaryotes can be divided into several groups. In the first group, the genetic crossover occurs between two recombination substrates that are identical or nearly identical in sequence. The two product sites generated by the recombination event are therefore identical or similar to the substrate sites, and the recombination reaction is fully reversible. Notable examples from this group are Cre-*lox*, FLP-*FRT*, and R-*RS*, where Cre, FLP, and R are the *recombinases* and *lox*, *FRT*, and *RS* are the respective *recombination sites*. These recombination systems can perform the full range of recombination both within and between DNA molecules (Fig. 16.1). In a second group, the substrates and products of the recombination reaction are also identical or very similar to each other. However, only one of the two types of intramolecular recombination is possible, either deletion or inversion, and the reaction is not reversible. Examples of deletion systems from this group are β-*six*, ParA-*MRS*, and CinH-*RS2*, where β; ParA, and CinH are the recombinases (also known as *resolvases*) that catalyze DNA excision between pairs of sites known respectively as *six, MRS*, and *res*. In a third group, the recombinase (also known as *integrase*) recombines two substrates that do not share extensive similarity, typically known as bacterial and phage attachment

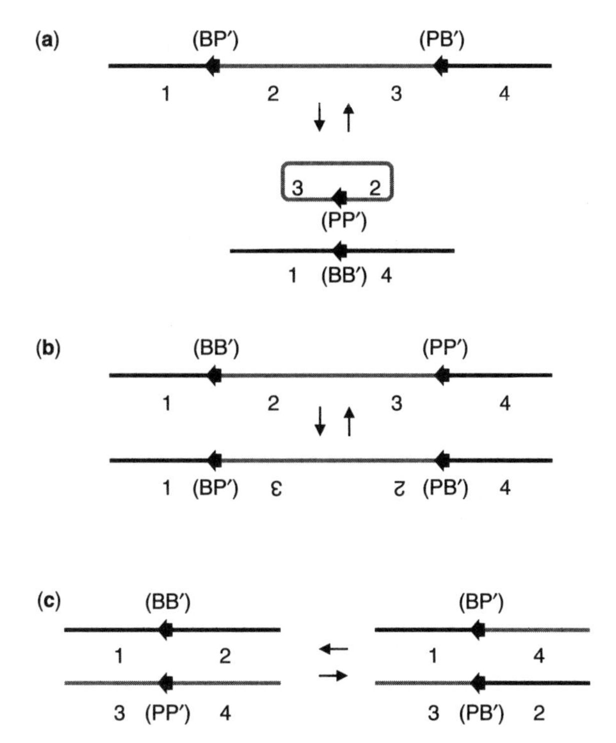

Figure 16.1. Recombination between recombination sites (arrowheads) leading to (a) deletion (excision of circular molecule 2,3 from molecule 1,2,3,4) or integration (insertion of molecule 2,3 into molecule 1,4); (b) inversion (of DNA segment 2,3 flanked by recombination sites of opposite orientation); or (c) translocation (of DNA of different molecules). Some recombination systems use recombination sites that differ in sequence generally known as *attB*, *attP*, *attL*, and *attR*, shown here as BB′, PP′, BP′, and PB′, respectively. In these systems, recombination between *attL* and *attR* requires an excisionase protein in addition to an integrase protein.

sites *attB* and *attP*, respectively, to yield product sites known as *attL* and *attR*. These were explained in detail in Chapter 7 and are the basis of the Gateway cloning system. The reaction can be excision, inversion, or integration, and is not reversible unless an additional protein, an excisionase, is provided. With some members of this group, the integrase can catalyze recombination without the need of a host protein; examples of these recombination systems include ϕC31, ϕBT1, and Bxb1.

16.2.1. Removal of DNA from Transgenic Plants or Plant Parts

The reciprocal exchange of strategically placed recombination sites offers a variety of applications. For instance, the specific deletion of DNA provides a means to remove transgenic materials not needed in the commercial product, in the consumed portion of the commercial crop, or in plant pollen. The site-specific insertion of DNA into pre-defined sites in the genome could help minimize unwanted "position effects" caused by the random integration of introduced DNA. The rearrangement of DNA could also be used to alter gene expression, as in a genetic switch to turn on or off a large set of genes; and the specific recombination of sites within a genome could potentially lead to the large-scale restructuring of chromosomes. Of these potential applications, the removal of DNA by site-specific recombination has already been adopted for commercial product development. Renessen, a joint venture between Cargill and Monsanto, introduced a maize line, LY038, for use in animal feed, marketed under the trade name Mavera™. High-value corn with lysine, LY038, is derived from biolistic transformation of maize to incorporate a *cordapA* gene that directs seed-specific production of a lysine-insensitive dihydrodipicolinate synthase enzyme. The kanamycin resistance marker used for transformation, *nptII*, was subsequently removed by site-specific recombination, a process first described in model plants some 15 years earlier (Dale and Ow 1991; Russell et al. 1992). In the transformation vector, *nptII* was flanked by directly oriented *lox* sites from the Cre-*lox* recombination system (Fig. 16.2). On recovery of the desired transformant, a *cre* gene was introduced into the genome from a genetic cross. Cre recombinase excised away the *lox*-flanked DNA, and the *cre* gene was then removed by genetic segregation. With a decade-long development timeline for genetically modified (GM) maize, the decision to incorporate a recombinase-mediated marker removal step must have been made in the mid-1990s.

Another example is the ability to remove transgenes from pollen to delimit unwanted gene flow to wild relatives and nontransgenic plants that are sexually compatible. Not only are crop-wild transgenic hybrids undesirable from the perspective of transgene escape; repeated backcrossing of transgenes into wild relatives is of regulatory and consumer concern as well. This introgression of transgenes is already likely very rare (Stewart et al. 2003), but could probably be eliminated entirely by a "GM-gene-deletor system" (Luo et al. 2007). In such a system, transgenes of interest are located within dual fused recombination recognition sites with the recombinase under the control of a pollen-specific promoter. Only in pollen grains would the transgenes be deleted, and thus risks of hybridization and introgression would be greatly decreased (Fig. 16.3).

16.2.2. More Precise Integration of DNA

As commercial products improve over time through new innovations, it is likely that additional features will be incorporated into future generations of transgenic plants and

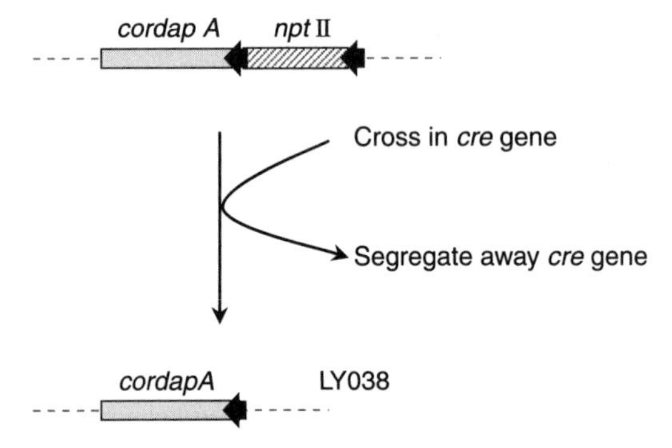

Figure 16.2. Renessen's high-lysine corn line LY038 used site-specific recombination to remove the transformation selectable marker, the kanamycin resistance gene *nptII*, after stable incorporation of *cordapA* that directs high-lysine production in seed. Cre recombinase, introduced from hybridization with a *cre* transgenic plant, excised the *nptII* marker flanked by directly oriented *lox* recombination sites. The *cre* gene was subsequently segregated away in the following generation.

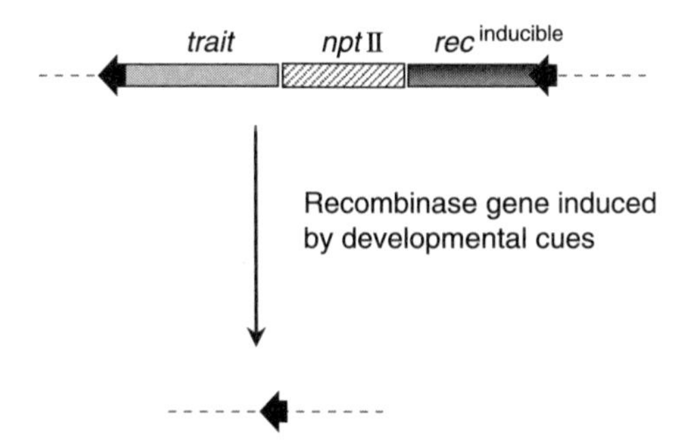

Figure 16.3. Recombination sites that flank the entire transgenic locus permits removal of transgenic DNA on induced expression of a recombinase gene. For instance, if the recombinase gene is placed under the control of sperm-specific or fruit-specific promoters, the excision of transgenic DNA may help reduce the outcross of transgenes, or minimize the production of transgene-encoded proteins needed elsewhere in the plant but not in the edible portions of food.

animals. Site-specific integration of DNA will likely be used in the next generation of agricultural products. The recombinase-mediated integration process is more efficient for delivering precise single-copy DNA into a host genome. More importantly, it has the potential to cluster the introduced DNA at a single locus in the genome. Clustering the transgenic traits provides an important benefit in the breeding process. Over time, more and more genetic improvements will be added or "stacked" to a transgenic plant or animal. Currently, this gene stacking is conducted by combining different transgenes from independent transformation events (integration sites). For example, the coassortment of 4 independent loci is 1 in 16 progeny, which can be found with ease among the typically hundreds of plant progeny (although it is already beyond the litter size limits of many farm animals). For 10 independent loci, however, this would be 1 out of 1024; and here we are not even considering the nontransgenic traits that also must be combined into the genome of the desired commercial product, or the possibility of linkage drag that makes the selective assortment of some of the genetic loci more difficult. If the transgenic traits were clustered, however, then there would be fewer genetic loci to assemble in a breeding program.

16.3. ZINC-FINGER NUCLEASES

In the high-flying world of human gene therapy, a technology that has captured the imagination of researchers is called *zinc-finger nucleases* (ZFNs). In 2005 researchers corrected a mutation in severe combined immune deficiency (SCID) in cultured cells (Urnov et al. 2005). ZFNs are similar to *restriction enzymes* in their action of digesting DNA at certain sequences, but, unlike restriction endonucleases, they can be made to be specific to single locations in a genome. These designer proteins are created from zinc finger and nuclease domains. *Zinc-finger domains* bind DNA at specific sequences, with each zinc finger recognizing primarily a three-nucleotide sequence (Porteus and Carroll 2005). Hence, a domain composed of three zinc fingers would recognize a specific 9-bp sequence. The theory behind this approach is that with the addition of a zinc-finger domain to a DNA endonuclease, the endonuclease would be directed to cut the DNA near the site where the zinc-finger peptide binds. The damaged DNA would then activate the host DNA repair

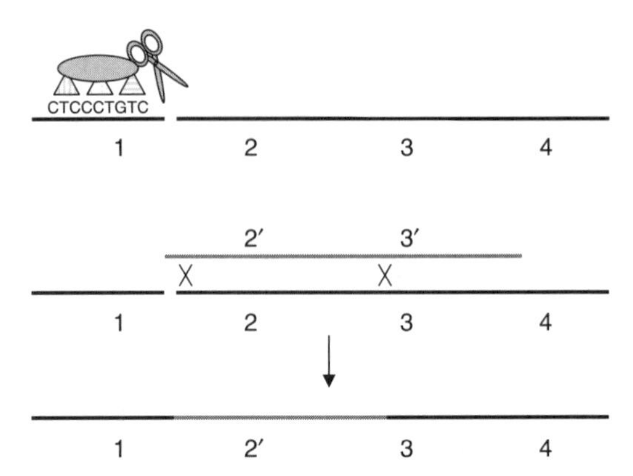

Figure 16.4. A promising approach for homologous gene replacement. Site-directed DNA cleavage by a zinc finger nuclease, with each zinc finger (triangles) recognizing a 3-nucleotide sequence, promotes DNA repair that leads to higher rate of homologous recombination. Example shows replacement of gene 2 by gene 2′.

process, in which the damaged DNA might then be replaced by the similar, but not identical, template introduced into the cell (Fig. 16.4). As with many biotechnologies, the relatively rich field of biomedical science often spills over into plant biology. A number of researchers agree that the idea seems promising, but this technology is relatively new and only time tell whether it will be effective for engineering crop plants. If it lives up to promise, there will be many applications for zinc-finger nucleases, such as correcting mutant sequences in plants ("plant gene therapy"), engineering herbicide resistance by point mutations, and applications similar to those proposed for site-specific recombination systems indicated above (Wright et al. 2005).

16.4. THE FUTURE OF FOOD (AND FUEL AND PHARMACEUTICALS)

One of the unexpected developments in plant biotechnology has been the concern of food companies about accidental transgene transgenic product "contamination" in foods—called *adventitious presence* or *admixture*. These companies often employ biotechnology in their own right, but are concerned about potential marketing problems and food safety issues, especially in export markets. For example, a beer company was not happy that transgenic rice for human pharmaceuticals was being grown in the same state that was home to its rice fields—they were worried about admixture that might complicate its product. And there was concern among farmers that admixture might harm export markets to "no GMO" countries. While industry and farmers are generally pro-innovation, this was a case in which they were more worried about potential harmful occurrence that might hamper economics of existing products.

What does the admixture problem mean for the future of plant biotechnology? It has been argued that using biotechnology to modify food plants for pharmaceuticals and now bioenergy production must take into consideration food uses first. For example, engineering a soybean for a better diesel fuel would have to first consider soybean oil for vegetable oil uses, since the latter is the predominant market. Because even if modified soybean oil were safe for consumption, the consumer would not be expected to be enthralled with the thought of consuming "diesel fuel"; we can expect consumer preferences to be increasingly important, which is to be expected in affluent societies. As was so poignantly discussed in the previous chapter, is the public is not only embracing choices in the marketplace but are also increasingly exercising their power to boycott products that do not meet their standards or ideals. Therefore, biotechnology products must be made desirable to consumers. The plant biotechnologist of tomorrow will be reminded of this fact repeatedly.

16.5. CONCLUSIONS

We cannot foresee how the story of plant biotechnology will end. Some people have predicted that genomics will render plant biotechnology obsolete. That is, the more we understand about the genomic makeup of crops and use genomics in conventional breeding programs, the less we will need to introduce foreign DNA into plants. This prediction seems limited by inherent genetic variation in a particular crop and not reasonable. Although plant breeding will continue to be a valuable tool, it is still highly limited by native genetic variation. Some people have predicted that nearly all economically important plants will be genetically engineered a few decades from now. This prediction would likely be hampered, at least, by regulatory statutes and practices, if nothing else. It takes millions of dollars to move a transgenic event into deregulated status, and therefore, mainly large companies can afford the regulatory costs, and so they have only engineered crops that are planted over large areas. The issue of public acceptance is also at large. For example, would people accept transgenic ornamentals? This will depend on how much value (aesthetic and otherwise) a transgenic plant might possess over its nontransgenic counterpart and whether biosafety can be assured. Finally, it also depends on how much value ornamentals have and how badly people want them. Recall the Dutch tulip industry where novel bulbs would sell for small fortunes. A valuable engineered houseplant could change the market and public perception paradigms. What about an engineered "wild" plant? One can imagine that a transgenic American chestnut tree that is resistant to chestnut blight could be a highly desirable wild plant that could be somewhat disruptive to current ecosystems by its reestablishment as a dominant tree. The American chestnut was decimated in the Appalachian Mountains by blight 100 years ago and has been reduced to a nondominant plant in the understory of forests. In the case of transgenic chestnut, the restoration ecologist would expect gene flow and alterations in biodiversity, which would alter the no-gene-flow paradigm (Stewart 2004).

The brief discussion above shows how the plant biotechnologist of tomorrow not only will need to imagine new technologies for innovation but should also have training in ecology among other scientific subjects, as well as an understanding of human nature and sociology. As we have seen in this book, regulatory affairs and patent law are also very important determinants of technological development. The future of plant biotechnology will be shaped my people who understand multiple disciplines and provide solutions to help feed and clothe the world, as well as provide fuel, pharmaceuticals, and yes, even fun.

LIFE BOX 16.1. TANIYA DHILLON

Taniya Dhillon, Graduate Student, Ohio State University

Taniya Dhillon

I have always lived in an environment very much influenced by agriculture—I was born and raised in Punjab, a major agricultural state of India, and frequently visited my father's family that still lives in a village and practices farming. Agricultural science became a part of my family because both my parents worked as professors in Punjab Agricultural University (PAU). In school, I enjoyed studying biology more than mathematics. However, since I did not have a desire to become a doctor of humans or animals, I decided to study the biology of plants. I started my Bachelors in Crop Science at PAU and studied a range of subjects related to field crops, livestock, poultry, economics of farm production and extension education.

My interest developed specifically in genetics and biotechnology. I was amazed how a small DNA molecule, invisible to the naked eye, was the essence of life. I still remember the day when I actually got to see this molecule after following the procedure for DNA extraction from wheat leaves. It was almost impossible to believe that a colorless, thread-like entity defined the morphology, structure, color, behavior and functionality of each individual. It was fascinating that this thread-like structure was composed of even smaller units or nucleotides, the arrangement of which determined the uniqueness of an individual.

Since biotechnology and molecular biology were not very extensively taught at my university, I started looking into US universities for pursuing a masters degree. With some luck and some hard work, I was accepted in the department of Horticulture and Crop Science at Ohio State University.

I currently work in a laboratory that focuses on plant transformation. Biotechnology has emerged as a very fast approach for genotype modifications/alterations. Several methods of gene introduction have been standardized (see Chapter 11). However, the biology of a plant is highly complex. Once a gene is introduced, it does not always have a uniform expression and this can happen due to a number of factors. In simple terms, the plant sometimes senses the introduced gene as an attack by a biological entity and can "silence" the transgene, which means that transcription or translation of the transgene is suppressed. However, certain genes from plant viruses can be used to stabilize the transgene expression. My research project involves the evaluation of six such viral genes or "suppressors of silencing" to stabilize transgene expression. But the introduction of viral genes in a plant can also lead to developmental abnormalities in plants. At present, I'm analyzing the structural differences recovered in the leaves of transgenic soybean plants due to the introduction of a silencing suppressor. Upon finishing my masters, I plan to start a Ph.D. that will focus on understanding the molecular mechanisms of cold tolerance in cereals like wheat, barley and rye.

Genes are generally introduced to study their function in heterologous systems. However, the biology of any living system is so complex that things don't always work according to our predictions. We generally end up dealing with the "side-effects", and this is how our knowledge about living systems increases. The more we try to explore, the more there is left unexplored. For instance, I introduced a silencing suppressor, *p19* in soybean plants with the aim to stabilize the expression of the *green fluorescent protein* (*gfp*) gene. However, introduction of *p19* did not help in stabilizing *gfp* expression, but instead resulted in an abnormal leaf phenotype in soybean plants. Some additional results have created doubts if such silencing suppressors can lead to genome instability in plants. In the near future, I would like to explore this possible function of *p19*. But I also hope that we can find a permanent solution to stabilize transgene expression in plants. Stable transformation of plants is of least value if the expression of the transgene in not uniform across clones.

In the long run, I hope to hear about and contribute towards another "green revolution" or may be a "biotechnology-based green revolution" in which the yield of cereal crops can be tremendously increased to serve the needs of present day world population. I also hope that government regulations on the public distribution of transgenic crops are eased, especially in developing countries where cereal crops like wheat, rice and corn form the staple food of a majority of people. All the scientific advances would be of no good unless they are accessible to the general public. I hope to use biotechnology to improve the quality of life of the people and to bridge the gap between the poor and the rich, since we all equally deserve the right to live and the right to the basic necessities of life, the foremost being food.

LIFE BOX 16.2. JOSHUA YUAN

Joshua Yuan, PhD Student and Genomics Scientist, University of Tennessee

Joshua Yuan by the microarray printer at the Gallo Center of UCSF.

I became interested in plant research during my college years when I worked as an undergraduate assistant with Prof. Pifang Zhang in Fudan University, Shanghai, China. After graduation, I was enrolled as a masters student in the University of Arizona, where I met many elite plant biologists including David Galbraith, who later became the advisor for my masters thesis studying the expression of ice plant water channel promoters in *Arabidopsis* and developing the MANTRA (Microarray Analysis of Nuclear TRAnscriptome) technology.

David was a great mentor who always gives a grace period to allow students to grow as scientists. After the masters degree, I took an adventure into industry to work at the new BASF Plant Sciences LLC, where I helped the company to established functional genomics platforms. I quickly found that my nature of curiosity and desire for free-style research doesn't fit an industry career well, and moved back to academia to work at UCSF (University of California, San Francisco) as a microarray manager, where I helped with different neuroscience projects. Regardless of being accepted by several top graduate programs, my attempt to go back to graduate school failed due to the complicated issue of permanent residency application. However, I was lucky enough to be offered a job at University of Tennessee to manage their genomics hub and pursue a PhD degree at the same time. I am supervised by a group of excellent researchers including Neil Rhodes, Neal Stewart and Feng Chen for my job and graduate study, respectively. UT turned out to be a promised land for me, where I had a chance to revive my love for plant biology research. Neal is a great mentor, who always encourages you to go beyond your limits to develop a multidisciplinary research interest emphasizing on both fundamental research and application. My research at UT covers a broad spectrum ranging from identifying volatile producing genes involved in tritrophic interaction, discovering the genes for low temperature germinability, genetic engineering of key cell wall genes for better bioenergy feedstock, to developing bioinformatics tools for genomics data analysis. I was lucky enough to be trained by scientists with strong background[s] in technology development as well as both basic and applied scientific research, which makes me believe that my research should be driven by new technology, scientific questioning and needs of the society. After wandering in different fields, I came to realize that plant biotechnology is emerging as a field with more and more significant impact on our society and lives. As a traditional source of food, energy and pharmaceuticals for mankind, the success in plant biotechnology research will enable more environmentally friendly energy supplies, more food, better nutrition, and cheaper healthcare products, all of which will contribute to the sustainable growth and peaceful development of human society.

REFERENCES

Dale EC, Ow DW (1991): Gene transfer with subsequent removal of the selection gene from the host genome. *Proc Natl Acad Sci USA* **88**:10558–10562.

Luo K, Duan H, Zhao D, Zheng X, Deng W, Chen Y, Stewart CN Jr, Wu Y, Jiang X, He A, McAvoy R. Pei Y, Li Y (2007): "GM-gene-deletor": Fused loxP-FRT recognition sequences dramatically improves efficiency of FLP or Cre recombinase on transgene excision from pollen and seed of tobacco plants. *Plant Biotechnol J* **5**:263–274.

Porteus MH, Carroll D (2005): Gene targeting using zinc finger nucleases. *Nat Biotechnol* **23**: 967–973.

Russell SH, Hoopes JL, Odell JT (1992): Directed excision of a transgene from the plant genome. *Mol Gen Genet* **234**:49–59.

Schaefer DG (2001): Gene targeting in *Physcomitrella patens*. *Curr Opin Plant Biol* **4**:143–150.

Stewart CN Jr (2004): *Genetically Modified Planet: Environmental Impacts of Genetically Engineered Plants*. Oxford Univ Press, New York.

Stewart CN Jr, Halfhill MD, Warwick SI (2003): Transgene introgression from genetically modified crops to their wild relatives. *Nat Rev Genet* **4**:806–817.

Urnov FD, Miller JC, Lee Y-L, Beausejour CM, Rock JM, Augustus S, Jamieson AC, Porteus MH, Gregory PD, Holmes MC (2005): Highly efficient endogenous human gene correction using designed zinc-finger nucleases. *Nature* **435**:646–651.

Wright DA, Townsend JA, Winfrey RJ Jr et al. (2005): High-frequency homologous recombination in plants mediated by zinc-finger nucleases. *Plant J* **44**:693–705.

INDEX

2,4-D, 117
35S promoter, *see* CaMV 35S promoter
5-enolpyruvylshikimate 3-phosphate synthase, *see* EPSPS

ABA, *see* abscisic acid
ABC model of flower development, 84, 99–101
Abscisic acid, 90–91, 97, 101, 103, 117
Admixture, *see* Adventitious presence of transgenes
Adventitious presence of transgenes, 364–365
Agrobacterium tumefaciens-mediated transformation,
  Analysis of transgenic plants, 279–288
  History, 18–19, 358
  Intellectual property, 331
  Methodologies, 222–274
  Vectors, 166–190
Agroinfiltration, 254
Amino acids, 76, 104, 136, 148–161, 170, 195–205, 295, 304, 348
Anther culture, 116, 123
Antibiotics resistance gene, *see* NptII and Hpt
Anticommons, 331–333
Antisense construct, 220, 223
Apomixis, 38–39, 77
Arabidopsis thaliana,
  Developmental biology, 89–106
  Floral dip transformation, 254–255, 261, 277, 282
  Genomics, 195
Arnold, Michael, 43–45
Asilomar conference, 293
AttR sites, 174, *see also* Gateway cloning
Autogamy, *see* Selfing
Autoradiography, 281, 285
Auxin, 89–104, 116–130, 166, 230

Bacillus thuringiensis, *see* Bt transgenes
Backcross breeding, 66–67
Baenziger, Stephen, 80–81
Bar gene, *see* PAT gene
Barnase, 230–231
Barstar, 230–231
Bartsch, Detlef, 321–323
Bayh-Dole Act, 330
Beta glucuronidase, *see* GUS reporter gene
Binary vector, *see Agrobacterium tumefaciens*, vectors
Biofuels, 110, 177, 209–210, 271
Bioinformatics, 196, 368
Biolistics, *see* Particle bombardment
Biosafety, 215, 218, 230, 291–354, 357–365
Borlaug, Norman, 15–16
Brassinosteriod, 105–106
Bt transgenes
  Biosafety, 307, 312–322, 349–353
  Cry genes, 2, 8, 18, 170, 185, 187
  Mechanisms of toxicity, 200–202
Bulk breeding, 65–70

CAAT box, 143, 168–169
Callus, 116–117, 120–126, 222, 251, 277
CAMBIA, 338
CaMV 35S promoter, 179, 183, 189, 220, 223–242, 332, 349
Carbon dioxide emissions, 12–13
Cauliflower mosaic virus, *see* CaMV 35S promoter
cDNA, 110, 154, 172–176, 196, 286
Cell suspension culture, 122–123
Central dogma, 136, 141
Centromere, 23–24, 29–32
Chilton, Mary-Dell, 17–18
Chloroplast transformation, 186–188, 227–230, 259
Chrispeels, Maarten, 153–154

Plant Biotechnology and Genetics: Principles, Techniques and Applications, Edited by Neal Stewart, Jr

Chromatin, 88, 136–140, 144–146
Cis-acting element, 144, 164, 186
Cleistogamy, Cleistogamous flowers, 35–36
Codons, 76, 145–152, 169–170, 178–179, 220
Collins, Glenn, 130–131
Complementary DNA, *see* cDNA
Conner, Tony, 155–156
Cotransformation, 235–236, 259
Cre-*lox* recombination system, 175–176, 360–362
Cytokinin, 100, 105, 114–116, 120, 125, 166, 230, 236

Delmer, Deborah, 108–110
Development of
  Flowers, 98–101
  Fruits, 88–89
  Leaves, 95–98
  Roots, 94–95
Dhillon, Taniya, 366–367
Disease-resistance 57, 64–5, 67, 202–205
DNA blot analysis, *see* Southern blot analysis
Doubled haploids, 75, 81, 123, 131

Electroporation, 162, 262, 267
ELISA, 277, 279–280, 288
Embryogenesis, 85–95
Enabling technologies, 326, 327, 332, 334–339
Enhancers, 143, 144, 180,
Enhancer trap, 225–226
Environmental impact quotient, 9–14
Environmental Protection Agency, US, *see* EPA
Environmental Risk Assessment, 312–321
Enzyme-linked immunosorbent assay, *see* ELISA
EPA, 293–296, 298–300
Epistasis, 55
EPSPS, 197–199, 207
ESTs, 196
Ethylene, 101–105, 117, 133
EU, 294, 299–301, 305, 312
European Union, *see* EU
Event, transgenic, 74, 77, 162, 198, 221–224, 227, 230, 255, 259, 269, 358–360, 363, 365
Exon, 146–147
Explant, 114–128, 222
Expressed sequence tags, *see* ESTs

FDA, 293–298
Flavr Savr tomato, 295, 345, 350, 352
FLP-*FRT* recombination system, 236, 360
Fluorescence *in situ* hybridization, 34
Food and Drug Administration, US, *see* FDA
Frankenstein, 343–344
Freedom to operate, *see* FTO

GA, *see* Giberellic acid
Gametogenesis, 85–88
Gateway cloning, 172–181, 361
Gene, definition, 22–23
Gene gun, *see* particle bombardment
Gene silencing, *see* RNAi, *see also* antisense construct
Genetic linkage, 28–29, 76, 236, 363
Genome, 22–24, 32–35, 40
Genomics, 194–197
Genotype, 25–28, 51–52
GFP reporter gene, 178, 233–234, 255, 260, 268
Gibberellic acid, 91, 101–103, 116, 185
Glyphosate resistance, 7, 9, 14, 197–199, 229, *see also* herbicide resistance
Golden Rice, 205–207, 209, 210, 212–215
Gonsalves, 211–212
Green fluorescent protein, *see* GFP
GUS reporter gene, 95, 222–224, 231–232, 268, 272

Hemizygous, 277
Herbicide resistance (or tolerance), 7, 9–14, 182, 197–200
Heterologous, 224, 332, 334
Hinchee, Maud, 270–271
Histone acetyltransferase, 145
Histones, 136, 138, 145
HPT selectable marker gene, 228–229, 332
Hybrids, 26–29, 39–44, 54–55, 61–72
Hybrid varieties, 58, 72–74
Hygromycin phophotransferase, *see* HPT selectable marker gene

IAA, 104, 114, 116
IBA, 117
Indole acetic acid, *see* IAA
Indole butyric acid, *see* IBA
Insect resistance, 7, 9–14, 200–202
Intellectual property, *see* Patents
Intron, 146–147
IPT gene, 230, 236
Isopentyl transferase gene, *see* IPT gene

Jasmonic acid, 102–103

Khush, Gurdev, 78–80
Klein, Ted, 271–272
Kozak sequence, 150, 170

Laser micropuncture, 263
Left border, T-DNA, 166–167, 282–283
Ligase, 160, 165
Luciferase reporter gene, 231, 233, 240

Male sterility, 38, 61, 230
Marker-assisted selection, 56, 74–76
Marker-free strategies, 218, 233–237
Marker gene, selectable, 167, 182, 218–222, 226–231, 277–279, 331, 332, 362
Material transfer agreement, 328, 334
Matrix attachment region, 186
Meiosis, 30–34
Mendel, Gregor, 25, 28
Mendelian
  Segregation, 25–30, 50, 54, 222, 269, 277
  Trait, 25–30
Meristems
  Root apical, 84, 85, 90, 94–95
  Shoot apical, 84, 85, 90, 92–94
Messenger RNA, *see* mRNA
Methylation, DNA, 146, 160
Microarray analysis, 195–196
Mitosis, 30–34
Monarch butterfly, 315–319, 349–350
mRNA,
  Analysis, 284–286, *see also* Northern blot analysis and Real-time PCR
  Production, 140–142, 146–150
  Stability, 169–170, 195–196
MS medium, 114–116
Multiple cloning site, 170
Murashige and Skoog medium, *see* MS medium
Mutation breeding, 76, 77, 156

Nanofiber arrays, 263–264
National Institutes of Health, *see* NIH
Neomycin phosphotransferase, *see* NptII selectable marker gene
NIH, 293
Northern blot analysis, 284–288
NptII selectable marker gene, 228–229, 332
Nucleosome, 139–140, 145

OECD, 293–294
Oils, modified, 207–208
Oral vaccine, 208
Organization for Economic Cooperation and Development, *see* OECD
Organogenesis, 120, 125, 230
Origins of replication, 163, 166–168
Open access, 325–339
Open reading frame, 179–181
Open source, 338
ORF, *see* Open reading frame
Outcrossing, 35–39, 64, 69, 187
Ow, David, 238–240

Papaya ringspot virus, 204, 212
Parrott, Wayne, 189–190
Patents, 251, 257, 265, 325–339
PAT gene, 99–101
Pathogen resistance, 202–205
PCR
  Defined, 171
  Real-time, 280, 283, 286
  Use in cloning, 175
  Use in identifying transgenic plants, 278–279
Pedigree breeding, 64
PEG, 262
Pharmaceuticals, 208–209, 364–365
Phosphinothricin, *see* PPT
Phosphinothricin acetyltransferase gene, *see* PAT gene
Phosphoisomerase gene, *see* PMI gene
Photomorphogenesis, 91–92
PIPRA, 333, 337
Plant-produced pharmaceuticals, *see* Pharmaceuticals
Plasmids, 159–188
Plastid vector, *see* Chloroplast transformation
PMI gene, 229
Polyadenylation, *see* PolyA tail
PolyA tail, 147–147
Polyethylene glycol, *see* PEG
Polygenic traits, 26, 50–51, 185
Polylinker, *see* Multiple cloning site
Polymerase chain reaction, *see* PCR
Polypeptide, 136, 148–152
Polyploids, 38–41
Population genetics, 55
Position effects, 223, 224, 259, 361
Posttranscriptional gene silencing, *see* RNAi
Posttranslational modification, 152–153, 186, 196
Potrykus, Ingo, 213–215
PPT, 229, 309
Prior art, 334–336

Promoter trap, 225–226
Protein blot analysis, *see* Western blot analysis
Protoplasts, 123–124, 261–262
Public Sector Intellectual Property Resources for Agriculture, *see* PIPRA
Punnett square, 27–29
Putative transformants, 277, 279, 288

Quantitative PCR, *see* Real-time PCR
Quantitative trait loci (QTL), 56, 75–76

Raikel, Natasha, 106–108
Reading frame, 148–150, 178–179
Real-time PCR, *see* PCR, Real-time
Recombination
  Homologous, 186–187, 259, 359–360, 364
  Site-specific, 172–181, 236, 359–361, *see also* Gateway cloning and Cre-*lox*
Recurrent selection, 58, 61, 63, 67, 70–72
Reporter genes, 170, 220, 222–227, 231–235
Restriction endonuclease, 160–161, 165, 170, 280–284
Restriction enzyme, *see* Restriction endonuclease
R gene, 202–203
Ribonuclease, 230–231
Right border, T-DNA, 166–167, 282–283
RNA blot analysis, *see* Northern blot analysis
RNAi, 179, 259
RNA interference, *see* RNAi
RNA polymerase II, 140–144, 162, 188, 227
RNases, 146–147, 284
Root apical meristem, 80, 82, 84, 90, 94–95
Roundup, Roundup-Ready, *see* glyphosate

Seed germination, 85, 91
Selectable marker genes, 167–168, 182, 187, 218–222, 226–231, 234–237, 332
Self-fertilization, 27–29, 35–36, 58, 63–64, 60–74
Self-incompatibility, 35–38
Self-pollination, *see* Self-fertilization
Shelton, Anthony, 306–307
Shifting balance theory, 60
Shillito, Ray, 308–309
Signal transduction, 90–93, 97, 101–106
Silicon carbide whiskers, 262
Single-seed descent, 64–65
Somaclonal variation, 122, 125–126, 349
Somatic embryos, 120, 121, 123, 126–128, 188, 221
Southern blot analysis, 280–284
Splicesome, 147
Stacked genes, 2, 62, 185, 202, 363
Start codon, 150, 169–170
Stop codon, 148, 152, 178–179
Surface sterilization of explants, 118–119
Synthetic varieties, 71–72
Systems biology, 107, 196

Tapetum, 230–231
Taq DNA polymerase, 278, 286
TATA box, 148, 168–169
T-DNA, 18, 166–170, 178, 185–190, 249–254, 283
Ti Plasmid, 18, 166–168, 225, 228, 230, 249, 252
Totipotency, Totipotent, 113–114, 248
Traits
  Input, 62, 197–205
  Output, 205–210
Transcription
  Factors, 92, 94, 98, 140–146, 169, 182
  Process, 140–147
  Start site, 169
Transfer DNA, *see* T-DNA
Transfer RNA, *see* tRNA
Translation, 148–152
Translation start site, *see* start codon
tRNA, 150–151
Tumor-inducing plasmid, *see* Ti Plasmid

UidA gene, *see* GUS reporter gene
United States Department of Agriculture Animal and Plant Health Inspection Service
  Biotechnology Regulation Service, *see* USDA APHIS BRS
Untranslated region, *see* UTR
Upregulated gene, 195
USDA APHIS BRS, 293–297
UTR, 142, 169

Viral vectors, 263
Vir genes, *see* Virulence genes
Virulence genes, 166, 250, 263

Western blot analysis, 286–287
World Trade Organization, *see* WTO
Wright, Martha, 129–130
WTO, 300

Yuan, Joshua, 367–368

Zinc finger nuclease, 363–364

CUSTOMER NOTE: IF THIS BOOK IS ACCOMPANIED BY SOFTWARE, PLEASE READ THE FOLLOWING BEFORE OPENING THE PACKAGE.

This software contains files to help you utilize the models described in the accompanying book. By opening the package, you are agreeing to be bound by the following agreement:

This software product is protected by copyright and all rights are reserved by the author and John Wiley & Sons, Inc. You are licensed to use this software on a single computer. Copying the software to another medium or format for use on a single computer does not violate the U.S. Copyright Law. Copying the software for any other purpose is a violation of the U.S. Copyright Law.

This software product is sold as is without warranty of any kind, either express or implied, including but not limited to the implied warranty of merchantability and fitness for a particular purpose. Neither Wiley nor its dealers or distributors assumes any liability of any alleged or actual damages arising from the use of or the inability to use this software. (Some states do not allow the exclusion of implied warranties, so the exclusion may not apply to you.)

WILEY